T0398875

HYBRID CENSORING KNOW-HOW

HYBRID CENSORING KNOW-HOW

Designs and Implementations

N. BALAKRISHNAN
McMaster University
Department of Mathematics and Statistics
Hamilton, ON, Canada

ERHARD CRAMER
RWTH Aachen
Institute of Statistics
Aachen, Germany

DEBASIS KUNDU
Indian Institute of Technology Kanpur
Department of Mathematics and Statistics
Kanpur, India

Academic Press is an imprint of Elsevier
125 London Wall, London EC2Y 5AS, United Kingdom
525 B Street, Suite 1650, San Diego, CA 92101, United States
50 Hampshire Street, 5th Floor, Cambridge, MA 02139, United States
The Boulevard, Langford Lane, Kidlington, Oxford OX5 1GB, United Kingdom

Notices

Knowledge and best practice in this field are constantly changing. As new research and experience broaden our understanding, changes in research methods, professional practices, or medical treatment may become necessary.

Practitioners and researchers must always rely on their own experience and knowledge in evaluating and using any information, methods, compounds, or experiments described herein. In using such information or methods they should be mindful of their own safety and the safety of others, including parties for whom they have a professional responsibility.

To the fullest extent of the law, neither the Publisher nor the authors, contributors, or editors, assume any liability for any injury and/or damage to persons or property as a matter of products liability, negligence or otherwise, or from any use or operation of any methods, products, instructions, or ideas contained in the material herein.

ISBN: 978-0-12-398387-9

For information on all Academic Press publications
visit our website at https://www.elsevier.com/books-and-journals

Publisher: Mara Conner
Editorial Project Manager: Susan Dennis
Production Project Manager: Paul Prasad Chandramohan
Cover Designer: Matthew Limbert

Typeset by VTeX

In the loving memory of my mother

N. B.

Für Katharina, Anna, Jakob & Johannes

E. C.

Dedicated to my mother

D. K.

Contents

Preface

Though the notion of hybrid censoring was introduced in 1950s, very little work had been done on it until 2000. During the last two decades, however, the literature on hybrid censoring has exploded, with many new censoring plans having been introduced and several inferential methods having been developed for numerous lifetime distributions. A concise synthesis of all these developments were provided in the Discussion paper by N. Balakrishnan and D. Kundu in 2013 (*Computational Statistics & Data Analysis*, Vol. 57, pages 166–209). Since the publication of this Discussion paper in 2013, more than 300 papers have been published in this topic. The current book is an expanded version of the Discussion paper covering all different aspects of hybrid censoring, especially focusing on recent results and developments.

As research in this area is still intensive, with many papers being published every year (see Chapter 1 for pertinent details), we have tried our best to make the bibliography as complete and up-to-date as possible. We have also indicated a number of unresolved issues and problems that remain open, and these should be of interest to any researcher who wishes to engage in this interesting and active area of research.

Our sincere thanks go to all our families who provided constant support and great engagement through out the course of this project. Thanks are also due to Ms. Susan Ikeda (Senior Editorial Project Manager, Elsevier) for her keen interest in the book project and also in its progress, Ms. Debbie Iscoe (McMaster University, Canada) for her help with typesetting some parts of the manuscript, and Dr. Faisal Khamis for his diligent work on literature survey in the beginning stage of the project. Without all their help and cooperation, this book would not have been completed!

We enjoyed working on this project and in bringing together all the results and developments on hybrid censoring methodology. It is our sincere hope that the readers of this book will find it to be a useful resource and guide while doing their research!

N. Balakrishnan (McMaster University, Hamilton, ON, Canada)
Erhard Cramer (RWTH, Aachen, Germany)
Debasis Kundu (IIT, Kanpur, India)

CHAPTER 1

Introduction

Contents

1.1. Historical perspectives

Reliability of many manufactured items has increased substantially over time due to the ever-growing demands of the customers, heavy competition from numerous producers across the world, and strict requirements on quality assurance, as stipulated by ISO9000, for example. As a result, providing information on the reliability of items (such as mean lifetime, median lifetime, $100p^{th}$ percentile of the lifetime distribution, and so on), within a reasonable period of time, becomes very difficult under a traditional life-test. So, because of cost and time considerations, life-tests in these situations necessarily become censored life-testing experiments (often, with heavy censoring). The development of accurate inferential methods based on data obtained from such censored life-tests becomes a challenging task to say the least. This issue, therefore, has attracted the attention of numerous researchers over the last several decades!

Early scenarios requiring the consideration of censoring seemed to have originated in the context of survival data. For example, Boag (1949) discussed the estimation of proportion of cancer patients in UK who were cured following a treatment, and while doing so, did consider censoring corresponding to all those cancer patients who were alive at the end of the clinical trial; one may also refer to Harris et al. (1950), Berkson and Gage (1952) and Littel (1952) for further detailed discussions in this regard. Note that the form of censoring considered in this context is time-censoring, meaning the trial ended at a certain pre-fixed time; consequently, individuals who were alive at the end of the trial ended up getting censored at this specific termination time. However, according to David (1995), the word *censoring* explicitly appeared for the first time in the works of Hald (1949) and Gupta (1952). In fact, Hald (1949) made a clear distinction between *truncation* and *censoring*, depending on whether the population from which the sample is drawn is truncated or the sample itself is truncated, respectively. Yet, in many of the early works, the word truncation got used in place of censoring; see, for example,

Hybrid Censoring Know-How
https://doi.org/10.1016/B978-0-12-398387-9.00009-X

Cohen (1949, 1950). It was Gupta (1952) who pointed out that censoring could arise in two different ways, one when the observation gets terminated at a pre-fixed time (with all observations up till that time point being observed and all those after being censored as in the study of Boag (1949) mentioned above) and another when the observation gets terminated when a certain number of smaller observations is achieved (with the remaining larger observations being censored). Then, for the purpose of distinguishing between these two different cases, he referred to them as *Type-I censoring* and *Type-II censoring*, respectively.

What is clear from the descriptions and explanations of Hald (1949) and Gupta (1952) is that truncation is a feature of the population distribution while censoring is inherently a feature of the sample observed. In spite of this clear distinction, still today, some authors mistakenly refer to "censored distributions" and "truncated samples".

Following the work of Gupta (1952), Epstein and Sobel (1953, 1954) published pioneering results on censored data analysis from a life-testing and reliability viewpoint, basing their results on exponential lifetime distribution. They utilized some interesting distributional properties of order statistics (more specifically, on spacings) from the exponential distribution to develop exact inferential results for the case of Type-II censored samples. Though their distributional results on spacings from the exponential distribution were already known from the earlier works of Sukhatme (1937) and Rényi (1953), the approach taken by Epstein and Sobel (1954) in developing exact inferential methods based on censored data from a life-testing viewpoint attracted the attention of many researchers subsequently, resulting in numerous publications in the following years; the works of Deemer and Votaw (1955), Cohen (1955), and Bartholomew (1957) are some of the early noteworthy ones in this direction.

During the last six decades or so, since the publication of these early works, the literature on censored data analysis has expanded phenomenally, by considering varying forms of censored data, dealing with a wide range of lifetime distributions, and developing many different methods of inference. For a detailed overview of various developments on inferential methods for truncated distributions and based on censored samples, interested readers may refer to Nelson (1982), Schneider (1986), Cohen (1991), and Balakrishnan and Cohen (1991).

1.2. Type-I and Type-II censoring

In the preceding section, while describing different forms of censoring that had been considered, it becomes evident that are two basic forms of censoring:

(i) *Type-I censoring* and

(ii) *Type-II censoring*.

As the historical details in the last section clearly reveal, both these forms of censoring have been discussed extensively in the literature, and this continues on to date. Though the form of censoring schemes in these two cases may look somewhat similar, with slight variation in the resulting likelihood functions, there is a significant difference when it comes to development of inferential methods based on these two forms of censoring, as will be seen in the subsequent chapters.

Type-I censoring would naturally arise in life-testing experiments when there is a constraint on time allocated for the reliability experiment to be conducted. This may be due to many practical considerations such as limitations on the availability of test facility, cost of conducting the experiment, need to make reliability assessment in a timely manner, and so on. It is evident that the duration of the life-test is fixed in this censoring scheme, but the number of complete failure times to be observed will be random. These result in advantages as well as disadvantages, the fixed duration being a distinct advantage in the sense that the experimenter would know a priori how long the test is going to last, while the random number of failures to be observed being a clear disadvantage. For example, if the duration was fixed to be too small compared to the average (or median) lifetime of the product, then a rather small number of complete failures would be realized with a high probability, and this would in turn have a negative impact on the precision or accuracy of the inferential methods subsequently developed.

In a *Type-II censoring scheme*, on the other hand, the number of complete failures to be observed is fixed a priori; consequently, the experimenter can have a control on the amount of information (in the form of complete failures) to be collected from the life-testing experiment, thus having a positive impact on the precision of subsequent inferential methods. However, it has a clear disadvantage that the duration of the life-test is random, which would pose difficulty in the planning/conducting of the reliability experiment, and also has the potential to result in unduly long life-test (especially when the product under test is highly reliable).

For the purpose of illustration, let us consider the $Uniform(0, 1)$ distribution.[1] Let us further suppose we have $n = 10$ units available for the life-test, and that we choose $T = 0.4, 0.5, 0.6, 0.7, 0.8$ as the pre-fixed termination times under Type-I censoring scheme, and $m = 4, 5, 6, 7, 8$ as the pre-fixed number of complete failures to be observed under Type-II censoring scheme. In this situation, the following facts are evident:

(a) The random number of failures that would occur until the pre-fixed time T, say D, will have a $bin(10, T)$-distribution;

(b) Consequently, the number of failures that would be expected to occur by time T will simply be $E(D) = np = 10T$, under the Type-I censoring scheme;

[1] The uniform distribution is not a suitable model for lifetimes, but is used here for illustrating the idea behind the two censoring schemes, and their primary difference. The uniform distribution, however, can be thought of as a nonparametric model, obtained after performing a probability integral transformation on the observed lifetimes, as will be explained later on in the book!

(c) Under the Type-II censoring scheme, as stated above, the termination time will be random, say Y,[2] and will in fact equal the m-th order statistic from a sample of size n from the *Uniform*$(0, 1)$-distribution, and hence is known to have a *Beta*$(m, n - m + 1)$-distribution.

To get a clear idea about the difference between the two censoring schemes, we have presented, in Table 1.1, the exact values of the following quantities:

(i) Termination time T, expected number of failures $E(D)$, and $\Pr(D \geq m)$, corresponding to Type-I censoring;

(ii) Expected termination time $E(Y)$, number of failures m, and $\Pr(Y \geq T)$, corresponding to Type-II censoring.

From binomial probabilities, it is then easy to verify that the cumulative distribution function (cdf) of Y is, for $0 < y < 1$,

$$F^Y(y) = \sum_{i=m}^{n} \binom{n}{i} y^i (1-y)^{n-i} = I_y(m, n-m+1)$$
$$= \frac{1}{\mathrm{B}(m, n-m+1)} \int_0^y t^{m-1}(1-t)^{n-m} dt, \qquad (1.1)$$

where $I_y(a, b)$ denotes the incomplete beta ratio and $\mathrm{B}(a, b) = \frac{\Gamma(a)\Gamma(b)}{\Gamma(a+b)}$ (for $a, b > 0$) denotes the complete beta function. It is also easy to see that $E(Y) = \frac{m}{n+1}$. Furthermore, we may also observe in this situation that

$$\Pr(D \geq m) = \sum_{i=m}^{n} \binom{n}{i} T^i (1-T)^{n-i} = 1 - \Pr(Y > T), \qquad (1.2)$$

a fact that is readily seen in the last column of Table 1.1.

1.3. Need for hybrid censoring

From the results presented in Table 1.1, the following essential points need to be emphasized, which should justify the coverage/exposition of this book being focused on the importance and need for hybrid censoring methodology:

(1) First issue relating to Type-I censoring is concerning the occurrence of no failures. This would especially be the case if the termination time T had been chosen to be unduly small (in other words, if the length of the test compared to mean life time of the product under test is too small). For example, in the case of $n = 10$, had T been fixed as 0.2, then there is almost a 11% chance that the life-test would not result any complete failure. However, if the termination time T gets increased to

[2] Keep in mind that the notation used here is a preliminary one, and an unified notation that will be used throughout the book will be presented later at the end of this chapter.

Table 1.1 Comparison of Type-I and Type-II censoring schemes based on termination time and number of failures for the uniform distribution when the number of test units is $n = 10$.

Scheme parameters		Type-I censoring			Type-II censoring		
T	m	T	$E(D)$	$\Pr(D \geq m)$	$E(Y)$	m	$\Pr(Y > T)$
0.4	4	0.4	4.0	0.6177	0.3636	4	0.3823
0.5	4	0.5	5.0	0.8281	0.3636	4	0.1719
0.6	4	0.6	6.0	0.9452	0.3636	4	0.0548
0.7	4	0.7	7.0	0.9894	0.3636	4	0.0106
0.8	4	0.8	8.0	0.9991	0.3636	4	0.0009
0.4	5	0.4	4.0	0.3669	0.4545	5	0.6331
0.5	5	0.5	5.0	0.6231	0.4545	5	0.3770
0.6	5	0.6	6.0	0.8338	0.4545	5	0.1662
0.7	5	0.7	7.0	0.9527	0.4545	5	0.0474
0.8	5	0.8	8.0	0.9936	0.4545	5	0.0064
0.4	6	0.4	4.0	0.1662	0.5455	6	0.8338
0.5	6	0.5	5.0	0.3770	0.5455	6	0.6231
0.6	6	0.6	6.0	0.6331	0.5455	6	0.3669
0.7	6	0.7	7.0	0.8497	0.5455	6	0.1503
0.8	6	0.8	8.0	0.9672	0.5455	6	0.0328
0.4	7	0.4	4.0	0.0548	0.6363	7	0.9452
0.5	7	0.5	5.0	0.1719	0.6363	7	0.8281
0.6	7	0.6	6.0	0.3823	0.6363	7	0.6177
0.7	7	0.7	7.0	0.6496	0.6363	7	0.3504
0.8	7	0.8	8.0	0.8791	0.6363	7	0.1209
0.4	8	0.4	4.0	0.0123	0.7273	8	0.9877
0.5	8	0.5	5.0	0.0547	0.7273	8	0.9453
0.6	8	0.6	6.0	0.1673	0.7273	8	0.8327
0.7	8	0.7	7.0	0.3828	0.7273	8	0.6172
0.8	8	0.8	8.0	0.6778	0.7273	8	0.3222

0.3, then the chance of zero failures being observed would decrease to about 3%. Moreover, if the number of units tested becomes larger, say $n = 20$ then even with the termination time being $T = 0.2$, the chance of zero failures being observed would be as small as 1%;

(2) Another issue with the case of Type-I censoring is that if the termination time T is pre-fixed to be large, though one has a certain required number of complete failures m in mind (at the planning stage of the experiment), then the corresponding values of $\Pr(D \geq m)$ suggest that, with high probability, the actual number of failures observed would be at least m. For example, had the experimenter chosen the termination time of the experiment to be $T = 0.7$ and in fact had $m = 6$ number of complete failures in mind, then there is almost a 85% chance that the number of failures observed would be at least 6. Similarly, if the termination time had been chosen to be $T = 0.8$ and had $m = 7$ preliminarily in mind, then there is almost a 88% chance that the number of failures observed would be at least 7. This suggests that if the experimenter has an idea on the number of failures to be observed and chooses the termination time T to be large, then a Type-I censoring scheme would end up result in an unnecessarily long life-test with high probability;

(3) Furthermore, in the case of Type-I censoring, it can also be observed that, with the preliminary value of m one has in mind, if the termination time T had been chosen to be small, then with high probability the actual number of complete failures that will be observed will end up being less than m. For example, if the experimenter had in mind $m = 8$ but had chosen the termination time to be $T = 0.7$, then there is almost a 62% chance that the number of failures observed would be at most 7, being less than $m = 8$ that was in the mind of the experimenter prior to conducting the experiment. Thus, it is more likely that the life-test would be concluded by time T in this case;

(4) The final point worth mentioning is that in the case of Type-II censoring, the test duration would be long if one were to choose m to be large enough in comparison to n, but still may not exceed a large value of T the experimenter may preliminarily have in mind before conducting the experiment with a high probability. For example, suppose $n = 10$ units are under test and the number of failures to be observed has been fixed to be $m = 8$. If the experimenter had an idea of having the duration of the test to be at least $T = 0.8$, for example, the chance that the Type-II censoring scheme would result in a test exceeding 0.8 would be only about 32%.

Let us consider the first two points above and discuss their ramifications in statistical as well as pragmatic terms. With regard to Point (1) above, it becomes evident that, at least when the number of test units n is small, there will be a non-negligible probability that one may not observe any complete failure at all from the life-test in the case of Type-I censoring. In such a situation, it is clear that a meaningful inferential method can not be developed (whether it is point/interval estimation or test of hypothesis) unconditionally and, therefore, all pertinent inferential methods in this case need to be developed only conditionally, conditioned on the event that at least one complete failure is observed.[3] In fact, this is the basis for the comment that "*there is a significant difference when it comes to development of inferential methods based on these two forms of censoring*" made in the beginning of last section!

With regard to Point (2) that if the termination time T is pre-fixed to be large under Type-I censoring scheme, then with a high probability, the actual number of

[3] The condition may sometimes require the number of complete failures observed to be even more than one, depending on the assumed model and the number of parameters involved in it!

failures observed would be larger than the preliminary number of failures (say, m) the experimenter would have had in mind, which is what formed the basis for the original proposal of hybrid censoring by Epstein (1954). It is for this reason he defined the hybrid termination time of $T_I = \min\{Y, T\} = Y \wedge T$ in order to terminate the life-test as soon as the preliminary number of failures the experimenter had in mind is achieved, and otherwise terminate at the pre-fixed time T. We refer to this censoring scheme here as Type-I hybrid censoring scheme, adding the phrase Type-I to emphasize that this scheme is based on a time-based guarantee (viz., not to exceed a pre-fixed time T). It is then evident that

$$T_I = \begin{cases} Y & \text{if } Y \leq T \\ T & \text{if } Y > T \end{cases}$$

where Y has its cumulative distribution function as given in (1.1), which readily yields the mixture representation for the hybrid termination time T_I as

$$T_I \stackrel{d}{=} \begin{cases} Y & \text{with probability } \pi \\ T & \text{with probability } 1 - \pi \end{cases} \tag{1.3}$$

with the mixture probability $\pi = \Pr(Y \leq T)$ as in (1.2). Strictly speaking, in the above mixture form, T may be viewed as a degenerate random variable at time T. With time T_I, it is of interest to associate a count random variable D_I (analogous to D) corresponding to the number of complete failures observed in the life-test. It is then clear that

$$D_I = \begin{cases} D & \text{if } Y > T \\ m & \text{if } Y \leq T \end{cases} \tag{1.4}$$

which is in fact a *clumped binomial random variable*, with all the binomial probabilities for m to n being clumped at the value m; see, for example, Johnson et al. (2005) for details on this clumped binomial distribution. From (1.3) and (1.4), the values of $E(T_I)$ and $E(D_I)$ can be readily computed, and these are presented in Table 1.2 for the purpose of comparing Type-I censoring and Type-I hybrid censoring schemes in terms of termination time and expected number of failures.

From Table 1.2, we observe the following points:

(5) The intended purpose of Type-I hybrid censoring scheme, as introduced by Epstein (1954), is clearly achieved in certain circumstances. For example, if the experimenter was planning to conduct the life-test for a period of $T = 0.7$ and had an interest in observing 7 complete failures out of a total of $n = 10$ units under test, then under the Type-I hybrid censoring scheme, the test on an average would have lasted for a period of 0.61, and would have resulted in 6.44 complete failures on an average (instead of 7 the experimenter had in mind). Instead, had the experimenter planned to conduct the life-test for a period of $T = 0.8$ in the same setting,

Table 1.2 Comparison of Type-I censoring and Type-I hybrid censoring based on termination time and number of failures for the uniform distribution when the number of test units is $n = 10$.

Scheme parameters		Type-I censoring			Type-I HCS		
T	m	T	$E(D)$	$\Pr(D \geq m)$	$E(T_{\text{I}})$	$E(D_{\text{I}})$	$\Pr(Y>T)$
0.4	4	0.4	4.0	0.6177	0.3228	3.3960	0.3823
0.5	4	0.5	5.0	0.8281	0.3498	3.7618	0.1719
0.6	4	0.6	6.0	0.9452	0.3604	3.9312	0.0548
0.7	4	0.7	7.0	0.9894	0.3632	3.9877	0.0106
0.8	4	0.8	8.0	0.9991	0.3636	3.9991	0.0009
0.4	5	0.4	4.0	0.3669	0.3653	3.7649	0.6331
0.5	5	0.5	5.0	0.6231	0.4158	4.3848	0.3770
0.6	5	0.6	6.0	0.8338	0.4422	4.7649	0.1662
0.7	5	0.7	7.0	0.9527	0.4522	4.9403	0.0474
0.8	5	0.8	8.0	0.9936	0.4544	4.9927	0.0064
0.4	6	0.4	4.0	0.1662	0.3877	3.9312	0.8338
0.5	6	0.5	5.0	0.3770	0.4612	4.7617	0.6231
0.6	6	0.6	6.0	0.6331	0.5107	5.3980	0.3669
0.7	6	0.7	7.0	0.8497	0.5360	5.7901	0.1503
0.8	6	0.8	8.0	0.9672	0.5422	5.9599	0.0328
0.4	7	0.4	4.0	0.0548	0.3967	3.9859	0.9452
0.5	7	0.5	5.0	0.1719	0.4861	4.9337	0.8281
0.6	7	0.6	6.0	0.3823	0.5592	5.7803	0.6177
0.7	7	0.7	7.0	0.6496	0.6077	6.4397	0.3504
0.8	7	0.8	8.0	0.8791	0.6305	6.8390	0.1209
0.4	8	0.4	4.0	0.0123	0.3994	3.9982	0.9877
0.5	8	0.5	5.0	0.0547	0.4964	4.9883	0.9453
0.6	8	0.6	6.0	0.1673	0.5861	5.9476	0.8327
0.7	8	0.7	7.0	0.3828	0.6595	6.8225	0.6172
0.8	8	0.8	8.0	0.6778	0.7068	7.5168	0.3222

then the test would have lasted on an average for a period of 0.63, and would have resulted in 6.84 complete failures on an average;

(6) However, if the time T had been chosen to be too small for the value of the number of complete failures interested in observing, then the life-test, with high probability, would end by time T. For example, if the experimenter had chosen to conduct the life-test for a period of $T = 0.5$, but had an interest in observing possibly 7 complete failures from the life-test, then the test on an average would have lasted for a period of 0.49, and would have resulted in 4.93 complete failures on an average (instead of 7 the experimenter had in mind). This follows intuitively from the fact that, in this case, there is about 83% chance that the 7th failure would occur after time $T = 0.5$ (see the value reported in the last column of Table 1.2);

(7) A final point worth noting is that, like in Type-I censoring scheme, the case of no failures is a possibility in the case of Type-I hybrid censoring scheme as well! This then means that inferential procedures can be developed only conditionally, conditioned on the event that at least one complete failure is observed, just as in the case of Type-I censoring!

While Point (5) above highlights the practical utility of Type-I hybrid censoring scheme, Point (6) indicates a potential shortcoming in Type-I hybrid censoring scheme exactly as Point (3) indicated earlier in the case of Type-I censoring, viz., that if the test time T is chosen to be too small in comparison to the number of complete failures the experimenter wishes to observe from the life-test, then with high probability, the test would terminate by time T, in which case few failures will be observed from the life-test, leading to possibly imprecise inferential results.

It is precisely this point that led Childs et al. (2003) to propose another form of hybrid censoring based on the hybrid termination time of $T_{\text{II}} = \max\{Y, T\} = Y \vee T$, called Type-II hybrid censoring scheme. It is of interest to mention that the phrase Type-II is incorporated here in order to emphasize the fact that this censoring scheme provides a guarantee for the number of failures to be observed (viz., that the observed number of complete failures would be at least m). It is then clear in this case that

$$T_{\text{II}} = \begin{cases} Y & \text{if } Y > T \\ T & \text{if } Y \leq T \end{cases} \tag{1.5}$$

where Y has its cdf as given in (1.1), which readily yields the mixture representation for the hybrid termination time T_{II} as

$$T_{\text{II}} \stackrel{d}{=} \begin{cases} Y & \text{with probability } 1 - \pi \\ T & \text{with probability } \pi \end{cases} \tag{1.6}$$

with the mixture probability $\pi = \Pr(Y \leq T)$ as in (1.2). Here again, T may be viewed as a degenerate random variable at time T. With time T_{II}, it will be useful to associate a count random variable D_{II} (analogous to D) corresponding to the number of complete failures observed in the life test, with support $\{m, m+1, \ldots, n\}$ and probability mass function as

$$\Pr(D_{\text{II}} = d) = \begin{cases} \sum_{\ell=0}^{m} \binom{n}{\ell} \gamma^{\ell}(1-\gamma)^{n-\ell} & \text{for } d = m \\ \binom{n}{d} \gamma^{d}(1-\gamma)^{n-d} & \text{for } d = m+1, \ldots, n, \end{cases} \tag{1.7}$$

which is in fact a *clumped binomial random variable*, with all the binomial probabilities for 0 to m being clumped at the value m. Observe the difference in the two clumped binomial distributions that arise in the cases of Type-I hybrid censoring scheme and Type-II hybrid censoring scheme here; in the former, it is clumped on the right at the value m and in the latter, it is clumped on the left at the value m. It is instructive to note here that the event $\{D = m\}$ would occur in both cases listed in (1.5): in the first case when $Y > T$, the termination of the life-test would occur at Y resulting in exactly m complete failures, and in the second case when $Y \leq T$, the termination would occur at T, but if no failure occurs in the interval (Y, T) then also exactly m complete failures would be realized. Now, from (1.6) and (1.7), we can readily compute the values of $E(T_{\text{II}})$ and $E(D_{\text{II}})$. These are presented in Table 1.3 for the purpose of comparing Type-II censoring and Type-II hybrid censoring schemes in terms of termination time and expected number of failures.

Thing to note with Type-II hybrid censoring scheme is that all pertinent inferential methods based on it will be unconditional, just like in the case of Type-II censoring, due to the fact that at least m failures are guaranteed to be observed. Additional advantages and the intended purpose of Type-II hybrid censoring scheme, as stated originally by Childs et al. (2003), become clear from Table 1.3. If the experimenter had chosen the test time T to be small compared to the number of complete failures to be observed from the life-test, then the life-test in all likelihood would proceed until that many complete failures are observed. For example, for the choice of $m = 8$, if the experimenter had chosen the time T to be 0.4, then there is about 99% chance that the 8th failure would occur after $T = 0.4$ and so the expected duration of the test becomes 0.728 and consequently the expected number of complete failures observed becomes 8.002, as seen in Table 1.3. On the other hand, had the experimenter chosen the test time T to be large compared to the number of complete failures to be observed from the life-test, then the life-test in all likelihood would proceed until time T. For example, for the choice of $m = 6$, if the experimenter had chosen the time T to be 0.8, then there is about 3% chance that the 6th failure would occur before $T = 0.8$, in which case we see that the expected duration of the test is 0.801 and the expected number of failures observed becomes 8.04. Thus, in either case, the Type-II hybrid censoring scheme provides guarantee for observing enough complete failures from the test to facilitate the development of precise inferential results, whether it is point/interval estimation or hypothesis tests. Of course, this advantage naturally comes at a price of having a longer life-test than under any of Type-I censoring, Type-I hybrid censoring, and Type-II censoring.

1.4. Antecedents

As mentioned earlier, Epstein (1954) was the first one to introduce a hybrid censoring scheme to facilitate early termination of a life-test as soon as a certain number of failures the experimenter had in mind is achieved, instead of carrying on with the test

Table 1.3 Comparison of Type-II censoring and Type-II hybrid censoring based on termination time and number of failures for the uniform distribution when the number of test units is $n = 10$.

Scheme parameters		Type-II censoring		Type-II HCS	
T	m	$E(Y)$	$\Pr(Y>T)$	$E(T_{II})$	$E(D_{II})$
0.4	4	0.3636	0.3823	0.4408	4.6020
0.5	4	0.3636	0.1719	0.5138	5.2384
0.6	4	0.3636	0.0548	0.6032	6.0688
0.7	4	0.3636	0.0106	0.7004	7.0123
0.8	4	0.3636	0.0009	0.8000	8.0010
0.4	5	0.4545	0.6331	0.4893	5.2351
0.5	5	0.4545	0.3770	0.5388	5.6154
0.6	5	0.4545	0.1662	0.6123	6.2351
0.7	5	0.4545	0.0474	0.7024	7.0597
0.8	5	0.4545	0.0064	0.8002	8.0073
0.4	6	0.5455	0.8338	0.5577	6.0688
0.5	6	0.5455	0.6231	0.5842	6.2384
0.6	6	0.5455	0.3669	0.6347	6.6020
0.7	6	0.5455	0.1503	0.7095	7.2100
0.8	6	0.5455	0.0328	0.8013	8.0401
0.4	7	0.6363	0.9452	0.6396	7.0141
0.5	7	0.6363	0.8281	0.6502	7.0665
0.6	7	0.6363	0.6177	0.6772	7.2197
0.7	7	0.6363	0.3504	0.7286	7.5603
0.8	7	0.6363	0.1209	0.8058	8.1610
0.4	8	0.7273	0.9877	0.7279	8.0017
0.5	8	0.7273	0.9453	0.7308	8.0118
0.6	8	0.7273	0.8327	0.7412	8.0524
0.7	8	0.7273	0.6172	0.7678	8.1776
0.8	8	0.7273	0.3222	0.8205	8.4832

until the pre-fixed time T. He then considered the case of exponential lifetimes and derived expressions for the mean termination time as well as the expected number of failures under the Type-I hybrid censoring scheme form that he introduced. In addition, he also considered a "replacement case" in which failed units are replaced a once by new units drawn from the same exponential population, and derived explicit expressions for the same quantities so that a comparison could be made between the two cases. Later, Epstein (1960b) developed hypothesis tests concerning the exponential mean parameter, while Epstein (1960a,c) discussed the construction of confidence intervals (one-sided and two-sided) for the mean lifetime of an exponential distribution based on a Type-I hybrid censored data, using chi-square distribution for the pivotal quantity and using a chi-square percentage point approximately even in the case of no failures. These procedures were subsequently adopted as reliability qualification tests and reliability acceptance tests based on exponential lifetimes as standard test plans in MIL-STD-781-C (1977), wherein the performance requirement is specified through mean-time-between-failure (MTBF). Harter (1978) evaluated the performance of these confidence bounds for the MTTF through Monte Carlo simulations. A formal rule for obtaining a two-sided confidence interval for the MTTF, in the exponential case, was given by Fairbanks et al. (1982) who demonstrated that their rule is very close to the approximation provided earlier by Epstein (1960c) and also provided a validation for their rule.

In the paper by Bartholomew (1963), the exact conditional distribution of the maximum likelihood estimator of the mean of an exponential distribution under a time censored life-test (i.e., under Type-I censoring) was derived through conditional moment generating function (Conditional MGF) approach, conditioned on the event that at least one complete failure is observed. This method was adopted by Chen and Bhattacharyya (1988) to develop the exact distribution theory for the maximum likelihood estimation of the exponential mean lifetime under Type-I hybrid censoring scheme, and the conditional moment generating function approach has since become a standard tool for developing exact distribution theory for maximum likelihood estimators of parameters under various forms of hybrid censored data, as will be seen in the ensuing chapters.

The prediction of times of future failures, based on a Type-I hybrid censoring scheme, for the case of exponential distribution, was discussed by Ebrahimi (1992) for both cases when the failed units are not replaced and replaced by new units. All the works mentioned so far dealt with a scaled exponential distribution, involving only the mean lifetime parameter. A two-parameter exponential distribution, consisting of a threshold parameter (interpreted as guarantee period in reliability literature) and a scale parameter (relating to the residual mean lifetime parameter), was considered by Ebrahimi (1986) who then developed point and interval estimation methods as well as hypothesis tests for both cases when the hybrid life-tests involved without and with replacement of failed units.

All these early developments were on various inferential aspects based on data observed from Type-I hybrid censoring scheme, as introduced by Epstein (1954). But, as mentioned in the preceding section, the Type-II hybrid censoring scheme, guaranteeing at least a pre-specified number of complete failures to be observed in the life-test, was introduced by Childs et al. (2003) to overcome some of the shortcomings of Type-I hybrid censoring scheme; but, this advantage comes at a price of having a longer life-test, as mentioned earlier. Since then, the literature on hybrid censoring has exploded with varying forms of hybrid schemes, for many different lifetime distributions, and the development of a wide range of inferential methods. The following section gives an account of the recent growth in this area of research!

Table 1.4 Publication record on hybrid censoring for the period 1954–2022 for given time intervals.

Period	No. of Publications
1954–1963	3
1964–1973	4
1974–1983	26
1984–1993	45
1994–1998	47
1999–2003	31
2004–2008	61
2009–2013	131
2014–2018	227
2019–2022	103

1.5. Burgeoning literature

As mentioned above, the literature on hybrid censoring has grown significantly in recent years. For example, a quick search on *zbmath*, using *hybrid censoring*, *hybrid censored*, and *truncated life test*, *truncated life tests*, *truncated life testing* as keywords, produced the frequency table (Table 1.4) and the histogram for the publication record[4] (see Fig. 1.1).

1.6. Scope of the book

The primary objectives in preparing this book have been to produce an up-to-date volume, with emphasis on both theory and applications that will form as a reference guide for practitioners involved in the design of a life-testing experiment as well as in the analysis of lifetime data observed from such experiments. The models and methods described in the book would provide the reader a know-how regarding the designs and implementations of various hybrid censoring schemes and their merits and demerits. A central aspect in our presentation and analysis of the different hybrid censoring schemes is to identify the key shared features as well as the structural elements. In particular, this enables a structured and efficient approach to complex hybrid censoring models and is intended to support both the design and analysis of new, possibly even more complex models and the development of associated statistical procedures. A complete treatment of basic theory, including the derivation of all associated characteristics, properties and inferential results, is presented for all the different hybrid censoring schemes considered in the literature. To facilitate a better understanding and appreciation of these develop-

[4] Retrieved on June 1, 2022 from http://zbmath.org/.

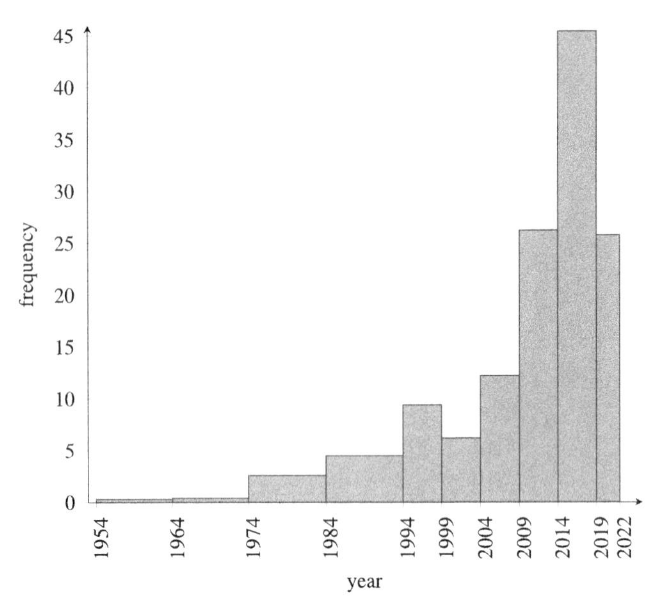

Figure 1.1 Histogram of publication record data on hybrid censoring for the period 1954–2022 for given time intervals given in Table 1.4.

ments, a comprehensive review of all pertinent results on the conventional censoring schemes, viz., Type-II, Type-I, and progressive censoring, is also given.

Throughout the book, many numerical examples as well as examples based on real-life data sets have been presented in order to demonstrate practicability and usefulness of the theoretical results discussed. In doing so, due recognition has been given to all the published works on hybrid censoring methodology by citing them appropriately. Thus, the Bibliography at the end of the book contains an exhaustive list of publications on the subject to-date and it should be valuable for any researcher interested in working on this topic of research.

The coverage of the rest of the book, intended to give a detailed overview of all pertinent developments, is as follows. Chapter 2 presents some preliminary results on order statistics, progressively Type-II right censored order statistics, generalized order statistics and sequential order statistics, and they get used repeatedly in the ensuing chapters. In Chapter 3, pertinent inferential methods and results are described for the classical censoring schemes, viz., Type-II, Type-I, and progressive censoring, and they become foundational for the corresponding results for hybrid censoring schemes developed subsequently. Chapter 4 discusses various models and distributional properties of hybrid censoring designs and inter-relationships between different forms of hybrid censoring. While Chapter 5 presents inferential results for the case of exponential lifetime distribution, Chapter 6 deals with inferential results for some other important lifetime distributions such as Weibull, log-normal, Birnbaum-Saunders, generalized exponential, (log-)Laplace, and uniform. The generalization to progressive hybrid censoring situation is handled in Chapter 7. Chapter 8 deals with the derivation of information measures for different forms of hybrid censored data. These include Fisher information, entropy, Kullback-Leibler information, and Pitman closeness. Next, applications of hybrid censoring to reliability problems are discussed in Chapters 9 and 10 with Chapter 9 focussing on step-stress accelerated life-tests and Chapter 10 handling competing risks analysis, stress-strength analysis, acceptance sampling plans, and optimal design problems. Chapters 11 and 12 deal with model validation methods and prediction (both point and interval prediction) issues, respectively. Finally, adaptive progressive hybrid censoring schemes are addressed in Chapter 13.

Even though the coverage of the topic this way is quite elaborate and detailed, some gaps that exist in the literature and also some problems that will be worthy of further study are pointed out at different places in the book. These could serve as convenient starting points for any one interested in engaging in this area of research.

1.7. Notation

The following (incomplete) list provides notation that will be used throughout the book. In selected cases, page numbers are provided in order to refer to pertinent definition.

Notation	**Explanation**	**Page**
Random variables		
X, Y, Z	random variables	
$\boldsymbol{X}, \boldsymbol{X}_n$	vector of random variables $X_1, \dots, X_n$	
$X_{i:n}$	i-th order statistic in a sample of size n	
$U_{i:n}$	i-th uniform order statistic in a sample of size n	
$S_{i,n}$	i-th normalized spacing in a sample of size n	
$X^{\mathscr{R}}_{i:m:n}, X_{i:m:n}$	i-th progressively Type-II censored order statistic based on censoring plan $\mathscr{R}$	
$U^{\mathscr{R}}_{i:m:n}, U_{i:m:n}$	i-th uniform progressively Type-II censored order statistic based on censoring plan $\mathscr{R}$	
$\mathbf{X}^{\mathscr{R}}, \mathbf{U}^{\mathscr{R}}, \mathbf{Z}^{\mathscr{R}}$	vector of progressively Type-II censored order statistics	
D	random counter $\sum_{j=1}^{n} \mathbb{1}_{(-\infty,\tau]}(X_{j:n})$ (for order statistics) or $\sum_{j=1}^{n^*} \mathbb{1}_{(-\infty,\tau]}(X_{j:n^*:n})$ (for progressively Type-II censored order statistics).	
$D_{\mathsf{HCS}}, D_{\mathsf{I}}, D_{\mathsf{II}}, D_{\mathsf{gI}}, D_{\mathsf{gII}}, D_{\mathsf{uI}}, D_{\mathsf{uII}}, D_{\mathsf{uIII}}, D_{\mathsf{uIV}}$	random counter for corresponding hybrid censoring scheme	
Censoring plans, coefficients, constants		
$\mathscr{R}, \mathscr{S}$	censoring plan $\mathscr{R} = (R_1, \dots, R_m)$, etc.	
$\gamma_i, \gamma_i(\mathscr{R})$	$\gamma_i = \sum_{j=i}^{m}(R_j + 1)$, $1 \le i \le m$, for a censoring plan $\mathscr{R} = (R_1, \dots, R_m)$	
c_{r-1}	$\prod_{j=1}^{r} \gamma_j$	

Notation	Explanation	Page
$a_{j,r}$	$\prod_{\substack{i=1\\i\neq j}}^{r} \frac{1}{\gamma_i-\gamma_j}$, $1 \leq j \leq r$	
$a_{j,k_2}^{(k_1)}$	$\prod_{\substack{\nu=k_1+1\\\nu\neq j}}^{k_2} \frac{1}{\gamma_\nu-\gamma_j}$, $k_1+1 \leq j \leq k_2$	
Density functions, cumulative distribution functions		
f, f^X, F, F^X	density function/cumulative distribution function (of a random variable X)	
$F^{\leftarrow}$	quantile function of F	
$\overline{F}$	survival/reliability function $\overline{F} = 1 - F$	
h, h^X	hazard rate $h = f/(1-F)$ of F, F^X	
$F_{Exp}, F_{Exp(\vartheta)}$	cumulative distribution function of a (standard) exponential distribution (with mean ϑ)	
φ, Φ, Φ^{-1}	density function, cumulative distribution function, and quantile function of a standard normal distribution $N(0, 1)$	
$f_{\chi^2(d)}, F_{\chi^2(d)}$	density function/cumulative distribution function of a χ^2-distribution with d degrees of freedom	
$f_{\Gamma(\vartheta,\beta)}, F_{\Gamma(\vartheta,\beta)}$	density function/cumulative distribution function of a $\Gamma(\vartheta, \beta)$-distribution with scale parameter ϑ and shape parameter β	
$f_{j:n}, F_{j:n}$	density function/cumulative distribution function of $X_{j:n}$	
$f_{j:m:n}, f_{j:m:n}^{\mathscr{R}}$	density function of $X_{j:m:n}^{\mathscr{R}}$	
$F_{j:m:n}, F_{j:m:n}^{\mathscr{R}}$	cumulative distribution function of $X_{j:m:n}^{\mathscr{R}}$	
$f_{1,\ldots,m:n}, F_{1,\ldots,m:n}$	joint density function/cumulative distribution function of $X_{1:n}, \ldots, X_{m:n}$	
$f_{1,\ldots,m:m:n}, F_{1,\ldots,m:m:n}$	joint density function/cumulative distribution function of $X_{1:m:n}^{\mathscr{R}}, \ldots, X_{m:m:n}^{\mathscr{R}}$	
$\mathscr{L}(\boldsymbol{\theta};\text{data}), \mathscr{L}(\boldsymbol{\theta};\boldsymbol{x}_d, d)$	likelihood function for parameter $\boldsymbol{\theta} = (\theta_1, \ldots, \theta_p)$ given some data or $\boldsymbol{x}_d, d$	
$\mathscr{L}^*(\boldsymbol{\theta};\text{data}), \mathscr{L}^*(\boldsymbol{\theta};\boldsymbol{x}_d, d)$	log-likelihood function for parameter $\boldsymbol{\theta} = (\theta_1, \ldots, \theta_p)$ given some data or $\boldsymbol{x}_d, d$	
$\mathscr{I}(\boldsymbol{X};\boldsymbol{\theta})$	Fisher information about $\boldsymbol{\theta} = (\theta_1, \ldots, \theta_p)$ (in $\boldsymbol{X}$)	
Distributions		
$Uniform(a, b)$	uniform distribution	354
$Beta(\alpha, \beta)$	beta distribution	354
$Power(\alpha)$	power distribution	354
$RPower(\beta)$	reflected power distribution	354
$Exp(\mu, \vartheta)$, $Exp(\vartheta)$	exponential distribution	354
$Weibull(\vartheta, \beta)$	Weibull distribution	355
$\Gamma(\vartheta, \beta)$	gamma distribution	355
$N(\mu, \sigma^2)$	normal distribution	356
χ_n^2	χ^2-distribution	355
$F_{n,m}$	F-distribution	355
$Pareto(\alpha)$	Pareto distribution	355
$Laplace(\mu, \vartheta)$	Laplace distribution	356
$bin(n, p)$	binomial distribution	356

Notation	Explanation	Page
Special functions		
$\exp(t)$, e^t	exponential function	
$\log(t)$	natural logarithm	
$\Gamma(\alpha)$	gamma function	
$\mathsf{IG}(t;\alpha)$	incomplete gamma function ratio defined as $\mathsf{IG}(t;\alpha) = \frac{1}{\Gamma(\alpha)}\int_0^t z^{\alpha-1}e^{-z}dz,\ t \geq 0$	
$\mathsf{B}(\alpha,\beta)$	beta function	
$I_t(\alpha,\beta)$	incomplete beta function ratio defined as $I_t(\alpha,\beta) = \frac{1}{\mathsf{B}(\alpha,\beta)}\int_0^t x^{\alpha-1}(1-x)^{\beta-1}dx,\ 0 < t < 1$	
$n!$	n factorial defined by $\prod_{j=1}^n j,\ n \in \mathbb{N}_0$, where $\prod_{j=1}^0 j = 1$	
$\binom{n}{k}$	binomial coefficient $\frac{n!}{k!(n-k)!},\ k, n \in \mathbb{N}_0,\ k \leq n$	
$\binom{n}{k_1,\ldots,k_r}$	multinomial coefficient $\frac{n!}{k_1!\cdot\ldots\cdot k_r!},\ k_i, n \in \mathbb{N}_0,\ \sum_{i=1}^r k_i = n$	
$B_k(\cdot \mid a_k,\ldots,a_1)$	univariate B-Spline B_k of degree k with knots $a_k,\ldots,a_1$	357
$\mathbb{1}_A$	indicator function on the set A	
$[x]_+ = \max(x,0)$	positive part of x	
$[x]_- = \min(x,0)$	negative part of x	
$\lfloor x \rfloor$	largest integer k satisfying $k \leq x$	
$\mathrm{sgn}(x)$	sign of x	
$[x_j,\ldots,x_\nu]h$	divided differences of order $\nu - j$ at $x_\nu > \cdots > x_j$ for function h	358
Sets		
$\mathbb{N}$	integers $\{1,2,3,\ldots\}$	
$\mathbb{N}_0$	$\mathbb{N} \cup \{0\} = \{0,1,2,3,\ldots\}$	
$\mathbb{R}$	real numbers	
$\mathbb{R}^n$	n-fold Cartesian product of $\mathbb{R}$	
$\mathbb{R}^n_\leq$	$\{(x_1,\ldots,x_n) \in \mathbb{R}^n \mid x_1 \leq \cdots \leq x_n\}$	
$\mathbb{R}^n_{\leq T}$	$\{(x_1,\ldots,x_n) \in \mathbb{R}^n \mid x_1 \leq \cdots \leq x_n \leq T\}$	
$\mathbb{R}^{k\times n}$	set of $(k \times n)$-matrices	
$\mathfrak{S}_n$	set of all permutations of $(1,\ldots,n)$	
$(\Omega,\mathfrak{A})$	measurable space with σ-algebra $\mathfrak{A}$	
$(\Omega,\mathfrak{A},P)$	probability space with σ-algebra $\mathfrak{A}$ and probability measure P	
Symbols		
$X \sim F$	X is distributed according to a cumulative distribution function F	
iid	independent and identically distributed	
$X_1,\ldots,X_n \overset{\text{iid}}{\sim} F$	$X_1,\ldots,X_n$ are iid random variables from a cumulative distribution function F	
$\overset{d}{=}$	equality in distribution	
$\overset{d}{\to}$	convergence in distribution	
$\overset{\text{a.e.}}{\longrightarrow}$	convergence almost everywhere	
$\alpha(F)$	left endpoint of the support of F	
$\omega(F)$	right endpoint of the support of F	
$\xi_p = F^{\leftarrow}(p),\ p \in (0,1)$	p-th quantile of F	

Notation	Explanation	Page
z_p	p-th quantile of a standard normal distribution	
$\chi^2_{\nu,p}$	p-th quantile of a χ^2-distribution with ν degrees of freedom	
$\mathbb{\lambda}^r$	r-dimensional Lebesgue measure	
δ_x	one-point distribution in x/Dirac measure in x	
$\mathrm{tr}(A)$	trace of a matrix A	
$\det(A)$	determinant of a matrix A	
$\mathrm{med}(F)$	median of cumulative distribution function F	
$\boldsymbol{a}_k$	$(a_1, \ldots, a_k) \in \mathbb{R}^k$	
(a^{*k})	$(a, \ldots, a) \in \mathbb{R}^k$	
$\mathbf{1}$	$\mathbf{1} = (1, \ldots, 1) = (1^{*k})$	
$g(t-)$	$\lim_{x \nearrow t} g(x)$	
$g(t+)$	$\lim_{x \searrow t} g(x)$	
const	represents all additive terms of a function which do not contain the variable of the function	
Operations		
$x \wedge y$	$\min\{x, y\}$, $x, y \in \mathbb{R}$	
$x \vee y$	$\max\{x, y\}$, $x, y \in \mathbb{R}$	
$\boldsymbol{x} \wedge \boldsymbol{y}$	$(x_1 \wedge y_1, \ldots, x_n \wedge y_n)^T$, $\boldsymbol{x}, \boldsymbol{y} \in \mathbb{R}^n$	
$\boldsymbol{x} \vee \boldsymbol{y}$	$(x_1 \vee y_1, \ldots, x_n \vee y_n)^T$, $\boldsymbol{x}, \boldsymbol{y} \in \mathbb{R}^n$	
$a_{\bullet r}$	$\sum_{j=1}^r a_j$ for $\boldsymbol{a}_r = (a_1, \ldots, a_r) \in \mathbb{R}^r$	
$\boldsymbol{x}^T$	transposed vector of vector $\boldsymbol{x} \in \mathbb{R}^n$	
Σ^{-1}	inverse matrix of $\Sigma \in \mathbb{R}^{n \times n}$	
Abbreviations		
a.s.	almost surely	
a.e.	almost everywhere	
e.g.	for example / exempli gratia	
i.e.	id est	
w.r.t.	with respect to	
IFR/DFR	increasing/decreasing failure rate	
BLUE	best linear unbiased estimator	
BLIE	best linear invariant estimator	
MLE	maximum likelihood estimator	
AMLE	approximate maximum likelihood estimator	
UMVUE	uniformly minimum variance unbiased estimator	
BUP	best unbiased predictor	
MLP	maximum likelihood predictor	
AMLP	approximate maximum likelihood predictor	
MUP	median unbiased predictor	
CMP	conditional median predictor	
BLUP	best linear unbiased predictor	
BLEP	best linear equivariant predictor	
PLF	predictive likelihood function	
HPD	highest probability density	
MSE	mean squared error	

CHAPTER 2

Preliminaries

Contents

2.1. Introduction

Inferential methods for handling various forms of hybrid censored data, to be discussed in detail in all ensuing chapters, will naturally require a thorough understanding and sound knowledge of distributional properties, dependence characteristics and moments associated with different forms of ordered data. These include order statistics, progressively Type-II censored order statistics, sequential order statistics, and generalized order statistics. The aim of this chapter is to provide a brief survey of various important distributional results and properties for all these forms of ordered data. These will get used repeatedly in the ensuing chapters.

In Section 2.2, we provide a basic introduction to the theory of usual order statistics and first present some general distributional results. We then present some specific results for order statistics arising from uniform, exponential, Weibull and Laplace distributions. Next, in Section 2.3, we provide a basic introduction to the theory of progressively Type-II censored order statistics and describe some general distributional results. We

Hybrid Censoring Know-How
https://doi.org/10.1016/B978-0-12-398387-9.00010-6

then present some specific results for progressively Type-II censored order statistics arising from uniform, exponential, Weibull and Laplace distributions. In Section 2.4, we introduce the notion of generalized order statistics which includes different models of ordered data as special cases and present some general distributional results. Finally, Section 2.5, we introduce briefly the notion of sequential order statistics and present some general distributional results, and mention how it is associated to proportional hazards family of distributions.

2.2. Order statistics

Let $X_1, \ldots, X_n$ be a random sample of size n from a continuous population with absolutely continuous cumulative distribution function F^X with density function f^X, defined over a support $S \subseteq \mathbb{R}$. For short, we write f and F instead of f^X and F^X, respectively, if the connection of the random variables and its distribution is obvious. Then, due to the iid assumption on X_i's, we readily have the joint density function of X_i's to be[1]

$$f^{X_1,\ldots,X_n}(x_1, \ldots, x_n) = \prod_{i=1}^{n} f^{X_i}(x_i), \quad x_1, \ldots, x_n \in \mathbb{R}. \tag{2.1}$$

There exists a large body of literature on order statistics and their applications. Interested readers may refer to the books by Arnold et al. (1992, 2008), Balakrishnan and Rao (1998b,c), and David and Nagaraja (2003) for elaborate discussions on various issues concerning order statistics. Recently, Cramer (2021) has provided an illustrative review of different types of lifetime data and censoring mechanisms. Here, we describe some key results that are essential for the developments in subsequent chapters.

2.2.1 Joint and marginal distributions

From (2.1), by considering the transformation from $(X_1, \ldots, X_n)$ to the set of order statistics $(X_{1:n}, \ldots, X_{n:n})$ and recognizing that it is a $n!$-to-1 transformation, it is readily observed that the joint density function of all n order statistics is as presented below.

Theorem 2.1 (Joint density of order statistics). *Let $X_1, \ldots, X_n$ be iid random variables with absolutely continuous density function f and cumulative distribution function F. Then, the joint density function of $(X_{1:n}, \ldots, X_{n:n})$ is*

$$f_{1,\ldots,n:n}(x_1, \ldots, x_n) = n! \prod_{i=1}^{n} f(x_i), \quad x_1 < x_2 < \cdots < x_n. \tag{2.2}$$

[1] For an iid sample $X_1, \ldots, X_n$ with common cumulative distribution function F or density function f we write subsequently for short $X_1, \ldots, X_n \overset{\text{iid}}{\sim} F$ and $X_1, \ldots, X_n \overset{\text{iid}}{\sim} f$, respectively.

Now, for a fixed $d \in \{1, \ldots, n-1\}$, let $(X_{1:n}, \ldots, X_{d:n})$ be a Type-II right censored sample of size d from a sample of size n, with the largest $n-d$ observations having been censored. Then, from (2.2), by integrating out the variables $x_{d+1}, \ldots, x_n$, over the range $x_d < x_{d+1} < \cdots < x_n < \infty$, we readily obtain the likelihood function (i.e., the joint density function) of the above Type-II right censored sample as presented in (2.3).

Theorem 2.2 (Likelihood function of a Type-II right censored sample). *For a fixed $d \in \{1, \ldots, n-1\}$, let $(X_{1:n}, \ldots, X_{d:n})$ be a Type-II right censored sample from a population with absolutely continuous cumulative distribution function F and density function f. Then, the corresponding joint density function is*

$$f_{1,\ldots,d:n}(x_1, \ldots, x_d) = \frac{n!}{(n-d)!}(1 - F(x_d))^{n-d} \prod_{i=1}^{d} f(x_i), \quad x_1 < \cdots < x_d. \tag{2.3}$$

Next, for a fixed $r \in \{0, 1, \ldots, n-1\}$, let $(X_{r+1:n}, X_{r+2:n}, \ldots, X_{n:n})$ be a Type-II left censored sample of size $n-r$ from a sample of size n, with the smallest r observations having been censored. Then, from (2.2), by integrating out the variables $x_1, \ldots, x_r$, over the range $-\infty < x_1 < \cdots < x_r < x_{r+1}$, we readily obtain the likelihood function (i.e., the joint density function) of the above Type-II left censored sample as presented in (2.4).

Theorem 2.3 (Likelihood function of a Type-II left censored sample). *For a fixed $r \in \{0, 1, \ldots, n-1\}$, let $(X_{r+1:n}, \ldots, X_{n:n})$ be a Type-II left censored sample from a population with absolutely continuous cumulative distribution function F and density function f. Then, the corresponding joint density function is*

$$f_{r+1,\ldots,n:n}(x_{r+1}, \ldots, x_n) = \frac{n!}{r!} F^r(x_{r+1}) \prod_{i=r+1}^{n} f(x_i), \quad x_{r+1} < \cdots < x_n. \tag{2.4}$$

Finally, for fixed r and d (with $1 \leq d - r \leq n - 1$), let $(X_{r+1:n}, \ldots, X_{d:n})$ be a Type-II doubly censored sample from a sample of size n, with the smallest r and largest $n-d$ observations having been censored. Then, from (2.2), by integrating out the variables $(X_{1:n}, \ldots, X_{r:n})$ and $(X_{d+1:n}, \ldots, X_{n:n})$, over the ranges $-\infty < x_1 < \cdots < x_r < x_{r+1}$ and $x_d < x_{d+1} < \cdots < x_n < \infty$, respectively, we obtain the likelihood function of the above Type-II doubly censored as presented in (2.5).

Theorem 2.4 (Likelihood function of a Type-II doubly censored sample). *For fixed r and d (with $1 \leq d - r \leq n - 1$), let $(X_{r+1:n}, \ldots X_{d:n})$ be a Type-II doubly censored sample from a population with absolutely continuous cumulative distribution function F and density function f. Then, the corresponding likelihood function is*

$$f_{r+1,\ldots,d:n}(x_{r+1}, \ldots, x_d) = \frac{n!}{r!(n-d)!} F^r(x_{r+1})(1 - F(x_d))^{n-d} \prod_{i=r+1}^{d} f(x_i),$$
$$x_{r+1} < \cdots < x_d. \tag{2.5}$$

Remark 2.5. The likelihood functions in (2.3)–(2.5) are commonly used while developing likelihood estimates of model parameters under these three forms of Type-II censored data, but often without considering the normalizing constants in them. Of course, that will have no impact on the maximum likelihood estimation itself. This is the reason why (2.3)–(2.5) may be viewed as joint densities of the sets of order statistics involved in the censored samples which necessarily would require the normalizing constants!

Proceeding in a similar manner from (2.1), by using judicious integration, we can arrive at the following general distribution for a set of order statistics (see, e.g., Balakrishnan and Rao, 1998a, (3.5); David and Nagaraja, 2003, (2.2.2)). A subsample $(X_{d_1:n}, X_{d_2:n}, \dots, X_{d_r:n})$ of order statistics with $1 \le d_1 < d_2 < \cdots < d_r \le n$ is also called multiply Type-II censored sample (see, e.g., Kong, 1998 and Section 7.2.8).

Theorem 2.6 (Joint density function of a set of order statistics). *Consider the set of order statistics $(X_{d_1:n}, X_{d_2:n}, \dots, X_{d_r:n})$, for $1 \le d_1 < d_2 < \cdots < d_r \le n$, from a sample of size n from a population with absolutely continuous cumulative distribution function F and density function f. Then, their joint density function is*

$$f_{d_1,\dots,d_r:n}(x_{d_1}, \dots, x_{d_r}) = c\prod_{i=1}^{r+1}\left(F(x_{d_i}) - F(x_{d_{i-1}})\right)^{d_i - d_{i-1} - 1}\prod_{i=1}^{r} f(x_i) \tag{2.6}$$

for $x_{d_1} < \cdots < x_{d_r}$, with $d_0 \equiv 0$, $d_{r+1} \equiv n+1$, $x_{d_0} \equiv -\infty$ and $x_{d_{r+1}} \equiv \infty$ so that $F(x_{d_0}) = 0$ and $F(x_{d_{r+1}}) = 1$, where

$$c = \frac{n!}{\prod_{i=1}^{r+1}(d_i - d_{i-1} - 1)!}.$$

An important particular case of the above density function is needed in prediction problems. Here, we need the joint density function of $X_{1:n}, \dots, X_{d+1:n}, X_{d+s:n}$ with $d \in \{1, \dots, n-2\}$ and $s \in \{2, \dots, n-d\}$.

Theorem 2.7. *Consider order statistics $X_{1:n}, \dots, X_{d+1:n}, X_{d+s:n}$ with $d \in \{1, \dots, n-2\}$ and $s \in \{2, \dots, n-d\}$ from a sample of size n from a population with absolutely continuous cumulative distribution function F and density function f. Then, their joint density function is*

$$\begin{aligned} &f_{1,\dots,d+1,d+s:n}(x_1, \dots, x_d, x_{d+s}) \\ &\quad = c\,(F(x_{d+s}) - F(x_{d+1}))^s\,(1 - F(x_{d+s}))^{n-d-s-1}\left(\prod_{i=1}^{d+1} f(x_i)\right) f(x_{d+s}) \end{aligned} \tag{2.7}$$

for $x_1 < \cdots < x_{d+1} < x_{d+s}$, where

$$c = \frac{n!}{(s-2)!(n-d-s)!}.$$

In the special case of $r = 2$, (2.6) readily yields the joint density function of two order statistics as presented in (2.8).

Theorem 2.8 (Joint density function of two order statistics). *The joint density function of order statistics $X_{d_1:n}$ and $X_{d_2:n}$, for $1 \leq d_1 < d_2 \leq n$, from a sample of size n from a population with absolutely continuous cumulative distribution function F and density function f, is*

$$f_{d_1,d_2:n}(x_{d_1}, x_{d_2}) = \frac{n!}{(d_1-1)!(d_2-d_1-1)!(n-d_2)!} F^{d_1-1}(x_{d_1}) \\ \times (F(x_{d_2}) - F(x_{d_1}))^{d_2-d_1-1}(1-F(x_{d_2}))^{n-d_2} f(x_{d_1}) f(x_{d_2}), \ x_{d_1} < x_{d_2}. \quad (2.8)$$

Finally, in the special case of $r = 1$ in Theorem 2.6, we simply get the marginal density function of $X_{d:n}$, for $1 \leq d \leq n$, from (2.6) as follows.

Theorem 2.9 (Marginal density function of a single order statistic). *The marginal density function of the order statistic $X_{d:n}$, for $1 \leq d \leq n$, from a sample of size n from a population with absolutely continuous cumulative distribution function F and density function f, is*

$$f_{d:n}(x_d) = \frac{n!}{(d-1)!(n-d)!} F^{d-1}(x_d)(1-F(x_d))^{n-d} f(x_d), \qquad x_d \in \mathbb{R}. \quad (2.9)$$

From the marginal density function in (2.9), through direct integration, we can obtain the corresponding cumulative distribution function as presented below. Using the quantile representation in Theorem 2.36 it follows that (2.10) holds for any cumulative distribution function F. Absolute continuity of F is not necessary for the representation to hold.

Theorem 2.10 (Cumulative distribution function of a single order statistic). *The cumulative distribution function of the order statistic $X_{d:n}$, for $1 \leq d \leq n$, from a sample of size n from a population with cumulative distribution function F, is given by*

$$F_{d:n}(x_d) = \int_{-\infty}^{x_d} \frac{n!}{(d-1)!(n-d)!} F^{d-1}(x)(1-F(x))^{n-d} f(x)dx \\ = \sum_{i=d}^{n} \binom{n}{i} F^i(x_d)(1-F(x_d))^{n-i}, \ x_d \in \mathbb{R}. \quad (2.10)$$

Remark 2.11. The expression in (2.10) can be derived directly by carrying out the required integration by parts repeatedly. Alternatively, it can be easily obtained through the use of Pearson's identity between incomplete beta function and binomial survival function (see (1.1)).

Furthermore, it needs to be mentioned that the expression of the cumulative distribution function in (2.10) can also be established simply by binomial argument that the event $\{X_{d:n} \leq x_d\}$ is exactly the same as the event of getting at least d successes out of n

iid Bernoulli trials each with probability of success $\Pr(X \le x_d) = F(x_d)$ (see David and Nagaraja, 2003, Section 2.1).

Finally, it is important to point out that the binomial argument mentioned above applies not only for the case of continuous distributions, but also for discrete distributions, as it is based only on the probability of success $F(x_d)$ of Bernoulli trials. However, the density function and joint density functions presented in all the above theorems do not hold for discrete distributions due to the possibility of ties; consequently, suitable modifications have to be made in their forms. Interested readers may refer to Chapter 3 of Arnold et al. (1992) for pertinent details. We abstain here, however, from a description of these results for order statistics from discrete distributions due to the fact that the focus of this book is on various life-testing experiments and that the lifetimes of units under test are assumed to arise from a continuous distribution.

2.2.2 Conditional distributions

We now present some important results concerning conditional distributions of order statistics.

Theorem 2.12 (Conditional distribution in terms of left truncation). *The conditional density function of $X_{d_2:n}$, given $X_{d_1:n} = x_{d_1}$, for $1 \le d_1 < d_2 \le n$, is*

$$f_{d_2|d_1:n}(x_{d_2}|x_{d_1}) = \frac{(n-d_1)!}{(d_2-d_1-1)!(n-d_2)!}\left\{\frac{F(x_{d_2})-F(x_{d_1})}{1-F(x_{d_1})}\right\}^{d_2-d_1-1} \times \left\{\frac{1-F(x_{d_2})}{1-F(x_{d_1})}\right\}^{n-d_2}\frac{f(x_{d_2})}{1-F(x_{d_1})}, \qquad x_{d_2} > x_{d_1}, \tag{2.11}$$

that is, the conditional distribution of $X_{d_2:n}$, given $X_{d_1:n} = x_{d_1}$, for $d_1 < d_2 \le n$, is the same as the distribution of the (d_2-d_1)-th order statistic from a random sample of size $n-d_1$ from the left-truncated distribution at x_{d_1}, with density function $\frac{f}{1-F(x_{d_1})}$ and cumulative distribution function $\frac{F(\cdot)-F(x_{d_1})}{1-F(x_{d_1})}$ on (x_{d_1}, ∞).

Theorem 2.13 (Conditional distribution in terms of right truncation). *The conditional density function of $X_{d_1:n}$, given $X_{d_2:n} = x_{d_2}$, for $1 \le d_1 < d_2$, is*

$$f_{d_1|d_2:n}(x_{d_1} \mid x_{d_2}) = \frac{(d_2-1)!}{(d_1-1)!(d_2-d_1)!}\left\{\frac{F(x_{d_1})}{F(x_{d_2})}\right\}^{d_1-1}\left\{\frac{F(x_{d_2})-F(x_{d_1})}{F(x_{d_2})}\right\}^{d_2-d_1}\frac{f(x_{d_1})}{F(x_{d_2})}, \qquad x_{d_1} < x_{d_2},$$

that is, the conditional distribution of $X_{d_1:n}$, given $X_{d_2:n} = x_{d_2}$, for $1 \le d_1 < d_2$, is the same as the distribution of the d_1-th order statistic from a random sample of size $d_2 - 1$ from the right-truncated distribution at x_{d_2}, with density function $\frac{f}{F(x_{d_2})}$ and cumulative distribution function $\frac{F}{F(x_{d_2})}$, on $(-\infty, x_{d_2})$.

As a generalization of the above two theorems, the following result can be presented for the conditional distribution of an order statistic, given two order statistics.

Theorem 2.14 (Conditional distribution in terms of double truncation). *The conditional density function of $X_{d:n}$, given $X_{d_1:n} = x_{d_1}$ and $X_{d_2:n} = x_{d_2}$, for $d_1 + 1 \le d \le d_2 - 1$, is*

$$f_{d|d_1,d_2:n}(x_d \mid x_{d_1}, x_{d_2}) = \frac{(d_2 - d_1 - 1)!}{(d - d_1 - 1)!(d_2 - 1 - d)!} \left\{ \frac{F(x_d) - F(x_{d_1})}{F(x_{d_2}) - F(x_{d_1})} \right\}^{d-d_1-1} \times \left\{ \frac{F(x_{d_2}) - F(x_d)}{F(x_{d_2}) - F(x_{d_1})} \right\}^{d_2-1-d} \frac{f(x_d)}{F(x_{d_2}) - F(x_{d_1})},$$

$$x_{d_1} < x_d < x_{d_2},$$

that is, the conditional distribution of $X_{d:n}$, given $X_{d_1:n} = x_{d_1}$ and $X_{d_2:n} = x_{d_2}$, for $d_1 + 1 \le d \le d_2 - 1$, is the same as the distribution of the $(d - d_1)$-th order statistic from a random sample of size $d_2 - d_1 - 1$ from the doubly-truncated distribution, with density function $\frac{f}{F(x_{d_2})-F(x_{d_1})}$ and cumulative distribution function $\frac{F(\cdot)-F(x_{d_1})}{F(x_{d_2})-F(x_{d_1})}$, on (x_{d_1}, x_{d_2}).

Remark 2.15. From the general result in Theorem 2.14, the results presented in Theorems 2.12 and 2.13 can be readily deduced by setting $d_2 = n + 1$ and $d_1 = 0$ (or, equivalently, by letting $x_{d_2} \to \infty$ and $x_{d_1} \to -\infty$, respectively.

Remark 2.16. In the viewpoint of censoring schemes to be discussed in this book, the conditional distributions of order statistics presented in Theorems 2.12–2.14 may be thought of as distributions of unobserved lifetimes of units under a life-test, having been given the observed lifetime(s) of one (two) units(s).

In this regard, some more conditional distributional results are also of great interest, and these are as follows. For convenience, we introduce the indicator function $\mathbb{1}_{(-\infty,\tau]}$ with $\tau \in \mathbb{R}$ as defined by

$$\mathbb{1}_{(-\infty,\tau]}(x) = \begin{cases} 1, & \text{when } x \in (-\infty, \tau] \\ 0, & \text{otherwise} \end{cases}.$$

Theorem 2.17 (Distribution of the number of failures up to a fixed time). *Suppose n units, with independent life times $X_1, \ldots, X_n$ and common cumulative distribution function F, are placed on a life-test. Let D denote the number of units that failed on or before a pre-fixed time τ, that is,*

$$D = \sum_{j=1}^{n} \mathbb{1}_{(-\infty,\tau]}(X_j). \tag{2.12}$$

Then, the distribution of D is readily seen to be a binomial distribution with n as the number of trials and $F(\tau)$ as the probability of success; that is, $D \sim bin(n, F(\tau))$ and the probability mass *function is given by*

$$\Pr(D=d) = \binom{n}{d} F^d(\tau)(1-F(\tau))^{n-d}, \quad d=0,1,\ldots,n. \tag{2.13}$$

Remark 2.18. **(a)** It is important to realize here that the number of failures D could be 0 with probability

$$\Pr(D=0) = (1-F(\tau))^n.$$

In case this is a realization of D in a life-testing experiment, then the development of inferential methods would not be possible. For this reason, inferential results may be developed conditionally, conditioned on $D \geq 1$, and in this case we need to use the zero-truncated distribution of D with its probability mass function as

$$\Pr(D=d \mid D \geq 1) = \frac{1}{1-(1-F(\tau))^n} \binom{n}{d} F^d(\tau)(1-F(\tau))^{n-d}, \ d=1,\ldots,n.$$

This will be encountered repeatedly in the ensuing chapters.

(b) From (2.12), it is clear that D can also be expressed via the order statistics $X_{1:n}, \ldots, X_{n:n}$ in the form

$$D = \sum_{j=1}^{n} \mathbb{1}_{(-\infty,\tau]}(X_{j:n}).$$

Theorem 2.19 (Conditional joint density functions and independence). *Under the setting of Theorem 2.17, suppose $D=d$ has been observed in a life-testing experiment. Then, the ordered lifetimes $(X_{1:n}, \ldots, X_{d:n})$ will have their conditional joint density function as*

$$f_{1,\ldots,d:n}(x_1,\ldots,x_d \mid D=d) = d! \prod_{i=1}^{d} \frac{f(x_i)}{F(\tau)}, \quad x_1 < \cdots < x_d < \tau, \tag{2.14}$$

and similarly the ordered lifetimes $(X_{d+1:n}, \ldots, X_{n:n})$ will have their conditional joint density function as

$$f_{d+1,\ldots,n:n}(x_{d+1},\ldots,x_n \mid D=d) = (n-d)! \prod_{i=d+1}^{n} \frac{f(x_i)}{1-F(\tau)},$$

$$\tau < x_{d+1} < \cdots < x_n, \tag{2.15}$$

with the cases of $d=0$ and $d=n$ being obviously excluded from (2.14) and (2.15), respectively. Moreover, the two sets of order statistics, conditionally on $D=d$, are mutually independent.

Remark 2.20. The above theorem can be readily proved by using (2.2) and (2.13) and then appealing to factorization theorem. However, a more general result, called *conditional block independence*, that includes this as a special case, is described in Section 2.2.4.

The conditional distribution of order statistics presented in Theorems 2.12–2.14 are all conditional on lifetimes(s) of one (two) unit(s) having been observed at specific time points (see Remark 2.16). Instead, if the conditioning event is that a unit has not failed or failed by a specific time point, then the conditional distributions of order statistics become a bit more complex. But, they do possess some nice properties and interpretations as presented in the following theorems.

Theorem 2.21. *Suppose from a life-test of n independent and identical units, it is known that the d_1-th ordered unit lifetime is larger than a specified time τ, i.e., $X_{d_1:n} > \tau$. Then, the conditional cumulative distribution function of the d-th order statistic, for $2 \le d_1 + 1 \le d \le n$, is given by*

$$\begin{aligned} F_{d:n}(x_d \mid X_{d_1:n} > \tau) &= \Pr(X_{d:n} \le x_d \mid X_{d_1:n} > \tau) \\ &= \sum_{i=0}^{d_1-1} \frac{\Pr(D=i)}{\Pr(D \le d_1 - 1)} \sum_{j=d-i}^{n-i} \binom{n-i}{j} \left\{ \frac{F(x_d) - F(\tau)}{1 - F(\tau)} \right\}^j \left\{ \frac{1 - F(x_d)}{1 - F(\tau)} \right\}^{n-i-j}, \quad x_d > \tau, \end{aligned}$$

and the corresponding conditional density function of the d-th order statistic, for $d_1 + 1 \le d \le n$, is given by

$$\begin{aligned} & f_{d:n}(x_d \mid X_{d_1:n} > \tau) \\ & \quad = \sum_{i=0}^{d_1-1} \frac{\Pr(D=i)}{\Pr(D \le d_1 - 1)} \frac{(n-i)!}{(d-i-1)!(n-d)!} \left\{ \frac{F(x_d) - F(\tau)}{1 - F(\tau)} \right\}^{d-i-1} \\ & \qquad \times \left\{ \frac{1 - F(x_d)}{1 - F(\tau)} \right\}^{n-d} \frac{f(x_d)}{1 - F(\tau)}, \quad x_d > \tau. \end{aligned}$$

In the above two equations, $\Pr(D=i)$ *is as given in* (2.13).

Proof. First, we observe that the given event $\{X_{d_1:n} > \tau\}$ is equivalent to the event $\{D \le d_1 - 1\}$ for $d_1 \in \{1, \ldots, n\}$. Next, conditional on the event $\{D=i\}$, for $i = 0, 1, \ldots, d_1 - 1$, we also note that $X_{d:n}$ is distributed exactly as the $(d-i)$-th order statistic from a random sample of size $n-i$ from the left-truncated distribution at τ, with density function $\frac{f}{1-F(\tau)}$ and cumulative distribution function $\frac{F(\cdot)-F(\tau)}{1-F(\tau)}$, on (τ, ∞). Hence,

$$F_{d:n}(x_d \mid X_{d_1:n} > \tau) = \sum_{i=0}^{d_1-1} \frac{\Pr(D=i)}{\Pr(D \le d_1 - 1)} F^{LT}_{d-i:n-i|\tau}(x_d \mid \tau)$$

for $x_d > \tau$, where $\Pr(D=i)$ is as given in (2.13) and $F^{LT}_{d-i:n-i|\tau}(\cdot \mid \tau)$ denotes the cumulative distribution function of the $(d-i)$-th order statistic from a random sample of size $n-i$ from the left-truncated distribution at τ given by (see (2.10))

$$F^{LT}_{d-i:n-i|\tau}(x \mid \tau) = \sum_{j=d-i}^{n-i} \binom{n-i}{j} \left\{ \frac{F(x)-F(\tau)}{1-F(\tau)} \right\}^{j} \left\{ \frac{1-F(x)}{1-F(\tau)} \right\}^{n-i-j}, \quad x > \tau. \tag{2.16}$$

Then, evidently, the corresponding conditional density function of $X_{d:n}$ is simply

$$f_{d:n}(x_d \mid X_{d_1:n} > \tau) = \sum_{i=0}^{d_1-1} \frac{\Pr(D=i)}{\Pr(D \le d_1-1)} f^{LT}_{d-i:n-i|\tau}(x_d \mid \tau)$$

for $x_d > \tau$, where

$$f^{LT}_{d-i:n-i|\tau}(x \mid \tau) = \frac{(n-i)!}{(d-i-1)!(n-d)!} \left\{ \frac{F(x)-F(\tau)}{1-F(\tau)} \right\}^{d-i-1} \left\{ \frac{1-F(x)}{1-F(\tau)} \right\}^{n-d} \frac{f(x)}{1-F(\tau)}, \quad x > \tau. \tag{2.17}$$

Hence, the theorem. □

In an analogous manner, we can present the following result for the conditional distribution of an order statistic given that a unit has failed by a specific time point.

Theorem 2.22. *Suppose from a life-test of n independent and identical units, it is known that the d_1-th ordered unit lifetime is no more than a specified time τ, i.e., $X_{d_1:n} \le \tau$. Then, the conditional cumulative distribution function of the d-th order statistic, for $2 \le d_1 + 1 \le d \le n$, is given by*

$$\begin{aligned} F_{d:n}(x_d \mid X_{d_1:n} \le \tau) &= \Pr(X_{d:n} \le x_d \mid X_{d_1:n} \le \tau) \\ &= \sum_{i=d_1}^{d-1} \frac{\Pr(D=i)}{\Pr(D \ge d_1)} F^{LT}_{d-i:n-i|\tau}(x_d \mid \tau) + \sum_{i=d}^{n} \frac{\Pr(D=i)}{\Pr(D \ge d_1)} F^{RT}_{d:i|\tau}(x_d \mid \tau), \end{aligned} \tag{2.18}$$

where $\Pr(D=i)$ is as given in (2.13), $F^{LT}_{d-i:n-i|\tau}(x \mid \tau)$ is as given in (2.16), and $F^{RT}_{d:i|\tau}(\cdot \mid \tau)$ denotes the cumulative distribution function of the d-th order statistic from a random sample of size i from the right-truncated distribution at τ given by

$$F^{RT}_{d:i|\tau}(x \mid \tau) = \sum_{j=d}^{i} \binom{i}{j} \left\{ \frac{F(x)}{F(\tau)} \right\}^{j} \left\{ \frac{F(\tau)-F(x)}{F(\tau)} \right\}^{i-j}, \quad x < \tau.$$

Then, evidently, the corresponding conditional density function of $X_{d:n}$ is simply

$$f_{d:n}(x_d \mid X_{d_1:n} \le \tau) = \sum_{i=d_1}^{d-1} \frac{\Pr(D=i)}{\Pr(D \ge d_1)} f^{LT}_{d-i:n-i|\tau}(x_d \mid \tau) + \sum_{i=d}^{n} \frac{\Pr(D=i)}{\Pr(D \ge d_1)} f^{RT}_{d:i|\tau}(x_d \mid \tau), \quad (2.19)$$

where $f^{LT}_{d-i:n-i|\tau}(x \mid \tau)$ *is as given in* (2.17) *and*

$$f^{RT}_{d:i|\tau}(x \mid \tau) = d\binom{i}{d}\left\{\frac{F(x)}{F(\tau)}\right\}^{d-1}\left\{\frac{F(\tau)-F(x)}{F(\tau)}\right\}\frac{f(x)}{F(\tau)}, \quad x \le \tau.$$

Remark 2.23. It should be noted that the first term on the right hand side of (2.18) and (2.19) both apply for $x_d > \tau$, while the second term on the right hand side of (2.18) and (2.19) both apply for $x_d \le \tau$.

Remark 2.24. The conditional density function of the d-th order statistic in Theorem 2.21 has its support as (τ, ∞), while that in Theorem 2.22 has its support as $(0, \infty)$ (the same as that of X), say. This is so because in the case of former, as it is given that $X_{d_1:n} > \tau$, $X_{d:n}$ (for $d_1 + 1 \le d \le n$) must necessarily be larger than τ. But, in the case of latter, as it is given that $X_{d_1:n} \le \tau$, $X_{d:n}$ (for $d_1 + 1 \le d \le n$) may either be larger than τ or be at most τ, thus resulting in two terms.

2.2.3 Markov dependence

An important property of order statistics arising from a population with an absolutely continuous cumulative distribution function F with density function f is that their dependence structure is Markovian in nature, as stated below.

Theorem 2.25 (Markov dependence of order statistics). *Let $X_1, \dots, X_n$, $n \in \mathbb{N}$, be iid random variables with absolutely continuous cumulative distribution function F and density function f. Then, the corresponding order statistics $X_{1:n}, \dots, X_{n:n}$ form a Markov chain.*

Proof. From (2.6), for any n, we have the joint density function of $(X_{d_1:n}, \dots, X_{d_{r-1}:n})$, for $1 \le d_1 < \cdots < d_{r-1} < n$, as

$$\begin{aligned} &f_{d_1,\dots,d_{r-1}:n}(x_{d_1}, \dots, x_{d_{r-1}}) \\ &\quad = \frac{n!}{(d_1-1)!(d_2-d_1-1)!\cdots(d_{r-1}-d_{r-2}-1)!(n-d_{r-1})!} \\ &\quad \times F^{d_1-1}(x_{d_1})f(x_{d_1})\left(\prod_{\ell=2}^{r-1}\left(\left[F(x_{d_\ell})-F(x_{d_{\ell-1}})\right]^{d_\ell-d_{\ell-1}-1}f(d_\ell)\right)\right) \\ &\quad \times (1-F(x_{d_{r-1}}))^{n-d_{r-1}-1}f(x_{d_{r-1}}), \quad x_{d_1} < \cdots < x_{d_{r-1}}, \end{aligned} \quad (2.20)$$

and the joint density function of $(X_{d_1:n}, \ldots X_{d_{r-1}:n}, X_{d_r:n})$, for $1 \leq d_1 < \cdots < d_{r-1} < d_r \leq n$, as

$$\begin{aligned}
&f_{d_1,\ldots,d_{r-1},d_r:n}(x_{d_1}, \ldots, x_{d_{r-1}}, x_{d_r}) \\
&= \frac{n!}{(d_1-1)!(d_2-d_1-1)!\cdots(d_r-d_{r-1}-1)!(n-d_r)!} F^{d_1-1}(x_{d_1}) f(x_{d_1}) \\
&\times (F(x_{d_1}))^{d_1-1} f(x_{d_1}) \left(\prod_{\ell=2}^{r-1} \left(\left[F(x_{d_\ell}) - F(x_{d_{\ell-1}}) \right]^{d_\ell - d_{\ell-1}-1} f(d_\ell) \right) \right) \\
&\times (F(x_{d_r}) - F(x_{d_{r-1}}))^{d_r - d_{r-1}-1} f(x_{d_{r-1}}) (1 - F(x_{d_r}))^{n-d_r} f(x_{d_r}), \qquad (2.21) \\
&\qquad\qquad x_{d_1} < \cdots < x_{d_{r-1}} < x_{d_r}.
\end{aligned}$$

From (2.20) and (2.21), we readily obtain the conditional density function of $X_{d_r:n}$, given $(X_{d_1:n} = x_{d_1}, \ldots, X_{d_{r-1}:n} = x_{d_{r-1}})$, for $d_{r-1} < d_r \leq n$, to be

$$\begin{aligned}
&f_{d_r|(d_1,\ldots,d_{r-1}):n}(x_{d_r}|x_{d_1}, \ldots, x_{d_r}) \\
&= \frac{(n-d_{r-1})!}{(d_r-d_{r-1}-1)!(n-d_r)!} \left\{ \frac{F(x_{d_r}) - F(x_{d_{r-1}})}{1 - F(x_{d_{r-1}})} \right\}^{d_r - d_{r-1}-1} \\
&\times \left\{ \frac{1 - F(x_{d_r})}{1 - F(x_{d_{r-1}})} \right\}^{n-d_r} \frac{f(x_{d_r})}{1 - F(x_{d_{r-1}})} \qquad (2.22)
\end{aligned}$$

for $x_{d_r} > x_{d_{r-1}}$. The desired result follows readily from the fact that the conditional density function in (2.22) depends only on $x_{d_{r-1}}$ and not on $x_{d_1}, \ldots, x_{d_{r-2}}$. □

Remark 2.26. In view of Theorem 2.12, (2.22) is just the conditional density function of $X_{d_r:n}$, given $X_{d_{r-1}:n} = x_{d_{r-1}}$.

Remark 2.27. Earlier, in Remark 2.11, a comment was made about the difference in results for order statistics from continuous and discrete distributions. One important difference is that the order statistics from discrete distributions do not form a Markov chain in general, in contrast to the Markovian property for the case of continuous distributions established in Theorem 2.25 (see, e.g., Arnold et al., 1984; Rüschendorf, 1985; Cramer and Tran, 2009). One may refer to Chapter 3 of Arnold et al. (1992) for elaborate details on various properties and results pertaining to order statistics from discrete distributions.

2.2.4 Conditional block independence

In Theorem 2.19, we had seen earlier that the two sets of order statistics $(X_{1:n}, \ldots, X_{d:n})$ and $(X_{d+1:n}, \ldots, X_{n:n})$, conditional on the number of failures (till a fixed time τ) $D = d$, are mutually independent. This result has been established in a general form by Iliopoulos and Balakrishnan (2009), referred to as *conditional block independence* of order statistics. For stating this result, let us introduce the following notation:

(1) $-\infty \equiv \tau_0 < \tau_1 < \cdots < \tau_p < \infty$ are pre-fixed time points,

(2) D_i equals the number of measurements in the interval $(\tau_{i-1}, \tau_i]$, $i = 1, \ldots, p$, that is,

$$D_i = \sum_{j=1}^{n} \mathbb{1}_{(\tau_{i-1},\tau_i]}(X_j) = \sum_{j=1}^{n} \mathbb{1}_{(\tau_{i-1},\tau_i]}(X_{j:n}), \qquad i = 1, \ldots, p,$$

(3) $(d_1, \ldots, d_p)$ be a fixed value of $(D_1, \ldots, D_p)$, and

(4) $d_{\bullet i}$ is the partial sum $d_{\bullet i} = \sum_{j=1}^{i} d_j$, with $d_{\bullet 0} \equiv 0$.

(5) $d_{\bullet p+1} = n - d_{\bullet p}$ denotes the number of measurements exceeding the largest threshold τ_p.

Then, the conditional block independence result established by Iliopoulos and Balakrishnan (2009) is as follows.

Theorem 2.28 (Conditional block independence). *Conditional on the event $(D_1, \ldots, D_p) = (d_1, \ldots, d_p)$, the vectors of order statistics*

$$(X_{1:n}, \ldots, X_{d_{\bullet 1}:n}), (X_{d_{\bullet 1}+1:n}, \ldots, X_{d_{\bullet 2}:n}), \ldots, (X_{d_{\bullet p}+1:n}, \ldots, X_{n:n})$$

are mutually independent. Furthermore,

$$(X_{d_{\bullet i-1}+1:n}, \ldots, X_{d_{\bullet i}:n}) \stackrel{d}{=} (Y^{(i)}_{1:d_i}, \ldots, Y^{(i)}_{d_i:d_i}), \qquad i = 1, \ldots, p,$$
$$(X_{d_{\bullet p}+1:n}, \ldots, X_{n:n}) \stackrel{d}{=} (Z^{(p+1)}_{1:n-d_{\bullet p}}, \ldots, Z^{(p+1)}_{n-d_{\bullet p}:n-d_{\bullet p}}),$$

where

(1) $Y^{(i)}_{1:d_i}, \ldots, Y^{(i)}_{d_i:d_i}$, *for $i = 1, \ldots, p$, are order statistics from a random sample of size d_i from the doubly-truncated distribution in the interval $(\tau_{i-1}, \tau_i]$ with density function $\frac{f}{F(\tau_i)-F(\tau_{i-1})}$ (on $(\tau_{i-1}, \tau_i]$),*

(2) $Z^{(p+1)}_{1:n-d_{\bullet p}}, \ldots, Z^{(p+1)}_{n-d_{\bullet p}:n-d_{\bullet p}}$ *are order statistics from a random sample of size $n - d_{\bullet p}$ from the left-truncated distribution at τ_p with density function $\frac{f}{1-F(\tau_p)}$ (on (τ_p, ∞)).*

Remark 2.29. From Theorem 2.28, we find the joint density function of $X_{1:n}, \ldots, X_{n:n}$ and D (w.r.t. the n dimensional Lebesgue measure and the counting measure). For $d \in \{1, \ldots, n-1\}$, we get for $x_1 \le \cdots \le x_n$

$$\begin{aligned} f^{X_{1:n},\ldots,X_{n:n},D}(\boldsymbol{x}_n, d) &= \Pr(D = d) f^{X_{1:n},\ldots,X_{d:n}|D}(\boldsymbol{x}_n \mid d) f^{X_{d+1:n},\ldots,X_{n:n}|D}(\boldsymbol{x}_n \mid d) \\ &= \binom{n}{d} F^d(\tau)(1 - F(\tau))^{n-d} \cdot d! \prod_{i=1}^{d} \frac{f(x_i)}{F(\tau)} \mathbb{1}_{(-\infty,\tau]}(x_d) \\ &\quad \times (n-d)! \prod_{i=d+1}^{n} \frac{f(x_i)}{1 - F(\tau)} \mathbb{1}_{(\tau,\infty)}(x_{d+1}) \end{aligned}$$

$$
\begin{aligned}
&= n!\left(\prod_{j=1}^{n} f(x_j)\right)\mathbb{1}_{(-\infty,\tau]}(x_d)\mathbb{1}_{(\tau,\infty)}(x_{d+1}) \\
&= f_{1,\ldots,n;n}(\boldsymbol{x}_n)\mathbb{1}_{(-\infty,\tau]}(x_d)\mathbb{1}_{(\tau,\infty)}(x_{d+1})
\end{aligned}
$$

(see (2.13), (2.14), (2.15)). With $x_0 = -\infty$ and $x_{n+1} = \infty$, this formula holds also for $d \in \{0, n\}$. Therefore, the joint density function depends on the value of D only via the indicator function.

Remark 2.30. The conditional independence result in Theorem 2.28 turns out to be quite useful in deriving the exact distributions of maximum likelihood estimators of model parameters based on various forms of hybrid censored data, as will be seen in subsequent chapters.

2.2.5 Results for uniform distribution

Order statistics from uniform distribution possess several useful and interesting properties. In this section, we present a brief description of these properties. It needs to be mentioned first that, because of probability integral transformation, these results for uniform distribution can be used to produce results for other continuous distributions as well.

Specifically, if X is a random variable with absolutely continuous cumulative distribution function F and density function f, the *probability integral transformation* $U = F(X)$ results in a *Uniform*$(0, 1)$-distribution, that is, $F(X) \sim$ *Uniform*$(0, 1)$. As the considered transformation is monotone increasing, it is then evident that the d-th order statistic $X_{d:n}$ arising from a random sample of size n from any continuous cumulative distribution function F gets transformed to the d-th order statistic $U_{d:n}$ arising from a random sample of size n from *Uniform*$(0, 1)$-distribution, through this probability integral transformation; that is, $F(X_{d:n}) \stackrel{d}{=} U_{d:n}$. Indeed, more generally, we have

$$
(F(X_{1:n}), \ldots, F(X_{n:n})) \stackrel{d}{=} (U_{1:n}, \ldots, U_{n:n}) \tag{2.23}
$$

(cf. the quantile representation given in Theorem 2.36).

Then, from (2.2), we have the joint density function of uniform order statistics $(U_{1:n}, \ldots, U_{n:n})$ as

$$
f_{1,\ldots,n:n}(u_1, \ldots, u_n) = n!, \quad 0 < u_1 < \cdots < u_n < 1. \tag{2.24}
$$

We then readily observe the following result (see Wilks, 1962; David and Nagaraja, 2003, p. 134).

Theorem 2.31. *Let $W_i = U_{i:n} - U_{i-1:n}$, for $i = 1, \ldots, n$, with the convention that $U_{0:n} \equiv 0$. Then, $(W_1, \ldots, W_n)$ has a Dirichlet(1^{*n+1})-distribution.*

The following result, due to Malmquist (1950), provides a product representation of uniform order statistics, that is, it relates the joint distribution of uniform order statistics to the distribution of a product of independent beta random variables (see (2.26)).

Theorem 2.32 (Malmquist's result). *Let $U_{1:n} < \cdots < U_{n:n}$ be order statistics from a random sample of size n from Uniform$(0,1)$-distribution. Then,*

$$V_1 = \frac{U_{1:n}}{U_{2:n}}, \quad V_2 = \frac{U_{2:n}}{U_{3:n}}, \ldots, \quad V_{n-1} = \frac{U_{n-1:n}}{U_{n:n}}, \quad V_n = U_{n:n} \tag{2.25}$$

are stochastically independent, with $V_i \sim$ Beta$(i, 1)$, for $i = 1, \ldots, n$. Consequently, the powers V_i^i (for $i = 1, \ldots, n$) are iid Uniform$(0,1)$-variables.

From (2.25), we get the representation

$$U_{d:n} = \prod_{i=d}^{n} V_i, \quad 1 \le d \le n, \quad \text{with } V_1^1, \ldots, V_n^n \overset{\text{iid}}{\sim} \mathit{Uniform}(0,1). \tag{2.26}$$

Corollary 2.33. *By using the independence result in Theorem 2.32, we readily obtain from (2.26) that*

$$E(U_{d:n}) = E\left(\prod_{i=d}^{n} V_i\right) = \prod_{i=d}^{n} E(V_i) = \prod_{i=d}^{n} \frac{i}{i+1} = \frac{d}{n+1} \tag{2.27}$$

for $d = 1, \ldots, n$. Similarly, we get

$$E(U_{d:n}^2) = E\left(\prod_{i=d}^{n} V_i^2\right) = \prod_{i=d}^{n} E(V_i^2) = \prod_{i=d}^{n} \frac{i}{i+2} = \frac{d(d+1)}{(n+1)(n+2)}$$

using which we obtain

$$\text{Var}(U_{d:n}) = \frac{d(n-d+1)}{(n+1)^2(n+2)}, \quad d = 1, \ldots, n. \tag{2.28}$$

Furthermore, from Theorem 2.32, we also find

$$\begin{aligned} E(U_{d_1:n} U_{d_2:n}) &= E\left(\prod_{i=d_1}^{d_2-1} V_i \prod_{i=d_2}^{n} V_i^2\right) \\ &= \prod_{i=d_1}^{d_2-1} \frac{i}{i+1} \prod_{i=d_2}^{n} \frac{i}{i+2} = \frac{d_1(d_2+1)}{(n+1)(n+2)} \end{aligned}$$

for $1 \le d_1 < d_2 \le n$, *using which we readily obtain*

$$\mathrm{Cov}(U_{d_1:n}, U_{d_2:n}) = \frac{d_1(n-d_2+1)}{(n+1)^2(n+2)}, \quad 1 \le d_1 < d_2 \le n. \tag{2.29}$$

From (2.24) (or from the symmetry of the Dirichlet(1^{*n+1})-distribution), it is evident that

$$(1-U_{n:n}, 1-U_{n-1:n}, \ldots, 1-U_{1:n}) \stackrel{d}{=} (U_{1:n}, U_{2:n}, \ldots, U_{n:n}),$$

and in particular,

$$U_{d:n} \stackrel{d}{=} 1-U_{n-d+1:n}, \quad d=1,\ldots,n.$$

Thus, using (2.25), we have the following analogue of Malmquist's result.

Theorem 2.34 (Analogue of Malmquist's result). *Let* $U_{1:n} < \cdots < U_{n:n}$ *be order statistics from a random sample of size n from Uniform*$(0,1)$*-distribution. Then,*

$$V_1' = \frac{1-U_{n:n}}{1-U_{n-1:n}}, \quad V_2' = \frac{1-U_{n-1:n}}{1-U_{n-2:n}}, \ldots, \quad V_{n-1}' = \frac{1-U_{2:n}}{1-U_{1:n}}, \quad V_n' = 1-U_{1:n}$$

are stochastically independent, with $V_i' \sim$ *Beta*$(i,1)$, *for* $i=1,\ldots,n$, *and so* $(V_i')^i$ *(for* $i=1,\ldots,n$*) are iid Uniform*$(0,1)$*-variables.*

Next, from (2.9), *we also readily find the density function of* $U_{d:n}$ *as*

$$f_{d:n}(u_d) = \frac{n!}{(d-1)!(n-d)!} u_d^{d-1}(1-u_d)^{n-d}, \quad 0 < u_d < 1, \tag{2.30}$$

i.e., $U_{d:n} \sim$ *Beta*$(d, n-d+1)$, *for* $d=1,\ldots,n$. *Similarly, from* (2.8), *we also find the joint density function of* $(U_{d_1:n}, U_{d_2:n})$, *for* $1 \le d_1 < d_2 \le n$, *as*

$$f_{d_1,d_2:n}(u_{d_1}, u_{d_2}) = \frac{n!}{(d_1-1)!(d_2-d_1-1)!(n-d_2)!} u_{d_1}^{d_1-1}(u_{d_2}-u_{d_1})^{d_2-d_1-1}(1-u_{d_2})^{n-d_2},$$
$$0 < u_{d_1} < u_{d_2} < 1. \tag{2.31}$$

Means, variances and covariances of order statistics can also be derived from (2.30) and (2.31) by means of integration, and these coincide with their expressions derived in (2.27), (2.28) and (2.29), respectively.

Remark 2.35. The independent beta decompositions of joint distribution of uniform order statistics presented in Theorems 2.32 and 2.34, can be used together with the recent results of Jones and Balakrishnan (2021) to produce some other simple functions of order statistics having independent beta distributions.

Finally, uniform order statistics may also be used to provide quantile representations for order statistics from any continuous distribution F as given below.

Theorem 2.36 (Quantile representation of order statistics). *Let $X_{1:n}, \ldots, X_{n:n}$ be the order statistics from any continuous cumulative distribution function F, with $F^{\leftarrow}$ being the quantile function of F (or, of the underlying random variable X). Then,*

$$(X_{1:n}, \ldots, X_{n:n}) \stackrel{d}{=} (F^{\leftarrow}(U_{1:n}), \ldots, F^{\leftarrow}(U_{n:n})),$$

and in particular,

$$X_{d:n} \stackrel{d}{=} F^{\leftarrow}(U_{d:n}), \qquad d = 1, \ldots, n.$$

Remark 2.37. The quantile representation of order statistics in Theorem 2.36 can be stated in more general terms than for the case of continuous distributions done here. But, as mentioned before, since we are only concerned with continuous lifetime data arising from life-testing experiments of different forms here in this book, the statement in Theorem 2.36 is made only for the case of continuous distributions.

Furthermore, note, in contrast to Theorem 2.36, the probability transformation in (2.23) is true only for continuous cumulative distribution function F.

This quantile representation will get used repeatedly in the following chapters!

2.2.6 Results for exponential distribution

Let $Z_{1:n}, \ldots, Z_{n:n}$ be order statistics from a random sample of size n from the standard exponential distribution with density function and cumulative distribution function as

$$f(z) = e^{-z}, \ z > 0, \qquad \text{and} \qquad F(z) = 1 - e^{-z}, \ z > 0, \tag{2.32}$$

respectively (see Definition B.5).

Then, from (2.2), we have the joint density function of exponential order statistics $(Z_{1:n}, \ldots, Z_{n:n})$ as

$$f_{1,\ldots,n:n}(z_1, \ldots, z_n) = n! e^{-\sum_{i=1}^{n} z_i}, \qquad 0 < z_1 < \cdots < z_n < \infty. \tag{2.33}$$

From (2.33), it can be immediately seen that the differences of consecutive order statistics will form independent random variables by the use of *factorization theorem*. This was formally established by Sukhatme (1937), in terms of *normalized spacings*, as stated below.

Theorem 2.38 (Sukhatme's result on exponential spacings). *Let $Z_{1:n}, \ldots, Z_{n:n}$ be order statistics from a random sample of size n from the standard exponential distribution in* (2.32). *Further, let $S_{i,n} = (n - i + 1)(Z_{i:n} - Z_{i-1:n})$, $i = 1, \ldots, n$, with $Z_{0:n} \equiv 0$, be the normalized spacings. Then, $S_{1,n}, \ldots, S_{n,n}$ are all iid standard exponential random variables.*

Remark 2.39. This property of exponential spacings, also established later by Rényi (1953) and Epstein and Sobel (1954), has played a key role in the development of many exact inferential methods for exponential distribution, as seen in the book by Balakrishnan and Basu (1995). Incidentally, this result will also get repeatedly used in the subsequent chapters while developing exact likelihood inferential results for different forms of hybrid censored data.

Now, upon using the probability integral transformation $U = F(Z) = 1 - e^{-Z}$ and the fact that $U_{d:n} \stackrel{d}{=} F(Z_{d:n}) = 1 - e^{-Z_{d:n}}$, for $d = 1, \dots, n$, in Theorem 2.25, we find that

$$\begin{aligned}
\frac{1 - U_{n:n}}{1 - U_{n-1:n}} &\stackrel{d}{=} e^{-(Z_{n:n} - Z_{n-1:n})} = e^{-S_{n,n}}, \\
\left(\frac{1 - U_{n-1:n}}{1 - U_{n-2:n}}\right)^2 &\stackrel{d}{=} e^{-2(Z_{n-1:n} - Z_{n-2:n})} = e^{-S_{n-1,n}}, \\
&\vdots \\
\left(\frac{1 - U_{2:n}}{1 - U_{1:n}}\right)^{n-1} &\stackrel{d}{=} e^{-(n-1)(Z_{2:n} - Z_{1:n})} = e^{-S_{2,n}}, \\
(1 - U_{1:n})^n &\stackrel{d}{=} e^{-nZ_{1:n}} = e^{-S_{1,n}}
\end{aligned}$$

are iid *Uniform*$(0, 1)$-distributed random variables. This is equivalent to stating that the normalized spacings $S_{1,n}, \dots, S_{n,n}$ are iid standard exponential random variables, as stated in Theorem 2.38.

Next, by using the same transformation in Malmquist's result in Theorem 2.32, we find alternatively that

$$\left(\frac{U_{j:n}}{U_{j+1:n}}\right)^j \stackrel{d}{=} \left(\frac{1 - e^{-Z_{j:n}}}{1 - e^{-Z_{j+1:n}}}\right)^j, j = 1, \dots, n-1, \quad U_{n:n}^n \stackrel{d}{=} (1 - e^{-Z_{n:n}})^n$$

are iid *Uniform*$(0, 1)$ random variables. Then, it is evident that

$$j\{-\ln(1 - e^{-Z_{j:n}}) + \ln(1 - e^{-Z_{j+1:n}})\}, j = 1, \dots, n-1, \quad -n\ln(1 - e^{-Z_{n:n}})$$

are all iid standard exponential random variables.

From the exponential spacings result in Theorem 2.38, we readily obtain the following result giving an additive Markov chain representation for exponential order statistics (see Rényi, 1953).

Theorem 2.40 (Additive Markov chain representation). *Let $Z_{1:n}, \dots, Z_{n:n}$ be order statistics from a random sample of size n from the standard exponential distribution in* (2.32). *Then,*

$$Z_{d:n} \stackrel{d}{=} \sum_{i=1}^{d} \frac{S_i}{n - i + 1}, \quad d = 1, 2, \dots, n, \; n \in \mathbb{N}, \tag{2.34}$$

where $S_1, \dots, S_n$ are iid standard exponential random variables, and thus form an additive Markov chain.

From (2.34), expressions for moments of order statistics, such as means, variances and covariances, can be obtained easily, as presented in the following corollary.

Corollary 2.41. *Let $S_1, \dots, S_n \overset{iid}{\sim} Exp(1)$ as in Theorem 2.40. Using the facts that $E(S_i) = \text{Var}(S_i) = 1$, for $i = 1, 2, \dots, n$, we get from (2.34) that*

$$E(Z_{d:n}) = \sum_{i=1}^{d} \frac{1}{n-i+1}, \quad d = 1, \dots, n, \tag{2.35}$$

$$\text{Var}(Z_{d:n}) = \sum_{i=1}^{d} \frac{1}{(n-i+1)^2}, \quad d = 1, \dots, n,$$

$$\text{Cov}(Z_{d_1:n}, Z_{d_2:n}) = \text{Var}(Z_{d_1:n}) = \sum_{i=1}^{d_1} \frac{1}{(n-i+1)^2}, \quad 1 \le d_1 < d_2 \le n.$$

The final result to mention concerning exponential distribution is about its closure property to formation of series systems stated below.

Theorem 2.42 (Closure property to formation of series systems). *For $i \in \{1, \dots, n\}$, let X_i, representing the lifetime of the i-th component, have a scaled-exponential distribution, say $Exp(\vartheta_i)$, with density function*

$$f^{X_i}(x) = \frac{1}{\vartheta_i} e^{-x/\vartheta_i}, \quad x > 0, \ \vartheta_i > 0.$$

Further, suppose a n-component series system is formed with independent components whose lifetimes are $X_1, \dots, X_n$. Then, the lifetime of that series system, T_{ser} has a scaled-exponential distribution with $\vartheta = 1 \Big/ \left(\frac{1}{\vartheta_1} + \dots + \frac{1}{\vartheta_n} \right)$.

Proof. Consider the survival function of T_{ser} given by

$$\begin{aligned}
\overline{F}_{T_{\text{ser}}}(t) &= \Pr(T_{\text{ser}} > t) = \Pr(\min(X_1, \dots, X_n) > t) \\
&= \Pr\left(\bigcap_{i=1}^{n} \{X_i > t\} \right) \\
&= \prod_{i=1}^{n} \Pr(X_i > t) \quad \text{(due to the independence of components)} \\
&= \prod_{i=1}^{n} e^{-t/\vartheta_i} = \exp\left\{ -t \Big/ \left(\frac{1}{\vartheta_1} + \dots + \frac{1}{\vartheta_n} \right) \right\}, \quad \text{for } t > 0,
\end{aligned}$$

thus establishing the required result. □

2.2.7 Results for Weibull distribution

Let $X_{1:n}, \ldots, X_{n:n}$ be order statistics from a random sample of size n from the standard *Weibull*$(1, \alpha)$-distribution with density function and cumulative distribution function as

$$f(x) = \alpha x^{\alpha-1} e^{-x^{\alpha}}, \ x > 0, \quad \text{and} \quad F(x) = 1 - e^{-x^{\alpha}}, \ x > 0, \tag{2.36}$$

where $\alpha > 0$ is the shape parameter (see Definition B.6). Note that this reduces to the exponential distribution in (2.32) when $\alpha = 1$. From (2.36), we immediately find the hazard function (or failure rate function) to be

$$h(x) = \frac{f(x)}{1 - F(x)} = \alpha x^{\alpha-1}, \qquad x, \alpha > 0,$$

from which we observe that Weibull distribution has *Increasing Failure Rate (IFR)* when $\alpha > 1$, *Decreasing Failure Rate (DFR)* when $\alpha < 1$, and *Constant Failure Rate* (i.e., exponential) when $\alpha = 1$. Because of these different shape characteristics, Weibull distribution is commonly used as a lifetime model; consequently, inferential results for Weibull distribution based on different forms of hybrid censored data are described in detail in the ensuing chapters.

Now, from (2.9) and (2.36), we have the density function of $X_{d:n}$, for $d = 1, \ldots, n$, as

$$f_{d:n}(x_d) = d\binom{n}{d} \alpha x_d^{\alpha-1} (1 - e^{-x_d^{\alpha}})^{d-1} e^{-(n-d+1)x_d^{\alpha}} \tag{2.37}$$

for $x_d > 0$ and $\alpha > 0$. From (2.37), we readily obtain the following result (see also Lieblein, 1955; Cramer and Kamps, 1998).

Theorem 2.43 (Moments of Weibull order statistics). *Let $X_{1:n}, \ldots, X_{n:n}$ be order statistics from a random sample of size n from the standard Weibull$(1, \alpha)$-distribution in* (2.36). *Then, for $d = 1, \ldots, n$, we have*

$$E(X_{d:n}) = \frac{n!}{(d-1)!(n-d)!} \Gamma\left(1 + \frac{1}{\alpha}\right) \sum_{i=0}^{d-1} (-1)^i \binom{d-1}{i} \Big/ (n-d+i+1)^{1+\frac{1}{\alpha}} \tag{2.38}$$

and

$$E(X_{d:n}^2) = \frac{n!}{(d-1)!(n-d)!} \Gamma\left(1 + \frac{2}{\alpha}\right) \sum_{i=0}^{d-1} (-1)^i \binom{d-1}{i} \Big/ (n-d+i+1)^{1+\frac{2}{\alpha}}, \tag{2.39}$$

where $\Gamma(\cdot)$ denotes the complete gamma function. The variance of $X_{d:n}$ can be obtained immediately from (2.38) *and* (2.39).

Proof. From (2.37), for $d=1,\ldots,n$ and $k\geq 1$, we find

$$E(X_{d:n}^k)=\frac{n!}{(d-1)!(n-d)!}\int_0^\infty x_d^k\alpha x_d^{\alpha-1}(1-e^{-x_d^\alpha})^{d-1}e^{-(n-d+1)x_d^\alpha}\,dx_d$$
$$=\frac{n!}{(d-1)!(n-d)!}\sum_{i=0}^{d-1}(-1)^i\binom{d-1}{i}\int_0^\infty \alpha x_d^{k+\alpha-1}e^{-(n-d+i+1)x_d^\alpha}\,dx_d. \quad (2.40)$$

Upon setting $w=x_d^\alpha$ in (2.40), we get

$$E(X_{d:n}^k)=\frac{n!}{(d-1)!(n-d)!}\sum_{i=0}^{d-1}(-1)^i\binom{d-1}{i}\int_0^\infty w^{\frac{k}{\alpha}}e^{-w(n-d+i+1)}dw$$
$$=\frac{n!}{(d-1)!(n-d)!}\Gamma\left(1+\frac{k}{\alpha}\right)\sum_{i=0}^{d-1}(-1)^i\binom{d-1}{i}\Big/(n-d+i+1)^{1+\frac{k}{\alpha}},$$

from which the expressions in (2.38) and (2.39) readily follow. □

Similarly, from (2.8) and (2.36), we have the joint density function of $X_{d_1:n}$ and $X_{d_2:n}$, for $1\leq d_1<d_2\leq n$, as

$$f_{d_1,d_2:n}(x_{d_1},x_{d_2})=\frac{n!}{(d_1-1)!(d_2-d_1-1)!(n-d_2)!}(1-e^{-x_{d_1}^\alpha})^{d_1-1}(e^{-x_{d_1}^\alpha}-e^{-x_{d_2}^\alpha})^{d_2-d_1-1}$$
$$\times\alpha^2x_1^{\alpha-1}x_2^{\alpha-1}e^{-x_{d_1}^\alpha}e^{-(n-d_2+1)x_{d_2}^\alpha},\quad 0<x_{d_1}<x_{d_2}<\infty. \quad (2.41)$$

From (2.41) and proceeding as in the proof of Theorem 2.43, by using linear differential equation technique, an explicit expression for the product moment $E(X_{d_1:n}X_{d_2:n})$, for $1\leq d_1<d_2\leq n$, can be derived as done in Lieblein (1955) and Balakrishnan and Kocherlakota (1985); see also Balakrishnan and Cohen (1991, Section 3.8) and Cramer and Kamps (1998) for relevant details.

Similar to the result in Theorem 2.42, the Weibull distribution also possesses the closure property to formation of series systems, as established in the following theorem.

Theorem 2.44 (Closure property to formation of series systems). *For $i\in\{1,\ldots,n\}$, let X_i, representing the lifetime of the i-th component, have a two-parameter Weibull distribution, say Weibull($\vartheta_i^\alpha,\alpha$), with density function f^{X_i} and survival function $\overline{F}^{X_i}=1-F^{X_i}$ as*

$$f^{X_i}(x)=\frac{\alpha}{\vartheta_i^\alpha}x^{\alpha-1}e^{-(x/\vartheta_i)^\alpha},\quad x>0,\ \vartheta_i>0,\ \alpha>0, \quad (2.42)$$
$$\overline{F}^{X_i}(x)=e^{-(x/\vartheta_i)^\alpha},\quad x>0,\ \vartheta_i>0,\ \alpha>0, \quad (2.43)$$

respectively. Further, suppose a n-component series system is formed with independent components whose lifetimes are $X_1,\ldots,X_n$. Then, the lifetime of that series system, T_{ser}, has a two-parameter

Weibull distribution in (2.42) *with* α *as the shape parameter and* $\vartheta = 1/(\frac{1}{\vartheta_1^\alpha} + \cdots + \frac{1}{\vartheta_n^\alpha})^{1/\alpha}$ *as the scale parameter.*

Proof. Consider the survival function of T_{ser} given by

$$\begin{aligned}
\overline{F}_{T_{\text{ser}}}(t) &= \Pr(T_{\text{ser}} > t) = \Pr(\min(X_1, \ldots, X_n) > t) \\
&= \Pr\left(\bigcap_{i=1}^{n}(X_i < t)\right) \\
&= \prod_{i=1}^{n} \Pr(X_i > t) \quad \text{(due to the independence of components)} \\
&= \prod_{i=1}^{n} e^{-(t/\vartheta_i)^\alpha} = e^{-(t/\vartheta)^\alpha}, \qquad \text{for } t > 0,
\end{aligned}$$

where ϑ is as given above. □

Remark 2.45. The result in Theorem 2.42 is a particular case of Theorem 2.44 when $\alpha = 1$.

Remark 2.46. Suppose the lifetimes of n independent devices (or systems), $X_1, \ldots, X_n$, follow *Weibull*$(\vartheta_i^\alpha, \alpha)$-distributions, for $i = 1, \ldots, n$, respectively, and that the i-th device is tested at stress level s_i, for $i = 1, \ldots, n$. Then, from (2.42) and (2.43), it is evident that the hazard function of the i-th device is

$$h^{X_i}(x) = \frac{f^{X_i}(x)}{\overline{F}^{X_i}(x)} = \frac{\alpha x^{\alpha-1}}{\vartheta_i^\alpha}, \tag{2.44}$$

and the ratio of hazard functions of devices at two different stress levels s_i and s_j then becomes

$$\left(\frac{\alpha x^{\alpha-1}}{\vartheta_i^\alpha}\right) \Big/ \left(\frac{\alpha x^{\alpha-1}}{\vartheta_j^\alpha}\right) = \left(\frac{\vartheta_j}{\vartheta_i}\right)^\alpha,$$

a constant. Thus, the Weibull distribution belongs to the *proportional hazards model*, which is one of the most commonly used models in reliability studies. In this case, it is also usual to link the scale parameter ϑ in (2.44) to the stress variables (i.e., the covariates) by the use of a log-linear link function as ϑ itself is positive.

Remark 2.47. It is important to bear in mind that, in Remark 2.46, the Weibull distribution belongs to the proportional hazards model only when all the devices (or systems) have the same shape parameter α. If the shape parameters vary among the devices, then it is easy to verify that the corresponding Weibull distributions do not belong to the proportional hazards model. In some practical situations, however, the shape parameter α may also vary with stress factors, as pointed out by Meeter and Meeker (1994) and Balakrishnan and Ling (2013).

2.2.8 Results for symmetric distributions

Suppose $X_{1:n}, \ldots, X_{n:n}$ are order statistics arising from a random sample of size n from a continuous distribution F that is symmetric about 0, without loss of generality. Let Y be the corresponding folded random variable, i.e., $Y \stackrel{d}{=} |X|$; then, the density function and cumulative distribution function of Y are

$$f^Y(y) = 2f(y), \; y > 0, \quad \text{and} \quad F^Y(y) = 2F(y) - 1, \; y > 0. \tag{2.45}$$

Further, let $Y_{1:m}, \ldots, Y_{m:m}$ denote order statistics from a random sample of size m from the folded distribution F^Y in (2.45). Then, Govindarajulu (1963) was the first to establish some relationships between the moments of these two sets of order statistics; one may also refer to Balakrishnan (1989b) and Balakrishnan et al. (1993) for different extensions and generalizations.

Remark 2.48. It should be noted that symmetric distribution will not be useful while modeling lifetime data which will usually be skewed. However, it is somewhat common to transform a given lifetime data, through a logarithmic transformation, for example, and then carry out an analysis on such a log-transformed data by assuming a symmetric distribution. For instance, the assumption of log-normal, log-logistic and log-Laplace distributions for the original lifetime data would lead to the consideration of normal, logistic and Laplace distributions, respectively, for the log-transformed lifetime data.

To begin with, it is important to note the following facts:

(a) $\Pr(X \le 0) = \Pr(X > 0) = \frac{1}{2}$;

(b) $X \mid \{X > 0\} \stackrel{d}{=} Y$ and $X \mid \{X \le 0\} \stackrel{d}{=} -Y$.

Then, from the conditional block independence result in Theorem 2.28 and by using the above facts, we readily obtain the following mixture result.

Theorem 2.49 (Mixture representation for order statistics from symmetric distributions). *Let $X_{1:n}, \ldots, X_{n:n}$ be order statistics from a random sample of size n from a continuous cumulative distribution function F that is symmetric about 0. Let Y denote the corresponding folded random variable with distribution F^Y as in (2.45), and $Y_{1:n}, \ldots, Y_{n:n}$ denote the corresponding order statistics from a random sample of size n from the distribution F^Y. Further, let $D = \sum_{j=1}^{n} \mathbb{1}_{(-\infty,0)}(X_j)$ be the random variable denoting the number of X's that are non-positive; evidently, $D \sim bin(n, \frac{1}{2})$. Then:*

(a) *Given $D = d$, we have*

$$X_{r:n} \stackrel{d}{=} \begin{cases} Y_{r-d:n-d} & \text{for } r = d+1, \ldots, n \\ -Y_{d-r+1:d} & \text{for } r = 1, \ldots, d \end{cases}. \tag{2.46}$$

(b) *Similarly, given $D=d$, we have*

$$(X_{r:n}, X_{s:n}) \stackrel{d}{=} \begin{cases} (Y_{r-d:n-d}, Y_{s-d:n-d}) & \text{for } d < r < s \leq n \\ (-Y_{d-s+1:d}, -Y_{d-r+1:d}) & \text{for } 1 \leq r < s \leq d \\ (-Y^{(I)}_{d-r+1:d}, Y^{(II)}_{s-d:n-d}) & \text{for } 1 \leq r \leq d < s \leq n \end{cases},$$

where $Y^{(I)}_{k:d}$ and $Y^{(II)}_{\ell:n-d}$ are order statistics from two mutually independent random samples of sizes d and $n-d$, respectively, from the folded distribution F^Y in (2.45).

Remark 2.50. Theorem 2.49 would facilitate easy handling of distributions and moments associated with order statistics from a symmetric distribution provided the corresponding results are conveniently available for order statistics from its folded distribution. Thus, for example, all the results presented in Section 2.2.6 for the exponential distribution would allow us to readily develop results for the Laplace distribution. As an illustration, from (2.46), we will have

$$\begin{aligned} E(X_{r:n}) &= \sum_{d=0}^{r-1} \Pr(D=d) E(Y_{r-d:n-d}) - \sum_{d=r}^{n} \Pr(D=d) E(Y_{d-r+1:d}) \\ &= \frac{1}{2^n} \left[\sum_{d=0}^{r-1} \binom{n}{d} \sum_{i=1}^{r-d} \frac{1}{n-d-i+1} - \sum_{d=r}^{n} \binom{n}{d} \sum_{i=1}^{d-r+1} \frac{1}{d-i+1} \right] \end{aligned}$$

upon using (2.35), for $r=1,2,\ldots,n$. These types of conditional distributional arguments and results will get used later on while developing exact likelihood inferential results for Laplace distribution based on different forms of hybrid censored data.

2.3. Progressively Type-II right censored order statistics

Consider the following life-testing experiment: n independent and identical units are placed on test with absolutely continuous lifetime cumulative distribution function F and density function f, and

(1) at the time of the first failure, R_1 of the $n-1$ surviving units are randomly withdrawn from the life-test;

(2) at the time of the next failure, R_2 of the remaining $n-2-R_1$ surviving units are randomly withdrawn from the life-test, and so on;

(3) finally, at the time of the m-th failure, all remaining $R_m = n-m-R_1-\cdots-R_{m-1}$ surviving units are withdrawn from the life-test.

Here, m is the fixed number of complete failures to be observed, and $R_1, R_2, \ldots, R_m$ are the pre-fixed numbers of live units to be censored at the times of the first, second, ..., m-th failures, respectively, with $R_i \geq 0$ $(i=1,\ldots,m)$ and $\sum_{i=1}^m R_i = n-m$ (see Fig. 2.1). The censoring plan $\mathscr{R} = (R_1,\ldots,R_m)$ is referred to as *progressive Type-II right censor-*

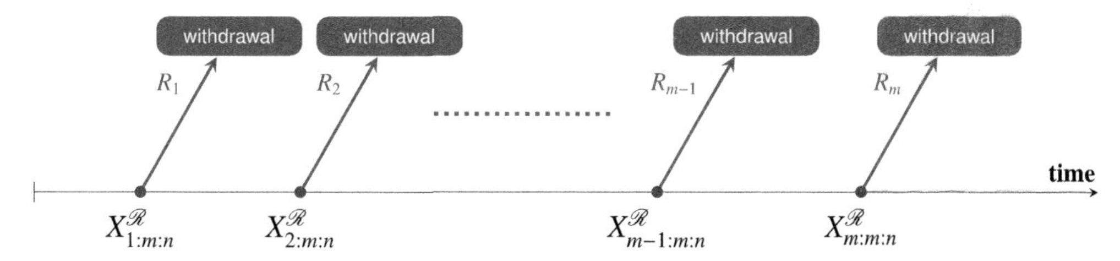

Figure 2.1 Progressive Type-II censoring with failure times $X^{\mathscr{R}}_{1:m:n}, \ldots, X^{\mathscr{R}}_{m:m:n}$ and censoring plan $\mathscr{R} = (R_1, \ldots, R_m)$.

ing plan and $(X^{\mathscr{R}}_{1:m:n}, \ldots, X^{\mathscr{R}}_{m:m:n})$ are referred to as *progressively Type-II right censored order statistics*. There exists a vast literature on these progressively Type-II right censored order statistics and associated theory and applications. Interested readers may refer to the books by Balakrishnan and Aggarwala (2000) and Balakrishnan and Cramer (2014), and the elaborate review articles by Balakrishnan (2007) and Balakrishnan and Cramer (2023). Here, we present only some key results which are essential for relevant developments in the subsequent chapters.

2.3.1 Joint and marginal distributions

From here on, for ease of presentation, we will drop the superscript $\mathscr{R}$ from the notation of progressively Type-II right censored order statistics, and retain it only when there is a specific need to emphasize the progressive censoring plan involved. Thus, we write $(X_{1:m:n}, \ldots, X_{m:m:n})$ for short. Otherwise, the underlying progressive censoring plan is to be taken as $\mathscr{R} = (R_1, \ldots, R_m)$.

The joint density function of $(X_{1:m:n}, \ldots, X_{m:m:n})$ is

$$f_{1,\ldots,m:m:n}(x_1, \ldots, x_m) = c_{m-1} \prod_{i=1}^{m} f(x_i)(1 - F(x_i))^{R_i}, \quad x_1 < \cdots < x_m, \tag{2.47}$$

where

$$\begin{aligned} c_{m-1} &= n(n - R_1 - 1)(n - R_1 - R_2 - 2) \cdots (n - R_1 - \cdots - R_{m-1} - m + 1) \\ &= \prod_{j=1}^{m} \gamma_j, \end{aligned} \tag{2.48}$$

and

$$\gamma_j = n - \sum_{i=1}^{j-1} (R_i + 1) = \sum_{i=j}^{m} (R_i + 1), \quad 1 \leq j \leq m,$$

denotes the number of units still in the life test immediately before the j-th failure.

From (2.47), we readily obtain the joint density function of a subset of progressively Type-II right censored order statistics as presented in the following theorem.

Theorem 2.51 (Joint density function of the first d progressively Type-II right censored order statistics). *The joint density function of* $(X_{1:m:n}, \dots, X_{d:m:n})$, *for* $d \leq m$, *is given by*

$$f_{1,\dots,d:m:n}(x_1, \dots, x_d) = c_{d-1}\left(\prod_{i=1}^{d-1} f(x_i)(1-F(x_i))^{R_i}\right) f(x_d)(1-F(x_d))^{\gamma_d - 1},$$

$$x_1 < \cdots < x_d, \quad (2.49)$$

where $c_{d-1} = \prod_{j=1}^{d} \gamma_j$.

Proof. The required joint density function can be derived from (2.47) by integrating out the variables $x_m, x_{m-1}, \dots, x_{d+1}$. Alternatively, the joint density function in (2.49) can be directly thought of as corresponding to the progressively Type-II censored order statistics arising from the progressive censoring plan $\mathscr{R}' = (R_1, \dots, R_{d-1}, \gamma_d - 1)$, where $\gamma_d - 1$ corresponds to the number of surviving units at x_d that are all withdrawn from the life-test. □

Remark 2.52. As has been pointed out in Balakrishnan and Cramer (2014, Theorem 2.1.1), progressively Type-II censored order statistics from a cumulative distribution function F exhibit a quantile representation similar to that valid for order statistics presented in Theorem 2.36 (see also Balakrishnan and Dembińska, 2008, 2009; Kamps and Cramer, 2001). Thus,

$$(X_{1:m:n}, \dots, X_{m:m:n}) \stackrel{d}{=} (F^{\leftarrow}(U_{1:m:n}), \dots, F^{\leftarrow}(U_{m:m:n})),$$

where $U_{1:m:n}, \dots, U_{m:m:n}$ are progressively Type-II censored order statistics from a standard uniform distribution and $F^{\leftarrow}$ denotes the quantile function of F (for details see Theorem 2.66). This implies that subsequently presented expressions for marginal cumulative distribution functions hold for any population cumulative distribution function F. The absolute continuity is only necessary to ensure the existence of the respective density functions. Furthermore, as mentioned in Cramer (2006), continuity of F is a sufficient condition to have a density function w.r.t. the measure P^F, so that, for a continuous cumulative distribution function F, the density function of a selection of k progressively Type-II censored order statistics always exists when the dominating measure is appropriately chosen as the product measure $\bigotimes_{j=1}^{k} P^F$.

The following theorem gives the marginal distribution of $X_{d:m:n}$, for $d = 1, \dots, m$ (see Kamps and Cramer, 2001).

Theorem 2.53 (Marginal distribution of a progressively Type-II right censored order statistic). *Let $(X_{1:m:n}, \ldots, X_{m:m:n})$ be progressively Type-II right censored order statistics from an absolutely continuous cumulative distribution function F with density function f. Then, for $d = 1, \ldots, m$, the density function $f_{d:m:n}$ of $X_{d:m:n}$ is*

$$f_{d:m:n}(x_d) = \left(\prod_{j=1}^{d} \gamma_j\right) f(x_d) \sum_{i=1}^{d} a_{i,d}(1 - F(x_d))^{\gamma_i - 1}, \quad x_d \in \mathbb{R}, \tag{2.50}$$

where $a_{i,d} = \prod_{\substack{k=1\\k\neq i}}^{d} \frac{1}{\gamma_k - \gamma_i}$, for $1 \le i \le d \le n$. The corresponding cumulative distribution function $F_{d:m:n}$ of $X_{d:m:n}$ (for $d = 1, \ldots, m$) is

$$F_{d:m:n}(x) = 1 - \left(\prod_{j=1}^{d} \gamma_j\right) \sum_{i=1}^{d} \frac{1}{\gamma_i} a_{i,d}(1 - F(x))^{\gamma_i}, \quad x \in \mathbb{R}.$$

Clearly, the cumulative distribution function can be written in terms of cumulative distribution functions of minima with sample size γ_j, that is,

$$F_{d:m:n}(t) = c_{d-1} \sum_{j=1}^{d} \frac{1}{\gamma_j} a_{j,d} F_{1:\gamma_j}(t), \quad t \in \mathbb{R}, \tag{2.51}$$

(see, e.g., Balakrishnan and Cramer, 2014, (2.27)). The representation in (2.51) shows that $F_{d:m:n}$ is a *generalized mixture* of $F_{1:\gamma_j}$, $j = 1, \ldots, m$ (see Cramer and Navarro, 2016; Navarro, 2021).

Similarly, the joint distribution of $X_{d_1:m:n}$ and $X_{d_2:m:n}$, for $1 \le d_1 < d_2 \le m$, can be presented as follows (see also Balakrishnan et al., 2002a, 2005).

Theorem 2.54 (Joint distribution of two progressively Type-II right censored order statistics). *Let $(X_{1:m:n}, \ldots, X_{m:m:n})$ be progressively Type-II right censored order statistics from an absolutely cumulative distribution function F with density function f. Then, the joint density function of $X_{d_1:m:n}$ and $X_{d_2:m:n}$, for $1 \le d_1 < d_2 \le m$, is*

$$\begin{aligned} f_{d_1,d_2:m:n}(x_{d_1}, x_{d_2}) &= \left(\prod_{j=1}^{d_2} \gamma_j\right) f(x_{d_1}) f(x_{d_2}) \\ &\times \sum_{i=1}^{d_1} \sum_{k=d_1+1}^{d_2} a_{i,d_1} a_{k,d_2}^{(d_1)} (1 - F(x_{d_1}))^{\gamma_i - 2} \left(\frac{1 - F(x_{d_2})}{1 - F(x_{d_1})}\right)^{\gamma_k - 1}, \quad x_{d_1} < x_{d_2}, \end{aligned}$$

where $a_{i,d}$ *is as defined in Theorem 2.53 and* $a_{k,d_2}^{(d_1)} = \prod_{\substack{\ell=d_1+1 \\ \ell\neq k}}^{d_2} \frac{1}{\gamma_\ell - \gamma_k}$. *The corresponding joint distribution function of* $X_{d_1:m:n}$ *and* $X_{d_2:m:n}$, *for* $1 \leq d_1 < d_2 \leq m$, *is*

$$F_{d_1,d_2:m:n}(x_{d_1}, x_{d_2}) = F_{d_1:m:n}(x_{d_1}) \\ - \left(\prod_{j=1}^{d_2} \gamma_j\right) \sum_{i=1}^{d_1} \sum_{k=d_1+1}^{d_2} \frac{a_{i,d_1} a_{k,d_2}^{(d_1)}}{\gamma_k(\gamma_i - \gamma_k)} \left\{1 - (1 - F(d_1))^{\gamma_i - \gamma_k}\right\} (1 - F(d_2))^{\gamma_k}, \ x_{d_1} < x_{d_2}.$$

Similar expressions for the joint density function of ℓ (selected) progressively Type-II censored order statistics have been presented by Cramer (2006).

As an example and for later use, we provide a closed form expression for the cumulative distribution function $F_{1,2:m:n}$. First, from (2.49), we have

$$f_{1,2:m:n}(x_1, x_2) = \gamma_1\gamma_2 f(x_1)(1 - F(x_1))^{R_1} f(x_2)(1 - F(x_2))^{\gamma_2 - 1}, \quad x_1 < x_2.$$

Then, for $t_1 > t_2$, we get

$$\begin{aligned} F_{1,2:m:n}(t_1, t_2) &= \Pr(X_{1:m:n} \leq t_1, X_{2:m:n} \leq t_2) \\ &= \Pr(X_{2:m:n} \leq t_2) - \Pr(X_{1:m:n} > t_1, X_{2:m:n} \leq t_2) = F_{2:m:n}(t_2) \\ &= 1 + \frac{\gamma_2}{\gamma_1 - \gamma_2}(1 - F(t_2))^{\gamma_1} - \frac{\gamma_1}{\gamma_1 - \gamma_2}(1 - F(t_2))^{\gamma_2}. \end{aligned}$$

For $t_1 \leq t_2$, we find

$$\begin{aligned} F_{1,2:m:n}(t_1, t_2) &= \int_{-\infty}^{t_1} \int_{x_1}^{t_2} \gamma_1\gamma_2 f(x_1)(1 - F(x_1))^{R_1} f(x_2)(1 - F(x_2))^{\gamma_2 - 1} dx_2 dx_1 \\ &= \int_{-\infty}^{t_1} \gamma_1 f(x_1)(1 - F(x_1))^{R_1} \int_{F(x_1)}^{F(t_2)} \gamma_2 (1 - z)^{\gamma_2 - 1} dz dx_1 \\ &= \int_{-\infty}^{t_1} \gamma_1 f(x_1)(1 - F(x_1))^{\gamma_2 + R_1} dx_1 \\ &\quad - (1 - F(t_2))^{\gamma_2} \int_{-\infty}^{t_1} \gamma_1 f(x_1)(1 - F(x_1))^{R_1} dx_1 \\ &= 1 - (1 - F(t_1))^{\gamma_1} - \frac{\gamma_1}{\gamma_1 - \gamma_2}(1 - F(t_2))^{\gamma_2} \\ &\quad + \frac{\gamma_1}{\gamma_1 - \gamma_2}(1 - F(t_1))^{\gamma_1 - \gamma_2}(1 - F(t_2))^{\gamma_2} \end{aligned} \tag{2.52}$$

since $\gamma_2 + R_1 = \gamma_1 - 1$.

2.3.2 Conditional distributions and dependence structure

Earlier, in Theorem 2.25, the Markovian property of order statistics got established. The following theorem extends this result to the case of progressively Type-II right censored order statistics. It can be extended to continuous cumulative distribution functions F (see Cramer and Tran, 2009).

Theorem 2.55 (Markov dependence of progressively Type-II right censored order statistics). *For $m, n \in \mathbb{N}$ with $m \leq n$, let $X_{1:m:n}, \ldots, X_{m:m:n}$ be progressively Type-II right censored order statistics with censoring plan $\mathscr{R}$ from an absolutely continuous cumulative distribution function F with density function f. Then, they form a Markov chain.*

Proof. From (2.49), we have the joint density function of $(X_{1:m:n}, \ldots, X_{d+1:m:n})$, for $1 \leq d \leq m-1$, is

$$f_{1,\ldots,d+1:m:n}(x_1, \ldots, x_{d+1}) = c_d\left(\prod_{i=1}^{d} f(x_i)(1-F(x_i))^{R_i}\right) f(x_{d+1})(1-F(x_{d+1}))^{\gamma_{d+1}-1} \quad (2.53)$$

for $x_1 < \cdots < x_d < x_{d+1}$, where $c_d = \prod_{j=1}^{d+1} \gamma_j$. Thus, from (2.49) and (2.53), we readily obtain the conditional density function of $X_{d+1:m:n}$, given $(X_{1:m:n} = x_1, \ldots, X_{d:m:n} = x_d)$, as

$$\begin{aligned} f_{d+1|(1,\ldots,d):m:n}(x_{d+1}|x_1, \ldots, x_d) &= \frac{f_{1,\ldots,d+1:m:n}(x_1, \ldots, x_{d+1})}{f_{1,\ldots,d:m:n}(x_1, \ldots, x_d)} \\ &= \gamma_{d+1}\frac{f(x_{d+1})}{1-F(x_d)}\left\{\frac{1-F(x_{d+1})}{1-F(x_d)}\right\}^{\gamma_{d+1}-1} \end{aligned} \quad (2.54)$$

for $x_{d+1} > x_d$. Notice that $\gamma_d - 1 = \gamma_{d+1} + R_d$. The desired result follows immediately from the fact that the conditional density function in (2.54) depends only on x_d (and not on $x_1, \ldots, x_{d-1}$). □

From the above derivation, we observe the following extension of Theorem 2.12 to the case of progressively Type-II right censored order statistics.

Theorem 2.56 (Conditional distribution of progressively Type-II right censored order statistics in terms of left truncation). *The conditional density function of $X_{d+1:m:n}$, given $X_{d:m:n} = x_d$, for $1 \leq d \leq m-1$, is*

$$f_{d+1|d:m:n}(x_{d+1} \mid x_d) = \gamma_{d+1}\frac{f(x_{d+1})}{1-F(x_d)}\left\{\frac{1-F(x_{d+1})}{1-F(x_d)}\right\}^{\gamma_{d+1}-1}$$

for $x_{d+1} > x_d$; that is, the conditional distribution of $X_{d+1:m:n}$, given $X_{d:m:n} = x_d$, for $1 \leq d \leq m-1$, is the same as the distribution of the smallest order statistic from a random sample of *size* γ_{d+1} *from the left-truncated distribution at* x_d*, with density function* $\frac{f}{1-F(x_d)}$ *and cumulative distribution function* $\frac{F(\cdot)-F(x_d)}{1-F(x_d)}$*, on* (x_d, ∞).

In fact, the following more general result can be stated in this regard.

Theorem 2.57 (Conditional joint distribution of progressively Type-II right censored order statistics in terms of left truncation). *The conditional joint density function of* $(X_{d+1:m:n}, \dots, X_{m:m:n})$*, given* $X_{d:m:n} = x_d$*, for* $1 \le d \le m-1$*, is*

$$f_{d+1,\dots,m|d:m:n}(x_{d+1}, \dots, x_m \mid x_d) = \left(\prod_{j=d+1}^{m} \gamma_j\right) \prod_{i=d+1}^{m} \left(\frac{f(x_i)}{1-F(x_d)} \left\{\frac{1-F(x_i)}{1-F(x_d)}\right\}^{R_i}\right), \quad x_d < x_{d+1} < \cdots < x_m;$$

that is, the conditional distribution of $(X_{d+1:m:n}, \dots, X_{m:m:n})$*, given* $X_{d:m:n} = x_d$*, for* $1 \le d \le m-1$*, is the same as the distribution of the progressively Type-II right censored order statistics from a random sample of size* $n - d - \sum_{j=1}^{d} R_i = \gamma_{d+1}$ *with progressive censoring plan* $\mathscr{R}' = (R_{d+1}, \dots, R_m)$ *from the left-truncated distribution at* x_d*, with density function* $\frac{f}{1-F(x_d)}$ *and cumulative distribution function* $\frac{F(\cdot)-F(x_d)}{1-F(x_d)}$*, on* (x_d, ∞).

Remark 2.58. There exists an important distinction between order statistics and progressively Type-II right censored order statistics and it is concerning the conditional distribution in terms of right truncation. While this holds for order statistics as stated in Theorem 2.13, an analogue of that result does not hold for progressively Type-II right censored order statistics. This is evident from the fact that the observations censored before observation x_d could be larger than x_d.

2.3.3 Conditional block independence

The conditional block independence result of order statistics discussed earlier in Section 2.2.4 can be naturally extended to the case of progressively Type-II right censored order statistics, as demonstrated by Iliopoulos and Balakrishnan (2009).

Now, let τ be a fixed time and D denote the number of progressively Type-II right censored order statistics that are at most τ, i.e.,

$$D = \sum_{j=1}^{m} \mathbb{1}_{(-\infty,\tau]}(X_{j:m:n}), . \tag{2.55}$$

Then, Xie et al. (2008) have derived the probability mass function of D as follows.

Theorem 2.59 (Distribution of the number of progressively Type-II right censored order statistics up to a fixed time). *Let* $(X_{1:m:n}, \dots, X_{m:m:n})$ *be progressively Type-II right* *censored order statistics, with progressive censoring plan $\mathscr{R} = (R_1, \ldots, R_m)$, from an absolutely continuous cumulative distribution function F with density function f. Let τ and D be as in (2.55). Then, the probability mass function of D is given by the probabilities*

$$\Pr(D = d) = \sum_{k=0}^{d} \frac{(-1)^k (1 - F(\tau))^{\gamma_{d-k+1}} \prod_{j=1}^{d} \gamma_j}{\left\{ \prod_{j=1}^{d-k} (\gamma_j - \gamma_{d-k+1}) \right\} \left\{ \prod_{j=d-k+1}^{d} (\gamma_{d-k+1} - \gamma_{j+1}) \right\}}$$

for $d = 0, \ldots, m$.

Then, using the fact that the conditional joint density function of $(X_{1:m:n}, \ldots, X_{m:m:n})$, given $D = d$, is given by

$$f_{1,\ldots,m:m:n}(x_1, \ldots, x_m \mid D = d) = \frac{f_{1,\ldots,m:m:n}(x_1, \ldots, x_m)}{\Pr(D = d)} \mathbb{1}_{(-\infty,\tau]}(x_d) \mathbb{1}_{(\tau,\infty)}(x_{d+1}),$$

Iliopoulos and Balakrishnan (2009) established the following block independence result.

Theorem 2.60 (Conditional block independence of progressively Type-II right censored order statistics). *Let $(X_{1:m:n}, \ldots, X_{m:m:n})$ be progressively Type-II right censored order statistics, with progressive censoring plan $\mathscr{R} = (R_1, \ldots, R_m)$, from an absolutely continuous cumulative distribution function F with density function f. Let τ and D be as in (2.55). Then, conditional on $D = d$, the vectors $(X_{1:m:n}, \ldots, X_{d:m:n})$ and $(X_{d+1:m:n}, \ldots, X_{m:m:n})$ are mutually independent with*

$$(X_{1:m:n}, \ldots, X_{d:m:n}) \stackrel{d}{=} \left(V_{1:d:d+K_{\bullet d}}^{(K_1,\ldots,K_d)}, \ldots, V_{d:d:d+K_{\bullet d}}^{(K_1,\ldots,K_d)} \right)$$

and

$$(X_{d+1:m:n}, \ldots, X_{m:m:n}) \stackrel{d}{=} \left(W_{1:m-d:n-d-R_{\bullet d}}^{(R_{d+1},\ldots,R_m)}, \ldots, V_{m-d:m-d:n-d-R_{\bullet d}}^{(R_{d+1},\ldots,R_m)} \right),$$

where the underlying variables $V_1, \ldots, V_{d+K_{\bullet d}}$ and $W_1, \ldots, W_{n-d-R_{\bullet d}}$ are sets of iid random variables, the first from F right-truncated at τ and the second from F left-truncated at τ, and $R_{\bullet d} = \sum_{i=1}^{d} R_i$ and $K_{\bullet d} = \sum_{i=1}^{d} K_i$. Further, $(K_1, \ldots, K_d)$ are discrete random variables with joint probability mass function

$$\Pr(K_1 = k_1, \ldots, K_d = k_d) = \frac{1}{\Pr(D = d)} \prod_{i=1}^{d} \frac{1}{\sum_{j=i}^{d} (k_j + 1)} \binom{R_i}{k_i} F^{k_i+1}(\tau)(1 - F(\tau))^{R_i - k_i}$$

for $0 \le k_i \le R_i$, $i = 1, \ldots, d$.

Remark 2.61. Observe in Theorem 2.60 that, conditional on $D = d$,

$$\left(W_{1:m-d:n-d-R_{\bullet d}}^{(R_{d+1},\ldots,R_m)}, \ldots, W_{m-d:m-d:n-d-R_{\bullet d}}^{(R_{d+1},\ldots,R_m)}\right)$$

do form a progressively Type-II right censored sample, with progressive censoring plan $\mathscr{R}' = (R_{d+1}, \ldots, R_m)$, from the distribution F truncated on the left at τ, with density function $\frac{f}{1-F(\tau)}$ and cumulative distribution function $\frac{F(\cdot)-F(\tau)}{1-F(\tau)}$, on (τ, ∞). However, $\left(V_{1:d:d+K_{\bullet d}}^{(K_1,\ldots,K_d)}, \ldots, V_{d:d:d+K_{\bullet d}}^{(K_1,\ldots,K_d)}\right)$ do not form a progressively Type-II right censored sample in the usual sense (see Remark 2.58), but form a progressively Type-II right censored sample of random sample size, with a random progressive censoring scheme, from the distribution F, truncated on the right at τ with density function $\frac{f}{F(\tau)}$ and cumulative distribution function $\frac{F}{F(\tau)}$, on $(\infty, \tau]$.

Remark 2.62. The block independence result in Theorem 2.60 can also be presented for the case of multiple cut-points $\tau_1, \ldots, \tau_m$, as done in Iliopoulos and Balakrishnan (2009).

2.3.4 Mixture representation

It may be easily observed from the very nature of the progressive Type-II right censored life-testing experiment that

$$X_{1:m:n} \stackrel{d}{=} X_{1:n},$$

where $X_{1:n}$ denotes the smallest order statistic among the n iid lifetimes of the units under test. Then, at the time of this failure, R_1 of the $n-1$ surviving units (i.e., the remaining $n-1$ order statistics) are randomly removed from the life-test. Thus, $X_{2:m:n}$ could be one of $X_{2:n}, X_{3:n}, \ldots, X_{R_1+2:n}$ depending on which R_1 order statistics are removed, and hence its distribution is a mixture of distributions of these order statistics with corresponding probabilities of removal as mixing weights. This feature extends to all other progressively Type-II right censored order statistics $X_{d:m:n}$ (for $d \le m$) as well. Such a general mixture representation, first given by Thomas and Wilson (1972), is as presented below (see also Fischer et al., 2008; Cramer and Lenz, 2010).

Theorem 2.63 (Mixture representation of progressively Type-II right censored order statistics). *Let $(X_{1:m:n}, \ldots, X_{m:m:n})$ be progressively Type-II right censored order statistics, with progressive censoring plan $\mathscr{R} = (R_1, \ldots, R_m)$, from an absolutely continuous cumulative distribution function F with density function f. Then, the joint cumulative distribution function of $(X_{1:m:n}, X_{2:m:n}, \ldots, X_{m:m:n})$ is, for $x_1 < x_2 < \cdots < x_m$,*

$$\begin{aligned}
F_{1,\ldots,m:m:n}(\boldsymbol{x}_m) &= \Pr(X_{1:m:n} \le x_1, X_{2:m:n} \le x_2, \ldots, X_{m:m:n} \le x_m) \\
&= \sum_{1=d_1<d_2<\cdots<d_m \le n-R_m} \pi(d_2, \ldots, d_m) \Pr(X_{d_1:n} \le x_1, X_{d_2:n} \le x_2, \ldots, X_{d_m:n} \le x_m)
\end{aligned}$$

$$= \sum_{1=d_1<d_2<\cdots<d_m\leq n-R_m} \pi(d_2,\ldots,d_m)\, F_{d_1,\ldots,d_m:n}(\boldsymbol{x}_m),$$

where

$$\pi(d_2,\ldots,d_m) = \prod_{j=2}^{m} \frac{\binom{n-d_j}{n-\gamma_j-d_j+1}}{\binom{n-d_{j-1}}{n-\gamma_j-d_{j-1}}} \tag{2.56}$$

and $X_{d_1:n}(\equiv X_{1:n}), X_{d_2:n}, \ldots, X_{d_m:n}$ *(for* $1 \leq d_2 < \cdots < d_m \leq n - R_m$*) are order statistics from a random sample of size n from f. Note that* $\pi(d_2,\ldots,d_m)$ *in (2.56) is precisely the probability that* $(X_{2:n},\ldots,X_{d_2-1:n})$, $(X_{d_2+1:n},\ldots,X_{d_3-1:n}),\ldots,(X_{d_{m-1}+1:n},\ldots,X_{d_m-1:n})$ *and* $(X_{d_m+1:n},\ldots,X_{n:n})$ *are the order statistics removed from the progressive censoring involved in the life-testing experiment.*

The mixture representation in Theorem 2.63 has been utilized by Guilbaud (2001, 2004) for developing exact nonparametric inferential methods based on progressively Type-II right censored order statistics. Fischer et al. (2008) generalized Theorem 2.63 to the case of independent and non-identical underlying variables, while Rezapour et al. (2013a,b) presented generalizations to the case when the underlying variables are dependent. A representation of the density function is presented in Balakrishnan and Cramer (2008) (see also Cramer et al., 2009 for representations in terms of permanents).

2.3.5 General progressive Type-II censoring

The concept of general progressive Type-II censoring, introduced originally by Balakrishnan and Sandhu (1996), involves Type-II left censoring in addition to progressive Type-II right censoring. Specifically, the associated data observation mechanism from the life-testing experiment is as follows: n independent and identical units are placed on test with lifetime density function f and cumulative distribution function F; the first r failures to occur are not observed (resulting in Type-II left censoring) and the time of the $(r+1)$-th failure is observed at which time R_{r+1} of the $n-r-1$ surviving units are randomly withdrawn from the life-test; at the time of the next failure, R_{r+2} of the remaining $n-r-R_{r+2}-2$ surviving units are randomly withdrawn from the life-test, and so on; finally, at the time of $(m-r)$-th failure, all remaining $R_m = n-m-R_{r+1}-\cdots-R_{m-1}$ surviving units are withdrawn from the life-test. Here, $m-r$ is the fixed number of failures observed, $R_{r+1},\ldots,R_m$ are the pre-fixed numbers of live units to be censored at the times of the $(r+1)$-th, $(r+2)$-th, $\ldots$, m-th failures, respectively, with $R_i \geq 0$ $(i = r+1,\ldots,m)$ and $\sum_{i=r+1}^{m} R_i = n-m$. Such a general progressively Type-II censored sample is denoted by $(X^{\mathscr{R}'}_{r+1:m:n},\ldots,X^{\mathscr{R}'}_{m:m:n})$, where $\mathscr{R}' = (R_{r+1},\ldots,R_m)$ is the adopted progressive Type-II censoring plan.

Then, the joint density function (or the likelihood function) of $(X^{\mathscr{R}'}_{r+1:m:n}, \ldots, X^{\mathscr{R}'}_{m:m:n})$ is given by

$$
\begin{aligned}
f_{r+1,\ldots,m:m:n}(x_{r+1}, \ldots, x_m) &= \frac{n!}{r!(n-r-1)!} \\
&\quad \times (n-r-R_{r+1}-1)(n-r-R_{r+1}-R_{r+2}-2)\cdots\left(n-r-\sum_{j=r+1}^{m-1}(R_j+1)\right) \\
&\quad \times F^r(x_{r+1}) \prod_{i=r+1}^{m} f(x_i)(1-F(x_i))^{R_i} \\
&= \frac{n!}{r!(n-r-1)!}\left(\prod_{j=r+2}^{m}\gamma_j\right) F^r(x_{r+1}) \prod_{i=r+1}^{m} f(x_i)(1-F(x_i))^{R_i}, \qquad (2.57) \\
&\qquad x_{r+1} < x_{r+2} < \cdots < x_m.
\end{aligned}
$$

Evidently, when we set $r=0$, the joint density function in (2.57) simply reduces to the joint density function of progressively Type-II right censored order statistics in (2.47). As pointed out by Balakrishnan and Cramer (2014), the general progressively Type-II censored sample $(X^{\mathscr{R}'}_{r+1:n}, \ldots, X^{\mathscr{R}'}_{m:n})$ may be thought of as a Type-II left-censored sample from the progressively Type-II censored sample $(X^{\mathscr{R}}_{1:m:n}, \ldots, X^{\mathscr{R}}_{m:m:n})$ with progressive censoring plan

$$\mathscr{R} = (R_1, \ldots, R_m) = (0^{*r}, R_{r+1}, \ldots, R_m)$$

where the first r (progressively Type-II right censored) order statistics having been censored. Notice that, with $R_1 = \cdots = R_r = 0$ and $r + \sum_{i=r+1}^{m}(R_i+1) = n$, we can write

$$\gamma_j = \sum_{i=j}^{m}(R_i+1) = \begin{cases} n-j+1, & j \in \{1, \ldots, r+1\} \\ n-j+1-\sum_{i=r+1}^{j-1} R_i, & j \in \{r+2, \ldots, m\} \end{cases}$$

(see also Eq. (2.57)).

2.3.6 Results for uniform distribution

Let us assume that the progressively Type-II censored sample $(X_{1:m:n}, \ldots, X_{m:m:n})$, with progressive censoring plan $\mathscr{R} = (R_1, \ldots, R_m)$, is from an absolutely continuous lifetime cumulative distribution function F with density function f. Then, as in Section 2.2.5, the probability integral transformation $U = F(X)$ transforms the above progressively Type-II right censored sample to the progressively Type-II right censored sample $(U_{1:m:n}, \ldots, U_{m:m:n})$ from the *Uniform*(0, 1)-distribution, from the same progressive censoring plan $\mathscr{R}$.

Now, from (2.47), we have the joint density function of $(U_{1:m:n}, \ldots, U_{m:m:n})$ as

$$f_{1,\ldots,m:m:n}(u_1, \ldots, u_m) = \prod_{j=1}^{m} \gamma_j (1-u_j)^{R_j}, \quad 0 < u_1 < \cdots < u_m < 1.$$

We then have the following result, due to Balakrishnan and Sandhu (1995), which presents a direct generalization of Theorem 2.34.

Theorem 2.64 (Generalization of Malmquist's result to progressive Type-II right censoring). *Let $(U_{1:m:n}, \ldots, U_{m:m:n})$ be progressively Type-II right censored order statistics from Uniform$(0,1)$-distribution, with progressive censoring plan $\mathscr{R} = (R_1, \ldots, R_m)$. Then,*

$$V_1 = 1 - U_{1:m:n}, \ V_2 = \frac{1-U_{2:m:n}}{1-U_{1:m:n}}, \ldots, \ V_m = \frac{1-U_{m:m:n}}{1-U_{m-1:m:n}} \tag{2.58}$$

are statistically independent random variables, with $V_i \sim$ Beta$(\gamma_i, 1)$, for $i = 1, 2, \ldots, m$, and so $V_i^{\gamma_i}$ (for $i = 1, \ldots, m$) are iid Uniform$(0,1)$-variables.

Corollary 2.65. *By using the independence result in Theorem 2.64, we readily obtain from* (2.58) *that*

$$U_{d:m:n} \stackrel{d}{=} 1 - V_1 V_2 \cdots V_d, \quad \text{for } d = 1, \ldots, m,$$

where $V_i \sim$ Beta$(\gamma_i, 1)$ independently. Hence, we have

$$E(U_{d:m:n}) = 1 - \prod_{j=1}^{d} \frac{\gamma_j}{\gamma_j + 1}, \quad d = 1, \ldots, m,$$

$$\text{Var}(U_{d:m:n}) = \prod_{j=1}^{d} \frac{\gamma_j}{\gamma_j + 2} - \prod_{j=1}^{d} \frac{\gamma_j^2}{(\gamma_j + 1)^2}, \quad d = 1, \ldots, m,$$

$$\text{Cov}(U_{d_1:m:n}, U_{d_2:m:n}) = \left(\prod_{j=d_1+1}^{d_2} \frac{\gamma_j}{\gamma_j + 1} \right) \text{Var}(U_{d_1:m:n}), \ 1 \le d_1 \le d_2 \le n.$$

Next, analogous to Theorem 2.36, we can provide quantile representations for progressively Type-II right censored order statistics from any continuous cumulative distribution function F, based on progressively Type-II right censored uniform order statistics, as follows (see Remark 2.52).

Theorem 2.66 (Quantile representation of progressively Type-II right censored order statistics). *Let $X_{1:m:n}, \ldots, X_{m:m:n}$ be progressively Type-II censored order statistics from any continuous cumulative distribution function F, with $F^{\leftarrow}$ being the quantile function of F. Then,*

$$(X_{1:m:n}, \ldots, X_{m:m:n}) \stackrel{d}{=} (F^{\leftarrow}(U_{1:m:n}), \ldots, F^{\leftarrow}(U_{m:m:n})), \tag{2.59}$$

and, in particular,

$$X_{d:m:n} \stackrel{d}{=} F^{\leftarrow}(U_{d:m:n}), \quad d = 1, \ldots, m.$$

This quantile representation will get used repeatedly in the following chapters for the purpose of simplifying the involved algebraic derivations.

2.3.7 Results for exponential distribution

Let $(Z_{1:m:n}, \ldots, Z_{m:m:n})$ be progressively Type-II right censored order statistics, with progressive censoring plan $\mathscr{R} = (R_1, \ldots, R_m)$, from the standard exponential distribution in (2.32).

Then, from (2.47), we have the joint density function of $(Z_{1:m:n}, \ldots, Z_{m:m:n})$ as

$$f_{1,\ldots,m:m:n}(z_1, \ldots, z_m) = \prod_{j=1}^{m} \gamma_j e^{-(R_j+1)z_j} = \left(\prod_{j=1}^{m} \gamma_j\right) e^{-\sum_{j=1}^{m}(R_j+1)z_j}$$

for $0 < z_1 < \cdots < z_m < \infty$.

Then, we have the following result (see Viveros and Balakrishnan, 1994), which presents a direct generalization of Sukhatme's result in Theorem 2.38.

Theorem 2.67 (Spacings result for progressively Type-II right censored order statistics). *Let* $(Z^{\mathscr{R}}_{1:m:n}, \ldots, Z^{\mathscr{R}}_{m:m:n})$ *be progressively Type-II right censored order statistics from the standard exponential distribution in* (2.32). *Further, let*

$$S^{\mathscr{R}}_{1,m} = nZ^{\mathscr{R}}_{1:m:n}, \; S^{\mathscr{R}}_{2,m} = \gamma_2(Z^{\mathscr{R}}_{2:m:n} - Z^{\mathscr{R}}_{1:m:n}), \; \ldots,$$
$$S^{\mathscr{R}}_{m,m} = \gamma_m(Z^{\mathscr{R}}_{m:m:n} - Z^{\mathscr{R}}_{m-1:m:n}),$$

i.e., $S^{\mathscr{R}}_{i,m} = \gamma_i(Z^{\mathscr{R}}_{i:m:n} - Z^{\mathscr{R}}_{i-1:m:n})$, *with* $Z^{\mathscr{R}}_{0:m:n} \equiv 0$, *be the normalized spacings. Then,* $S^{\mathscr{R}}_{1,m}, \ldots, S^{\mathscr{R}}_{m,m}$ *are all iid standard exponential random variables.*

From the above result, we then obtain the following generalization of the additive Markov chain representation in Theorem 2.40.

Theorem 2.68 (Additive Markov chain representation of progressively Type-II right censored order statistics). *Let* $(Z^{\mathscr{R}}_{1:m:n}, \ldots, Z^{\mathscr{R}}_{m:m:n})$ *be progressively Type-II right censored order statistics from the standard exponential distribution in* (2.32). *Then, for* $d = 1, \ldots, m$,

$$Z^{\mathscr{R}}_{d:m:n} \stackrel{d}{=} \sum_{i=1}^{d} \frac{S_i}{\gamma_i}, \tag{2.60}$$

where $S_1, \ldots, S_m$ *are iid standard exponential random variables, and thus form an additive Markov chain.*

From (2.60), expressions for moments of progressively Type-II right censored order statistics, such as means, variances and covariances, can be obtained easily, as presented in the following corollary.

Corollary 2.69. *Using the facts that* $E(S_i) = \mathrm{Var}(S_i) = 1$, *for* $i = 1, \ldots, m$, *we get from* (2.60) *that*

$$E(Z^{\mathscr{R}}_{d:m:n}) = \sum_{i=1}^{d} \frac{1}{\gamma_i}, \quad 1 \leq d \leq m,$$

$$\mathrm{Var}(Z^{\mathscr{R}}_{d:m:n}) = \sum_{i=1}^{d} \frac{1}{\gamma_i^2}, \quad 1 \leq d \leq m,$$

$$\mathrm{Cov}(Z^{\mathscr{R}}_{d_1:m:n}, Z^{\mathscr{R}}_{d_2:m:n}) = \mathrm{Var}(Z^{\mathscr{R}}_{d_1:m:n}) = \sum_{i=1}^{d_1} \frac{1}{\gamma_i^2}, \quad 1 \leq d_1 \leq d_2 \leq m.$$

Now, upon using the probability integral transformation $U = F(Z) = 1 - e^{-Z}$ and the quantile representation of progressively Type-II right censored order statistics in (2.59), we readily obtain from the generalization of Malmquist's result in Theorem 2.64 the following result:

$$V_1^{\gamma_1} \stackrel{d}{=} \left(e^{-Z^{\mathscr{R}}_{1:m:n}}\right)^{\gamma_1} = e^{-\gamma_1 Z^{\mathscr{R}}_{1:m:n}} = e^{-S^{\mathscr{R}}_{1,m}},$$

$$V_2^{\gamma_2} \stackrel{d}{=} \left(e^{-(Z^{\mathscr{R}}_{2:m:n} - Z^{\mathscr{R}}_{1:m:n})}\right)^{\gamma_2} = e^{-\gamma_2(Z^{\mathscr{R}}_{2:m:n} - Z^{\mathscr{R}}_{1:m:n})} = e^{-S^{\mathscr{R}}_{2,m}},$$

$$\vdots$$

$$V_m^{\gamma_m} \stackrel{d}{=} \left(e^{-(Z^{\mathscr{R}}_{m:m:n} - Z^{\mathscr{R}}_{m-1:m:n})}\right)^{\gamma_m} = e^{-\gamma_m(Z^{\mathscr{R}}_{m:m:n} - Z^{\mathscr{R}}_{m-1:m:n})} = e^{-S^{\mathscr{R}}_{m,m}},$$

are all iid *Uniform*$(0, 1)$-random variables; that is, $S^{\mathscr{R}}_{1,m}, \ldots, S^{\mathscr{R}}_{m,m}$ are all iid standard exponential random variables. This is precisely the generalized spacings result for progressively Type-II right censored order statistics from exponential distribution stated in Theorem 2.67.

2.3.8 Progressive Type-I right censoring

In contrast to progressive Type-II right censored life-testing experiment described at the beginning of this section, progressive Type-I right censored experiment proceeds as follows. A set of k times $\tau_1 < \cdots < \tau_k$ is pre-fixed. Here, time τ_k corresponds to the maximum experimental time. Note that τ_k will be the termination time only when some units under test are still alive at time τ_k, at which time all those units are removed from the test. Then, a progressive censoring plan $\mathscr{R} = (R_1, \ldots, R_{k-1})$ is fixed with the intention of removing R_i surviving units at random at time τ_i, for $i = 1, \ldots, k - 1$. It

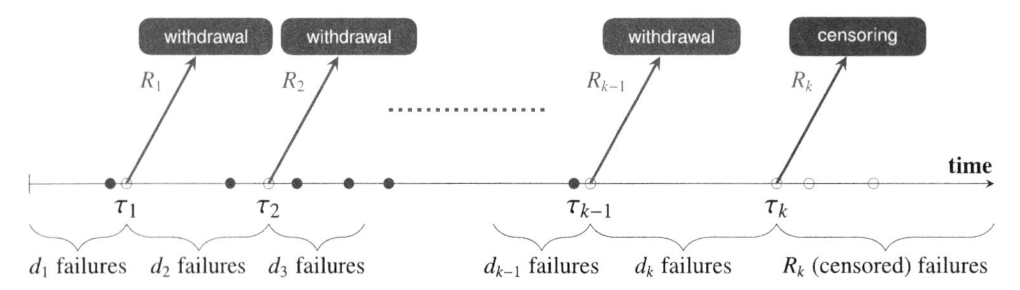

Figure 2.2 Progressive Type-I right censoring with progressive censoring times $\tau_1, \ldots, \tau_{k-1}$, right censoring time τ_k and initially planned censoring plan $(R_1, \ldots, R_{k-1})$.

is important to observe that the number of observations that will be made m (say) is random, with m being in $\{0, 1, \ldots, n\}$. Furthermore, the censoring number R_i at time τ_i would depend on the history of the life-testing experiment and will be random, in fact. Clearly, if the number of surviving units at time τ_i, say R_i', is at most R_i, then all surviving units will be withdrawn from the life-test and consequently the experiment will terminate at that time τ_i. The censoring scheme is illustrated in Fig. 2.2.

It is evident in this case that inferential methods based on such a progressively Type-I censored sample need to be conditional on the pattern of failures observed. In the case when R_k (the random number of surviving units at time τ_k that are withdrawn from the life-test, thus terminating the experiment) is positive, then withdrawal of R_i surviving units would have certainly been possible at time τ_i, for $i = 1, \ldots, k-1$. Then, in this situation, conditional on $R_k \geq 1$ and on observing d_1 failures in $(0, \tau_1]$ and d_i failures in $(\tau_{i-1}, \tau_i]$, for $i = 2, \ldots, k-1$, the likelihood function is evidently given by (see Laumen and Cramer, 2019; Balakrishnan and Cramer, 2023 for further details)

$$\mathscr{L} \propto \left(\prod_{i=1}^{d_{\bullet 1}} f(x_i)\right)(1 - F(\tau_1))^{R_1}\left(\prod_{i=d_{\bullet 1}+1}^{d_{\bullet 2}} f(x_i)\right)(1 - F(\tau_2))^{R_2}$$
$$\times \cdots \times \left(\prod_{i=d_{\bullet k-1}+1}^{m} f(x_i)\right)(1 - F(\tau_k))^{R_k} \quad (2.61)$$

for $0 < x_1 < \cdots < x_{d_{\bullet 1}} < \tau_1 < x_{d_{\bullet 1}+1} < \cdots < x_{d_{\bullet 2}} < \tau_2 < \cdots < \tau_{k-1} < x_{d_{\bullet k-1}+1} < \cdots$ $\cdots < x_m < \tau_k$, where $m = \sum_{i=1}^{k} d_i = d_{\bullet k}$, $n = \sum_{i=1}^{k}(d_i + R_i)$, and $d_{\bullet j} = \sum_{i=1}^{j} d_i$, $j = 1, \ldots, k$.

In the subsequent chapters, while dealing with hybrid censoring schemes involving progressive Type-I right censoring, conditional likelihood function like the one presented in (2.61) will be used to develop likelihood inferential methods. In this regard, one result that will be very useful in developing exact distributional results is the conditional block independence result due to Iliopoulos and Balakrishnan (2009), which is as follows.

Theorem 2.70 (Conditional block independence of progressively Type-I right censored order statistics). *Let $(X_{1:m:n}, \dots, X_{m:m:n})$ be progressively Type-I right censored order statistics from an absolutely continuous cumulative distribution function F with density function f, with progressive censoring plan $(R_1, \dots, R_{k-1})$ occurring at pre-fixed times $\tau_1, \dots, \tau_{k-1}$, and at time τ_k, all remaining surviving units R_k are censored.*

Let $D_1 = \sum_{i=1}^{m} \mathbb{1}_{(0,\tau_1]}(X_{i:m:n})$ and $D_j = \sum_{i=1}^{m} \mathbb{1}_{(\tau_{j-1},\tau_j]}(X_{i:m:n})$, for $j = 2, \dots, k-1$. Furthermore, let the corresponding progressively Type-I right censored order statistics observed in the time interval $(\tau_{j-1}, \tau_j]$ be denoted by $(X_{D_{\bullet j-1}+1:m:n}, \dots, X_{D_{\bullet j}:m:n})$, where $\tau_0 = 0$, $D_{\bullet j} = \sum_{i=1}^{j} D_i$ with $D_0 = 0$ and $D_{\bullet k} = m$, for $j = 1, \dots, k$. Then, conditional on $(D_1, \dots, D_k) = (d_1, \dots, d_k)$, the random vectors

$$(X_{1:m:n}, \dots, X_{d_{\bullet 1}:m:n}),\ (X_{d_{\bullet 1}+1:m:n}, \dots, X_{d_{\bullet 2}:m:n}),\ \dots,\ (X_{d_{\bullet k-1}+1:m:n}, \dots, X_{m:m:n})$$

are mutually independent, with

$$\left(X_{d_{\bullet j-1}+1:m:n}, \dots, X_{d_{\bullet j}:m:n}\right) \stackrel{d}{=} \left(V_{1:d_j}^{(j)}, \dots, V_{d_j:d_j}^{(j)}\right),$$

where the underlying variables $V_1^{(j)}, \dots, V_{d_j}^{(j)}$ are iid from the cumulative distribution function F doubly truncated in the interval $(\tau_{j-1}, \tau_j]$, with density function $\frac{f}{F(\tau_j)-F(\tau_{j-1})}$, for $j = 1, \dots, k$.

Remark 2.71. Of course, it is possible that, for some j, $D_j = D_{\bullet j} - D_{\bullet j-1} = 0$ (i.e., no failure occurs in the time interval $(\tau_{j-1}, \tau_j]$). In this case, the corresponding random vector $(X_{D_{\bullet j-1}+1:m:n}, \dots, X_{D_{\bullet j}:m:n})$ will be a vector of dimension 0.

A multi-sample version of progressive Type-I censoring has been discussed in Cramer et al. (2021).

2.4. Generalized order statistics

In his pioneering work, Kamps (1995a,b) introduced the concept of *generalized order statistics*, through a joint density function of n ordered variables, denoted by $X(1, n, \mathbf{m}, k), \dots, X(n, n, \mathbf{m}, k)$. The proposed joint density function is given by

$$f^{X(1,n,\mathbf{m},k),\dots,X(n,n,\mathbf{m},k)}(x_1, \dots, x_n) = k\left(\prod_{i=1}^{n-1} \gamma_i\right)\left\{\prod_{i=1}^{n-1}(1 - F(x_i))^{m_i} f(x_i)\right\}(1 - F(x_n))^{k-1} f(x_n),$$

$$x_1 < x_2 < \dots < x_n; \quad (2.62)$$

here, $\mathbf{m} = (m_1, \dots, m_{n-1})$ and k are the model parameters; $\gamma_i = k + n - i + \sum_{j=i}^{n-1} m_j$ $(i = 1, \dots, n-1)$. One may also refer to Kamps (2016), Cramer and Kamps (2003), and Cramer (2003) for more elaborate details.

Now, a comparison of the joint density function in (2.47) with the one in (2.62) above readily reveals that the case of progressively Type-II right censored order statistics is included as a particular case in generalized order statistics. Indeed, the model parameters in (2.62) need not be non-negative integers, but the parameters R_i's involved in the progressively Type-II right censored samples in (2.47) must all be non-negative integers satisfying the condition that $\sum_{i=1}^{m} R_i = n - m$. In fact, it is this difference that corresponds to an underlying sampling framework for progressively Type-II right censored order statistics, while the generalized order statistics, in a more general form, remains as a mathematical construct (with no sampling framework) based on the joint density function in (2.62). In fact, the restrictions on the parameters in the generalized order statistics model in (2.62) corresponding to the case of progressively Type-II right censored order statistics are

(a) $\gamma_j \in \mathbb{N}$ for $j = 1, \ldots, m$,

(b) $1 \le \gamma_m < \gamma_{m-1} < \cdots < \gamma_1 = n$.

Further, the progressively Type-II right censored order statistics, with progressive censoring plan $\mathscr{R} = (R_1, \ldots, R_m)$, are precisely the generalized order statistics

$$X(\underset{\substack{\downarrow\\ n}}{1}, m, \underbrace{(R_1, \ldots, R_{m-1})}_{\substack{\downarrow\\ \mathbf{m}}}, \underset{\substack{\downarrow\\ k}}{R_m + 1}), \ldots, X(\underset{\substack{\downarrow\\ n}}{m}, \underset{\substack{\downarrow\\ n}}{m}, \underbrace{(R_1, \ldots, R_{m-1})}_{\substack{\downarrow\\ \mathbf{m}}}, \underset{\substack{\downarrow\\ k}}{R_m + 1}).$$

It is important to mention here that due to the general form and nature of the joint distribution of generalized order statistics, it may become convenient sometimes to use some of the corresponding results while dealing with inferential results based on progressively Type-II right censored samples and also with associated hybrid censoring schemes. This may be seen in some instances in subsequent chapters.

2.5. Sequential order statistics

The notion of *sequential order statistics*, also introduced by Kamps (1995b), is to extend the well-known k-out-of-n structures in reliability theory. These sequential order statistics, denoted by $X_*^{(1)}, \ldots, X_*^{(m)}$, requiring the specification of cumulative distribution functions $F_1, \ldots, F_m$, has the following quantile representation:

$$X_*^{(r)} = F_r^{\leftarrow}\left(1 - V_r(1 - F_r(X_*^{(r-1)}))\right)$$

for $r = 1, \ldots, m$, where $X_*^{(0)} = -\infty$, $F_1^{\leftarrow}(1) \le \cdots \le F_m^{\leftarrow}(1)$, and $V_1, \ldots, V_m$ are independent with $V_r \sim \mathit{Beta}(m - r + 1, 1)$, for $1 \le r \le m$. One may refer to Cramer and Kamps (2001b, 2003), Burkschat (2009), and Cramer (2016) for further details. In fact, the sequential order statistics $X_*^{(1)}, \ldots, X_*^{(m)}$ are generalized order statistics based on a baseline cumulative distribution function F and that the cumulative distribution functions

$F_1, \ldots, F_m$ are such that

$$1 - F_i(x) = (1 - F(x))^{\alpha_i}, \quad x \in \mathbb{R},$$

for $i = 1, \ldots, m$, with $\alpha_i > 0$ for all i. Observe that this means that $F_1, \ldots, F_m$ belong to the proportional hazards family mentioned earlier in Remark 2.46. From a reliability viewpoint, inferential methods for these proportionality parameters α_i have been discussed by a number of authors including Cramer and Kamps (1996, 2001a) and Balakrishnan et al. (2008a, 2012).

CHAPTER 3

Inference for Type-II, Type-I, and progressive censoring

Contents

3.1. Introduction

In this chapter, we consider the conventional Type-II, Type-I and progressive censoring schemes and describe some common inferential methods under these schemes. These will include best linear unbiased estimation, best linear invariant estimation, maximum likelihood estimation, approximate maximum likelihood estimation, and Bayesian estimation methods. Even though all these estimation methods can be demonstrated for many lifetime distributions, we shall furnish here the details only for a few key lifetime distributions such as exponential, Weibull, half logistic, log-normal and log-Laplace distributions; but, we shall give due reference to works on other lifetime distributions as we see fit. Finally, it is important to mention that the inferential methods described in this chapter will form a foundational basis for analogous inferential results to be developed in the subsequent chapters for different hybrid censoring schemes, providing even the luxury to remain brief in the corresponding derivations and details.

3.2. Type-II censoring

Suppose we have observed, from a life-testing experiment of n independent and identical units, the first r failures to occur and that the experiment gets terminated immediately after the r-th failure. Here, r $(1 \leq r \leq n)$ is a pre-fixed number of complete failures

Hybrid Censoring Know-How
https://doi.org/10.1016/B978-0-12-398387-9.00011-8

to be observed. This censoring scheme, as mentioned earlier in Chapter 2, is the conventional *Type-II censoring*. Now, with $X_1, \ldots, X_n$ denoting the lifetimes of the n units under test, what we observe in this case will be

$$(X_{1:n}, X_{2:n}, \ldots, X_{r:n}), \tag{3.1}$$

and all we know about the last $n - r$ observations that got censored is that they are all larger than the termination time, $X_{r:n}$.

Let us further assume that the lifetimes of units arise from an absolutely continuous cumulative distribution function $F_{\boldsymbol{\theta}}$ with density function $f_{\boldsymbol{\theta}}$ and survival function $\overline{F}_{\boldsymbol{\theta}}$, where $\boldsymbol{\theta} \in \Theta \subseteq \mathbb{R}^p$ is the model parameter. In this section, we will especially focus on scale- and (log-) location-scale-families of distributions, that is, for some given cumulative distribution function F (standard member),

$$\begin{aligned} \mathscr{F} &= \{F_\vartheta\}_{\vartheta>0} = \Big\{F_\vartheta \mid F_\vartheta(t) = F\Big(\frac{t}{\vartheta}\Big), t \in \mathbb{R}, \vartheta > 0\Big\}, \\ \mathscr{F} &= \{F_{\mu,\vartheta}\}_{\mu\in\mathbb{R},\vartheta>0} = \Big\{F_{\mu,\vartheta} \mid F_{\mu,\vartheta}(t) = F\Big(\frac{t-\mu}{\vartheta}\Big), t \in \mathbb{R}, \mu \in \mathbb{R}, \vartheta > 0\Big\}, \end{aligned} \tag{3.2}$$

and then describe different methods of inference based on censored samples, and also illustrate the methods with some examples involving specific members of these families. In particular, one- and two-parameter exponential distributions are discussed in details since they lead to explicit representations of the estimators in many cases and allow for exact (conditional) inference.

3.2.1 Best linear unbiased estimation

3.2.1.1 Scale family

Suppose the lifetime density function f_ϑ belongs to the scale-family in (3.2), and is of the form

$$f_\vartheta(x) = \frac{1}{\vartheta} f\left(\frac{x}{\vartheta}\right), \quad x > 0, \ \vartheta > 0. \tag{3.3}$$

Then, $Z = \frac{X}{\vartheta}$ will be the corresponding standard variable with density function $f^Z = f$, which is simply the standard member of the scale family $\{f_\vartheta\}_{\vartheta>0}$ corresponding to $\vartheta = 1$. In this case, with $Z_{i:n} = \frac{X_{i:n}}{\vartheta}$,

$$(Z_{1:n}, Z_{2:n}, \ldots, Z_{r:n})$$

will be the corresponding Type-II censored sample of $Z_1, \ldots, Z_n$ from the standard density function f.

Now, let us denote

$$E(Z_{i:n}) = \alpha_{i:n}, \ \text{Var}(Z_{i:n}) = \sigma_{i,i:n}, \ \text{and} \ \text{Cov}(Z_{i:n}, Z_{j:n}) = \sigma_{i,j:n},$$

for $1 \leq i, j \leq n$, and further

$$\boldsymbol{Z} = (Z_{1:n}, \ldots, Z_{r:n})^T, \quad \boldsymbol{\alpha} = E(\boldsymbol{Z}) = (\alpha_{1:n}, \ldots, \alpha_{r:n})^T$$

as well as

$$\boldsymbol{\Sigma} = \mathrm{Var}(\boldsymbol{Z}) = \left((\sigma_{i,j:n})\right)_{i,j=1}^{n}.$$

It is evident that, with $\boldsymbol{X} = (X_{1:n}, \ldots, X_{r:n})^T$, we have

$$E(\boldsymbol{X}) = \vartheta \boldsymbol{\alpha} \quad \text{and} \quad \mathrm{Var}(\boldsymbol{X}) = \vartheta^2 \boldsymbol{\Sigma}.$$

We then have the following theorem with regard to the best linear unbiased estimator (BLUE) of ϑ based on the given Type-II right censored sample $\boldsymbol{X}$.

Theorem 3.1 (BLUE for the scale-family). *Based on a Type-II right censored sample* $\boldsymbol{X}$ *from a scale-family of density functions* $\{f_\vartheta\}_{\vartheta > 0}$, *the BLUE of* ϑ *is given by*

$$\vartheta^* = \frac{\boldsymbol{\alpha}^T \boldsymbol{\Sigma}^{-1} \boldsymbol{X}}{\boldsymbol{\alpha}^T \boldsymbol{\Sigma}^{-1} \boldsymbol{\alpha}} \tag{3.4}$$

and its variance is given by

$$\mathrm{Var}(\vartheta^*) = \frac{\vartheta^2}{\boldsymbol{\alpha}^T \boldsymbol{\Sigma}^{-1} \boldsymbol{\alpha}}.$$

Proof. Let us consider an arbitrary linear estimator of ϑ, based on $\boldsymbol{X}$, as $\vartheta^* = \boldsymbol{a}^T \boldsymbol{X}$, where $\boldsymbol{a} = (a_1, \ldots, a_r)^T$ is a vector of coefficients that need to be suitably determined. Then, it is clear that, for $\vartheta > 0$,

$$E(\vartheta^*) = \vartheta \boldsymbol{a}^T \boldsymbol{\alpha} \quad \text{and} \quad \mathrm{Var}(\vartheta^*) = \vartheta^2 \boldsymbol{a}^T \boldsymbol{\Sigma} \boldsymbol{a}.$$

So, for ϑ^* to be an unbiased estimator of ϑ, we require $\boldsymbol{a}^T \boldsymbol{\alpha} = 1$. Hence, for minimizing the variance, by using Lagrangian method, we need to minimize the objective function

$$Q(\boldsymbol{a}) = \boldsymbol{a}^T \boldsymbol{\Sigma} \boldsymbol{a} - 2\lambda(\boldsymbol{a}^T \boldsymbol{\alpha} - 1),$$

where λ is the Lagrangian multiplier that needs to be suitably determined. Differentiating $Q(\boldsymbol{a})$ with respect to $\boldsymbol{a}$, we get

$$2\boldsymbol{\Sigma} \boldsymbol{a} - 2\lambda \boldsymbol{\alpha} = \boldsymbol{0},$$

which yields $\boldsymbol{a} = \lambda \boldsymbol{\Sigma}^{-1} \boldsymbol{\alpha}$. Now, by using the constraint $\boldsymbol{a}^T \boldsymbol{\alpha} = 1$, we obtain $\lambda \boldsymbol{\alpha}^T \boldsymbol{\Sigma}^{-1} \boldsymbol{\alpha} = 1$, which yields $\lambda = \frac{1}{\boldsymbol{\alpha}^T \boldsymbol{\Sigma}^{-1} \boldsymbol{\alpha}}$. With this choice of λ, we then find the BLUE of ϑ to be

$$\vartheta^* = \boldsymbol{a}^T \boldsymbol{X} = \frac{\boldsymbol{\alpha}^T \boldsymbol{\Sigma}^{-1} \boldsymbol{X}}{\boldsymbol{\alpha}^T \boldsymbol{\Sigma}^{-1} \boldsymbol{\alpha}}.$$

Moreover, we readily find with $\boldsymbol{a} = \lambda \boldsymbol{\Sigma}^{-1} \boldsymbol{\alpha}$

$$\mathrm{Var}(\vartheta^*) = \boldsymbol{a}^T \,\mathrm{Var}(\boldsymbol{X}) \boldsymbol{a} = \vartheta^2 \boldsymbol{a}^T \boldsymbol{\Sigma} \boldsymbol{a} = \frac{\vartheta^2}{\boldsymbol{\alpha}^T \boldsymbol{\Sigma}^{-1} \boldsymbol{\alpha}} .$$

□

Remark 3.2. The BLUE ϑ^* in (3.4) was originally derived by Lloyd (1952) by minimizing the generalized variance

$$(\boldsymbol{X} - \vartheta \boldsymbol{\alpha})^T \boldsymbol{\Sigma}^{-1} (\boldsymbol{X} - \vartheta \boldsymbol{\alpha})$$

(see Balakrishnan and Cohen, 1991; Arnold et al., 1992, and David and Nagaraja, 2003 for further details).

Special case of scaled-exponential distribution

Let us consider the special case when the lifetimes have an *Exp*(ϑ)-distribution with density function

$$f_\vartheta(x) = \frac{1}{\vartheta} e^{-\frac{x}{\vartheta}}, \qquad x > 0,\ \vartheta > 0,$$

in which case the standard density function (of $Z = \frac{X}{\vartheta}$) is given by

$$f(z) = e^{-z}, \qquad z > 0.$$

Then, from Corollary 2.41, it is known that

$$\alpha_{i:n} = \sum_{k=1}^{i} \frac{1}{n-k+1}, \qquad \sigma_{i,i:n} = \sum_{k=1}^{i} \frac{1}{(n-k+1)^2}, \qquad \sigma_{i,j:n} = \sigma_{i,i:n}$$

for $1 \le i \le j \le n$. In this case, with $\boldsymbol{\Sigma}$ being

$$\boldsymbol{\Sigma} = \begin{bmatrix} \sigma_{1,1:n} & \sigma_{1,1:n} & \cdots & \sigma_{1,1:n} \\ \sigma_{1,1:n} & \sigma_{2,2:n} & \cdots & \sigma_{2,2:n} \\ \vdots & \vdots & \ddots & \vdots \\ \sigma_{1,1:n} & \sigma_{2,2:n} & \cdots & \sigma_{r,r:n} \end{bmatrix}_{r \times r},$$

it can be shown that its inverse is given by (see Graybill, 1983, p. 187)

$$\boldsymbol{\Sigma}^{-1} = \begin{bmatrix} n^2+(n-1)^2 & -(n-1)^2 & 0 & \cdots & 0 \\ -(n-1)^2 & (n-1)^2+(n-2)^2 & -(n-2)^2 & \cdots & \vdots \\ 0 & -(n-2)^2 & & \ddots & \vdots \\ \vdots & \ddots & \ddots & \ddots & 0 \\ \vdots & & \ddots & (n-r+1)^2+(n-r)^2 & -(n-r+1)^2 \\ 0 & \cdots & \cdots & -(n-r+1)^2 & (n-r+1)^2 \end{bmatrix}_{r \times r}.$$

With these explicit expressions of $\boldsymbol{\alpha}$ and $\boldsymbol{\Sigma}^{-1}$, we find

$$\boldsymbol{\alpha}^T\boldsymbol{\Sigma}^{-1} = (1, 1, \ldots, 1, n-r+1)^T = (1^{*r-1}, n-r+1)^T$$

and

$$\boldsymbol{\alpha}^T\boldsymbol{\Sigma}^{-1}\boldsymbol{\alpha} = r;$$

consequently, we obtain the BLUE of ϑ to be

$$\vartheta^* = \frac{\boldsymbol{\alpha}^T\boldsymbol{\Sigma}^{-1}\boldsymbol{X}}{\boldsymbol{\alpha}^T\boldsymbol{\Sigma}^{-1}\boldsymbol{\alpha}} = \frac{1}{r}\left\{\sum_{i=1}^{r} X_{i:n} + (n-r)X_{r:n}\right\} \tag{3.5}$$

and its variance to be

$$\mathrm{Var}(\vartheta^*) = \frac{\vartheta^2}{\boldsymbol{\alpha}^T\boldsymbol{\Sigma}^{-1}\boldsymbol{\alpha}} = \frac{\vartheta^2}{r}. \tag{3.6}$$

Remark 3.3. In this special case, it can be readily verified that the Fisher information in $\boldsymbol{X}$ about ϑ is, in fact, $\mathscr{I}(\boldsymbol{X};\vartheta) = \frac{r}{\vartheta^2}$. Thus, from (3.6), we observe that the BLUE of ϑ in (3.5) attains the Rao-Cramér lower bound and is, therefore, the best unbiased estimator of ϑ.

Note that

$$\mathsf{TTT}_r = r \cdot \vartheta^* = \sum_{i=1}^{r} X_{i:n} + (n-r)X_{r:n}$$

denotes the *total time on test*, that is, the cumulated time objects are under study.

Example 3.4. Consider the following Type-II right censored data on times between successive failures of air-conditioning equipments in Boeing 720 airplanes, given by Proschan (1963, Table 1, Plane 7910), with the largest two observations of 15 having been censored[1]:

$$12, 21, 26, 27, 29, 48, 57, 59, 70, 74, 153, 326.$$

In this case, with $n = 15$ and $r = 13$, we find the BLUE of ϑ to be

$$\vartheta^* = \frac{1}{13}\left\{\sum_{i=1}^{13} X_{i:15} + 2X_{13:15}\right\} = 121.769$$

and its standard error to be

$$\mathrm{SE}(\vartheta^*) = \frac{\vartheta^*}{\sqrt{13}} = 33.773.$$

[1] The censored measurements are 386, 502.

Exact distribution of ϑ^* and associated inference

From the form of ϑ^* in (3.5), it is evident that

$$\frac{\vartheta^*}{\vartheta} = \frac{1}{r}\left\{\sum_{i=1}^{r} Z_{i:n} + (n-r)Z_{r:n}\right\} = \frac{1}{r}\sum_{i=1}^{r} S_{i,n},$$

where $S_{i,n}$ are the normalized spacings as defined in Theorem 2.38 of Chapter 2. Also, it is known from Sukhatme's result presented in that theorem that $S_{i,n}$ are all iid standard exponential random variables. Hence, we find

$$\frac{2r\vartheta^*}{\vartheta} \stackrel{d}{=} \sum_{i=1}^{r} V_i \sim \chi^2_{2r} \quad \text{where } V_1, \ldots, V_r \stackrel{\text{iid}}{\sim} \chi^2_2, \tag{3.7}$$

where χ^2_ν denotes a central chi-square random variable with ν degrees of freedom. Using this exact distributional result, inferential results, such as confidence intervals and hypothesis tests, can be readily developed.

For example, from the probability statement

$$\Pr_\vartheta\left(\chi^2_{2r,\frac{\alpha}{2}} \le \frac{2r\vartheta^*}{\vartheta} \le \chi^2_{2r,1-\frac{\alpha}{2}}\right) = 1-\alpha, \quad \vartheta > 0,$$

where $\chi^2_{\nu,a}$ denotes the a-th percentage point of a central chi-square distribution with ν degrees of freedom, we obtain a $100(1-\alpha)\%$ equi-tailed confidence interval for ϑ to be

$$\left(\frac{2r\vartheta^*}{\chi^2_{2r,1-\frac{\alpha}{2}}}, \frac{2r\vartheta^*}{\chi^2_{2r,\frac{\alpha}{2}}}\right).$$

One can also start from the probability statement

$$\Pr_\vartheta\left(\frac{2r\vartheta^*}{\vartheta} \le \chi^2_{2r,1-\alpha}\right) = 1-\alpha, \quad \vartheta > 0,$$

to obtain a $100(1-\alpha)\%$ lower confidence bound for ϑ to be

$$\frac{2r\vartheta^*}{\chi^2_{2r,1-\alpha}}.$$

In a similar manner, we can carry out exact tests of hypotheses concerning the parameter ϑ. For example, if we wish to test $H_0 : \vartheta = \vartheta_0$ against $H_1 : \vartheta > \vartheta_0$, where ϑ_0 is some specified value, it is clear from (3.7) that the exact null distribution of $\frac{2r\vartheta^*}{\vartheta_0}$ is central chi-square distribution with $2r$ degrees of freedom. Hence, at α level of significance, we would use the critical region

$$\left\{ \frac{2r\vartheta^*}{\vartheta_0} > \chi^2_{2r,1-\alpha} \right\}$$

to find evidence against $H_0 : \vartheta = \vartheta_0$, and in favor of $H_1 : \vartheta > \vartheta_0$, where $\chi^2_{2r,1-\alpha}$ is the $100(1-\alpha)\%$ point of chi-square distribution with $2r$ degrees of freedom. In fact, with the use of Neyman-Pearson theorem, it can be shown that this is indeed an uniformly most powerful test of level α.

Example 3.5. Let us consider the Type-II right censored data on times between successive failures of air-conditioning equipments presented earlier in Example 3.4. In this case, we get equi-tailed 95% confidence interval for ϑ to be

$$\left(\frac{3165.994}{41.923}, \frac{3165.994}{13.844} \right) = (75{,}519, 228.691),$$

with a lower 95% confidence bound for ϑ to be

$$\frac{3165.994}{38.885} = 81.419.$$

Instead, if we wish to test the hypothesis $H_0 : \vartheta = 100$ against the alternative $H_1 : \vartheta > 100$, we find the observed value of the test statistic to be

$$\frac{3165.994}{100} = 31.65994.$$

At 5% level of significance, as the critical region is $(38.885, \infty)$, we conclude that the data do not provide enough evidence against the null hypothesis $H_0 : \vartheta = 100$.

Special case of scaled half-logistic distribution

Let us consider the special case when the lifetimes have *HalfLogist*(ϑ)-distribution with density function

$$f_\vartheta(x) = \frac{2e^{-\frac{x}{\vartheta}}}{\vartheta(1 + e^{-\frac{x}{\vartheta}})^2}, \quad x > 0, \ \vartheta > 0,$$

and with corresponding standard density function f given by

$$f(z) = \frac{2e^{-z}}{(1 + e^{-z})^2}, \quad z > 0.$$

In this case, unlike in the exponential case, simple explicit expressions are not available for means, variances, and covariances of order statistics. But, they can be numerically computed in an efficient recursive manner, as demonstrated by Balakrishnan

(1985). Then, using these numerically computed values of mean vector $\boldsymbol{\alpha}$ and variance-covariance matrix $\boldsymbol{\Sigma}$, the BLUE of ϑ can be numerically determined; see, for example, Balakrishnan and Chan (1992) who have presented a table for the BLUE of ϑ for various choices of n and r.

Example 3.6. Consider the following Type-II right censored data on failure times, in minutes, of a specific type of electrical insulation that was subjected to a continuously increasing voltage stress, given by Lawless (2003, p. 208), with the largest observation of 12 having been censored:

$$12.3, 21.8, 24.4, 28.6, 43.2, 46.9, 70.7, 75.3, 95.5, 98.1, 138.6.$$

In this case, with $n = 12$ and $r = 11$, from the tables of Balakrishnan and Chan (1992), we readily find the BLUE of ϑ to be

$$\vartheta^* = 50.499$$

and its standard error to be

$$\mathrm{SE}(\vartheta^*) = \vartheta^* \sqrt{0.0630} = 12.675.$$

Using the above presented results, we also obtain the BLUE of the expected failure time of the insulation to be

$$\vartheta^*(\log 4) = 70.006$$

and its standard error to be

$$\mathrm{SE}(\vartheta^*)(\log 4) = 17.571.$$

3.2.1.2 Location-scale family

Suppose the lifetime density function $f_{\mu,\vartheta}$ belongs to the location-scale-family in (3.2), and is of the form

$$f_{\mu,\vartheta}(x) = \frac{1}{\vartheta} f\left(\frac{x-\mu}{\vartheta}\right), \quad x \in \mathbb{R},\ \mu \in \mathbb{R}, \vartheta > 0. \tag{3.8}$$

Then, $Z = \frac{X-\mu}{\vartheta}$ will be the corresponding standard variable with density function $f^Z = f$, which is simply the standard member of the family $\{f_{\mu,\vartheta}\}_{\mu\in\mathbb{R},\vartheta>0}$ corresponding to $\mu = 0$ and $\vartheta = 1$. In this case, with $Z_{i:n} = \frac{X_{i:n}-\mu}{\vartheta}$, we will have $\boldsymbol{Z} = (Z_{1:n}, \ldots, Z_{r:n})^T$ as the corresponding Type-II right censored sample from the standard density function f. As in Section 3.2.1.1, let $\boldsymbol{\alpha}$ and $\boldsymbol{\Sigma}$ denote the mean vector and the variance-covariance matrix of $\boldsymbol{Z}$, respectively. It is then evident that, with $\boldsymbol{X} = (X_{1:n}, \ldots, X_{r:n})^T$, we have

$$E(\boldsymbol{X}) = \mu\mathbf{1} + \vartheta\boldsymbol{\alpha} \quad \text{and} \quad \mathrm{Var}(\boldsymbol{X}) = \vartheta^2\boldsymbol{\Sigma},$$

where $\mathbf{1} = (1^{*r})$ is a column vector of 1's of dimension r.

We then have the following theorem with regard to the optimal linear estimation of parameters μ and ϑ based on the given Type-II right censored sample $\boldsymbol{X}$.

Theorem 3.7 (BLUEs for the location-scale-family). *Based on a Type-II right censored sample from a location-scale family of density functions* $\{f_{\mu,\vartheta}\}_{\mu\in\mathbb{R},\vartheta>0}$, *the BLUEs of* μ *and* ϑ *are given by*

$$\mu^* = \frac{1}{\Delta}\left\{(\boldsymbol{\alpha}^T\boldsymbol{\Sigma}^{-1}\boldsymbol{\alpha})\mathbf{1}^T\boldsymbol{\Sigma}^{-1} - (\boldsymbol{\alpha}^T\boldsymbol{\Sigma}^{-1}\mathbf{1})\boldsymbol{\alpha}^T\boldsymbol{\Sigma}^{-1}\right\}\mathbf{X} \tag{3.9}$$

and

$$\vartheta^* = \frac{1}{\Delta}\left\{(\mathbf{1}^T\boldsymbol{\Sigma}^{-1}\mathbf{1})\boldsymbol{\alpha}^T\boldsymbol{\Sigma}^{-1} - (\mathbf{1}^T\boldsymbol{\Sigma}^{-1}\boldsymbol{\alpha})\mathbf{1}^T\boldsymbol{\Sigma}^{-1}\right\}\mathbf{X}, \tag{3.10}$$

and their variances and covariance are given by

$$\begin{aligned} \mathrm{Var}(\mu^*) &= \frac{\vartheta^2}{\Delta}\boldsymbol{\alpha}^T\boldsymbol{\Sigma}^{-1}\boldsymbol{\alpha}, \quad \mathrm{Var}(\vartheta^*) = \frac{\vartheta^2}{\Delta}\mathbf{1}^T\boldsymbol{\Sigma}^{-1}\mathbf{1}, \\ \mathrm{Cov}(\mu^*, \vartheta^*) &= -\frac{\vartheta^2}{\Delta}\boldsymbol{\alpha}^T\boldsymbol{\Sigma}^{-1}\mathbf{1}, \end{aligned} \tag{3.11}$$

where

$$\Delta = (\boldsymbol{\alpha}^T\boldsymbol{\Sigma}^{-1}\boldsymbol{\alpha})(\mathbf{1}^T\boldsymbol{\Sigma}^{-1}\mathbf{1}) - (\boldsymbol{\alpha}^T\boldsymbol{\Sigma}^{-1}\mathbf{1})^2.$$

Proof. Let us consider arbitrary linear estimators of μ and ϑ, based on $\boldsymbol{X}$, as $\mu^* = \boldsymbol{a}^T\boldsymbol{X}$ and $\vartheta^* = \boldsymbol{b}^T\mathbf{X}$, where $\boldsymbol{a} = (a_1, \ldots, a_r)^T$ and $\boldsymbol{b} = (b_1, \ldots, b_r)^T$ are vectors of coefficients that need to be suitably determined. Then, it is clear that

$$E(\mu^*) = \mu\boldsymbol{a}^T\mathbf{1} + \vartheta\boldsymbol{a}^T\boldsymbol{\alpha} \quad \text{and} \quad E(\vartheta^*) = \mu\boldsymbol{b}^T\mathbf{1} + \vartheta\boldsymbol{b}^T\boldsymbol{\alpha}.$$

So, for μ^* and ϑ^* to be unbiased estimators of μ and ϑ, respectively, we require the constraints

$$\boldsymbol{a}^T\mathbf{1} = 1, \ \boldsymbol{a}^T\boldsymbol{\alpha} = 0, \ \boldsymbol{b}^T\mathbf{1} = 0, \ \boldsymbol{b}^T\boldsymbol{\alpha} = 1. \tag{3.12}$$

Thus, to obtain '*trace-efficient linear unbiased estimators*' of μ and ϑ, we need to minimize

$$\mathrm{Var}(\mu^*) + \mathrm{Var}(\vartheta^*) = \vartheta^2\left[\boldsymbol{a}^T\boldsymbol{\Sigma}\boldsymbol{a} + \boldsymbol{b}^T\boldsymbol{\Sigma}\boldsymbol{b}\right],$$

subject to the unbiasedness conditions in (3.12). Thus, by employing the Lagrangian method, we need to minimize the objective function

$$\begin{aligned} Q(\boldsymbol{a}, \boldsymbol{b}) &= \boldsymbol{a}^T\boldsymbol{\Sigma}\boldsymbol{a} + \boldsymbol{b}^T\boldsymbol{\Sigma}\boldsymbol{b} - 2\lambda_1(\boldsymbol{a}^T\mathbf{1} - 1) - 2\lambda_2(\boldsymbol{a}^T\boldsymbol{\alpha}) \\ &\quad - 2\lambda_3(\boldsymbol{b}^T\mathbf{1}) - 2\lambda_4(\boldsymbol{b}^T\boldsymbol{\alpha} - 1), \end{aligned} \tag{3.13}$$

where λ_1, λ_2, λ_3 and λ_4 are the Lagrangian multipliers that need to be suitably determined. Differentiating $Q(\boldsymbol{a}, \boldsymbol{b})$ in (3.13) with respect to $\boldsymbol{a}$ and $\boldsymbol{b}$, we obtain the equations

$$2\boldsymbol{\Sigma a} - 2\lambda_1\mathbf{1} - 2\lambda_2\boldsymbol{\alpha} = \mathbf{0},$$
$$2\boldsymbol{\Sigma b} - 2\lambda_3\mathbf{1} - 2\lambda_4\boldsymbol{\alpha} = \mathbf{0},$$

respectively. These equations yield $\boldsymbol{a}$ and $\boldsymbol{b}$ to be

$$\boldsymbol{a} = \boldsymbol{\Sigma}^{-1}(\lambda_1\mathbf{1} + \lambda_2\boldsymbol{\alpha}) \quad \text{and} \quad \boldsymbol{b} = \boldsymbol{\Sigma}^{-1}(\lambda_3\mathbf{1} + \lambda_4\boldsymbol{\alpha}). \tag{3.14}$$

Now, by using the unbiased conditions in (3.12), we can determine the Lagrangian multipliers λ_1, λ_2, λ_3, and λ_4 readily from (3.14), using which we can find the coefficient vectors $\boldsymbol{a}$ and $\boldsymbol{b}$ as

$$\boldsymbol{a}^T = \frac{1}{\Delta}\left[(\boldsymbol{\alpha}^T\boldsymbol{\Sigma}^{-1}\boldsymbol{\alpha})\mathbf{1}^T\boldsymbol{\Sigma}^{-1} - (\boldsymbol{\alpha}^T\boldsymbol{\Sigma}^{-1}\mathbf{1})\boldsymbol{\alpha}^T\boldsymbol{\Sigma}^{-1}\right]$$

and

$$\boldsymbol{b}^T = \frac{1}{\Delta}\left[(\mathbf{1}^T\boldsymbol{\Sigma}^{-1}\mathbf{1})\boldsymbol{\alpha}^T\boldsymbol{\Sigma}^{-1} - (\mathbf{1}^T\boldsymbol{\Sigma}^{-1}\boldsymbol{\alpha})\mathbf{1}^T\boldsymbol{\Sigma}^{-1}\right].$$

These agree with the estimators presented in (3.9) and (3.10). Moreover, we find

$$\text{Var}(\mu^*) = \text{Var}(\boldsymbol{a}^T\boldsymbol{X}) = \vartheta^2\boldsymbol{a}^T\boldsymbol{\Sigma a},$$
$$\text{Var}(\vartheta^*) = \text{Var}(\boldsymbol{b}^T\boldsymbol{X}) = \vartheta^2\boldsymbol{b}^T\boldsymbol{\Sigma b},$$

and

$$\text{Cov}(\mu^*, \vartheta^*) = \text{Cov}(\boldsymbol{a}^T\boldsymbol{X}, \boldsymbol{b}^T\boldsymbol{X}) = \vartheta^2\boldsymbol{a}^T\boldsymbol{\Sigma b},$$

which upon simplification, reduce to the expressions presented in (3.11). □

Remark 3.8. The BLUEs of μ and ϑ in (3.9) and (3.10) were originally derived by Lloyd (1952) by minimizing the generalized variance

$$(\boldsymbol{X} - \mu\mathbf{1} - \vartheta\boldsymbol{\alpha})^T\boldsymbol{\Sigma}^{-1}(\boldsymbol{X} - \mu\mathbf{1} - \vartheta\boldsymbol{\alpha})$$

(see Balakrishnan and Cohen, 1991; Arnold et al., 1992, and David and Nagaraja, 2003 for further details).

Remark 3.9. (a) The estimators derived in Theorem 3.7, by minimizing the trace of the variance-covariance matrix of the estimators, as shown by Balakrishnan and Rao (1997, 2003), also hold as the optimal linear unbiased estimators if the determinant of the variance-covariance matrix of the estimators gets minimized instead.

(b) In fact, Balakrishnan and Rao (2003) have established more generally that these estimators possess the '*complete covariance matrix dominance property*' among the class of all linear unbiased estimators. The trace-efficiency and determinant-efficiency property of the estimators are indeed included as special cases of this complete covariance matrix dominance property.

(c) The BLUE of ϑ in (3.10) has been derived in an alternate simpler form as a linear function of spacings by Balakrishnan and Papadatos (2002). They then used it to establish a sufficient condition for the non-negativity of the estimator of ϑ, and also to present necessary and sufficient conditions for the BLUE ϑ^* in (3.10) to be a constant multiple of the sample range, $X_{n:n} - X_{1:n}$, in the case of complete samples.

Special case of two-parameter exponential distribution

Let us consider the special case when the lifetimes have an $Exp(\mu, \vartheta)$-distribution with density function

$$f_{\mu,\vartheta}(x) = \frac{1}{\vartheta}\, e^{-\frac{x-\mu}{\vartheta}}, \qquad x > \mu,\ \vartheta > 0, \tag{3.15}$$

in which case the standard density function (of $Z = \frac{X-\mu}{\vartheta}$) is given by $f(z) = e^{-z}$, $z > 0$. Then, by substituting the formulas of $\boldsymbol{\alpha}$, $\boldsymbol{\Sigma}$ and $\boldsymbol{\Sigma}^{-1}$ presented in the last section into (3.9) and (3.10) and simplifying the resulting expressions, we obtain the coefficient vectors $\boldsymbol{a}$ and $\boldsymbol{b}$ to be

$$a_i = \begin{cases} 1 + \dfrac{n-1}{n(r-1)}, & i = 1 \\ -\dfrac{1}{n(r-1)}, & 2 \le i \le r-1\,, \\ -\dfrac{n-r+1}{n(r-1)}, & i = r \end{cases} \tag{3.16}$$

and

$$b_i = \begin{cases} -\dfrac{n-1}{r-1}, & i = 1 \\ \dfrac{1}{r-1}, & 2 \le i \le r-1\,. \\ \dfrac{n-r+1}{r-1}, & i = r \end{cases} \tag{3.17}$$

From (3.17), we find the BLUE of ϑ to be

$$\vartheta^* = \sum_{i=1}^{r} b_i X_{i:n} = \frac{1}{r-1}\left\{\sum_{i=2}^{r-1} X_{i:n} + (n-r+1)X_{r:n} - (n-1)X_{1:n}\right\}$$

$$= \frac{1}{r-1} \sum_{i=2}^{r} (n-i+1)\,(X_{i:n} - X_{i-1:n})\,. \tag{3.18}$$

Thence, from (3.16), we find the BLUE of μ to be

$$\mu^* = \sum_{i=1}^{r} a_i X_{i:n} = X_{1:n} - \frac{\vartheta^*}{n}\,. \tag{3.19}$$

Moreover, upon substituting the expressions of $\boldsymbol{\alpha}$ and $\boldsymbol{\Sigma}^{-1}$ presented in the last section into (3.11) and simplifying the resulting expressions, we obtain the variances and covariance of the above BLUEs to be

$$\text{Var}(\mu^*) = \vartheta^2 \frac{r}{n^2(r-1)}, \quad \text{Var}(\vartheta^*) = \frac{\vartheta^2}{r-1}, \quad \text{and } \text{Cov}(\mu^*, \vartheta^*) = -\frac{\vartheta^2}{n(r-1)}. \tag{3.20}$$

Remark 3.10. In this special case, it can be shown that $X_{1:n}$ and ϑ^* are jointly sufficient for (μ, ϑ) (see, e.g., Balakrishnan et al., 2001 who also discussed a multi-sample scenario). However, the two-parameter exponential density function in (3.15) is an irregular density function as its support depends on the parameter μ. Hence, the Fisher information does not exist in this case; but, Tukey's linear sensitivity measure could be found and could be used to determine the efficiency of the BLUEs μ^* and ϑ^* (see Tukey, 1965 and Nagaraja, 1994). As proved by Chandrasekar and Balakrishnan (2002), the variance-covariance matrix of (μ^*, ϑ^*) in (3.20) is indeed the inverse of Tukey's linear sensitivity measure, in analogy to the property of BLUE mentioned in Remark 3.3 for the scaled-exponential distribution.

Example 3.11. Let us consider the Type-II right censored data on failure times of electrical insulation that was considered earlier in Example 3.6. In this case, with $n = 12$ and $r = 11$, we find the BLUEs of μ and ϑ to be

$$\mu^* = 6.913 \quad \text{and} \quad \vartheta^* = 64.640.$$

With these, we readily find the BLUE of the mean lifetime, $\mu + \vartheta$, to be

$$\mu^* + \vartheta^* = 6.913 + 64.640 = 71.553.$$

Also, its standard error could be found from (3.20) as

$$\sqrt{\widehat{\text{Var}(\mu^*)} + \widehat{\text{Var}(\vartheta^*)} + 2\widehat{\text{Cov}(\mu^*, \vartheta^*)}} = 19.497.$$

Exact distributions of μ^* and ϑ^* and associated inference

From the form of ϑ^* in (3.18), it is evident that

$$\frac{\vartheta^*}{\vartheta} = \frac{1}{r-1}\left\{\sum_{i=2}^{r}(n-i+1)(Z_{i:n} - Z_{i-1:n})\right\} = \frac{1}{r-1}\sum_{i=2}^{r} S_{i,n}, \tag{3.21}$$

where $S_{i,n}$ are the normalized spacings as defined in Theorem 2.38 of Chapter 2. Also, it is known from there that $S_{i,n}$ are all iid standard exponential random variables,

$$\frac{2(r-1)\vartheta^*}{\vartheta} \stackrel{d}{=} \sum_{i=2}^{r} V_i \sim \chi^2_{2r-2} \quad \text{with } V_2, \ldots, V_r \stackrel{\text{iid}}{\sim} \chi^2_2, \tag{3.22}$$

where, as before, χ^2_ν denotes a central chi-square random variable with ν degrees of freedom. We can use this exact distributional result to construct a $100(1-\alpha)\%$ equi-tailed confidence interval for ϑ, for example, as

$$\left(\frac{2(r-1)\vartheta^*}{\chi^2_{2r-2,1-\frac{\alpha}{2}}}, \frac{2(r-1)\vartheta^*}{\chi^2_{2r-2,\frac{\alpha}{2}}}\right), \tag{3.23}$$

where $\chi^2_{\nu,a}$ denote the a-th percentage point of a central chi-square distribution with ν degrees of freedom. We can similarly construct a $100(1-\alpha)\%$ lower confidence bound for ϑ to be

$$\frac{2(r-1)\vartheta^*}{\chi^2_{2r-2,1-\alpha}}.$$

Likewise, we can also carry out exact tests of hypotheses concerning the scale parameter ϑ. If, for example, we wish to test $H_0 : \vartheta = \vartheta_0$ against $H_1 : \vartheta > \vartheta_0$, where ϑ_0 is some specified value, it is evident from (3.22) that the exact null distribution of $\frac{2(r-1)\vartheta^*}{\vartheta_0}$ is central chi-square distribution with $2r-2$ degrees of freedom. So, at α level of significance, we would use the critical region

$$\left\{\frac{2(r-1)\vartheta^*}{\vartheta_0} > \chi^2_{2r-2,1-\alpha}\right\}$$

to find evidence against $H_0 : \vartheta = \vartheta_0$, and in favor of $H_1 : \vartheta > \vartheta_0$.

Next, for developing inference about the location parameter μ, we observe from (3.19) that

$$\frac{\mu^* - \mu}{\vartheta} = \left(\frac{X_{1:n} - \mu}{\vartheta}\right) - \frac{\vartheta^*}{n\vartheta},$$

and that

$$n\left(\frac{X_{1:n} - \mu}{\vartheta}\right) = S_{1,n}, \tag{3.24}$$

where $S_{1,n}$ is a standard exponential random variable. Thence, we have

$$\frac{2n(X_{1:n}-\mu)}{\vartheta} \sim \chi_2^2. \tag{3.25}$$

Now, from (3.22) and (3.25), we observe that

$$\frac{\frac{2n(X_{1:n}-\mu)}{2\vartheta}}{\frac{2(r-1)\vartheta^*}{(2r-2)\vartheta}} = \frac{n(X_{1:n}-\mu)}{\vartheta^*} \sim F_{2,2r-2}, \tag{3.26}$$

where F_{ν_1,ν_2} denotes a central F random variable with (ν_1, ν_2) degrees of freedom. Note that the required independence between the two chi-square random variables in the numerator and the denominator follows readily from the fact that the former is based on the spacing $S_{1,n}$ (see (3.24)) while the latter is based on the spacings $S_{2,n}\ldots, S_{r,n}$ (see (3.21)), and that these two sets of spacings are mutually independent. We can then use this exact distributional result to develop exact inferential results for the location parameter μ.

For example, from the probability statement

$$\Pr_{\mu,\vartheta}\left(F_{2,2r-2,\frac{\alpha}{2}} \leq \frac{n(X_{1:n}-\mu)}{\vartheta^*} \leq F_{2,2r-2,1-\frac{\alpha}{2}}\right) = 1-\alpha, \quad \mu \in \mathbb{R}, \vartheta > 0$$

where $F_{\nu_1,\nu_2,a}$ denotes the a-th percentage point of a central F-distribution with (ν_1, ν_2) degrees of freedom, we obtain a $100(1-\alpha)\%$ equi-tailed confidence interval for μ to be

$$\left(X_{1:n} - \frac{\vartheta^*}{n}F_{2,2r-2,1-\frac{\alpha}{2}}, X_{1:n} - \frac{\vartheta^*}{n}F_{2,2r-2,\frac{\alpha}{2}}\right). \tag{3.27}$$

One can also similarly obtain a $100(1-\alpha)\%$ lower confidence bound for μ to be

$$X_{1:n} - \frac{\vartheta^*}{n}F_{2,2r-2,1-\alpha}.$$

Joint confidence regions can be constructed along the lines of Wu (2010) using the independence of $X_{1:n}$ and ϑ^* (see also the survey in Balakrishnan and Cramer, 2014, Section 17.1).

Proceeding in an analogous manner, we can also carry out exact tests of hypotheses concerning the location parameter μ. If we wish to test $H_0 : \mu = \mu_0$ against $H_1 : \mu > \mu_0$, where μ_0 is some specified value, for example, it is evident from (3.26) that the exact null distribution of $\frac{n}{\vartheta^*}(X_{1:n} - \mu_0)$ is a central F-distribution with $(2, 2r-2)$ degrees of freedom. So, at α level of significance, we would use the critical region

$$\left\{\frac{n}{\vartheta^*}(X_{1:n} - \mu_0) > F_{2,2r-2,1-\alpha}\right\}$$

to find evidence against $H_0 : \mu = \mu_0$, and in favor of $H_1 : \mu > \mu_0$.

Example 3.12. Let us consider the Type-II right censored data on failure times of electrical insulation considered earlier in Example 3.11. In this case, with $n = 12$ and $r = 11$, we found earlier the BLUEs of μ and ϑ to be

$$\mu^* = 6.913 \quad \text{and} \quad \vartheta^* = 64.640.$$

Then, from (3.23), we find the 95% equi-tailed confidence interval for ϑ to be

$$\left(\frac{20 \times 64.640}{34.1696}, \frac{20 \times 64.640}{9.5908} \right) = (37.835, 134.796).$$

Similarly, from (3.27), we find the 95% equi-tailed confidence interval for μ to be

$$\left(12.3 - \frac{64.640 \times 4.4613}{12}, 12.3 - \frac{64.640 \times 0.0253}{12} \right) = (-11.732, 12.164).$$

As 0 is contained within this confidence interval for μ, we can conclude that the scaled-exponential distribution itself will be adequate as a model for these data.

Special case of two-parameter half-logistic distribution

Let us consider the special case when the lifetimes have $HalfLogist(\mu, \vartheta)$-distribution with density function

$$f_{\mu,\vartheta}(x) = \frac{2e^{-\frac{x-\mu}{\vartheta}}}{\vartheta(1 + e^{-\frac{x-\mu}{\vartheta}})^2}, \qquad \mu < x < \infty, \ \vartheta > 0,$$

for which the standard density function f is given by

$$f(z) = \frac{2e^{-z}}{(1 + e^{-z})^2}, \qquad z > 0.$$

In this case, by using the numerically computed values of mean vector $\boldsymbol{\alpha}$ and the variance-covariance matrix $\boldsymbol{\Sigma}$ determined from the recursive computational procedure of Balakrishnan (1985), Balakrishnan and Wong (1994) tabulated the coefficients of the BLUEs μ^* and ϑ^* and also their variances and covariance for various choices of n and r.

Example 3.13. Let us consider the Type-II right censored data on failure times of electrical insulation considered earlier in Examples 3.6 and 3.11. In this case, with $n = 12$ and $r = 11$, from the tables of Balakrishnan and Wong (1994), we find the BLUEs of μ and ϑ to be

$$\mu^* = 4.848 \qquad \text{and} \qquad \vartheta^* = 47.433.$$

Moreover, from the tables of Balakrishnan and Wong (1994), we also find

$$\widehat{\text{Var}(\mu^*)} = (47.433)^2 \times 0.02440 = 54.897,$$

$$\widehat{\mathrm{Var}(\vartheta^*)} = (47.433)^2 \times 0.07282 = 163.837$$

and

$$\widehat{\mathrm{Cov}(\mu^*, \vartheta^*)} = -(47.433)^2 \times 0.01547 = -34.806.$$

Thence, we obtain the BLUE of the mean failure time of the insulation to be

$$\mu^* + \vartheta^* \log 4 = 4.848 + 47.433 \log 4 = 70.603$$

and its standard error to be

$$\begin{aligned} &\vartheta^* \{0.02440 + 0.07282(\log 4)^2 - 2 \times 0.01547(\log 4)\}^{1/2} \\ &\quad = 47.433 \times 0.349 = 16.530. \end{aligned}$$

A comparison of these results with those presented in Example 3.11, based on the two-parameter exponential distribution, reveals that the two estimates of mean failure time are quite close, but the standard error of the estimate is smaller under the two-parameter half-logistic distribution.

Example 3.14. Let us consider the Type-II right censored data on times between successive failures of air-conditioning equipments considered earlier in Examples 3.4 and 3.5. In this case, with $n = 15$ and $r = 13$, from the tables of Balakrishnan and Wong (1994), we find the BLUEs of μ and ϑ to be

$$\mu^* = -0.531 \quad \text{and} \quad \vartheta^* = 88.229.$$

Moreover, from the tables of Balakrishnan and Wong (1994), we also find

$$\begin{aligned} \widehat{\mathrm{Var}(\mu^*)} &= (88.229)^2 \times 0.01597 = 124.316, \\ \widehat{\mathrm{Var}(\vartheta^*)} &= (88.229)^2 \times 0.05997 = 466.828 \end{aligned}$$

and

$$\widehat{\mathrm{Cov}(\mu^*, \vartheta^*)} = -(88.229)^2 \times 0.01051 = -81.814.$$

Thence, we obtain the BLUE of the expected time between successive failures to be

$$\mu^* + \vartheta^* \log 4 = -0.531 + 88.229 \log 4 = 121.781$$

and its standard error to be

$$\begin{aligned} &\vartheta^* \{0.01597 + 0.05997(\log 4)^2 - 2 \times 0.01051(\log 4)\}^{1/2} \\ &\quad = 88.229 \times 0.320 = 28.189. \end{aligned}$$

A comparison of these results with those presented in Example 3.4, based on the scaled-exponential distribution, reveals that the two estimates of expected time between successive failures are quite close, but the standard error of the estimate is smaller under the two-parameter half-logistic distribution.

3.2.2 Best linear invariant estimation

3.2.2.1 Scale family

For the scale-family of distributions in (3.3), the BLUE of the scale parameter ϑ was derived earlier in Section 3.2.1.1 by imposing unbiasedness condition and then by minimizing the variance of the estimator. Instead, one can derive an estimator by minimizing the mean squared estimator which, as a criterion, takes into account both bias and variance; see Mann (1967) and Balakrishnan et al. (2008b).

Theorem 3.15 (Best linear invariant estimator for the scale-family). *Based on a Type-II right censored sample $\boldsymbol{X}$ from a scale-family of density functions $\{f_\vartheta\}_{\vartheta>0}$, the best linear invariant estimator (BLIE) of ϑ is given by*

$$\vartheta^{**} = \boldsymbol{a}^T\boldsymbol{X} \tag{3.28}$$

and its mean squared error is given by

$$\mathrm{MSE}(\vartheta^{**}) = \vartheta^2\left\{\boldsymbol{a}^T\boldsymbol{\Sigma}\boldsymbol{a} + (\boldsymbol{a}^T\boldsymbol{\alpha} - 1)^2\right\}, \tag{3.29}$$

where

$$\boldsymbol{a}^T = \boldsymbol{\alpha}^T(\boldsymbol{\Sigma} - \boldsymbol{\alpha}\boldsymbol{\alpha}^T)^{-1} = \boldsymbol{\alpha}^T\boldsymbol{\Sigma}^{-1}\left(\boldsymbol{I} - (\boldsymbol{\Sigma}^{-1/2}\boldsymbol{\alpha})(\boldsymbol{\Sigma}^{-1/2}\boldsymbol{\alpha})^T\right)^{-1}.$$

Proof. As in Section 3.2.1.1, let us consider an arbitrary linear estimator of ϑ, based on $\boldsymbol{X}$, as $\vartheta^{**} = \boldsymbol{a}^T\boldsymbol{X}$, where $\boldsymbol{a} = (a_1, \ldots, a_r)^T$ is a vector of coefficients that need to be suitably determined. It is clear that

$$E(\vartheta^{**}) = \vartheta\boldsymbol{a}^T\boldsymbol{\alpha}, \quad \mathrm{Var}(\vartheta^{**}) = \vartheta^2\boldsymbol{a}^T\boldsymbol{\Sigma}\boldsymbol{a},$$

so that

$$\mathrm{bias}(\vartheta^{**}) = E(\vartheta^{**}) - \vartheta = \vartheta(\boldsymbol{a}^T\boldsymbol{\alpha} - 1).$$

Hence, the mean squared error of ϑ^{**} is given by

$$\begin{aligned}\mathrm{MSE}(\vartheta^{**}) &= \mathrm{Var}(\vartheta^{**}) + \left[\mathrm{bias}(\vartheta^{**})\right]^2 \\ &= \vartheta^2\{\boldsymbol{a}^T\boldsymbol{\Sigma}\boldsymbol{a} + (\boldsymbol{a}^T\boldsymbol{\alpha} - 1)^2\}.\end{aligned} \tag{3.30}$$

Upon minimizing $\mathrm{MSE}(\vartheta^{**})$ in (3.30) with respect to the coefficient vector $\boldsymbol{a}$ (leaving out the constant multiple ϑ^2), we get the equation

$$2\boldsymbol{\Sigma}\boldsymbol{a} - 2\boldsymbol{\alpha}(\boldsymbol{\alpha}^T\boldsymbol{a} - 1) = 0,$$

which yields the optimal coefficient vector $\boldsymbol{a}$ to be

$$\boldsymbol{a} = (\boldsymbol{\Sigma} - \boldsymbol{\alpha}\boldsymbol{\alpha}^T)^{-1}\boldsymbol{\alpha}. \tag{3.31}$$

Evidently, (3.31) can be rewritten as

$$\boldsymbol{a}^T = \boldsymbol{\alpha}^T(\boldsymbol{\Sigma} - \boldsymbol{\alpha}\boldsymbol{\alpha}^T)^{-1} = \boldsymbol{\alpha}^T\boldsymbol{\Sigma}^{-1}\left(\boldsymbol{I} - (\boldsymbol{\Sigma}^{-1/2}\boldsymbol{\alpha})(\boldsymbol{\Sigma}^{-1/2}\boldsymbol{\alpha})^T\right)^{-1},$$

where $\boldsymbol{I}$ is an identity matrix of dimension r. Finally, upon substituting the above expression of $\boldsymbol{a}$ in (3.30) and simplifying the resulting expression, we obtain the mean squared error of the estimator ϑ^{**} as given in (3.29). □

3.2.2.2 Location-scale family

In the case of location-scale family, one can proceed similarly by considering linear estimators $\mu^{**} = \boldsymbol{a}^T\boldsymbol{X}$ and $\vartheta^{**} = \boldsymbol{b}^T\boldsymbol{X}$ and then by minimizing a norm on the mean squared error and product matrix. One could use either the trace or determinant of this matrix and both these norms result in the same BLIEs of μ and ϑ; for details, see Mann (1967) and Balakrishnan et al. (2008b). We abstain from presenting these details here.

It is of interest to mention here that optimal linear estimators have been discussed by a number of authors for a wide range of lifetime distributions!

3.2.3 Maximum likelihood estimation

3.2.3.1 Scale family

Let us consider the situation when the Type-II right censored sample in (3.1) is observed from the scale-family of distributions with density function f_ϑ, $\vartheta > 0$, as in (3.3). Then, it is evident that the likelihood function is

$$\mathscr{L}(\vartheta; x_{1:n}, \ldots, x_{r:n}) = \frac{n!}{(n-r)!}\prod_{i=1}^{r} f_\vartheta(x_{i:n})\,(1 - F_\vartheta(x_{r:n}))^{n-r}, \tag{3.32}$$

which can be re-expressed in terms of the standard density function f in (3.3) and the corresponding standardized order statistics $z_{i:n} = \frac{x_{i:n}}{\vartheta}$, $i = 1, 2, \ldots, r$, as

$$\mathscr{L}(\vartheta; z_{1:n}, \ldots, z_{r:n}) = \frac{n!}{(n-r)!\vartheta^r}\prod_{i=1}^{r} f(z_{i:n})\,(1 - F(z_{r:n}))^{n-r}.$$

Then, the log-likelihood function is simply given by

$$\begin{aligned}\mathscr{L}^*(\vartheta) &\equiv \log \mathscr{L}(\vartheta)\\ &= \text{const} - r\log\vartheta + \sum_{i=1}^{r}\log f(z_{i:n}) + (n-r)\log(1 - F(z_{r:n})).\end{aligned} \tag{3.33}$$

It is this function that needs to be maximized for determining the Maximum Likelihood Estimate (MLE) of ϑ!

Remark 3.16. For a few distributions such as exponential and uniform, the MLE of ϑ can be derived explicitly from (3.33). But, in most cases, numerical optimization methods need to be employed for maximizing the log-likelihood function in (3.33).

Due to the fact that $z_{i:n} = \frac{x_{i:n}}{\vartheta}$, we observe that

$$\frac{\partial z_{i:n}}{\partial \vartheta} = -\frac{z_{i:n}}{\vartheta}, \quad \frac{\partial}{\partial \vartheta} \log f(z_{i:n}) = -\frac{z_{i:n}}{\vartheta} \cdot \frac{\frac{\partial}{\partial z_{i:n}} f(z_{i:n})}{f(z_{i:n})}$$

and

$$\frac{\partial}{\partial \vartheta} \log (1 - F(z_{r:n})) = \frac{z_{r:n}}{\vartheta} \cdot \frac{f(z_{r:n})}{1 - F(z_{r:n})} = \frac{z_{r:n}}{\vartheta} h(z_{r:n}),$$

where $h(\cdot)$ is the hazard function of the standard random variable Z. Hence, from (3.33), by taking derivative with respect to ϑ, we obtain the likelihood equation for ϑ to be

$$\begin{aligned} \frac{\partial \mathcal{L}^*}{\partial \vartheta}(\vartheta) &\equiv \frac{\partial \log \mathcal{L}}{\partial \vartheta}(\vartheta) \\ &= -\frac{r}{\vartheta} - \frac{1}{\vartheta} \sum_{i=1}^{r} z_{i:n} \frac{\frac{\partial}{\partial z_{i:n}} f(z_{i:n})}{f(z_{i:n})} + \frac{n-r}{\vartheta} z_{r:n} h(z_{r:n}) = 0. \end{aligned} \quad (3.34)$$

Proceeding similarly, and differentiating once again with respect to ϑ, we find

$$\begin{aligned} \frac{\partial^2 \mathcal{L}^*}{\partial \vartheta^2}(\vartheta) &= \frac{r}{\vartheta^2} + \frac{2}{\vartheta^2} \sum_{i=1}^{r} z_{i:n} \frac{\frac{\partial}{\partial z_{i:n}} f(z_{i:n})}{f(z_{i:n})} + \frac{1}{\vartheta^2} \sum_{i=1}^{r} z_{i:n}^2 \frac{\frac{\partial^2}{\partial z_{i:n}^2} f(z_{i:n})}{f^Z(z_{i:n})} \\ &\quad - \frac{1}{\vartheta^2} \sum_{i=1}^{r} z_{i:n}^2 \frac{\left(\frac{\partial}{\partial z_{i:n}} f(z_{i:n})\right)^2}{\left(f(z_{i:n})\right)^2} - \frac{2(n-r)}{\vartheta^2} z_{r:n} h(z_{r:n}) \\ &\quad - \frac{n-r}{\vartheta} z_{r:n}^2 \frac{\partial}{\partial z_{r:n}} h(z_{r:n}). \end{aligned} \quad (3.35)$$

From the above expression, we can readily obtain the Fisher information as $\mathcal{I}(\mathbf{X}; \vartheta) = E\left(-\frac{\partial^2 \mathcal{L}^*}{\partial \vartheta^2}\right)$.

Remark 3.17. From (3.35), it is evident that there are some complicated functions of the variables $z_{1:n}, \ldots, z_{r:n}$ for which we need to find expectations in order to obtain the Fisher information. This may be achievable for some simple distributions as exponential, but will not be achievable for many other distributions. In such cases, it will be convenient to determine the observed Fisher information as $\left(-\frac{\partial^2 \mathcal{L}^*}{\partial \vartheta^2}\right)\big|_{\vartheta=\widehat{\vartheta}}$, where $\widehat{\vartheta}$ is the MLE of ϑ determined from (3.34). It should be realized that this is just an estimate of the Fisher information $\mathcal{I}(\mathbf{X}; \vartheta)$, albeit based on the observed Type-II censored data.

Remark 3.18. With the values of the MLE of ϑ and the Fisher information (or, observed Fisher information) so determined, and upon using the well-known asymptotic properties of the MLE subject to some regularity conditions (see, for example, Casella and Berger, 2002), we can readily construct asymptotic confidence intervals for ϑ or carry out asymptotic Wald-type hypothesis tests concerning ϑ. For instance, by using the property that $\frac{\widehat{\vartheta}-\vartheta}{\mathrm{SE}(\widehat{\vartheta})}$, where $\mathrm{SE}(\widehat{\vartheta})$ is the standard error of the MLE $\widehat{\vartheta}$, determined as $\frac{1}{\sqrt{\mathcal{I}(\mathbf{X};\vartheta)}}\Big|_{\vartheta=\widehat{\vartheta}}$, is asymptotically standard normal, we get an approximate $100(1-\alpha)\%$ confidence interval for ϑ to be

$$\left(\widehat{\vartheta} - z_{1-\alpha/2}\,\mathrm{SE}(\widehat{\vartheta}), \widehat{\vartheta} + z_{1-\alpha/2}\,\mathrm{SE}(\widehat{\vartheta})\right),$$

where $z_{1-\alpha/2}$ is the upper $\alpha/2$ percentage point of the standard normal distribution. In a similar vein, we can carry out a test for the hypothesis testing problem

$$H_0 : \vartheta = \vartheta_0 \quad \text{vs.} \quad H_1 : \vartheta > \vartheta_0,$$

for example, where ϑ_0 is some pre-specified value, using the test statistic

$$Z = \frac{\widehat{\vartheta} - \vartheta_0}{\mathrm{SE}(\widehat{\vartheta})},$$

and finding evidence against H_0 if the observed value of Z is larger than $z_{1-\alpha}$, at α level of significance.

Remark 3.19. It is important to mention that in the preceding discussion, we have assumed the likelihood function in (3.32) to be differentiable with respect to ϑ. However, in cases like *Uniform*$(0, \vartheta)$, this is not true. In such cases, we need to use direct monotonicity arguments to determine the MLE of ϑ, and then derive the corresponding distributional properties by using the form of the MLE directly.

Special case of scaled-exponential distribution

Let us now consider the special case when the lifetimes have $Exp(\vartheta)$-distribution with

$$f_\vartheta(x) = \frac{1}{\vartheta}e^{-x/\vartheta} \quad \text{and} \quad 1 - F_\vartheta(x) = e^{-x/\vartheta}, \quad x > 0,\ \vartheta > 0.$$

In this case, $Z \sim Exp(1)$ with

$$f(z) = 1 - F(z) = e^{-z} \quad \text{and} \quad h(z) = 1 \quad \text{for } z > 0,$$

and consequently the likelihood equation in (3.34) simply reduces to

$$\frac{\partial \mathcal{L}^*}{\partial \vartheta} = -\frac{r}{\vartheta} + \frac{1}{\vartheta}\sum_{i=1}^{r} z_{i:n} + \frac{n-r}{\vartheta} z_{r:n} = 0,$$

which readily yields the MLE of ϑ to be

$$\widehat{\vartheta} = \frac{1}{r}\left\{\sum_{i=1}^{r} X_{i:n} + (n-r)X_{r:n}\right\}.$$

Remark 3.20. We observe that this is exactly the same as the BLUE of ϑ derived earlier in (3.5). Hence, all the distributional properties and exact inferential methods discussed earlier in Section 3.2.1.1 based on the BLUE hold true here as well for the MLE $\widehat{\vartheta}$.

Furthermore, we obtain in this case from (3.35) that

$$\frac{\partial^2 \mathcal{L}^*}{\partial \vartheta^2} = \frac{r}{\vartheta^2} - \frac{2}{\vartheta^2}\sum_{i=1}^{r} z_{i:n} + \frac{1}{\vartheta^2}\sum_{i=1}^{r} z_{i:n}^2 - \frac{1}{\vartheta^2}\sum_{i=1}^{r} z_{i:n}^2 - \frac{2(n-r)}{\vartheta^2} z_{r:n},$$

so that the Fisher information becomes

$$\mathcal{I}(\mathbf{X};\vartheta) = E\left(-\frac{\partial^2 \mathcal{L}^*}{\partial \vartheta^2}\right) = -\frac{r}{\vartheta^2} + \frac{2}{\vartheta^2} E\left[\sum_{i=1}^{r} Z_{i:n} + (n-r)Z_{r:n}\right].$$

But, since

$$\sum_{i=1}^{r} Z_{i:n} + (n-r)Z_{r:n} = \sum_{i=1}^{r} S_{i,n} \sim \Gamma(1,r),$$

we readily obtain the Fisher information to be

$$\mathcal{I}(\mathbf{X};\vartheta) = -\frac{r}{\vartheta^2} + \frac{2r}{\vartheta^2} = \frac{r}{\vartheta^2}.$$

Hence, we have $\mathrm{Var}_\vartheta(\widehat{\vartheta}) = \frac{1}{\mathcal{I}(\mathbf{X};\vartheta)} = \frac{\vartheta^2}{r}$ which, in the present case, is an exact result (and not an asymptotic result). Note that this is identical to the formula derived earlier in (3.6) for the BLUE of ϑ.

Special case of scaled half-logistic distribution

Let us now consider the special case when the lifetimes have *HalfLogist*(ϑ)-distribution with

$$f_\vartheta(x) = \frac{2e^{-\frac{x}{\vartheta}}}{\vartheta(1+e^{-\frac{x}{\vartheta}})^2} \quad \text{and} \quad 1 - F_\vartheta(x) = \frac{2e^{-\frac{x}{\vartheta}}}{1+e^{-\frac{x}{\vartheta}}}, \quad x > 0,\ \vartheta > 0.$$

In this case, $Z \sim$ *HalfLogist*(1) with

$$f(z) = \frac{2e^{-z}}{(1+e^{-z})^2},\ 1 - F(z) = \frac{2e^{-z}}{1+e^{-z}},\ \text{and } h(z) = \frac{1}{1+e^{-z}} \text{ for } z > 0,$$

and consequently the likelihood equation in (3.34) simply reduces to

$$\frac{\partial \mathcal{L}^*}{\partial \vartheta} = -\frac{r}{\vartheta} + \frac{1}{\vartheta} \sum_{i=1}^{r} z_{i:n} \left(\frac{1 - e^{-z_{i:n}}}{1 + e^{-z_{i:n}}} \right) + \frac{(n-r)}{\vartheta} \frac{z_{r:n}}{1 + e^{-z_{r:n}}} = 0.$$

This non-linear equation needs to be solved for ϑ by some numerical method, as mentioned earlier. The Newton-Raphson procedure, for example, would prove to be useful for this purpose.

Approximate maximum likelihood estimation

As the MLE of ϑ, as the solution of the likelihood equation in (3.34), needs to be determined by employing numerical iterative methods, it would be convenient to develop some approximations to the MLE of ϑ by suitably approximating the likelihood equation in (3.34). By realizing that the presence of the functions $\frac{\frac{\partial}{\partial z_{i:n}} f(z_{i:n})}{f(z_{i:n})}$ and $h(z_{r:n})$ in (3.34) hinder in obtaining an explicit form for the MLE of ϑ, one possibility will be to approximate them linearly[2] as

$$\frac{\frac{\partial}{\partial z_{i:n}} f(z_{i:n})}{f(z_{i:n})} \doteq \alpha_i + \beta_i z_{i:n} \quad \text{and} \quad h(z_{r:n}) \doteq \gamma + \delta z_{r:n};$$

here, the coefficients α_i, β_i, γ and δ involved in these linear approximations can be determined by expanding the involved functions in a Taylor-series in $z_{i:n}$ and $z_{r:n}$, respectively, around

$$F^{\leftarrow}(p_i) = F^{\leftarrow}\left(\frac{i}{n+1}\right) = \log\left(\frac{1 + \frac{i}{n+1}}{1 - \frac{i}{n+1}}\right) = \log\left(\frac{n+1+i}{n+1-i}\right),$$

for $i = 1, \ldots, r$. Note that the quantile function of the *HalfLogist*(1)-distribution is given by

$$F^{\leftarrow}(t) = \log\left(\frac{1+t}{1-t}\right), \qquad t \in (0, 1).$$

Of course, this approximate method can be developed for any other lifetime distribution as well, with the corresponding form of the quantile function $F^{\leftarrow}$.

Remark 3.21. Use of these two linear approximations in (3.34) yields the likelihood equation to be a quadratic equation in ϑ. In many cases, it can be shown that only one of the two roots of the quadratic equation is positive and is therefore admissible as an estimate. Such an estimator is referred to as the Approximate Maximum Likelihood Estimator (AMLE).

[2] We use the notation $\doteq$ to express the linear approximation throughout.

Remark 3.22. Eventhough this AMLE is only an approximation to the MLE, it has been observed in many cases that the two estimates are very close numerically. Hence, it can either be used directly as an estimator of ϑ for inferential purposes or can be utilized as a starting value while numerically solving (3.34) for finding the MLE of ϑ.

Example 3.23. Let us consider the Type-II right censored data on failure times of electrical insulation considered earlier in Example 3.8. In this case, Balakrishnan and Chan (1992) employed the Newton-Raphson iterative procedure to determine numerically from (3.34) the MLE of ϑ to be $\widehat{\vartheta} = 47.416$ mins. Balakrishnan and Chan (1992) also computed the AMLE of ϑ, by using the approximation described above, to be $\tilde{\vartheta} = 47.416$ mins. Note that this AMLE is identical to the MLE numerically.

3.2.3.2 Location-scale family

Let us now consider the situation when the Type-II right censored sample in (3.1) is observed from the location-scale-family of distributions with density function $f_{\mu,\vartheta}$ as in (3.8). Then, it is evident that the corresponding likelihood function is

$$\mathscr{L}(\mu, \vartheta; x_{1:n}, \dots, x_{r:n}) = \frac{n!}{(n-r)!} \prod_{i=1}^{r} f_{\mu,\vartheta}(x_{i:n}) \left(1 - F_{\mu,\vartheta}(x_{r:n})\right)^{n-r}, \tag{3.36}$$

which can be re-expressed in terms of the standard density function f in (3.3) and the corresponding standardized order statistics $z_{i:n} = \frac{x_{i:n}-\mu}{\vartheta}$, $i = 1, 2, \dots, r$, as

$$\mathscr{L}(\mu, \vartheta; z_{1:n}, \dots, z_{r:n}) = \frac{n!}{(n-r)!\vartheta^r} \prod_{i=1}^{r} f(z_{i:n}) \left(1 - F(z_{r:n})\right)^{n-r}.$$

Then, the log-likelihood function is simply

$$\begin{aligned} \mathscr{L}^*(\mu, \vartheta) &\equiv \log \mathscr{L}(\mu, \vartheta) \\ &= \text{const} - r \log \vartheta + \sum_{i=1}^{r} \log f(z_{i:n}) + (n-r) \log\left(1 - F(z_{r:n})\right). \end{aligned} \tag{3.37}$$

This is the functions that needs to be maximized for determining the MLEs of μ and ϑ!

Remark 3.24. As mentioned earlier in Remark 3.16, the MLEs of μ and ϑ can be derived explicitly from (3.37) only for a few distributions such as exponential and uniform. For most lifetime distributions, however, numerical methods need to be employed for determining the MLEs from (3.37).

Due to the fact that $z_{i:n} = \frac{x_{i:n}-\mu}{\vartheta}$, we observe that

$$\frac{\partial z_{i:n}}{\partial \mu} = -\frac{1}{\vartheta}, \qquad \frac{\partial z_{i:n}}{\partial \vartheta} = -\frac{z_{i:n}}{\vartheta},$$

$$\frac{\partial \log(z_{i:n})}{\partial \mu} = -\frac{1}{\vartheta}\frac{\frac{\partial}{\partial z_{i:n}} f(z_{i:n})}{f(z_{i:n})}, \qquad \frac{\partial \log f(z_{i:n})}{\partial \vartheta} = -\frac{z_{i:n}}{\vartheta}\frac{\frac{\partial}{\partial z_{i:n}} f(z_{i:n})}{f(z_{i:n})},$$
$$\frac{\partial \log(1-F(z_{r:n}))}{\partial \mu} = \frac{1}{\vartheta}\frac{f(z_{r:n})}{1-F(z_{r:n})} = \frac{1}{\vartheta}h(z_{r:n})$$

and

$$\frac{\partial \log(1-F(z_{r:n}))}{\partial \vartheta} = \frac{z_{r:n}}{\vartheta}\frac{f(z_{r:n})}{1-F(z_{r:n})} = \frac{z_{r:n}}{\vartheta}h(z_{r:n}),$$

where $h(\cdot)$ is the hazard function of the standard random variable Z. Thence, by taking partial derivatives with respect to μ and ϑ in (3.37), we obtain the likelihood equations for μ and ϑ to be

$$\begin{aligned}\frac{\partial \mathscr{L}^*}{\partial \mu} &\equiv \frac{\partial \log \mathscr{L}(\mu,\vartheta)}{\partial \mu} \\ &= -\frac{1}{\vartheta}\sum_{i=1}^{r}\frac{\frac{\partial}{\partial z_{i:n}} f(z_{i:n})}{f(z_{i:n})} + \frac{n-r}{\vartheta}h(z_{r:n}) = 0\end{aligned} \tag{3.38}$$

and

$$\begin{aligned}\frac{\partial \mathscr{L}^*}{\partial \vartheta} &\equiv \frac{\partial \log \mathscr{L}(\mu,\vartheta)}{\partial \vartheta} \\ &= -\frac{r}{\vartheta} - \frac{1}{\vartheta}\sum_{i=1}^{r} z_{i:n}\frac{\frac{\partial}{\partial z_{i:n}} f(z_{i:n})}{f(z_{i:n})} + \frac{n-r}{\vartheta}z_{r:n}h(z_{r:n}) = 0,\end{aligned} \tag{3.39}$$

respectively.

Proceeding similarly, and differentiating once again partially with respect to μ and ϑ, we obtain the following expressions:

$$\begin{aligned}\frac{\partial^2 \mathscr{L}^*}{\partial \mu^2} &= \frac{1}{\vartheta^2}\sum_{i=1}^{r}\frac{\frac{\partial^2}{\partial z_{i:n}^2} f(z_{i:n})}{f(z_{i:n})} - \frac{1}{\vartheta^2}\sum_{i=1}^{r}\frac{\left(\frac{\partial}{\partial z_{i:n}} f(z_{i:n})\right)^2}{\left(f(z_{i:n})\right)^2} \\ &\quad - \frac{(n-r)}{\vartheta^2}\frac{\partial}{\partial z_{r:n}}h(z_{r:n}),\end{aligned} \tag{3.40}$$

$$\begin{aligned}\frac{\partial^2 \mathscr{L}^*}{\partial \mu \partial \vartheta} &= \frac{1}{\vartheta^2}\sum_{i=1}^{r}\frac{\frac{\partial}{\partial z_{i:n}} f(z_{i:n})}{f(z_{i:n})} + \frac{1}{\vartheta^2}\sum_{i=1}^{r} z_{i:n}\frac{\frac{\partial^2}{\partial z_{i:n}^2} f(z_{i:n})}{f(z_{i:n})} \\ &\quad - \frac{1}{\vartheta^2}\sum_{i=1}^{r} z_{i:n}\frac{\left(\frac{\partial}{\partial z_{i:n}} f(z_{i:n})\right)^2}{\left(f(z_{i:n})\right)^2} - \frac{(n-r)}{\vartheta^2}h(z_{r:n}) \\ &\quad - \frac{(n-r)}{\vartheta^2}z_{r:n}\frac{\partial}{\partial z_{r:n}}h(z_{r:n}),\end{aligned}$$

$$\frac{\partial^2 \mathscr{L}^*}{\partial \vartheta^2} = \frac{r}{\vartheta^2} + \frac{2}{\vartheta^2} \sum_{i=1}^{r} z_{i:n} \frac{\frac{\partial}{\partial z_{i:n}} f(z_{i:n})}{f(z_{i:n})} + \frac{1}{\vartheta^2} \sum_{i=1}^{r} z_{i:n}^2 \frac{\frac{\partial^2}{\partial z_{i:n}^2} f(z_{i:n})}{f(z_{i:n})}$$
$$- \frac{1}{\vartheta^2} \sum_{i=1}^{r} z_{i:n}^2 \frac{\left(\frac{\partial}{\partial z_{i:n}} f(z_{i:n})\right)^2}{\left(f(z_{i:n})\right)^2} - \frac{2(n-r)}{\vartheta^2} z_{r:n} h(z_{r:n})$$
$$- \frac{(n-r)}{\vartheta^2} z_{r:n}^2 \frac{\partial}{\partial z_{r:n}} h(z_{r:n}). \tag{3.41}$$

From the above expressions in (3.40) and (3.41), we can readily obtain the Fisher information matrix as

$$\mathscr{I}(\mathbf{X}; \mu, \vartheta) = \begin{pmatrix} E\left(-\frac{\partial^2 \mathscr{L}^*}{\partial \mu^2}\right) & E\left(-\frac{\partial^2 \mathscr{L}^*}{\partial \mu \partial \vartheta}\right) \\ E\left(-\frac{\partial^2 \mathscr{L}^*}{\partial \mu \partial \vartheta}\right) & E\left(-\frac{\partial^2 \mathscr{L}^*}{\partial \vartheta^2}\right) \end{pmatrix}.$$

Remark 3.25. From (3.40) and (3.41), it is apparent that there are some complicated functions of the variables $z_{1:n}, \ldots, z_{r:n}$ for which we need to find expectations in order to obtain the Fisher information. This is feasible only for some simple distributions like Weibull or extreme value (see Dahmen et al., 2012), but will not be achievable for many other lifetime distributions. In such cases, it will be convenient to find the observed Fisher information matrix as

$$\left.\begin{pmatrix} -\frac{\partial^2 \mathscr{L}^*}{\partial \mu^2} & -\frac{\partial^2 \mathscr{L}^*}{\partial \mu \partial \vartheta} \\ -\frac{\partial^2 \mathscr{L}^*}{\partial \mu \partial \vartheta} & -\frac{\partial^2 \mathscr{L}^*}{\partial \vartheta^2} \end{pmatrix}\right|_{(\mu=\widehat{\mu},\ \vartheta=\widehat{\vartheta})},$$

where $(\widehat{\mu}, \widehat{\vartheta})$ is the MLE of (μ, ϑ) determined by solving (3.38) and (3.39). Note that this is just an estimate of the Fisher information $\mathscr{I}(\mathbf{X}; \mu, \vartheta)$, albeit based on the observed Type-II censored data.

Remark 3.26. With the values of the MLEs of (μ, ϑ) and the Fisher information matrix (or, observed Fisher information matrix) so determined, and upon using the well-known asymptotic properties of the MLEs subject to some regularity conditions (see Casella and Berger, 2002), we can construct asymptotic confidence intervals for μ and ϑ or carry out asymptotic Wald-type hypothesis tests concerning the parameters μ and ϑ. For example, by using the marginal asymptotic distributional properties that $\frac{\widehat{\mu}-\mu}{\mathrm{SE}(\widehat{\mu})}$ and $\frac{\widehat{\vartheta}-\vartheta}{\mathrm{SE}(\widehat{\vartheta})}$, where $\mathrm{SE}(\widehat{\mu})$ and $\mathrm{SE}(\widehat{\vartheta})$ are the standard errors of the MLEs $\widehat{\mu}$ and $\widehat{\vartheta}$, respectively, determined as the square root of the diagonal elements of the matrix $(\mathscr{I}(\mathbf{X}; \mu, \vartheta))^{-1}|_{(\mu=\widehat{\mu}, \vartheta=\widehat{\vartheta})}$, are asymptotically standard normal, we can get approximate $100(1-\alpha)\%$ confidence intervals for μ and ϑ to be

$$\left(\widehat{\mu} - z_{1-\alpha/2}\,\mathrm{SE}(\widehat{\mu}),\ \widehat{\mu} + z_{1-\alpha/2}\,\mathrm{SE}(\widehat{\mu})\right)$$

and

$$\left(\widehat{\vartheta} - z_{1-\alpha/2}\,\mathrm{SE}(\widehat{\vartheta}),\ \widehat{\vartheta} + z_{1-\alpha/2}\,\mathrm{SE}(\widehat{\vartheta})\right),$$

respectively; here, $z_{1-\alpha/2}$ is the upper $\alpha/2$ percentage point of the standard normal distribution, as before. Similarly, we can carry out tests for the hypothesis testing problems

$$H_0 : \mu = \mu_0 \quad \text{vs.} \quad H_1 : \mu > \mu_0$$

and

$$H_0' : \vartheta = \vartheta_0 \quad \text{vs.} \quad H_1' : \vartheta > \vartheta_0,$$

where μ_0 and ϑ_0 are some pre-specified values, using the test statistics

$$Z = \frac{\widehat{\mu} - \mu_0}{\mathrm{SE}(\widehat{\mu})} \quad \text{and} \quad Z' = \frac{\widehat{\vartheta} - \vartheta_0}{\mathrm{SE}(\widehat{\vartheta})},$$

respectively, and finding evidence against $H_0(H_0')$ if the observed value of $Z(Z')$ is larger than $z_{1-\alpha}$, at α level of significance.

Remark 3.27. It should be mentioned that in the preceding discussion, we have assumed the likelihood function in (3.36) to be differentiable with respect to both parameters μ and ϑ. However, in some cases like *Exp*(μ, ϑ) and *Uniform*$(\mu, \mu + \vartheta)$, this is not true. In such cases, we need to use direct monotonicity arguments to determine the MLEs, and then derive their corresponding distributional properties by using their explicit forms.

Special case of two-parameter exponential distribution

Let us consider the special case when the lifetimes have *Exp*(μ, ϑ)-distribution with

$$f_{\mu,\vartheta}(x) = \frac{1}{\vartheta}\, e^{-\frac{x-\mu}{\vartheta}} \quad \text{and} \quad 1 - F_{\mu,\vartheta}(x) = e^{-\frac{x-\mu}{\vartheta}}, \quad x \geq \mu,\ \vartheta > 0.$$

In this case, the log-likelihood function in (3.37) is

$$\begin{aligned} \mathcal{L}^*(\mu, \vartheta) &= \text{const} - r \log \vartheta - \sum_{i-1}^{r} \frac{x_{i:n} - \mu}{\vartheta} - (n-r)\frac{x_{r:n} - \mu}{\vartheta} \\ &= \text{const} - r \log \vartheta - \frac{1}{\vartheta} \sum_{i=1}^{r} x_{i:n} - \frac{n-r}{\vartheta} x_{r:n} + \frac{n\mu}{\vartheta}, \qquad (3.42) \\ &\qquad \mu \leq x_{1:n} < \cdots < x_{r:n} < \infty,\ \vartheta > 0. \end{aligned}$$

Evidently, (3.42) is an increasing function of μ and due to the restriction that $\mu \leq x_{1:n}$, we immediately find the MLE of μ to be $\widehat{\mu} = x_{1:n}$. Upon replacing μ by $\widehat{\mu} = x_{1:n}$ in

(3.42), we get

$$\mathscr{L}^*(\mu, \vartheta) \le \mathscr{L}^*(x_{1:n}, \vartheta)$$

with equality iff $\mu = x_{1:n}$. Therefore, we get the likelihood function for ϑ to be

$$\mathscr{L}^*(x_{1:n}, \vartheta) = \text{const} - r \log \vartheta - \frac{1}{\vartheta} \sum_{i=1}^{r} x_{i:n} - \frac{n-r}{\vartheta} x_{r:n} + \frac{n x_{1:n}}{\vartheta},$$

which is indeed differentiable with respect to ϑ. So, we immediately obtain the likelihood equation for ϑ to be

$$\frac{\partial \mathscr{L}^*}{\partial \vartheta}(x_{1:n}, \vartheta) = -\frac{r}{\vartheta} + \frac{1}{\vartheta^2} \left\{ \sum_{i=1}^{r} x_{i:n} + (n-r) x_{r:n} - n x_{1:n} \right\},$$

which yields the MLE of ϑ to be

$$\begin{aligned} \widehat{\vartheta} &= \frac{1}{r} \left\{ \sum_{i=1}^{r} X_{i:n} + (n-r) X_{r:n} - n X_{1:n} \right\} \\ &= \frac{1}{r} \sum_{i=2}^{r} (n-i+1)(X_{i:n} - X_{i-1:n}). \end{aligned}$$

Using the same idea as in Remark 3.34 for Weibull distributions and applying the inequality $\log t \le t - 1$, $t > 0$, it can be shown that

$$\mathscr{L}^*(\mu, \vartheta) \le \mathscr{L}^*(\widehat{\mu}, \vartheta) \le \mathscr{L}^*(\widehat{\mu}, \widehat{\vartheta}) \quad \text{for all } \mu \in \mathbb{R}, \vartheta > 0,$$

with equality iff $\mu = \widehat{\mu}$ and $\vartheta = \widehat{\vartheta}$. This proves that $(\widehat{\mu}, \widehat{\vartheta})$ yields a global maximum of the log-likelihood function so that $(\widehat{\mu}, \widehat{\vartheta})$ is indeed the MLE.

Remark 3.28. We observe that $\widehat{\vartheta}$ is exactly $\frac{r-1}{r}\vartheta^*$, where ϑ^* is the BLUE of ϑ in (3.18). As ϑ^* is unbiased for ϑ, we immediately see that

$$\text{bias}_\vartheta(\widehat{\vartheta}) = E_\vartheta(\widehat{\vartheta}) - \vartheta = E_\vartheta\left(\frac{r-1}{r}\vartheta^*\right) - \vartheta = \frac{r-1}{r}\vartheta - \vartheta = -\frac{\vartheta}{r},$$

and so $\widehat{\vartheta}$ is negatively biased for ϑ. It does become asymptotically unbiased for ϑ when $r \to \infty$. Furthermore, we observe that $\widehat{\mu} = X_{1:n} = \mu^* + \frac{\vartheta^*}{n}$, where μ^* and ϑ^* are the BLUEs of μ and ϑ, respectively. Then, we find

$$\text{bias}_{\mu,\vartheta}(\widehat{\mu}) = E_{\mu,\vartheta}(\widehat{\mu}) - \mu = E_{\mu,\vartheta}\left(\mu^* + \frac{\vartheta^*}{n}\right) - \mu = \frac{\vartheta}{n},$$

and so $\widehat{\mu}$ is positively biased for μ. It does become unbiased for μ when $n \to \infty$.

Remark 3.29. As the MLEs $\widehat{\mu}$ and $\widehat{\vartheta}$ are same as the BLUEs μ^* and ϑ^*, except for the bias terms, all the distributional properties and exact inferential methods discussed earlier in Section 3.2.1.2 based on the BLUEs hold true here as well for the MLEs $\widehat{\mu}$ and $\widehat{\vartheta}$. Note that the MLEs are negatively correlated (and, thus, dependent), that is,

$$\mathrm{Cov}(\widehat{\mu}, \widehat{\vartheta}) = -\frac{r-1}{nr}\,\mathrm{Var}(\vartheta^*) = -\frac{\vartheta^2}{nr}.$$

Example 3.30. Let us consider the Type-II right censored data on failure times of electrical insulation considered earlier in Example 3.6. In this case, with $n = 12$ and $r = 11$, we find the MLEs of μ and ϑ to be

$$\widehat{\mu} = 12.3 \quad \text{and} \quad \widehat{\vartheta} = 58.764.$$

Upon comparing these with the BLUEs presented in Example 3.11, we immediately find that $\widehat{\mu} > \mu^*$ $(= 6.913)$ and $\widehat{\vartheta} < \vartheta^*$ $(= 64.640)$, supporting the facts that $\widehat{\mu}$ and $\widehat{\vartheta}$ are positively and negatively biased estimates of μ and ϑ, respectively, as pointed out in Remark 3.28. Further, we find the MLE of the mean lifetime, $\mu + \vartheta$, to be $\widehat{\mu} + \widehat{\vartheta} = 71.064$. Note that this estimate is close to 71.553, the BLUE of the mean lifetime.

Special case of two-parameter half-logistic distribution

Let us now consider the special case when the lifetimes have *HalfLogist*(μ, ϑ)-distribution with

$$f_{\mu,\vartheta}(x) = \frac{2e^{-\frac{x-\mu}{\vartheta}}}{\vartheta\left(1+e^{-\frac{x-\mu}{\vartheta}}\right)^2} \quad \text{and} \quad 1 - F_{\mu,\vartheta}(x) = \frac{2e^{-\frac{x-\mu}{\vartheta}}}{1+e^{-\frac{x-\mu}{\vartheta}}}, \quad x > \mu, \vartheta > 0.$$

In this case as well, the log-likelihood function in (3.37) is an increasing function of μ, and so we have the MLE of μ to be $\widehat{\mu} = x_{1:n}$. Then, the MLE of ϑ can be determined by maximizing the log-likelihood function in (3.37) after replacing μ by $\widehat{\mu} = x_{1:n}$. Further, by using linear approximations for the non-linear functions that are present in the likelihood equation for ϑ, Balakrishnan and Wong (1991) developed the AMLE of ϑ.

Example 3.31. Let us consider again the Type-II right censored data on failure times of electrical insulation considered earlier in Example 3.6. In this case, the MLEs of μ and ϑ were found by Balakrishnan and Wong (1991) to be

$$\widehat{\mu} = 12.3 \quad \text{and} \quad \widehat{\vartheta} = 42.7.$$

Also, they found the AMLE of ϑ to be $\tilde{\vartheta} = 42.7$. Then, the MLE of the mean lifetime, $\mu + \vartheta \log 4$, is

$$12.3 + (42.7 \times 1.386) = 71.482.$$

Special case of two-parameter log-normal distribution

Suppose the lifetimes have *log-N*(μ, ϑ)-distribution, so that the log-lifetimes have $N(\mu, \vartheta^2)$-distribution. In this case, we can assume that the Type-II censored data

$$x_{1:n}(=\log y_{1:n}), x_{2:n}(=\log y_{2:n}), \ldots, x_{r:n}(=\log y_{r:n})$$

have been observed from normal, $N(\mu, \vartheta^2)$, distribution. In this case, the standard variable $Z = \frac{X-\mu}{\vartheta}$ has a standard normal, $N(0, 1)$, distribution with

$$\varphi(z) = \frac{1}{\sqrt{2\pi}} e^{-\frac{z^2}{2}}, \quad \frac{\partial}{\partial z}\varphi(z) = -z\varphi(z), \qquad \text{and} \qquad h(z) = \frac{\varphi(z)}{1-\Phi(z)}, \qquad z \in \mathbb{R}.$$

The likelihood equations in (3.38) and (3.39) then simplify to

$$\frac{\partial \mathscr{L}^*}{\partial \mu} = \frac{1}{\vartheta}\sum_{i=1}^{r} z_{i:n} + \frac{n-r}{\vartheta}\frac{\varphi(z_{r:n})}{1-\Phi(z_{r:n})} = 0, \tag{3.43}$$

$$\frac{\partial \mathscr{L}^*}{\partial \vartheta} = -\frac{r}{\vartheta} + \frac{1}{\vartheta}\sum_{i=1}^{r} z_{i:n}^2 + \frac{n-r}{\vartheta}\frac{z_{r:n}\varphi(z_{r:n})}{1-\Phi(z_{r:n})} = 0. \tag{3.44}$$

These two equations can be solved numerically by the use of Newton-Raphson method.

By expanding the function $\frac{\varphi(z_{r:n})}{1-\Phi(z_{r:n})}$ in a Taylor-series around $\Phi^{-1}(\frac{r}{n+1})$, where Φ^{-1} is the standard normal quantile function, and using the linear approximation

$$\frac{\varphi(z_{r:n})}{1-\Phi(z_{r:n})} \doteq \gamma + \delta z_{r:n},$$

Balakrishnan (1989a) derived AMLEs of μ and ϑ from (3.43) and (3.44), respectively. These are explicit estimators, and have also been shown to be as efficient as the MLEs. Similar AMLEs of parameters of some other distributions have been developed by Balakrishnan and Varadan (1991), Balakrishnan (1990), Balakrishnan (1992), and Kang et al. (2001).

Example 3.32. The following Type-II right censored data, presented by Gupta (1952), give the number of days to death (y) of the first 7 mice to die in a sample of 10 mice after inoculation with uniform culture of human tuberculosis:

$y_{i:10}$	41	44	46	54	55	58	60
$x_{i:10} = \log y_{i:10}$	1.613	1.644	1.663	1.732	1.740	1.763	1.778

Upon making use of the tables of Cohen (1961), the MLEs of μ and ϑ, as solutions of (3.43) and (3.44), can be found to be

$$\widehat{\mu} = 1.742 \quad \text{and} \quad \widehat{\vartheta} = 0.079,$$

with their standard errors as

$$\mathrm{SE}(\widehat{\mu}) = 0.027 \quad \text{and} \quad \mathrm{SE}(\widehat{\vartheta}) = 0.023.$$

With $n = 10$ and $r = 7$, based on the above Type-II right censored data, Balakrishnan (1989a) determined the AMLEs (using the linear approximation in (3.43) and (3.44)) of μ and ϑ to be

$$\tilde{\mu} = 1.742 \quad \text{and} \quad \tilde{\vartheta} = 0.079,$$

with their standard errors as

$$\mathrm{SE}(\tilde{\mu}) = 0.027 \quad \text{and} \quad \mathrm{SE}(\tilde{\vartheta}) = 0.022.$$

Finally, the tables of BLUEs presented by Sarhan and Greenberg (1956) can be used to compute the BLUEs of μ and ϑ to be

$$\mu^* = 1.746 \quad \text{and} \quad \vartheta^* = 0.091,$$

with their standard errors as

$$\mathrm{SE}(\mu^*) = 0.31 \quad \text{and} \quad \mathrm{SE}(\vartheta^*) = 0.029.$$

Special case of two-parameter log-Laplace distribution

Suppose the lifetimes have a *log-Laplace*(μ, ϑ)-distribution, so that the log-lifetimes have a *Laplace*(μ, ϑ)-distribution with

$$f_{\mu,\vartheta}(x) = \frac{1}{2\vartheta} e^{-\frac{|x-\mu|}{\vartheta}}, \; x \in \mathbb{R}, \qquad \text{and } F_{\mu,\vartheta}(x) = \begin{cases} \frac{1}{2} e^{\frac{x-\mu}{\vartheta}}, & x < \mu \\ 1 - \frac{1}{2} e^{\frac{\mu - x}{\vartheta}}, & x \geq \mu \end{cases}$$

for $-\infty < \mu < \infty$ and $\vartheta > 0$ (see Definition B.15). Let $x_{1:n}, \ldots, x_{r:n}$ be a Type-II right censored sample from the above *Laplace*(μ, ϑ)-distribution. In this case, quite interestingly by using a combination of monotonicity argument and differentiation-based method, the MLEs of μ and ϑ have been derived in explicit forms as linear functions of the ordered Type-II censored data; see Balakrishnan and Cutler (1996), Childs and Balakrishnan (1997), Balakrishnan and Cramer (2014, Section 12.6), Zhu and Balakrishnan (2016), and Balakrishnan and Zhu (2016).

The explicit MLEs of μ and ϑ so derived are as follows:

Case 1: When $r < \frac{n}{2}$,

$$\widehat{\mu} = X_{r:n} + \widehat{\vartheta} \log\left(\frac{n}{2r}\right),$$

$$\widehat{\vartheta} = \frac{1}{r} \sum_{i=1}^{r-1} (X_{r:n} - X_{i:n});$$

Case 2: When $n = 2m + 1$ and $r \geq m + 2$,

$$\widehat{\mu} = X_{m+1:n},$$
$$\widehat{\vartheta} = \frac{1}{r}\left[\sum_{i=m+2}^{r} X_{i:n} - \sum_{i=1}^{m} X_{i:n} + (n-r)X_{r:n}\right];$$

Case 3: When $n = 2m + 1$ and $r = m + 1$,

$$\widehat{\mu} = X_{m+1:n},$$
$$\widehat{\vartheta} = \frac{1}{r}\sum_{i=1}^{m}(X_{m+1:n} - X_{i:n});$$

Case 4: When $n = 2m$ and $r > m + 1$,

$$\widehat{\mu} = \frac{1}{2}(X_{m:n} + X_{m+1:n}),$$
$$\widehat{\vartheta} = \frac{1}{r}\left[\sum_{i=m+1}^{r} X_{i:n} - \sum_{i=1}^{m} X_{i:n} + (n-r)X_{r:n}\right];$$

Case 5: When $n = 2m$ and $r = m + 1$,

$$\widehat{\mu} = \frac{1}{2}(X_{m:n} + X_{m+1:n}),$$
$$\widehat{\vartheta} = \frac{1}{r}\left[mX_{r:n} - \sum_{i=1}^{m} X_{i:n}\right];$$

Case 6: When $n = 2m$ and $r = m$,

$$\widehat{\mu} = X_{m:n},$$
$$\widehat{\vartheta} = \frac{1}{r}\left[(r-1)X_{r:n} - \sum_{i=1}^{r-1} X_{i:n}\right].$$

As the MLEs of μ and ϑ in all the above cases are linear functions of the order statistics, exact likelihood inferential procedures have been developed by Zhu and Balakrishnan (2016).

Example 3.33. Consider the following Type-II right censored data, given by Mann and Fertig (1973), on the lifetimes of 13 aeroplane components with the last 3 having been censored:

$$0.22, 0.50, 0.88, 1.00, 1.32, 1.33, 1.54, 1.76, 2.50, 3.00$$

Now, assuming *log-Laplace*(μ, ϑ) distribution for these Type-II right censored data, let us consider the following log-lifetimes:

$$-1.514, -0.693, -0.128, 0, 0.278, 0.285, 0.432, 0.565, 0.916, 1.099.$$

Let us now assume *Laplace*(μ, ϑ)-distribution for these Type-II right censored data on log-lifetimes. Then, in this case, we have $n = 13$ and $r = 10$ and so *Case 2* holds in the above list of cases. So, the MLEs of μ and ϑ are simply

$$\widehat{\mu} = x_{7:13} = 0.432$$

and

$$\widehat{\vartheta} = \frac{1}{10}\left[\sum_{i=8}^{10} x_{i:13} - \sum_{i=1}^{6} x_{i:13} + 3x_{10:13}\right] = 0.765.$$

In addition to these MLEs, Zhu and Balakrishnan (2016) have also presented exact inferential results, including exact confidence intervals, for population quantiles, reliability function and cumulative hazard function, based on these Type-II right censored data.

3.2.3.3 Scale-shape family

In reliability studies, it is more common and convenient to use some scale-shape family of distributions to model lifetime data due to the flexibility they provide in terms of shape characteristics. For example, Weibull, gamma, Birnbaum-Saunders, log-logistic, Pareto and log-normal distributions are some prominent members of this family. Yet, as seen in the last section with regard to log-normal distribution, a simple transformation (such as log) could transform a scale-shape family to a location-scale family of distributions. In fact, we utilized this transformation-based technique to transform lifetime data from log-normal and log-Laplace distributions to those from normal and Laplace distributions, respectively, and then estimate the distributional properties based on the results on location-scale family of distributions. However, this would not work while dealing with some other scale-shape distributions, such as gamma and Birnbaum-Saunders distributions.

Special case of two-parameter Weibull distribution

Let us now consider the case when the Type-II right censored data come from Weibull, *Weibull*(ϑ^β, β), distribution with

$$f_{\vartheta,\beta}(x) = \frac{\beta}{\vartheta^\beta} x^{\beta-1} e^{-(\frac{x}{\vartheta})^\beta} \quad \text{and} \quad 1 - F_{\vartheta,\beta}(x) = e^{-(\frac{x}{\vartheta})^\beta}, \quad x > 0, \tag{3.45}$$

for $\vartheta, \beta > 0$ (see Definition B.6). Upon substituting these expressions into the likelihood function, we obtain the explicit likelihood function as

$$\mathscr{L}(\vartheta, \beta; x_{1:n}, \ldots, x_{r:n}) = \frac{n!}{(n-r)!} \frac{\beta^r}{\vartheta^{r\beta}} \left(\prod_{i=1}^{r} x_{i:n}^{\beta-1} \right) \exp\left\{ -\frac{1}{\vartheta^{\beta}} \left(\sum_{i=1}^{r} x_{i:n}^{\beta} + (n-r)x_{r:n}^{\beta} \right) \right\},$$

from which we immediately get the log-likelihood function as

$$\begin{aligned} \mathscr{L}^*(\vartheta, \beta) &\equiv \log \mathscr{L}(\vartheta, \beta) \\ &= \text{const} + r \log \beta - r\beta \log \vartheta + (\beta - 1) \sum_{i=1}^{r} \log x_{i:n} \\ &\quad - \frac{1}{\vartheta^{\beta}} \left\{ \sum_{i=1}^{r} x_{i:n}^{\beta} + (n-r)x_{r:n}^{\beta} \right\}. \end{aligned} \tag{3.46}$$

From (3.46), by taking partial derivatives with respect to ϑ and β, we obtain the likelihood equations as follows:

$$\begin{aligned} \frac{\partial \mathscr{L}^*}{\partial \vartheta} &\equiv \frac{\partial \log \mathscr{L}(\vartheta, \beta)}{\partial \vartheta} \\ &= -\frac{r\beta}{\vartheta} + \frac{\beta}{\vartheta^{\beta+1}} \left\{ \sum_{i=1}^{r} x_{i:n}^{\beta} + (n-r)x_{r:n}^{\beta} \right\} = 0, \end{aligned} \tag{3.47}$$

$$\begin{aligned} \frac{\partial \mathscr{L}^*}{\partial \beta} &\equiv \frac{\partial \log \mathscr{L}(\vartheta, \beta)}{\partial \beta} \\ &= \frac{r}{\beta} - r \log \vartheta + \sum_{i=1}^{r} \log x_{i:n} + \frac{\log \vartheta}{\vartheta^{\beta}} \left\{ \sum_{i=1}^{r} x_{i:n}^{\beta} + (n-r)x_{r:n}^{\beta} \right\} \\ &\quad - \frac{1}{\vartheta^{\beta}} \left\{ \sum_{i=1}^{r} x_{i:n}^{\beta} \log x_{i:n} + (n-r)x_{r:n}^{\beta} \log x_{r:n} \right\} = 0. \end{aligned} \tag{3.48}$$

Observe that (3.47) yields an explicit MLE of ϑ as

$$\widehat{\vartheta} = \widehat{\vartheta}(\beta) = \left[\frac{1}{r} \left\{ \sum_{i=1}^{r} x_{i:n}^{\beta} + (n-r)x_{r:n}^{\beta} \right\} \right]^{1/\beta} \tag{3.49}$$

which, when substituted in (3.48), yields a simplified likelihood equation for β as

$$\begin{aligned} \frac{r}{\beta} - \frac{r}{\sum_{i=1}^{r} x_{i:n}^{\beta} + (n-r)x_{r:n}^{\beta}} &\left\{ \sum_{i=1}^{r} x_{i:n}^{\beta} \log x_{i:n} + (n-r)x_{r:n}^{\beta} \log x_{r:n} \right\} \\ &+ \sum_{i=1}^{r} \log x_{i:n} = 0. \end{aligned} \tag{3.50}$$

The MLE of β needs to be determined numerically by solving (3.50) through the use of Newton-Raphson method, for example.

Remark 3.34. Using the expression in (3.49) for $\widehat{\vartheta}(\beta)$ as well as the inequality $\log t \le t-1$, $t > 0$, the log-likelihood function can be bounded from above as follows:

$$\begin{aligned}\mathcal{L}^*(\vartheta,\beta) &= \text{const} + r\log\beta + (\beta-1)\sum_{i=1}^{r}\log x_{i:n} - r\beta\log\widehat{\vartheta}(\beta)\\ &\quad + r\log\left(\frac{\widehat{\vartheta}(\beta)}{\vartheta}\right)^{\beta} - r\left(\frac{\widehat{\vartheta}(\beta)}{\vartheta}\right)^{\beta}\\ &\le \text{const} + r\log\beta + (\beta-1)\sum_{i=1}^{r}\log x_{i:n} - r\beta\log\widehat{\vartheta}(\beta) - r = \mathcal{L}^*(\widehat{\vartheta}(\beta),\beta),\end{aligned}$$

with equality iff $\vartheta = \widehat{\vartheta}(\beta)$. Notice that (3.50) denotes the likelihood equation belonging to the log-likelihood function $\mathcal{L}^*(\widehat{\vartheta}(\beta),\beta)$. According to the discussion presented in Section 3.2.3.4, this shows that the bivariate optimization problem has a unique global maximum.

Example 3.35. Let us consider the following Type-II right censored data, presented by Dodson (2006), on failure times of grinders. The failure times of the first 12 grinders to fail, of the 20 grinders that were placed on a life-test, are as follows:

$$12.5, 24.4, 58.2, 68.0, 69.1, 95.5, 96.6, 97.0, 114.2, 123.2, 125.6, 152.7$$

Now, assuming *Weibull*$(\vartheta^{\beta},\beta)$-distribution for these Type-II right censored data, and solving the likelihood equation in (3.50) for β by Newton-Raphson method, we determine the MLE of β to be $\widehat{\beta} = 1.646$. Next, by using this estimate of β in (3.49), we immediately find the MLE of ϑ to be $\widehat{\vartheta} = 162.2$.

Though we have discussed the likelihood estimation of parameters ϑ and β of the Weibull distribution belonging to the scale-shape family of distributions, it is important to mention that it can be transformed to an extreme value distribution belonging to the location-scale family of distributions by means of logarithmic transformation. Specifically, if $X \sim$ *Weibull*$(\vartheta^{\beta},\beta)$ with density functions and survival functions as given in (3.45), then $Y = \log X$ has an extreme value, $EV_1(\mu',\vartheta')$, distribution with its density function and cumulative distribution function as

$$f_{\mu';\vartheta'}(y) = \frac{1}{\vartheta'}e^{\frac{y-\mu'}{\vartheta'}}e^{-e^{\frac{y-\mu'}{\vartheta'}}} \quad \text{and} \quad F_{\mu';\vartheta'}(y) = 1 - e^{-e^{\frac{y-\mu'}{\vartheta'}}} \tag{3.51}$$

for $y \in \mathbb{R}$, with parameters $\mu' = \log\vartheta \in \mathbb{R}$ and $\vartheta' = \frac{1}{\beta} > 0$ (see Definition B.13). Then, this logarithmic transformation, being a monotone increasing transformation, transforms the Type-II right censored data in (3.1) from *Weibull*$(\vartheta^{\beta},\beta)$-distribution in (3.45) to a Type-II right censored data

$$(Y_{1:n}, Y_{2:n}, \ldots, Y_{r:n})$$

from $EV_I(\mu', \vartheta')$-distribution in (3.51). In this case, the standardized variable $Z = \frac{Y-\mu'}{\vartheta'}$ has the standard extreme value, $EV_I(0,1)$, distribution with

$$f(z) = e^z e^{-e^z} \quad \text{and} \quad F(z) = 1 - e^{-e^z} \text{ for } z \in \mathbb{R}.$$

Then, using the facts that $z_{i:n} = \frac{y_{i:n}-\mu'}{\vartheta'}$, $\frac{\partial}{\partial z} f(z) = (1-e^z)f(z)$ and $h(z) = e^z$, the likelihood equations in (3.38) and (3.39) readily simplify to

$$\frac{\partial \mathcal{L}^*}{\partial \mu'} = -\frac{1}{\vartheta'} \sum_{i=1}^{r} (1 - e^{z_{i:n}}) + \frac{n-r}{\vartheta'} e^{z_{r:n}} = 0, \tag{3.52}$$

$$\frac{\partial \mathcal{L}^*}{\partial \vartheta'} = -\frac{r}{\vartheta'} - \frac{1}{\vartheta'} \sum_{i=1}^{r} z_{i:n}(1 - e^{z_{i:n}}) + \frac{n-r}{\vartheta'} z_{r:n} e^{z_{r:n}} = 0. \tag{3.53}$$

The MLEs of the location and scale parameters, μ' and ϑ', can then be determined numerically by solving the likelihood equations in (3.52) and (3.53).

Once the MLEs of μ' and ϑ', denoted by $\widehat{\mu}'$ and $\widehat{\vartheta}'$, have been so determined, the MLEs of the original Weibull scale and shape parameters (ϑ and β) can be readily found due to the simple explicit relationships between these two sets of parameters. To be specific, the MLEs of ϑ and η will simply be

$$\widehat{\vartheta} = e^{\widehat{\mu}'} \quad \text{and} \quad \widehat{\beta} = \frac{1}{\widehat{\vartheta}'}.$$

Thus, all the inferential procedures for μ' and ϑ', such as confidence intervals and hypothesis tests, can be readily transformed to the corresponding ones for the Weibull parameters ϑ and β!

Remark 3.36. Even though the logarithmic transformation transforms the scale-shape Weibull distribution to the location-scale extreme value distribution, it does not avert the need for the use of numerical iterative methods for the determination of MLEs of the distributional parameters!

Example 3.37. Let us consider the Type-II right censored data of Dodson (2006) on failure times of grinders, considered earlier in Example 3.35. Now, upon making log-transformation of these data, we can regard

$$2.5257, 3.1946, 4.0639, 4.2195, 4.2356, 4.5591, 4.5706, 4.5747, 4.7380,$$
$$4.8138, 4.8331, 5.0285$$

as a Type-II right censored data from $EV_{\mathrm{I}}(\mu', \vartheta')$-distribution with $n = 20$ and $r = 12$. Then, upon solving the likelihood equations in (3.52) and (3.53) by Newton-Raphson method, we find the MLEs of μ' and ϑ' to be

$$\widehat{\mu}' = 5.0888 \quad \text{and} \quad \widehat{\vartheta}' = 0.6075.$$

Observe that these do match the values of $\log(162.2)$ and $\frac{1}{1.646}$ found from the MLEs of Weibull parameters ϑ and β determined earlier in Example 3.35.

By expanding the function $e^{z_{i:n}}$ that is present in the likelihood equations in (3.52) and (3.53) in a Taylor-series around $F^{\leftarrow}(\frac{i}{n+1}) = \log\left(-\log\left(\frac{n-i+1}{n+1}\right)\right)$, for $i = 1, \ldots, r$, and then using the linear approximation

$$e^{z_{i:n}} \doteq \gamma_i + \delta_i z_{i:n},$$

Balakrishnan and Varadan (1991) derived the AMLEs of μ' and ϑ' from (3.52) and (3.53), respectively. These are explicit estimators, and have also been shown by these authors to be as efficient as the MLEs.

Example 3.38. Consider the Type-II right censored data of Mann and Fertig (1973) on lifetimes of aeroplane components, considered earlier in Example 3.33. Then, by treating the log-lifetimes presented there as a Type-II right censored data from $EV_{\mathrm{I}}(\mu', \vartheta')$ distribution with $n = 13$ and $r = 10$, the MLEs of μ' and ϑ' can be found by solving the likelihood equations in (3.52) and (3.53) to be $\widehat{\mu}' = 0.821$ and $\widehat{\vartheta}' = 0.706$; see Lawless (2003). By employing the above mentioned linear approximation in (3.52) and (3.53), Balakrishnan and Varadan (1991) determined the AMLEs of μ and ϑ to be 0.811 and 0.710, respectively. We do observe that these estimates are numerically quite close to the MLEs.

3.2.3.4 Existence and uniqueness

So far, we have discussed the maximum likelihood estimation of distributional parameters under Type-II right censored data. In a few instances, like in the cases of scaled-exponential, two-parameter exponential and log-Laplace distributions, the MLEs exist in explicit form. But, in other cases such as scaled half-logistic, two-parameter half-logistic, log-normal and Weibull distributions, the MLEs of distributional parameters needed to be determined by the use of numerical iterative methods. In such instances, it will be natural to ask the question whether the MLEs even exist and if they do, whether they are unique?

Though the answer to this question does not seem to be possible in general, it is certainly possible to do in certain cases. For example, let us consider the case of Weibull, $Weibull(\vartheta^{\beta}, \beta)$, distribution. In this case, we saw earlier that the MLE of ϑ exists uniquely and explicitly as in (3.49). But, the MLE of β does not exist explicitly and needs to be determined numerically by solving the corresponding likelihood equation in (3.50). However, as done by Balakrishnan and Kateri (2008), the existence and uniqueness of the MLE $\widehat{\beta}$ can be proved in the following simple manner.[3] For this purpose, let $2 \leq r \leq n$ and suppose $x_{1:n} < \cdots < x_{r:n}$ are strictly ordered (which holds with probability one). First rewrite the likelihood equation for β in (3.50) as

$$\begin{aligned} \frac{1}{\beta} &= \frac{1}{\sum_{i=1}^{r} x_{i:n}^{\beta} + (n-r)x_{r:n}^{\beta}} \left\{ \sum_{i=1}^{r} x_{i:n}^{\beta} \log x_{i:n} + (n-r)x_{r:n}^{\beta} \log x_{r:n} \right\} - \frac{1}{r} \sum_{i=1}^{r} \log x_{i:n} \\ &\equiv H(\beta; \boldsymbol{x}), \quad \text{say.} \end{aligned} \tag{3.54}$$

We now show that, for a given Type-II right censored data in (3.1), $H(\cdot; \boldsymbol{x})$ in (3.54) is a monotone increasing function of β with a finite positive limit as $\beta \to \infty$ (for any given $\boldsymbol{x}$). To do this, we need to check that the derivative

$$\frac{\partial H(\beta; \boldsymbol{x})}{\partial \beta} = \frac{H^*(\beta; \boldsymbol{x})}{\left\{ \sum_{i=1}^{r} x_{i:n}^{\beta} + (n-r)x_{r:n}^{\beta} \right\}^2} \geq 0,$$

that is, $H^*(\beta; \boldsymbol{x}) \geq 0$, where the function $H^*(\beta; \boldsymbol{x})$ is defined by

$$\begin{aligned} H^*(\beta; \boldsymbol{x}) &= \left\{ \sum_{i=1}^{r} x_{i:n}^{\beta} + (n-r)x_{r:n}^{\beta} \right\} \left\{ \sum_{i=1}^{r} x_{i:n}^{\beta} (\log x_{i:n})^2 + (n-r)x_{r:n}^{\beta} (\log x_{r:n})^2 \right\} \\ &\quad - \left\{ \sum_{i=1}^{r} x_{i:n}^{\beta} \log x_{i:n} + (n-r)x_{r:n}^{\beta} \log x_{r:n} \right\}^2. \end{aligned} \tag{3.55}$$

If we now set

$$a_i = \begin{cases} x_{i:n}^{\beta/2}, & \text{for } i = 1, \ldots, r \\ x_{r:n}^{\beta/2}, & \text{for } i = r+1, \ldots, n \end{cases} \quad \text{and} \quad b_i = \begin{cases} x_{i:n}^{\beta/2} \log x_{i:n}, & \text{for } i = 1, \ldots, r \\ x_{r:n}^{\beta/2} \log x_{r:n}, & \text{for } i = r+1, \ldots, n \end{cases}.$$

$H^*(\beta; \boldsymbol{x})$ in (3.55) simply becomes

$$H^*(\beta; \boldsymbol{x}) = \sum_{i=1}^{n} a_i^2 \sum_{i=1}^{n} b_i^2 - \left(\sum_{i=1}^{n} a_i b_i \right)^2,$$

which is clearly non-negative by Cauchy-Schwarz inequality. Moreover, we observe from (3.54) that

$$\lim_{\beta \to 0} H(\beta; \boldsymbol{x}) = \left(\frac{1}{r} - \frac{1}{n} \right) \sum_{i=1}^{r-1} \log \left(\frac{x_{r:n}}{x_{i:n}} \right) > 0,$$

[3] A similar approach had already been utilized in Cramer and Kamps (1996) in case of a multiple sample scenario of sequential order statistics from a *Weibull*$(1/\lambda, \beta)$-distribution.

which equals 0 only in the complete sample case (i.e., $r = n$), and

$$\lim_{\beta\to\infty} H(\beta;\mathbf{x}) = \frac{1}{r}\sum_{i=1}^{r-1}\log\left(\frac{x_{r:n}}{x_{i:n}}\right) > 0.$$

Thus, with the above established property that the function $H(\cdot;\mathbf{x})$ in (3.54) is a monotone increasing function of β with a finite positive limit as $\beta \to \infty$ and the fact that $\frac{1}{\beta}$ is a strictly decreasing function of β with right limit of $+\infty$ at $\beta = 0$, it is evident that the plots of the two functions $\frac{1}{\beta}$ and $H(\beta;\mathbf{x})$ would intersect, and intersect only once, at the MLE $\widehat{\beta}$. This establishes the existence and uniqueness of the MLE of β and also additionally provides a quick graphical way of determining the MLE $\widehat{\beta}$.

Example 3.39. Let us consider the Type-II right censored data of Dodson (2006) on failure times of grinders, considered earlier in Example 3.35. There, we fitted the Weibull, $Weibull(\vartheta^{\beta}, \beta)$, distribution for these data and found the MLEs by numerical iterative method to be $\widehat{\beta} = 1.646$ and $\widehat{\vartheta} = 162.2$. Using the above described simple graphical method, Balakrishnan and Kateri (2008) found the graphical estimate of β to be 1.647 which, when used in (3.49), yields an estimate of ϑ to be 162.223. Observe that these values are very nearly the same as the MLEs, which is not surprising as this method is just a graphical solution for the MLE of β.

Remark 3.40. The above discussion shows how the existence and uniqueness of the MLEs can be easily addressed for the case of Weibull distribution. However, the discussions become quite complex for many other distributions, especially in the presence of censored data as evidenced in the works of Balakrishnan and Mi (2003), Ghitany et al. (2013, 2014), Balakrishnan and Zhu (2014), and Zhu et al. (2019).

3.2.3.5 EM- and stochastic EM-algorithms

So far, we have discussed many situations wherein numerical solutions of likelihood equations became necessary. Alternatively, one could also make use of a general maximization algorithm (such as Broyden–Fletcher–Goldfarb–Shanno (BFGS) method) to directly maximize the log-likelihood function with respect to the distributional parameters for determining the MLEs. Such algorithms (like *maxnr*) are conveniently available in R-software, for example.

Yet another possibility is to exploit distributional and dependence properties of censored data to develop efficient numerical algorithms for the determination of MLEs in the presence of censoring. One such algorithm is the *EM-algorithm*, originally developed by Dempster et al. (1977), which proceeds by imputing the missing data (i.e., the censored observations) by their conditional expectations, conditioned on the given censored data and the current values of the parameters. This algorithm has now become a standard tool while carrying out analysis of data with missing values; one may refer to

McLachlan and Krishnan (2008) for an elaborate discussion on the EM-algorithm and its extensions and their diverse applications.

We shall now demonstrate this approach for the case of Type-II censored log-normally distributed data. As in Section 3.2.3.2, let the Type-II right censored data

$$(y_{1:n}, y_{2:n}, \ldots, y_{r:n})$$

be from lognormal, *log*-$N(\mu, \vartheta)$, distribution, so that the log-transformed data

$$x_{1:n} = \log y_{1:n}, x_{2:n} = \log y_{2:n}, \ldots, x_{r:n} = \log y_{r:n}$$

can be assumed to have come from normal, $N(\mu, \vartheta^2)$, distribution. For convenience in notation, let us now denote the censored ($n - r$ largest) log-lifetimes $x_{r+1:n}, \ldots, x_{n:n}$ by $w_1, \ldots, w_{n-r}$, respectively. Then, the likelihood function for the complete data, i.e., $(\boldsymbol{x}, \boldsymbol{w})$, is evidently

$$\mathcal{L}(\mu, \vartheta; \boldsymbol{x}, \boldsymbol{w}) = \frac{1}{(2\pi)^{n/2}\vartheta^n} e^{-\frac{1}{2\vartheta^2}\sum_{i=1}^{r}(x_{i:n}-\mu)^2} e^{-\frac{1}{2\vartheta^2}\sum_{j=1}^{n-r}(w_j-\mu)^2},$$

so that the complete log-likelihood function is

$$\mathcal{L}^*(\mu, \vartheta; \boldsymbol{x}, \boldsymbol{w}) = \text{const} - n\log\vartheta - \frac{1}{2\vartheta^2}\sum_{i=1}^{r}(x_{i:n} - \mu)^2 - \frac{1}{2\vartheta^2}\sum_{j=1}^{n-r}(w_j - \mu)^2. \tag{3.56}$$

Note that the last term on the right hand side of (3.56) is not computable as w_j are unobserved. So, the EM-algorithm proceeds in the following way, first by imputing the functions of missing values w_j (for $j = 1, 2, \ldots, n - r$) by their conditional expectations, conditioned on the observed Type-II right censored data $\boldsymbol{x}$ and the current value of the parameters μ and ϑ, say $\widehat{\mu}^{(k)}$ and $\widehat{\vartheta}^{(k)}$.

E-step: Upon realizing that, conditional on $\boldsymbol{x}$ and $(\widehat{\mu}^{(k)}, \widehat{\vartheta}^{(k)})$, the missing values w_j ($j = 1, 2, \ldots, n - r$) are iid observations from a random sample of size $n - r$ from $N(\mu, \vartheta^2)$-distribution left-truncated at $x_{r:n}$, it is known that (see, e.g., Johnson et al., 1994 and Cohen, 1991)

$$E\Big(W_j \mid W_j > x_{r:n}, \widehat{\mu}^{(k)}, \widehat{\vartheta}^{(k)}\Big) = \widehat{\mu}^{(k)} + \widehat{\vartheta}^{(k)} h(\xi^{(k)})$$

and

$$E\Big(W_j^2 \mid W_j > x_{r:n}, \widehat{\mu}^{(k)}, \widehat{\vartheta}^{(k)}\Big) = \widehat{\mu}^{(k)^2} + 2\widehat{\mu}^{(k)}\widehat{\vartheta}^{(k)} h(\xi^{(k)}) + \widehat{\vartheta}^{(k)^2}\left\{1 + \xi^{(k)} h(\xi^{(k)})\right\}$$

for $j = 1, 2, \ldots, n - r$, where $\xi^{(k)} = \frac{x_{r:n} - \widehat{\mu}^{(k)}}{\widehat{\vartheta}^{(k)}}$ and $h(\xi) = \frac{\varphi(\xi)}{1-\Phi(\xi)}$ is the hazard function of the standard normal distribution evaluated at ξ;

M-step: Now, upon replacing w_j and w_j^2 in (3.56) by the above conditional expectations, and then subsequently maximizing with respect to μ and ϑ, we obtain the next iterate of the estimates of μ and ϑ as follows:

$$\widehat{\mu}^{(k+1)} = \frac{1}{n}\left[\sum_{i=1}^{r} x_{i:n} + (n-r)\left\{\widehat{\mu}^{(k)} + \widehat{\vartheta}^{(k)} h(\xi^{(k)})\right\}\right],$$

$$\widehat{\vartheta}^{(k+1)} = \left\{\frac{1}{n}\left[\sum_{i=1}^{r} x_{i:n}^2 + (n-r)\left\{\widehat{\mu}^{(k+1)^2} + 2\widehat{\mu}^{(k+1)}\widehat{\vartheta}^{(k)} h(\tilde{\xi}^{(k)}) + \widehat{\vartheta}^{(k)^2}\left(1 + \tilde{\xi}^{(k)} h(\tilde{\xi}^{(k)})\right)\right\}\right] - \widehat{\mu}^{(k+1)^2}\right\}^{1/2},$$

where $\xi^{(k)} = \frac{x_{r:n} - \widehat{\mu}^{(k)}}{\widehat{\vartheta}^{(k)}}$, $\tilde{\xi}^{(k)} = \frac{x_{r:n} - \widehat{\mu}^{(k+1)}}{\widehat{\vartheta}^{(k)}}$ and $h(\cdot)$ denotes the hazard function of the standard normal distribution as before.

Then, the EM-algorithm proceeds by alternating between E- and M-steps until the estimates of μ and ϑ will converge to the desired level of accuracy. To start the iterative process, we may use the naive estimates

$$\widehat{\mu}^{(0)} = \frac{1}{n}\left[\sum_{i=1}^{r} x_{i:n} + (n-r)x_{r:n}\right]$$

and

$$\widehat{\vartheta}^{(0)} = \left\{\frac{1}{n}\left[\sum_{i=1}^{r}(x_{i:n} - \widehat{\mu}^{(0)})^2 + (n-r)(x_{r:n} - \widehat{\mu}^{(0)})^2\right]\right\}^{1/2}.$$

Alternatively, we could use the AMLEs of μ and ϑ based on the Type-II censored sample $\boldsymbol{x} = (x_{1:n}, \ldots, x_{r:n})$ as starting values $\widehat{\mu}^{(0)}$ and $\widehat{\vartheta}^{(0)}$ for the iterative process (see Balakrishnan, 1989a). In the presence of moderate and heavy censoring, the use of AMLEs as starting values hastens the convergence of the EM-algorithm.

It is important to mention that in the above case of normal (log-normal) distribution, the implementation of the EM-algorithm turns out to be rather easy because the conditional expectations required in the E-step have explicit simple expressions and the maximization needed in the M-step also have explicit solutions. This, however, is not the case with most other lifetime distributions. Often, the conditional expectations required in the E-step may not be available in explicit forms. In such a case, we may simulate a random sample of size N from the involved left-truncated distribution and then estimate the required conditional expectations by the empirical averages of those functions evaluated based on the N observations. An algorithm in which such simulation-based estimates are used in the E-step is called a *stochastic EM-algorithm*. Even though this is intuitively simple and easy to implement, we would need a very large N

(say, 10^5 or 10^6) to achieve reasonably close approximations to the required conditional expectations as the approximations used are based on weak law of large numbers. This would then result in a high computational effort, but would certainly provide a general approach for likelihood-based estimation; see Celeux and Dieboldt (1985), Celeux et al. (1996), and Ye and Ng (2014) for further details on stochastic EM-algorithm.

3.2.4 Bayesian estimation

The likelihood approach described in detail in the last section can be combined suitably with any prior information available on the distributional parameters to develop Bayesian inferential methods. This process is the usual process for Bayesian inference, with the only complication arising due to the presence of censoring in the observed data. However, this does not preclude us from using the well-known Bayesian computational tools, such as Metropolis-Hastings algorithm and more general Markov Chain Monte Carlo (MCMC) methods, for generating a sequence of random samples from a complex posterior distribution from which direct random sampling may be very difficult. This is indeed the case when the observed lifetime data involve censoring rendering the computation of the integral involved in the posterior distribution to be quite difficult, if not impossible. One may refer to Green et al. (2015) for an elaborate overview on Bayesian computational methods.

In order to demonstrate the Bayesian approach in the censoring set-up in a concise manner, let us consider the case when the observed Type-II censored data in (3.1) is from a scale-family of distributions. Then, with the prior knowledge on the scale parameter ϑ being represented by the prior density $\pi(\vartheta)$, $\vartheta \in \Theta$, we immediately have the posterior density of ϑ as

$$\begin{aligned} g(\vartheta \mid \boldsymbol{x}) &= \frac{g(\boldsymbol{x}, \vartheta)}{g(\boldsymbol{x})} = \frac{\pi(\vartheta)\mathscr{L}(\vartheta; \boldsymbol{x})}{g(\boldsymbol{x})} \\ &= \frac{\pi(\vartheta)\mathscr{L}(\vartheta; \boldsymbol{x})}{\int_\Theta \pi(\eta)\mathscr{L}(\eta; \boldsymbol{x})d\eta}, \end{aligned} \tag{3.57}$$

where $\mathscr{L}(\vartheta; \boldsymbol{x})$ is the likelihood function of the observed Type-II censored data. Then, with $\mathscr{L}(\vartheta; \boldsymbol{x})$ as given in (3.32), we readily have from (3.57) the posterior density of ϑ to be

$$g(\vartheta \mid \boldsymbol{x}) = \frac{\pi(\vartheta)\prod_{i=1}^r f_\vartheta(x_{i:n})\,(1 - F_\vartheta(x_{r:n}))^{n-r}}{\int_\Theta \pi(\eta)\prod_{i=1}^r f_\eta(x_{i:n})\,\left(1 - F_\eta(x_{r:n})\right)^{n-r} d\eta}. \tag{3.58}$$

Now, upon considering different loss functions, such as

(1) Quadratic loss: $L(\vartheta, a) = (a - \vartheta)^2 c$,

(2) Absolute loss: $L(\vartheta, a) = |a - \vartheta| c$,

(3) Exponential loss: $L(\vartheta, a) = e^{c(\vartheta - a) - 1}$,

(4) Linear-exponential (LINEX) loss: $L(\vartheta, a) = e^{c(\vartheta - a)} - 1 - c(\vartheta - a)$, and so on, the corresponding Bayesian estimates can be determined by minimizing the expected loss computed from the posterior density in (3.58). Of course, if the chosen loss function is the quadratic loss, then the posterior mean is the Bayesian estimator no matter what the choice of c is. It is then clearly evident that the integration needed to find the Bayesian estimate would be intractable in most cases (except in some simple and convenient cases), thus requiring the use of MCMC sampling as stated earlier.

Let us now consider the special case of scaled-exponential distribution considered earlier in Section 3.2.3.1. In this case, we have the posterior density of ϑ from (3.58) to be

$$g(\vartheta \mid \boldsymbol{x}) = \frac{\pi(\vartheta)\vartheta^{-r}e^{-\frac{1}{\vartheta}\{\sum_{i=1}^{r} x_{i:n}+(n-r)x_{r:n}\}}}{\int_{\Theta} \pi(\eta)\eta^{-r}e^{-\frac{1}{\eta}\{\sum_{i=1}^{r} x_{i:n}+(n-r)x_{r:n}\}}\,d\eta}. \tag{3.59}$$

Suppose the prior $\pi(\cdot)$ is chosen to be an inverse gamma, $\mathsf{I\Gamma}(\gamma, \delta)$, density, i.e.,

$$\pi(\vartheta) = \frac{\gamma^{\delta}}{\Gamma(\delta)}\frac{e^{-\frac{\gamma}{\vartheta}}}{\vartheta^{\delta+1}}, \qquad \vartheta > 0,\ \gamma, \delta > 0$$

(see Definition B.8). Note that this is indeed a natural conjugate prior for ϑ. Then, (3.59) readily reveals that

$$g(\vartheta \mid \boldsymbol{x}) \propto \frac{1}{\vartheta^{r+\delta+1}} \exp\left[-\frac{1}{\vartheta}\left\{\sum_{i=1}^{r} x_{i:n} + (n-r)x_{r:n} + \gamma\right\}\right],$$

that is, the posterior distribution of ϑ is $\mathsf{I\Gamma}(\gamma', \delta')$, where

$$\gamma' = \sum_{i=1}^{r} x_{i:n} + (n-r)x_{r:n} + \gamma \quad \text{and} \quad \delta' = r + \delta.$$

So, if we choose to use the quadratic loss function, then the Bayesian estimator is the posterior mean given by

$$\begin{aligned}\widehat{\vartheta}_{\mathsf{B}} &= \frac{\gamma'}{\delta'-1} \quad (\text{provided } \delta' > 1)\\ &= \frac{1}{r+\delta-1}\left[\sum_{i=1}^{r} x_{i:n} + (n-r)x_{r:n} + \gamma\right] \quad (\text{provided } r+\delta > 1).\end{aligned}$$

Observe that this Bayesian estimator can be expressed as

$$\widehat{\vartheta}_{\mathsf{B}} = \frac{r}{r+\delta-1}\left[\frac{1}{r}\left\{\sum_{i=1}^{r} x_{i:n} + (n-r)x_{r:n}\right\}\right] + \frac{\delta-1}{r+\delta-1}\left(\frac{\gamma}{\delta-1}\right). \tag{3.60}$$

Thus, $\widehat{\vartheta}_{\mathsf{B}}$ in (3.60) is a convex combination of the MLE $\widehat{\vartheta}$ and the prior mean $\frac{\gamma}{\delta-1}$ (provided $\delta > 1$), with the corresponding weights as $\frac{r}{r+\delta-1}$ and $\frac{\delta-1}{r+\delta-1}$. This is intuitively appealing for the reason that if $r \gg \delta$, then the first weight will tend to 1 (meaning that the Bayesian estimate will tend to the MLE), while when $\delta \gg r$, then the second weight will tend to 1 (meaning that the Bayesian estimate will tend to the prior mean). Furthermore, the explicit posterior density in this case (viz., the inverse gamma density) would also facilitate the construction of Highest Posterior Density (HPD) intervals for ϑ, for example.

Such Bayesian inferential results, either in explicit forms (if possible) or through computational methods mentioned before, have been developed for a wide range of distributions based on different forms of censored data. Some of these are mentioned in subsequent chapters of this book.

3.3. Type-I censoring

Suppose we have observed, from a life-testing experiment of n independent and identical units, the lifetimes of all units that failed up to a pre-fixed time T. Here, T is a time fixed by the reliability engineer prior to the experiment, and that the experiment gets terminated precisely at that time. This censoring scheme, as mentioned earlier in Chapter 1, is the conventional *Type-I censoring*. Note that, in this case, the number of units to fail by pre-fixed time T (denoted by D as defined in (2.12)) is a random variable; in fact, it has a binomial distribution with n as the number of trials and $p = F(T)$ as the probability of success. It is then obvious that the number of units to fail, D, could very well be 0 and this would occur with a probability of $(1 - F(T))^n$. Though this probability would tend to 0 as $n \to \infty$, for fixed T, it would be a positive value for finite n and may be even non-negligible if the mean lifetime of units under test is significantly larger than the pre-fixed time T.

For this reason, all likelihood inferential results need to be developed under the condition that the lifetime of at least one failed unit is observed (could be even more than one depending on the lifetime distribution assumed and the number of parameters it contains); that is, only conditional likelihood inferential methods are possible. Then, the observed censored data in this case will be of the form

$$(X_{1:n}, X_{2:n}, \dots, X_{D:n}), \tag{3.61}$$

with the remaining $n - D$ lifetimes that got censored are known to be only larger than the pre-fixed time T. The corresponding likelihood function is given by

$$\mathscr{L}(\boldsymbol{\theta}; \boldsymbol{x}, d) = \frac{n!}{(n-d)!} \prod_{i=1}^{d} f_{\boldsymbol{\theta}}(x_{i:n}) \, (1 - F_{\boldsymbol{\theta}}(T))^{n-d}, \tag{3.62}$$

conditional on $D \geq 1$. It is this conditional likelihood function that needs to be maximized with respect to $\boldsymbol{\theta}$ for determining the MLE of $\boldsymbol{\theta}$.

Remark 3.41. The MLEs of distributional parameters can be derived in explicit form only for a few simple distributions like uniform and exponential. In all other cases, however, the MLEs can only be determined by using some numerical methods to optimize the likelihood function in (3.62); see, for example, Balakrishnan and Cohen (1991) and Cohen (1991).

Remark 3.42. In the case of scale- and location-scale families of distributions, if we standardize the variables by $z_{i:n} = \frac{x_{i:n}}{\vartheta}$ and $z_{i:n} = \frac{x_{i:n}-\mu}{\vartheta}$, respectively, unlike in the Type-II censoring scenario, the distributions of $Z_{i:n}$ do not become free of the parameters as they are bounded (on the right) by $\frac{T}{\vartheta}$ and $\frac{T-\mu}{\vartheta}$, respectively. Because of this dependence of the distributions of $Z_{i:n}$ (and consequently their moments) on the distributional parameters, estimates such as BLUEs and BLIEs are not possible in this case. Therefore, the conditional likelihood and Bayesian methods seem to be the only viable options.

Now, let us assume that the available Type-I censored data, as in (3.61), are from the scaled-exponential, *Exp*(ϑ), distribution. Then, from (3.62), we have the conditional likelihood function to be

$$\mathscr{L}(\vartheta; \boldsymbol{x}, d) = \frac{n!}{(n-d)!\vartheta^d} \exp\left\{ -\frac{1}{\vartheta}\left(\sum_{i=1}^{d} x_{i:n} + (n-d)T \right) \right\},$$

$$0 < x_{1:n} < \cdots < x_{d:n} < T, \ \vartheta > 0, \quad (3.63)$$

conditional on $D \geq 1$. d denotes the observed value of D. Observe that, in the case of $D = 0$, the likelihood function is simply $e^{-nT/\vartheta}$ which is a monotone increasing function of ϑ, and so the MLE of ϑ does not exist in this case, confirming the statement made earlier. From (3.63), conditional on $D \geq 1$, we readily obtain the MLE of ϑ to be

$$\widehat{\vartheta} = \frac{1}{D}\left\{ \sum_{i=1}^{D} X_{i:n} + (n-D)T \right\}. \quad (3.64)$$

Remark 3.43. Even though the estimator in (3.64) looks similar in form to the MLE of ϑ in the Type-II censoring scenario, there is a significant difference in its distributional properties and characteristics. This is due to two reasons: firstly, the denominator D itself is random and also conditioned to be at least 1, and secondly, the order statistics $X_{i:n}$ are not the usual order statistics from *Exp*(ϑ)-distribution possessing nice properties (such as spacings property) as they are truncated on the right at time T. Nevertheless, the MLE $\widehat{\vartheta} = D \cdot \mathsf{TTT}_D$ can be written in terms of the total time on test

$$\mathsf{TTT}_D = \sum_{i=1}^{D} X_{i:n} + (n-D)T$$

resulting under Type-I censoring of the life test. Similar comments apply to hybrid censoring schemes as well be presented in ensuing chapters (see, e.g., Chapter 5).

Bartholomew (1963), in his pioneering work, discussed a method of deriving the exact conditional distribution of $\widehat{\vartheta}$ in (3.64). The technique he employed, based on moment generating function (MGF), is referred to as *Conditional MGF Method*. It has now become one of the commonly used tools in deriving the conditional distributions of MLE of ϑ under different censoring scenarios, as well be seen in the ensuing chapters of this book.

We now present the following exact results for the scaled-exponential distribution.

Theorem 3.44 (Nonparametric MLE of ϑ). *Suppose n independent and identical units, with Exp(ϑ)-distributed lifetimes, were placed on a life-testing experiment, and that the experiment got terminated at a pre-fixed time $T > 0$. Further, suppose exactly $D = d \geq 1$ failures are known to have occurred by time T. Then, the non-parametric MLE of ϑ is*

$$\widehat{\vartheta}_N = \frac{T}{-\log\left(1 - \frac{d}{n}\right)},$$

and the non-parametric MLE of the reliability at time T, $R(T)$, is

$$\widehat{R(T)}_N = 1 - \frac{d}{n}.$$

Proof. It is well-known that the MLE of the probability of success p, based on n iid Bernoulli trials, is the sample proportion of successes. Then, the MLE of $p = F_\vartheta(T) = 1 - e^{-\frac{T}{\vartheta}}$ is $\frac{d}{n}$, from which the stated results follow readily. □

Theorem 3.45 (Moment generating function of the parametric MLE of ϑ). *Suppose n independent and identical units, with Exp(ϑ)-distributed lifetimes, were placed on a life-testing experiment, and that the experiment got terminated at a pre-fixed time $T > 0$. Further, conditional on having observed at least one failure time (i.e., $D \geq 1$), and that the Type-I censored data of the form* (3.61) *have been observed, then the parametric MLE of ϑ, $\widehat{\vartheta}$, is as given in* (3.64). *Moreover, its exact conditional moment generating function is*

$$M_\vartheta^{D\geq 1}(s) = \frac{1}{1 - e^{-\frac{nT}{\vartheta}}} \sum_{d=1}^{n} \binom{n}{d} \frac{e^{(\frac{n-d}{d})Ts} e^{-\frac{(n-d)T}{\vartheta}} \{1 - e^{-\frac{T}{\vartheta}} e^{\frac{T}{d}s}\}^d}{(1 - \frac{\vartheta}{d}s)^d}, \quad s < \frac{1}{\vartheta}. \tag{3.65}$$

Proof. The derivation of the conditional MLE of ϑ in (3.64) has already been done. Next, to derive the conditional moment generating function of $\widehat{\vartheta}$, we use the following facts for the given set-up:

(a) The number of units failed until time T (viz., D) is distributed as $bin(n, F_\vartheta(T))$;

(b) Conditional on having observed at least one failure until time T, the random variable D is distributed as truncated binomial, truncated at 0. So, its probability mass function is

$$\begin{aligned}\Pr_\vartheta(D=d \mid D\geq 1) &= \frac{1}{1-(1-F_\vartheta(T))^n}\binom{n}{d}F_\vartheta^d(T)\,(1-F_\vartheta(T))^{n-d}\\ &= \frac{1}{1-e^{-\frac{nT}{\vartheta}}}\binom{n}{d}(1-e^{-\frac{T}{\vartheta}})^d e^{-\frac{(n-d)T}{\vartheta}}\end{aligned}$$

for $d=1,2,\ldots,n$;

(c) Conditional on $D=d\geq 1$, the ordered observed lifetimes $(x_{1:n},x_{2:n},\ldots,x_{d:n})$ are distributed as order statistics from a random sample of size d from the right-truncated exponential distribution truncated at T with density function given by $\frac{f_\vartheta(x)}{F_\vartheta(T)} = \frac{\frac{1}{\vartheta}e^{-\frac{x}{\vartheta}}}{1-e^{-\frac{T}{\vartheta}}}$, $0<x<T$; see Chapter 2 for pertinent details. So, the conditional joint density function of $(X_{1:n},X_{2:n},\ldots,X_{d:n})$, conditioned on $D=d\geq 1$, is

$$\frac{d!}{\vartheta^d}\frac{e^{-\frac{1}{\vartheta}\sum_{i=1}^d x_{i:n}}}{\left(1-e^{-\frac{T}{\vartheta}}\right)^d},\qquad 0<x_{1:n}<x_{2:n}<\cdots<x_{d:n}<T.$$

By making use of all the above facts, we have the conditional moment generating function of $\widehat{\vartheta}$, conditioned on $D\geq 1$, as

$$\begin{aligned}M_\vartheta^{D\geq 1}(s) &= E_\vartheta\left[e^{s\widehat{\vartheta}} \mid D\geq 1\right]\\ &= \sum_{d=1}^n \frac{\Pr_\vartheta(D=d)}{1-\Pr_\vartheta(D=0)}E_\vartheta\left[e^{\frac{s}{d}\{\sum_{i=1}^d x_{i:n}+(n-d)T\}} \mid D=d\right]\\ &= \frac{1}{1-e^{\frac{-nT}{\vartheta}}}\sum_{d=1}^n\binom{n}{d}\left(1-e^{-\frac{T}{\vartheta}}\right)^d e^{-\frac{(n-d)T}{\vartheta}+(\frac{n-d}{d})Ts}\\ &\quad\times \frac{d!}{\vartheta^d(1-e^{-\frac{T}{\vartheta}})^d}\int_0^T\int_0^{x_{d:n}}\cdots\int_0^{x_{2:n}} e^{-\alpha\sum_{i=1}^d x_{i:n}}dx_{1:n}\cdots dx_{d:n},\end{aligned}\tag{3.66}$$

where $\alpha=\frac{1}{\vartheta}-\frac{s}{d}=\frac{d-\vartheta s}{d\vartheta}>0$, $d=1,\ldots,n$. Now, upon using the integral identity that

$$\int_0^T\int_0^{x_{d:n}}\cdots\int_0^{x_{2:n}} e^{-\alpha\sum_{i=1}^d x_{i:n}}dx_{1:n}\cdots dx_{d:n} = \frac{(1-e^{-\alpha T})^d}{d!\alpha^d}$$

in (3.66) and simplifying the resulting expression, we readily obtain the conditional moment generating function of $\widehat{\vartheta}$, conditioned on $D=d\geq 1$, as presented in (3.65). Hence, the theorem. □

Inverting the conditional moment generating function yields the conditional density function of the MLE. In Figs. 5.3–5.6, plots of $f_{\vartheta}^{\widehat{\vartheta}|D\geq 1}$ are presented for $T=1$ and $\vartheta=1$ as well as $n\in\{2,3,4,5\}$.

Theorem 3.46 (Exact distribution of the parametric MLE of ϑ). *Under the setting of Theorem 3.45, the exact conditional density function of* $\widehat{\vartheta}$*, conditioned on* $D\geq 1$*, is*

$$f_{\vartheta}^{\widehat{\vartheta}|D\geq 1}(x)=\frac{1}{1-e^{-\frac{nT}{\vartheta}}}\sum_{d=1}^{n}\sum_{\ell=0}^{d}(-1)^{\ell}\binom{n}{d}\binom{d}{\ell}e^{-(n-d+\ell)\frac{T}{\vartheta}}f_{\Gamma(\vartheta/d,d)}\left(x-\left(\frac{n-d+\ell}{d}\right)T\right),\tag{3.67}$$

where $f_{\Gamma(b,a)}$ *is the gamma density function given in Definition B.7.*

Proof. To prove the required result, we first use binomial expansion for the term $\left\{1-e^{-\frac{T}{\vartheta}}e^{\frac{T}{d}s}\right\}^{d}$ in (3.65) to re-express the conditional moment generating function of $\widehat{\vartheta}$, conditioned on $D\geq 1$, as

$$M_{\vartheta}^{D\geq 1}(s)=\frac{1}{1-e^{-\frac{nT}{\vartheta}}}\sum_{d=1}^{n}\sum_{\ell=0}^{d}(-1)^{\ell}\binom{n}{d}\binom{d}{\ell}e^{-\frac{(n-d+\ell)T}{\vartheta}}\frac{e^{(\frac{n-d+\ell}{d})Ts}}{(1-\frac{\vartheta}{d}s)^{d}}\tag{3.68}$$

for $s<\frac{1}{\vartheta}$. Observe now that $\frac{1}{(1-\frac{s}{\gamma})^{p}}$ (for $s<\gamma$) is the moment generating function of the gamma, $\Gamma(1/\gamma,p)$, random variable with density function as in Definition B.7. It is then obvious that $\frac{e^{\delta s}}{(1-\frac{s}{\gamma})^{p}}$ (for $s<\gamma$) is the moment generating function of the gamma, $\Gamma(1/\gamma,p)$, random variable shifted by a constant δ. Upon using this fact in (3.68), the theorem follows readily. □

Remark 3.47. From the expression of the exact conditional density function of $\widehat{\vartheta}$, conditioned on $D\geq 1$, presented in (3.67) and then using the known moments of gamma distribution, we can immediately obtain the moments of the conditional MLE $\widehat{\vartheta}$. For example, since the expectation of the shifted-gamma variable with density function $f_{\Gamma(\vartheta/d,d)}\left(\cdot-\left(\frac{n-d+\ell}{d}\right)T\right)$ is given by $\vartheta+(\frac{n-d+\ell}{d})T$, we find the mean of the conditional MLE $\widehat{\vartheta}$, conditioned on $D\geq 1$, to be

$$E_{\vartheta}(\widehat{\vartheta}\mid D\geq 1)=\vartheta+\frac{1}{1-e^{-\frac{nT}{\vartheta}}}\sum_{d=1}^{n}\sum_{\ell=0}^{d}(-1)^{\ell}\binom{n}{d}\binom{d}{\ell}e^{-(n-d+\ell)\frac{T}{\vartheta}}\left(\frac{n-d+\ell}{d}\right)T.\tag{3.69}$$

The second term of the right hand side of (3.69) gives the bias in the conditional MLE $\widehat{\vartheta}$. We can also proceed similarly to derive other moments of $\widehat{\vartheta}$, including $\mathrm{Var}_{\vartheta}(\widehat{\vartheta})$.

Remark 3.48. From the discussion of Type-I hybrid censored data from exponential distributions, the more compact representation (5.14) of the conditional density function in (3.67) is obtained in terms of B-spline functions.

Remark 3.49. Balakrishnan and Iliopoulos (2009) established the stochastic monotonicity of the conditional distribution of $\widehat{\vartheta}$ in (3.67) with respect to ϑ which is useful in developing exact conditional inferential methods for ϑ, such as confidence intervals. The monotonicity result established by these authors also hold true for many other censoring scheme in addition to the Type-I censoring scheme discussed here (see also van Bentum and Cramer, 2019). Therefore, in the literature, the conditional moment generating function technique described here in detail and the stochastic monotonicity result of Balakrishnan and Iliopoulos (2009) has been used repeatedly for different hybrid censoring schemes! In the ensuing chapters, we will illustrate an alternative approach which is based on an appropriate decomposition of the (progressive) hybrid censoring schemes (see Górny and Cramer, 2018b).

Example 3.50. Let us consider the Type-I censored data of Barlow et al. (1968) based on 10 units placed on a time-censored life-testing experiment that got terminated at time 50. The lifetimes so observed were

$$4, 9, 11, 18, 27, 38.$$

So, in this case, with $n = 10$, $T = 50$ and $d = 6$, the MLE of ϑ is found from (3.64) to be

$$\widehat{\vartheta} = \frac{1}{6}\{107 + (4 \times 50)\} = \frac{307}{6} = 51.167.$$

While this is a parametric MLE of ϑ (based on the observed Type-I censored data), the non-parametric MLE of ϑ can be found from Theorem 3.44 to be

$$\widehat{\vartheta}_N = \frac{50}{-\log(1 - \frac{6}{10})} = \frac{50}{0.9163} = 54.567.$$

Similarly, we find the parametric MLE of the reliability at mission time 50 to be

$$\widehat{R(50)} = e^{-\frac{50}{\widehat{\vartheta}}} = e^{-\frac{50}{51.167}} = e^{-0.9772} = 0.376,$$

while the corresponding non-parametric MLE, found from Theorem 3.44, is

$$\widehat{R(50)}_N = 1 - \frac{6}{10} = 0.40.$$

Of course, by using the exact conditional distribution of $\widehat{\vartheta}$ in Theorem 3.46, the standard error of the estimate and exact confidence intervals for ϑ can be readily obtained.

Example 3.51. Let us consider the Type-I censored data of Bartholomew (1963) based on 20 items placed on a life-testing experiment that got terminated at time $T = 150$. The lifetimes so observed were

$$3, 19, 23, 26, 27, 37, 38, 41, 45, 58, 84, 90, 99, 109, 138.$$

So, in this case, with $n = 20$, $T = 150$ and $d = 15$, the parametric MLE of ϑ is found from (3.64) to be

$$\widehat{\vartheta} = \frac{1}{15}\{837 + (5 \times 150)\} = \frac{1587}{15} = 105.8.$$

The non-parametric MLE of ϑ, on the other hand, is found from Theorem 3.44 to be

$$\widehat{\vartheta}_N = \frac{150}{-\log(1 - \frac{15}{20})} = \frac{150}{1.3863} = 108.202.$$

Similarly, we find the parametric MLE of the reliability at mission time 150 to be

$$\widehat{R(150)} = e^{-\frac{150}{\widehat{\vartheta}}} = e^{-\frac{150}{105.8}} = e^{-1.4178} = 0.242,$$

while the corresponding non-parametric MLE, found from Theorem 3.44, is

$$\widehat{R(150)}_N = 1 - \frac{15}{20} = 0.25.$$

Here again, the exact conditional distribution of $\widehat{\vartheta}$ presented in Theorem 3.46 can be used to find the standard error the estimate as well as exact confidence intervals for ϑ.

3.4. Progressive Type-II censoring

In this section, we provide a review of some inferential results for progressive censoring. Needless to say, the inferential results for progressive Type-II censoring situation will be analogous to the corresponding ones for Type-II censoring case elaborated in detail in Section 3.2, while those for progressive Type-I censoring situation will be analogous to the corresponding results for Type-I censoring case discussed in Section 3.3. Due to the availability of extensive review articles by Balakrishnan (2007) and Balakrishnan and Cramer (2023) and the monographs by Balakrishnan and Aggarwala (2000) and Balakrishnan and Cramer (2014), this review is made quite brief detailing only some essential results pertaining to progressive censoring so that they form a basis for the related discussions in the ensuing chapters (see, e.g., Chapter 7).

3.4.1 Optimal linear estimation

Suppose $(X_{1:m:n}, \ldots, X_{m:m:n})$ is a progressively Type-II right censored sample from a life-testing experiment involving n independent and identical units and with progressive censoring plan $\mathscr{R} = (R_1, \ldots, R_m)$, where R_i denotes the number of units withdrawn from the life-test, from the $\gamma_i - 1 = n - i - R_1 - \cdots - R_{i-1}$ surviving units, at the time of the i-th failure. Then, if this progressively Type-II right censored sample arises from a scale family of distributions with density function f_ϑ, for $\vartheta \in \Theta$, as in (3.3), then the expressions of BLUE and BLIE of ϑ are exactly the same as in (3.4) and (3.28), respectively, with the only difference being that $\boldsymbol{\alpha}$ and $\boldsymbol{\Sigma}$ now correspond to the mean vector and variance-covariance matrix of progressively Type-II right censored order statistics from the standard density function f (see Balakrishnan and Cramer, 2014, Chapter 11).

In particular, in the special case of scaled-exponential, $Exp(\vartheta)$, distribution upon using the simple explicit expressions of means, variances and covariances of progressively Type-II right censored order statistics from standard exponential distribution presented earlier in Section 2.3, we can show that the BLUE of ϑ is

$$\vartheta^* = \frac{1}{m}\sum_{i=1}^{m}(R_i+1)X_{i:m:n} \tag{3.70}$$

and its variance is

$$\mathrm{Var}_\vartheta(\vartheta^*) = \frac{\vartheta^2}{m}. \tag{3.71}$$

Remark 3.52. It is of interest to observe from (3.71) that the precision of ϑ^* in (3.70) is just a function of m (the number of complete failures observed) and does not involve the progressive censoring plan $\mathscr{R} = (R_1, \dots, R_m)$.

Next, from (3.70), upon using the spacings result for progressively Type-II right censored exponential order statistics presented in Theorem 2.67, it can be easily shown that

$$\frac{m\vartheta^*}{\vartheta} \sim \Gamma(1, m) \quad \text{or} \quad \frac{2m\vartheta^*}{\vartheta} \sim \chi^2_{2m}.$$

From this exact distributional result, exact inferential methods, such as confidence intervals and hypothesis tests, can be readily developed.

Example 3.53. Times to breakdown of an insulating fluid in an accelerated test conducted at various test voltages were presented by Nelson (1982, p. 228). Viveros and Balakrishnan (1994) implemented a censoring plan $\mathscr{R} = (0, 0, 3, 0, 3, 0, 0, 5)$ on this data set with $n = 19$, and obtained the following progressively Type-II right censored sample of size $m = 8$:

$$0.19, 0.78, 0.96, 1.31, 2.78, 4.85, 6.50, 7.35.$$

Then, by assuming these data to have come from $Exp(\vartheta)$-distribution, we find from (3.70) the MLE of ϑ to be

$$\vartheta^* = \frac{72.69}{8} = 9.086$$

and its standard error from (3.71) to be

$$\mathrm{SE}(\vartheta^*) = \frac{\vartheta^*}{\sqrt{m}} = \frac{9.086}{\sqrt{8}} = 3.212.$$

It should be mentioned that Balakrishnan and Lin (2003) performed their goodness-of-fit test on these progressively Type-II right censored data to conclude that $Exp(\vartheta)$ is a suitable distribution for these data. One may also refer to Ahmadi et al. (2015) for some further analysis on these data.

Further, suppose the progressively Type-II right censored sample arises from a location-scale family of distributions with density function $f_{\mu,\vartheta}$, then the expressions of BLUEs are exactly the same as in (3.9) and (3.10), with $\boldsymbol{\alpha}$ and $\boldsymbol{\Sigma}$ being the mean vector and variance-covariance matrix of progressively Type-II right censored order statistics from the standard density function f.

In the special case of two-parameter exponential, $Exp(\mu, \vartheta)$, distribution, by using once again the simple explicit expressions of means, variances and covariances of progressively Type-II right censored order statistics from standard exponential distribution presented earlier in Section 2.3.7, we can obtain the BLUEs of μ and ϑ to be

$$\mu^* = X_{1:m:n} - \frac{\vartheta^*}{n} \quad \text{and} \quad \vartheta^* = \frac{1}{m-1}\sum_{i=2}^{m}(R_i+1)(X_{i:m:n} - X_{1:m:n}) \tag{3.72}$$

and their variances and covariance to be

$$\text{Var}(\mu^*) = \frac{m\vartheta^2}{n^2(m-1)}, \quad \text{Var}(\vartheta^*) = \frac{\vartheta^2}{m-1} \text{ and } \text{Cov}(\mu^*, \vartheta^*) = -\frac{\vartheta^2}{n(m-1)}. \tag{3.73}$$

Remark 3.54. Here again, it should be noted that the variances and covariance in (3.73) are all free of the progressive censoring plan $\mathscr{R} = (R_1, \dots, R_m)$.

Moreover, using the spacings result for progressively Type-II right censored exponential order statistics presented in Theorem 2.67, it can be shown that

$$\frac{2n(X_{1:m:n} - \mu)}{\vartheta} \sim \chi_2^2 \quad \text{and} \quad \frac{2(m-1)\vartheta^*}{\vartheta} \sim \chi_{2m-2}^2$$

are independent statistics (see, e.g., Balakrishnan et al., 2001). These exact distributional results can be used effectively to develop exact confidence intervals and hypothesis tests in this case.

Example 3.55. Let us reconsider the progressively Type-II right censored data in Example 3.53. Now, by assuming these data to have come from $Exp(\mu, \vartheta)$-distribution, we find from (3.72) the MLEs of ϑ and μ to be

$$\vartheta^* = \frac{69.08}{7} = 9.869 \quad \text{and} \quad \mu^* = 0.19 - \frac{9.869}{19} = -0.329$$

and their standard errors from (3.73) to be

$$\text{SE}(\vartheta^*) = \frac{9.869}{\sqrt{7}} = 3.730 \quad \text{and} \quad \text{SE}(\mu^*) = \frac{9.869\sqrt{8}}{19\sqrt{7}} = 0.555.$$

From the above estimate of μ and its standard error, it is evident that scaled-exponential distribution itself is adequate for these data.

It is of interest to mention here that these optimal linear estimators have been discussed by several authors for many different lifetime distributions!

3.4.2 Maximum likelihood estimation

Suppose the available progressively Type-II censored data are from the scale-family of distributions with density function f_ϑ, for $\vartheta \in \Theta$, as in (3.3). Then, from (2.47), the likelihood function is

$$\begin{aligned}\mathscr{L}(\vartheta; x_{1:m:n}, \ldots, x_{m:m:n}) &= c_{m-1} \prod_{i=1}^{m} f_\vartheta(x_{i:m:n})\,(1 - F_\vartheta(x_{i:m:n}))^{R_i} \\ &= \frac{c_{m-1}}{\vartheta^m} \prod_{i=1}^{m} f(z_{i:m:n})\,(1 - F(z_{i:m:n}))^{R_i}, \end{aligned} \tag{3.74}$$

where c_{m-1} is as in (2.48), $z_{i:m:n} = \frac{x_{i:m:n}}{\vartheta}$, and f is the standard density function in (3.3). Then, the log-likelihood function is

$$\mathscr{L}^*(\vartheta) \equiv \log \mathscr{L}(\vartheta) = \text{const} - m \log \vartheta + \sum_{i=1}^{m} \log f(z_{i:m:n}) + \sum_{i=1}^{m} R_i \log(1 - F(z_{i:m:n}))$$

which, upon differentiating with respect to ϑ, gives the likelihood equation for ϑ to be

$$\frac{\partial \mathscr{L}^*}{\partial \vartheta} = -\frac{m}{\vartheta} - \frac{1}{\vartheta} \sum_{i=1}^{m} z_{i:m:n} \frac{\frac{\partial}{\partial z_{i:m:n}} f(z_{i:m:n})}{f(z_{i:m:n})} + \frac{1}{\vartheta} \sum_{i=1}^{m} R_i z_{i:m:n} h(z_{i:m:n}) = 0, \tag{3.75}$$

where $h(\cdot)$ is the hazard function of the standard random variable Z. The MLE of ϑ, as a solution of (3.75), does not exist explicitly for most distributions. However, in the case of scaled-exponential, $Exp(\vartheta)$, distribution using the facts that $\frac{\partial}{\partial z} f(z) = -f(z)$, and $h(z) = 1$ for $z > 0$, (3.75) yields the MLE of ϑ to be

$$\widehat{\vartheta} = \frac{1}{m} \sum_{i=1}^{m} (R_i + 1) X_{i:m:n},$$

which is exactly the same as the BLUE of ϑ presented earlier.

Next, suppose the available progressively Type-II censored data are from the location-scale-family of distributions with density function $f_{\mu,\vartheta}$ as in (3.8). Then, it is evident from (2.47) that the likelihood function is

$$\mathscr{L}(\mu, \vartheta; x_{1:m:n}, \ldots, x_{m:m:n}) = c_{m-1} \prod_{i=1}^{m} f_{\mu,\vartheta}(x_{i:m:n}) \left(1 - F_{\mu,\vartheta}(x_{i:m:n})\right)^{R_i}$$

$$= \frac{c_{m-1}}{\vartheta^m} \prod_{i=1}^{m} f(z_{i:m:n}) \left(1 - F(z_{i:m:n})\right)^{R_i}, \tag{3.76}$$

where c_{m-1} is as in (2.48), $z_{i:m:n} = \frac{x_{i:m:n}-\mu}{\vartheta}$, and f is the standard density function in (3.3). Then, from (3.76), we can obtain the likelihood equations for μ and ϑ to be

$$\frac{\partial \mathscr{L}^*}{\partial \mu} = -\frac{1}{\vartheta} \sum_{i=1}^{m} \frac{\frac{\partial}{\partial z_{i:m:n}} f(z_{i:m:n})}{f(z_{i:m:n})} + \frac{1}{\vartheta} \sum_{i=1}^{m} R_i h(z_{i:m:n}) = 0 \tag{3.77}$$

and

$$\frac{\partial \mathscr{L}^*}{\partial \vartheta} = -\frac{m}{\vartheta} - \frac{1}{\vartheta} \sum_{i=1}^{m} z_{i:m:n} \frac{\frac{\partial}{\partial z_{i:m:n}} f(z_{i:m:n})}{f(z_{i:m:n})} + \frac{1}{\vartheta} \sum_{i=1}^{m} R_i z_{i:m:n} h(z_{i:m:n}) = 0, \tag{3.78}$$

where $h(\cdot)$ is the hazard function of the standard random variable. The MLEs of μ and ϑ, as solutions of (3.77) and (3.78), do not exist explicitly for most distributions. But, in the case of two-parameter, $Exp(\mu, \vartheta)$, distribution, we simply observe from (3.76) that the likelihood function is a monotone increasing function of μ and so the MLE of μ is $\widehat{\mu} = X_{1:m:n}$. Then, by differentiating the log-likelihood function with respect to ϑ, we obtain the MLE of ϑ to be

$$\widehat{\vartheta} = \frac{1}{m} \sum_{i=2}^{m} (R_i + 1)(X_{i:m:n} - X_{1:m:n}).$$

Observe that these MLEs of μ and ϑ are same as BLUEs of μ and ϑ presented earlier, except the former are biased.

As another example, let us assume that the observed progressively Type-II right censored sample are from Weibull, $Weibull(\vartheta^\beta, \beta)$, distribution. Then, as seen earlier in Section 3.2.3.1, $y_{i:m:n} = \log x_{i:m:n}$ $(i = 1, \ldots, m)$ can be assumed to arise from extreme value, $EV_I(\mu', \vartheta')$, distribution with density function as in (3.51). Then, using the facts that $z_{i:m:n} = \frac{y_{i:m:n}-\mu'}{\vartheta'}$, $f(z) = e^z e^{-e^z}$, $\frac{\partial}{\partial z} f(z) = (1 - e^z) f(z)$, and $h(z) = e^z$, we obtain from (3.77) and (3.78) the likelihood equations for μ' and ϑ' to be

$$\frac{\partial \mathscr{L}^*}{\partial \mu'} = -\frac{1}{\vartheta} \sum_{i=1}^{m} (1 - e^{z_{i:m:n}}) + \frac{1}{\vartheta} \sum_{i=1}^{m} R_i e^{z_{i:m:n}} = 0 \tag{3.79}$$

and

$$\frac{\partial \mathscr{L}^*}{\partial \vartheta'} = -\frac{m}{\vartheta} - \frac{1}{\vartheta} \sum_{i=1}^{m} z_{i:m:n} (1 - e^{z_{i:m:n}}) + \frac{1}{\vartheta} \sum_{i=1}^{m} R_i z_{i:m:n} e^{z_{i:m:n}} = 0. \tag{3.80}$$

These two likelihood equations need to be solved by the use of some numerical methods such as Newton-Raphson. Alternatively, by expanding the function $e^{z_{i:m:n}}$ that is present in (3.79) and (3.80) in a Taylor-series around

$$F^{\leftarrow}(E(U_{i:m:n})) = \log\left(-\log(1 - E(U_{i:m:n}))\right),$$

where $E(U_{i:m:n})$ is as given in Corollary 2.65, and then using the linear approximation

$$e^{z_{i:m:n}} \doteq \gamma_i + \delta_i z_{i:m:n},$$

AMLEs of μ' and ϑ' can be derived from (3.79) and (3.80), respectively; see Balakrishnan et al. (2004a) for pertinent details.

Example 3.56. Let us now assume Weibull, $Weibull(\vartheta^{\beta}, \beta)$, distribution for the progressively Type-II right censored data on breakdown times of an insulating fluid presented earlier in Example 3.53. Then, with $n = 19$, $m = 8$ and the progressive censoring plan as $\mathscr{R} = (0, 0, 3, 0, 3, 0, 0, 5)$, the log-lifetimes

$$-1.6608, -0.2485, -0.0409, 0.2700, 1.0224, 1.5789, 1.8718, 1.9947$$

can be assumed to be from extreme value, $EV_1(\mu', \vartheta')$, distribution. Then, the MLEs of μ' and ϑ' can be determined from (3.79) and (3.80) to be $\widehat{\mu}' = 2.222$ and $\widehat{\vartheta}' = 1.026$; see Viveros and Balakrishnan (1994) and Balakrishnan et al. (2004a). From these values, we then find the MLEs of Weibull parameters ϑ and β to be

$$\widehat{\vartheta} = e^{\widehat{\mu}'} = e^{2.222} = 9.226 \quad \text{and} \quad \widehat{\beta} = \frac{1}{\widehat{\vartheta}'} = \frac{1}{1.026} = 0.975.$$

Remark 3.57. The MLE $\widehat{\beta}$ of the shape parameter β (of the Weibull distribution) is quite close to 1, which does suggest that the scaled-exponential distribution itself is suitable for these data. This does agree with the same conclusion drawn by Balakrishnan and Lin (2003) based on an application of their goodness-of-fit to these progressively Type-II right censored data.

Instead of the preceding likelihood estimation based on extreme value distribution for log-transformed data, we can develop likelihood estimation directly based on Weibull, $Weibull(\vartheta^{\beta}, \beta)$, distribution, for the original lifetime data itself. In this case, with $f_{\vartheta,\beta}$ as given in (3.45), we have the likelihood function to be

$$\mathscr{L}(\vartheta, \beta; x_{1:m:n}, \ldots, x_{m:m:n}) = \frac{c_{m-1}}{\vartheta^{m\beta}} \beta^m \left(\prod_{i=1}^{m} x_{i:m:n}^{\beta-1} \right) \exp\left\{ -\frac{1}{\vartheta^{\beta}} \sum_{i=1}^{m} (R_i + 1) x_{i:m:n}^{\beta} \right\},$$

from which the log-likelihood function is obtained as

$$\mathscr{L}^*(\vartheta, \beta) \equiv \log \mathscr{L}(\vartheta, \beta)$$

$$= \text{const} + m \log \beta - m\beta \log \vartheta + (\beta - 1) \sum_{i=1}^{m} \log x_{i:m:n} - \frac{1}{\vartheta^{\beta}} \sum_{i=1}^{m} (R_i + 1) x_{i:m:n}^{\beta}. \tag{3.81}$$

From (3.81), upon taking partial derivatives with respect to ϑ and β, we obtain the likelihood equations to be

$$\frac{\partial \mathscr{L}^*}{\partial \vartheta} = -\frac{m\beta}{\vartheta} + \frac{\beta}{\vartheta^{\beta+1}} \sum_{i=1}^{m} (R_i + 1) x_{i:m:n}^{\beta} = 0 \tag{3.82}$$

and

$$\begin{aligned} \frac{\partial \mathscr{L}^*}{\partial \beta} = \frac{m}{\beta} - m \log \vartheta + \sum_{i=1}^{m} \log x_{i:m:n} + \frac{\log \vartheta}{\vartheta^{\beta}} \sum_{i=1}^{m} (R_i + 1) x_{i:m:n}^{\beta} \\ - \frac{1}{\vartheta^{\beta}} \sum_{i=1}^{m} (R_i + 1) x_{i:m:n}^{\beta} \log x_{i:m:n} = 0. \end{aligned} \tag{3.83}$$

Note that (3.82) yields an explicit MLE of ϑ as

$$\widehat{\vartheta} = \left\{ \frac{1}{m} \sum_{i=1}^{m} (R_i + 1) x_{i:m:n}^{\beta} \right\}^{1/\beta} \tag{3.84}$$

which, when substituted in (3.83), yields the simplified likelihood equation for β as

$$\frac{m}{\beta} - \frac{m}{\sum_{i=1}^{m} (R_i + 1) x_{i:m:n}^{\beta}} \sum_{i=1}^{m} (R_i + 1) x_{i:m:n}^{\beta} \log x_{i:m:n} + \sum_{i=1}^{m} \log x_{i:m:n} = 0. \tag{3.85}$$

The comments provided in Remark 3.34 for Type-II censored data from Weibull distribution apply here with obvious modifications, too. The MLE of β needs to be determined numerically by solving (3.85) by Newton-Raphson method, for example.

Example 3.58. By assuming the progressively Type-II right censored data on breakdown times of an insulating fluid presented in Example 3.53 to have come from Weibull, *Weibull*(ϑ^{β}, β), distribution and solving (3.85), we find the MLE of β to be $\widehat{\beta} = 0.975$. Then, by substituting this estimate in (3.84) for β, we find the MLE of ϑ to be $\widehat{\vartheta} = 9.226$. Observe that these are exactly the same as those found in Example 3.56.

Furthermore, by rewriting (3.85) as

$$\frac{1}{\beta} = \frac{1}{\sum_{i=1}^{m} (R_i + 1) x_{i:m:n}^{\beta}} \sum_{i=1}^{m} (R_i + 1) x_{i:m:n}^{\beta} \log x_{i:m:n} - \frac{1}{m} \sum_{i=1}^{m} \log x_{i:m:n} \tag{3.86}$$

and adopting an argument very similar to the one described in Section 3.2.3.3, Balakrishnan and Kateri (2008) have demonstrated that $\widehat{\beta}$, as a solution of (3.86), always exists and is unique (see also Remark 3.34).

EM- and stochastic EM-algorithms can also be utilized to determine the MLEs of distributional parameters very much along the lines of Section 3.2.3.5; as a matter of fact, the conditional expectations, conditioned on the observed progressively Type-II right censored data and the current values of the parameters, required here are exactly the same as those involved in the case of Type-II right censoring, at each stage of progressive censoring. One may refer to Ng et al. (2002, 2004) for further details in this regard.

3.4.3 Bayesian estimation

Bayesian inference can be readily developed in this case following the lines of Section 3.2.4. To this end, when the available progressively Type-II right censored data arise from a scale-family of distributions with density function f_ϑ, with the prior density of ϑ as $\pi(\vartheta)$, $\vartheta \in \Theta$, we have from (3.74) the posterior density of ϑ to be

$$g(\vartheta \mid \boldsymbol{x}) = \frac{\pi(\vartheta)\prod_{i=1}^{m} f_\vartheta(x_{i:m:n})\,(1 - F_\vartheta(x_{i:m:n}))^{R_i}}{\int_\Theta \pi(\eta)\prod_{i=1}^{m} f_\eta(x_{i:m:n})\,\big(1 - F_\eta(x_{i:m:n})\big)^{R_i}\,d\eta}. \tag{3.87}$$

For most distributions, Bayesian inference can be developed from the posterior density in (3.87) using MCMC method. In the special case of scaled-exponential, *Exp*(ϑ), distribution, the posterior density of ϑ in (3.87) takes on the simpler form

$$g(\vartheta \mid \boldsymbol{x}) = \frac{\pi(\vartheta)\vartheta^{-m} e^{-\frac{1}{\vartheta}\sum_{i=1}^{m}(R_i+1)x_{i:m:n}}}{\int_\Theta \pi(\eta)\eta^{-m} e^{-\frac{1}{\eta}\sum_{i=1}^{m}(R_i+1)x_{i:m:n}}\,d\eta}. \tag{3.88}$$

Thus, with the prior $\pi(\cdot)$ as inverse gamma, $\mathsf{I}\Gamma(\gamma, \delta)$, density function, for example, we can readily find the Bayesian estimate of ϑ from (3.88) to be

$$\widehat{\vartheta}_{\mathrm{B}} = \frac{\sum_{i=1}^{m}(R_i + 1)x_{i:m:n} + \gamma}{m + \delta - 1}$$

provided $m + \delta > 1$. It is evident that this is a direct generalization of the corresponding result for the case of Type-II censoring presented earlier in Section 3.2.4.

It is important to mention finally that similar Bayesian inferential methods have been discussed extensively for a wide range of lifetime distributions.

3.5. Progressive Type-I censoring

Progressive Type-I censoring is a direct generalization of Type-I censoring situation discussed earlier in Section 3.3. In this censoring scheme, inspections are to be performed at pre-fixed times $T_1, \ldots, T_{k-1}$ at which $R_1^0, \ldots, R_{k-1}^0$ of the surviving units are intended to be removed from the life-test, and finally at time T_k the test is to be terminated with all remaining surviving units getting censored at that time. However, as the numbers of surviving units at each stage of censoring are random, it is possible that planned numbers to censor may not be possible at some stages and that the experiment itself may end before the intended termination time T_k. Further, as in the case of Type-I censoring in Section 3.3, only conditional likelihood inference can be developed, conditioned on the assumption that at least one failure has occurred. These complications deter us from having explicit inferential results in this case for most lifetime distributions. In the exceptional case of scaled-exponential, $Exp(\vartheta)$, distribution, however, the MLE of ϑ and its exact sampling distribution become available in explicit forms. Then, with m as the total number of failures observed, $(R_1, \ldots, R_k)$ as the progressive censoring realized as compared to the intended censoring scheme $(R_1^0, \ldots, R_k^0)$ and $d_{\bullet j}$ as the number of failures observed till inspection time T_j $(j = 1, \ldots, k)$, with $d_{\bullet 0} \equiv 0$, the likelihood function in this case can be expressed as

$$\mathscr{L}(\vartheta; \boldsymbol{x}, \boldsymbol{d}_\bullet) \propto \frac{1}{\vartheta^m} \exp\left[-\frac{1}{\vartheta}\sum_{j=1}^{k}\left\{\sum_{i=d_{\bullet j-1}+1}^{d_{\bullet j}} x_{i:m:n} + R_j T_j\right\}\right]; \tag{3.89}$$

see Balakrishnan et al. (2011) and Cramer and Tamm (2014). The MLE of ϑ is readily obtained from (3.89) as

$$\widehat{\vartheta} = \frac{1}{m}\sum_{j=1}^{k}\left\{\sum_{i=d_{\bullet j-1}+1}^{d_{\bullet j}} X_{i:m:n} + R_j T_j\right\}. \tag{3.90}$$

In the above expressions, $x_{i:m:n}$ denotes the i-th progressively Type-I censored observation among the m complete failures observed from a life-test of n units (not to be confused with the notation of progressively Type-II right censored order statistics used before). For the MLE of ϑ in (3.90), Balakrishnan et al. (2011) have derived exact explicit expressions for its moments and the sampling distribution.

Example 3.59. Consider the progressively Type-I censored data on failure times of diesel engine fans given in Balakrishnan and Cramer (2014, p. 582); see also Nelson (1982), and Cohen (1991). In this case, $n = 70$ fans were tested, $m = 12$ complete failures were observed, and 58 units were censored at $k = 27$ different inspection times. Assuming a scaled-exponential, $Exp(\vartheta)$, distribution for this lifetime data, we compute the MLE of ϑ from (3.90) to be

$$\widehat{\vartheta} = \frac{1}{12}\sum_{j=1}^{27}\left\{\sum_{i=d_{\bullet j-1}+1}^{d_{\bullet j}} x_{i:12:70} + R_j T_j\right\} = \frac{344{,}440}{12} = 28{,}703.3 \text{ hours.}$$

We finally conclude this chapter by stating that all the methods and inferential results reviewed here for conventional Type-II, Type-I, progressive Type-II and progressive Type-I censoring schemes form a basis for all the developments in the ensuing chapters for different hybrid censoring scenarios!

CHAPTER 4

Models and distributional properties of hybrid censoring designs

Contents

4.1. Introduction

In this chapter, we introduce various hybrid censoring designs discussed in the literature and present the resulting sampling situations in terms of order statistics. It turns out that all these censoring schemes either face the problem of a possibly empty sample, that is, no failure may be observed within the test duration, or an unbounded experimental time. In order to provide the basis for (conditional) inference, we derive the likelihood functions under the discussed hybrid censoring schemes. For illustration, we present the derivation of the joint density functions for Type-I and Type-II hybrid censoring, respectively, in detail. Additionally, this will be done by applying the modularization approach proposed by Górny and Cramer (2018b) which is seen to provide a powerful tool to establish the corresponding results for even more complex hybrid censoring schemes. In fact, we demonstrate that the joint distribution of a hybrid censored sample can be obtained by considering four types of sampling scenarios called *module types O,*

https://doi.org/10.1016/B978-0-12-398387-9.00012-X

Table 4.1 Characteristics of Type-II censoring.

Type-II censoring	Expression	Properties
Scheme parameter	m	
Termination time	$X_{m:n}$	unbounded
Sample size D	m	$D \in \{m\}$
	The design guarantees exactly m observations.	
Experimental time W	$X_{m:n}$	

A, B, and AB.[1] This decomposition is explicitly provided for the generalized Type-I and Type-II hybrid censoring scheme, respectively. Finally, we present four hybrid censoring schemes called unified hybrid censoring where the application of the proposed decomposition method is only sketched.

It should already be mentioned here that the structure figured out for hybrid censoring schemes is independent of the underlying data (here order statistics). Therefore, we can apply the same approach by replacing the order statistics by the corresponding progressively Type-II censored order statistics when discussing progressive hybrid censoring schemes. Details are presented in Chapter 7.

4.2. Preliminaries

In this chapter, we consider order statistics $X_{1:n}, \ldots, X_{n:n}$ of an iid sample $X_1, \ldots, X_n$ with common cumulative distribution function F which is assumed to be absolutely continuous with density function f. Usually, F is supposed to be a lifetime distribution with support on the positive real line but, in most derivations, support restrictions are not of interest. Furthermore, we assume the sample to be Type-II censored, that is, the experiment is terminated at the m-th observed failure so that the data is given by

$$X_{1:n}, \ldots, X_{m:n} \tag{4.1}$$

where $m \in \{1, \ldots, n\}$ is prefixed. Characteristics of the Type-II censoring mechanism are summarized in Table 4.1, a flow chart with the sampling scenario is presented in Fig. 4.1.

Furthermore, time thresholds are present in many hybrid censoring models. In order to avoid trivialities, we assume throughout that such a threshold T is included in the support of F, that is, $0 < F(T) < 1$. For the threshold T, we introduce the random

[1] The module types O, A, B (and AB) are formally introduced in Section 4.5 but will be used earlier in the presentation of flow charts illustrating the various censoring mechanisms.

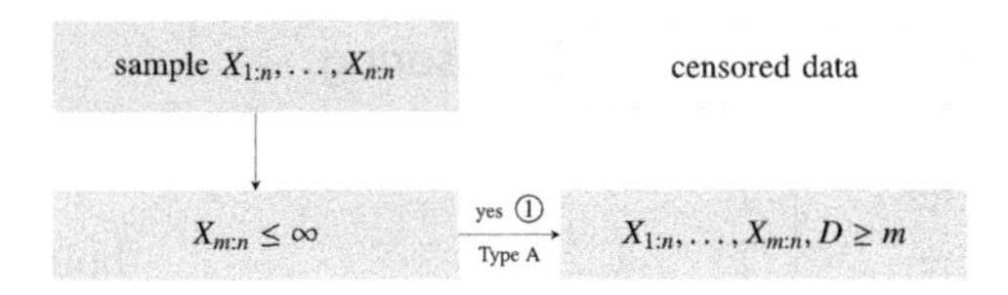

Figure 4.1 Sampling scenario in Type-II censoring.

counter $D = D(T)$ by (cf. (2.12))

$$D = \sum_{j=1}^{n} \mathbb{1}_{(-\infty,T]}(X_{j:n}). \tag{4.2}$$

For a threshold T_i, we use the notation $D_i = D(T_i)$.

For particular hybrid censoring schemes, we introduce a special notation for the involved counter D_{HCS}. For instance, D_{I} and D_{II} denote the corresponding random counters under Type-I and Type-II hybrid censoring, respectively. On the other hand, we will use the same notation for progressive hybrid censoring schemes (see (7.3)). This reflects the fact that the observed sample size D_{HCS} under a particular hybrid censoring scheme can be written as some function Υ_{HCS} of D (or D_i if more than one threshold are involved). The function Υ_{HCS} is independent of the particular ordered data and depends only on the design of the hybrid censoring scheme (see Remark 4.2). Such counters will play an important role in the probabilistic analysis of hybrid censoring schemes. In fact, the equivalence

$$D \geq m \iff X_{m:n} \leq T \tag{4.3}$$

will be very useful in the following derivations. According to Theorem 2.17, D has a binomial distribution with support $\{0, \ldots, n\}$ and probability mass function

$$\Pr(D = d) = \binom{n}{d} F^d(T)(1 - F(T))^{n-d}, \quad d \in \{0, \ldots, n\}. \tag{4.4}$$

In Type-I censoring, the experiment is terminated at T so that the data is given by

$$X_{1:n}, \ldots, X_{D:n}$$

with D as in (4.2) (see Section 3.3 and (3.61)). Characteristics of the censoring mechanism are summarized in Table 4.2, a flow chart with the different sampling scenarios is presented in Fig. 4.2.

Remark 4.1. In some papers on Type-I hybrid censoring, the random counter

$$D_{\mathsf{I}} = \sum_{j=1}^{m} \mathbb{1}_{(-\infty,T]}(X_{j:n}) \tag{4.5}$$

Table 4.2 Characteristics of Type-I censoring.

Type-I censoring	Expression	Properties
Scheme parameter	T	
Termination time	T	bounded by T
Sample size D	D	$D \in \{0, \dots, n\}$
	Sample size may be zero with positive probability.	
Experimental time W	$X_{n:n} \wedge T$	

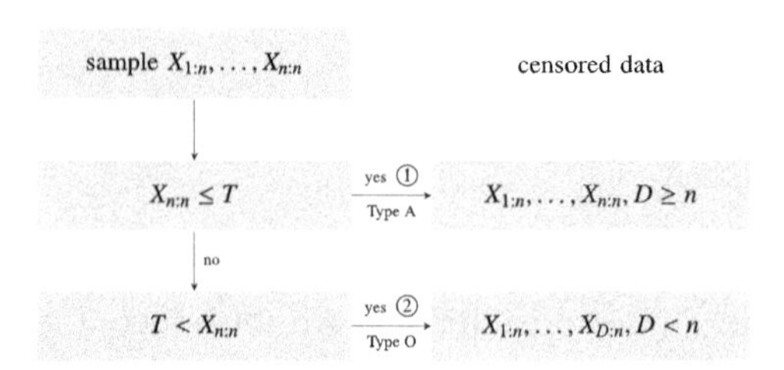

Figure 4.2 Sampling scenario in Type-I censoring.

is considered instead of D indicating that only the first m order statistics $X_{1:n}, \dots, X_{m:n}$ are really involved in the construction of the sample ($m \in \mathbb{N}$ with $m \leq n$). Thus, one has to check carefully which definition of the counter is used in the derivations. In the following, we use a definition in terms of D as in (4.2), that is,

$$D_{\mathsf{I}} = D \wedge m = \Upsilon_{\mathsf{I}}(D), \quad \text{say.}$$

Therefore, we can restrict the discussion to the counter D and some deterministic function Υ_{HCS} depending in the hybrid censoring scheme (for details, see Remark 4.2 and Table 5.1).

The probability mass function of D_{I} can be directly obtained from the distribution of D by using

$$\begin{aligned}
&\{D_{\mathsf{I}} = 0\} = \{D = 0\} = \{T < X_{1:n}\}, \\
&\{D_{\mathsf{I}} = d\} = \{D = d\} = \{X_{d:n} \leq T < X_{d+1:n}\}, \quad d \in \{1, \dots, m-1\}, \\
&\{D_{\mathsf{I}} = m\} = \{D \geq m\} = \{X_{m:n} \leq T\}.
\end{aligned}$$

Thus, the respective probabilities can be obtained from the distribution of order statistics based on F. Note that this approach can directly be extended to progressively Type-II censored order statistics (or other kinds of ordered data). Thus, we get

$$\begin{aligned}
\Pr(D_{\mathsf{I}} = 0) &= \Pr(X_{1:n} > T) = (1 - F(T))^n, \\
\Pr(D_{\mathsf{I}} = m) &= \Pr(D \geq m) = \Pr(X_{m:n} \leq T) = F_{m:n}(T).
\end{aligned}$$

For $d \in \{1, \ldots, m-1\}$, the cumulative distribution function of the order statistic $X_{j:n}$ is given in Theorem 2.10 so that

$$\begin{aligned}\Pr(D_{\mathsf{I}} = d) &= \Pr(X_{d:n} \leq T < X_{d+1:n}) = \Pr(X_{d+1:n} > T) - \Pr(X_{d:n} > T)\\ &= F_{d:n}(T) - F_{d+1:n}(T) = \binom{n}{d} F^d(T)(1 - F(T))^{n-d}.\end{aligned} \tag{4.6}$$

Notice that the right hand side of Eq. (4.6) can be expressed in terms of a density function (provided it exists and $f(T) > 0$) as

$$\Pr(D = d) = \frac{1 - F(T)}{(n-d)f(T)} f_{d+1:n}(T), \quad d \in \{0, \ldots, n-1\}. \tag{4.7}$$

Remark 4.2. From a structural point of view, hybrid censoring models depend on both the (complete) sample of order statistics $X_{1:n}, \ldots, X_{n:n}$ and a random counter D_{HCS} which determines the (observed) sample size under the particular hybrid censoring scheme, that is, the effectively observed sample is given by

$$X_{1:n}, \ldots, X_{D_{\mathsf{HCS}}:n}$$

with random sample size D_{HCS}. Thus, given $D_{\mathsf{HCS}} = d(>0)$, the first d failures times $X_{1:n}, \ldots, X_{d:n}$ are observed. The random counter D_{HCS} of some hybrid censoring schemes depend on the random variables $X_{1:n}, \ldots, X_{n:n}$ only via the random counter $D = \sum_{j=1}^{n} \mathbb{1}_{(-\infty, T]}(X_{j:n})$ (see (4.2)) and some deterministic function $\Upsilon_{\mathsf{HCS}} : \{0, \ldots, n\} \longrightarrow \{0, \ldots, n\}$, that is,

$$D_{\mathsf{HCS}} = \Upsilon_{\mathsf{HCS}}(D).$$

For hybrid censoring schemes incorporating several thresholds $T_1, \ldots, T_\ell$, a similar argument applies using the counters $D_i = \sum_{j=1}^{n} \mathbb{1}_{(-\infty, T_i]}(X_{j:n})$, $1 \leq i \leq \ell$, and a function $\Upsilon_{\mathsf{HCS}}(D_1, \ldots, D_\ell)$ depending on all these counters.

For instance, in Type-I hybrid censoring, $\Upsilon_{\mathsf{HCS}}(D) = D \wedge m$ where $1 \leq m \leq n$ is a fixed integer. For Type-I censoring, we have $\Upsilon_{\mathsf{HCS}}(D) = D$ whereas for Type-II censoring $\Upsilon_{\mathsf{HCS}}(D) = m$ holds. For other hybrid censoring schemes, the function Υ_{HCS} can be taken from Table 5.1.

In order to study distributions of the data and related statistics, the law of total probability will be applied to the hybrid censored sample as follows. Using that the events $\{D = d\}$, $d = 0, \ldots, n$, form a decomposition of the space Ω, we find

$$\Pr(X_{j:n} \leq t_j, 0 \leq j \leq D_{\mathsf{HCS}}) = \sum_{d=0}^{n} \Pr(X_{j:n} \leq t_j, 1 \leq j \leq \Upsilon_{\mathsf{HCS}}(d), D = d).$$

This illustrates that we can combine those probabilities (w.r.t. the values of D) where $\Upsilon_{\mathsf{HCS}}(\cdot)$ is constant. Therefore, the image[2] $\Upsilon_{\mathsf{HCS}}(\{0,\dots,n\}) \subseteq \{0,\dots,n\}$ can be used to identify the different data scenarios. Considering the particular functions Υ_{HCS} of the hybrid censoring scheme with a single threshold T, we find that the respective sets with equal values are of the form $\{D < m^\star\}$ and $\{D \geq m^\star\}$ for some $m^\star$ as well as $\{D = d\}$ for some d. Therefore, it is sufficient to study the conditional cumulative distribution functions

$$F^{X_{1:n},\dots,X_{d:n}|D=d}, \quad F^{X_{1:n},\dots,X_{m^\star:n}|D<m^\star}, \text{ and } F^{X_{1:n},\dots,X_{m^\star:n}|D\geq m^\star}$$

for some d and $m^\star$, respectively. In case of more than one threshold, the conditional cumulative distribution functions

$$F^{X_{1:n},\dots,X_{m^\star:n}|D_i<m^\star, D_{i+1}\geq m^\star}$$

have to be taken into account for each pair of successive thresholds $T_i < T_{i+1}$. This observation leads to the so-called *module types O, A, B, AB* (see p. 141 and p. 145). Details will be explained subsequently when the particular hybrid censoring schemes are discussed.

4.3. Type-I hybrid censoring

4.3.1 Censoring design

A Type-I hybrid censored sample can be considered as a Type-I censoring applied to Type-II censored data. In this regard, we consider the complete (ordered) sample $X_{1:n} \leq \cdots \leq X_{n:n}$ as baseline sample and

(a) first apply a Type-II censoring (with number m) and
(b) then a Type-I censoring (with threshold T)

to this sample (see Fig. 4.4). Given the Type-II censored sample in (4.1) and a threshold time T, the Type-I hybrid censored data is generated by the operation

$$\min(X_{j:n}, T) = X_{j:n} \wedge T, \quad 1 \leq j \leq m. \tag{4.8}$$

An illustration of the procedure is depicted in Fig. 4.3 for $m = 5 \leq n$ and a threshold T showing the two possible outcomes of the procedure. Depending on the value of D, the resulting data is illustrated in Fig. 4.4. In particular, we get two data scenarios:

① $X_{1:n},\dots,X_{m:n}, D \geq m$, (fixed sample size m)
② $X_{1:n},\dots,X_{D:n}, D < m$ (random sample size D).

[2] Notice that this image equals the support of D_{HCS}.

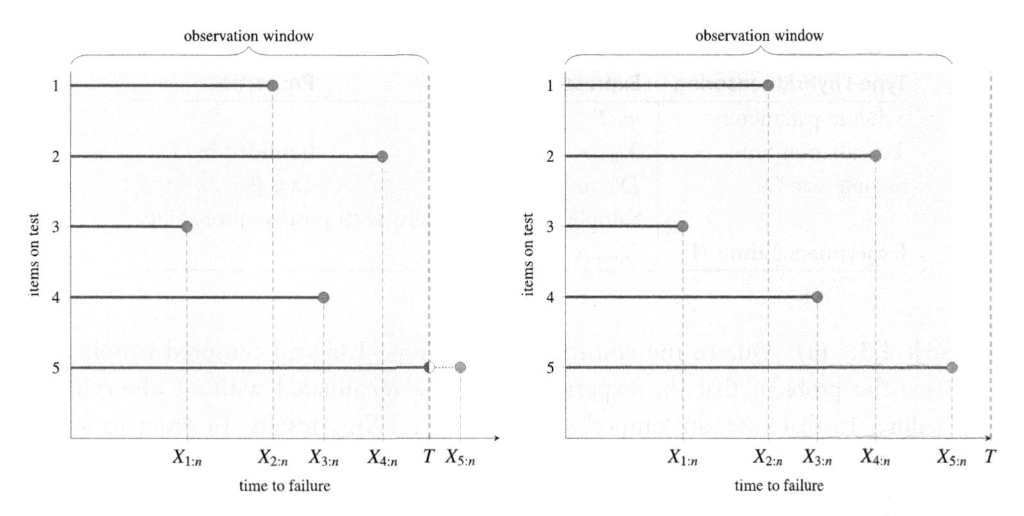

Figure 4.3 Type-I hybrid censoring scheme with $m = 5 \leq n$ and threshold T (left: case $T < X_{5:n}$ and $D_{\mathrm{I}} = 4$; right: case $T \geq X_{5:n}$ and $D_{\mathrm{I}} = 5$).

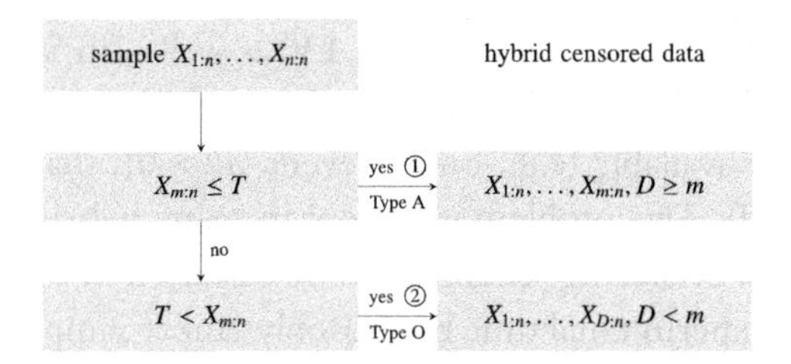

Figure 4.4 Sampling scenarios ①, ② in Type-I hybrid censoring.

The total number of observed failures, that is, the effective sample size is given by

$$D_{\mathrm{I}} = \min\{D, m\} = D \wedge m$$

as in (4.5), so that, from (4.8), the Type-I hybrid censored sample can be written in the form

$$X_{1:n}, \ldots, X_{D_{\mathrm{I}}:n}, \ T^{*m-D_{\mathrm{I}}}. \tag{4.9}$$

Notice that it may happen that the experiment terminates without observing a failure, that is, $D_{\mathrm{I}} = 0$ with probability $\Pr(D_{\mathrm{I}} = 0) = \Pr(D = 0) = (1 - F(T))^n > 0$ since $F(T) < 1$ (see Lemma 4.4). The termination time of the experiment (which equals the test duration) is given by $W = X_{m:n} \wedge T$. Characteristics of Type-I hybrid censoring are summarized in Table 4.3.

Table 4.3 Characteristics of Type-I hybrid censoring.

Type-I hybrid censoring	Expression	Properties
Scheme parameters	m, T	
Termination time	$X_{m:n} \wedge T$	bounded by T
Sample size D_{I}	$D \wedge m$	$D_{\mathsf{I}} \in \{0, \ldots, m\}$
	Sample size may be zero with positive probability.	
Experimental time W	$X_{m:n} \wedge T$	

Remark 4.3. **(a)** Due to the construction of a Type-I hybrid censored sample, we face the problem that the experiment may be terminated without observing a failure. In that case, an 'empty' sample $X_{1:n}, \ldots, X_{D_{\mathsf{I}}:n}$ results. In order to avoid this, the data can be generated as in (4.8) and (4.9) by adding observations with constant value T. Then, the sample size would always equal m and, for $D = D_{\mathsf{I}} = 0$, the sample is constant T^{*m}. Although this construction avoids the problem of an empty sample and will subsequently be used to avoid probabilities of not defined expressions, it does not help to solve the inferential problems caused by this behavior (see also the comment on p. 130 as well as in Section 3.3). Therefore, inference is usually conducted conditionally on the event that sufficiently many measurements are available (e.g., on the event $\{D > 0\}$, that is, at least one failure has been observed). This problem is present in many hybrid censored samples.
In fact, the hybrid censoring schemes can be classified in those with

- **(i)** bounded experimental time but possibly empty sample (see, e.g., Tables 4.2, 4.3, 4.6, 4.8, 4.11);
- **(ii)** unbounded experimental time but guaranteed minimum sample size (see, e.g., Tables 4.1, 4.4, 4.5, 4.9, 4.10).

The problem is also present for extensions to other kind of data like a progressively Type-II censored sample.

(b) We will see that a crucial quantity in the analysis of (progressive) hybrid censoring schemes is given by the observed sample size (generally denoted by D_{HCS}; for instance, we use the notation D_{I} under Type-I hybrid censoring). Then, the data is given by the ordered values $X_{(1)} \leq \cdots \leq X_{(D_{\mathsf{HCS}})}$. Interesting distributions are connected to conditional distributions of (progressively Type-II censored) order statistics $X_{(1)} \leq \cdots \leq X_{(d)}$ given $D_{\mathsf{HCS}} = d$.

4.3.2 Derivation of the joint density function

We start the analysis by presenting the probability mass function of the random sample size D_{I}. The derivation has already been provided on p. 123.

Lemma 4.4. *The random counter D_I has a discrete distribution with support $\{0, \ldots, m\}$ and probability mass function*

$$\Pr(D_I = d) = \begin{cases} \binom{n}{d} F^d(T)(1 - F(T))^{n-d}, & d \in \{0, \ldots, m-1\} \\ F_{m:n}(T), & d = m \end{cases}.$$

Thus, conditionally on $D_I = d$ with $d \in \{0, \ldots, m\}$, we have

$$X_{1:n} \wedge T, \ldots, X_{n:n} \wedge T \mid \{D_I = d\} \stackrel{d}{=} X_{1:n}, \ldots, X_{d:n}, T^{*m-d} \mid \{D_I = d\}. \tag{4.10}$$

This relation shows that, given $D_I = d \geq 1$, the data is determined by the first d order statistics $X_{1:n}, \ldots, X_{d:n}$ and $D_I = d$. For $d = 0$, the sample is deterministic and is given by T^{*m}. Furthermore, note that $\{D > 0\} = \{D_I > 0\}$.

Since the sample size as well as the experimental time are random, one may be interested in their means which are given in the following lemma. The mean of D_I follows directly from Lemma 4.4.

Lemma 4.5. (a) $ED_I = E(D \wedge m) = \sum_{d=0}^{m-1} d\binom{n}{d} F^d(T)(1 - F(T))^{n-d} + mF_{m:n}(T)$
(b) $EW = T(1 - F_{m:n}(T)) + \int_{-\infty}^{T} tf_{m:n}(t)\, dt$

Remark 4.6. Using (4.6), one gets for an *Exp*(ϑ)-distribution the identity

$$\begin{aligned} ED_I &= \sum_{d=0}^{m-1} d \Pr(D = d) + m \Pr(D \geq m) \\ &= \sum_{d=1}^{m-1} d\big(F_{d:n}(T) - F_{d+1:n}(T)\big) + mF_{m:n}(T) = \sum_{d=1}^{m} F_{d:n}(T), \end{aligned} \tag{4.11}$$

which has been established in Balakrishnan and Cramer (2014, p. 135) for progressive Type-I hybrid censoring.[3]

To establish representations for the joint cumulative distribution function, we need the following result.

Lemma 4.7. *Let $m, n \in \mathbb{N}$, $m \leq n$, and $F_{1,\ldots,m:n}$ denote the joint cumulative distribution function of order statistics $X_{1:n}, \ldots, X_{m:n}$. Then, for $\boldsymbol{t}_m \in \mathbb{R}^m$ and $T \in \mathbb{R}$:*

$$F_{1,\ldots,m:n}(\boldsymbol{t}_{m-1}, t_m \wedge T) = F_{1,\ldots,m:n}(\boldsymbol{t}_m \wedge T),$$

where $\boldsymbol{t}_m \wedge T = (t_1 \wedge T, \ldots, t_m \wedge T)$.

[3] In this representation, $F_{d:n}$ denotes the cumulative distribution function of the d-th order statistic based on an *Exp*(ϑ)-distribution. If $F_{d:n}$ denotes the respective cumulative distribution function from a standard exponential distribution, then we have $ED_I = \sum_{d=1}^{m} F_{d:n}(T/\vartheta)$.

Proof. The result follows directly from the property that, for $\boldsymbol{t}_m \in \mathbb{R}^m$,

$$F_{1,\dots,m:n}(\boldsymbol{t}_m) = F_{1,\dots,m:n}(t_1 \wedge \dots \wedge t_m, t_2 \wedge \dots \wedge t_m, \dots, t_{m-1} \wedge t_m, t_m)$$

using that $X_{1:n} \leq \dots \leq X_{m:n}$. □

In order to derive the joint distribution function of the Type-I hybrid censored data $X_{1:n}, \dots, X_{D_I:n}$, we use (4.10) and the data scenarios ① and ② illustrated in Fig. 4.4. These scenarios are examples of the module types O and A which will be explained in detail in Section 4.5.1.

Let $t_1, \dots, t_m \in \mathbb{R}$. Proceeding as in Cramer and Balakrishnan (2013) (see also Burkschat et al., 2016), we use the law of total probability to get

$$\begin{aligned}\Pr(X_{j:n} \leq t_j, 1 \leq j \leq D_I) &= \Pr(X_{j:n} \leq t_j, 1 \leq j \leq m, D \geq m) \\ &\quad + \sum_{d=0}^{m-1} \Pr(X_{j:n} \leq t_j, 1 \leq j \leq d, D = d). \qquad (4.12)\end{aligned}$$

Scenario ①: We get by (4.3) and Lemma 4.7

$$\begin{aligned}\Pr(X_{j:n} \leq t_j, 1 \leq j \leq m, D \geq m) &= \Pr(X_{j:n} \leq t_j, 1 \leq j \leq m-1, X_{m:n} \leq t_m \wedge T) \\ &= F_{1,\dots,m:n}(\boldsymbol{t}_{m-1}, t_m \wedge T) = F_{1,\dots,m:n}(\boldsymbol{t}_m \wedge T),\end{aligned}$$

Scenario ②: First, we get $(1 - F(T))^n \mathbb{1}_{[T,\infty)}(\min_{1 \leq j \leq m} t_j)$ for $d = 0$. Supposing $d \in \{1, \dots, m-1\}$, we find that

$$\begin{aligned}\Pr(X_{j:n} \leq t_j, 1 \leq j \leq d, D = d) &= \Pr(X_{j:n} \leq t_j, 1 \leq j \leq m, X_{d:n} \leq T < X_{d+1:n}) \\ &= \Pr(X_{j:n} \leq t_j, 1 \leq j \leq d-1, X_{d:n} \leq t_d \wedge T, X_{d+1:n} \geq T, T \leq \min_{d+1 \leq j \leq m} t_j) \\ &= \mathbb{1}_{[T,\infty)}(\min_{d+1 \leq j \leq m} t_j) \Big\{ \Pr(X_{j:n} \leq t_j, 1 \leq j \leq d-1, X_{d:n} \leq t_d \wedge T) \\ &\qquad - \Pr(X_{j:n} \leq t_j, 1 \leq j \leq d-1, X_{d:n} \leq t_d \wedge T, X_{d+1:n} \leq T) \Big\} \\ &= \mathbb{1}_{[T,\infty)}(\min_{d+1 \leq j \leq m} t_j) \Big\{ F_{1,\dots,d:n}(\boldsymbol{t}_{d-1}, t_d \wedge T) - F_{1,\dots,d+1:n}(\boldsymbol{t}_{d-1}, t_d \wedge T, T) \Big\}. \\ &= \mathbb{1}_{[T,\infty)}(\min_{d+1 \leq j \leq m} t_j) \Big\{ F_{1,\dots,d:n}(\boldsymbol{t}_d \wedge T) - F_{1,\dots,d+1:n}(\boldsymbol{t}_d \wedge T, T) \Big\}. \qquad (4.13)\end{aligned}$$

Let $v_1 \leq \dots \leq v_d \leq T$. Then, by using the marginal joint density function $f_{1,\dots,d:n}$ of order statistics and by introducing the notation $g(\boldsymbol{x}_d) = \frac{n!}{(n-d)!} \prod_{j=1}^{d} f(x_j)$, the joint cumulative distribution function $F_{1,\dots,d+1:n}(\boldsymbol{v}_d, T)$ can be expressed as

$$
\begin{aligned}
&F_{1,\ldots,d+1:n}(\boldsymbol{v}_d, T)\\
&\quad= \int_{-\infty}^{v_1} \cdots \int_{x_{d-1}}^{v_d} g(\boldsymbol{x}_d) \int_{x_d}^{T} (n-d) f(x_{d+1})(1-F(x_{d+1}))^{n-d-1} dx_{d+1} d\boldsymbol{x}_d\\
&\quad= F_{1,\ldots,d:m:n}(\boldsymbol{v}_d \wedge T) - (1-F(T))^{n-d} \int_{-\infty}^{v_1} \cdots \int_{x_{d-1}}^{v_d} g(\boldsymbol{x}_d) d\boldsymbol{x}_d
\end{aligned}
$$

Recalling that $\kappa_d^{-1} \cdot g$, with $\kappa_d = \frac{n!}{(n-d)!d!} = \binom{n}{d}$, can be interpreted as the joint density function of order statistics $X_{1:d}, \ldots, X_{d:d}$, this equals

$$
= F_{1,\ldots,d:m:n}(\boldsymbol{v}_d \wedge T) - \binom{n}{d}(1-F(T))^{n-d} F_{1,\ldots,d:d}(\boldsymbol{t}_d \wedge T).
$$

Summing up, we find

$$
\begin{aligned}
&\Pr(X_{j:n} \wedge T \leq t_j, 1 \leq j \leq m, D = d)\\
&\qquad= \mathbb{1}_{[T,\infty)}(\min_{d+1 \leq j \leq m} t_j)\binom{n}{d}(1-F(T))^{n-d} F_{1,\ldots,d:d}(\boldsymbol{t}_d \wedge T). \qquad (4.14)
\end{aligned}
$$

Notice that this expression also holds for $d = 0$ with $F_{1,\ldots,0:0} \equiv 1$. Combining the two cases as in (4.12), we arrive at the expression

$$
\begin{aligned}
&\Pr(X_{j:n} \wedge T \leq t_j, 1 \leq j \leq m) = \sum_{d=0}^{m} \Pr(X_{j:n} \wedge T \leq t_j, 1 \leq j \leq m, D_{\text{I}} = d)\\
&\quad= \sum_{d=0}^{m-1} \mathbb{1}_{[T,\infty)}(\min_{d+1 \leq j \leq m} t_j)\binom{n}{d}(1-F(T))^{n-d} F_{1,\ldots,d:d}(\boldsymbol{t}_d \wedge T)\\
&\qquad\qquad + F_{1,\ldots,m:n}(\boldsymbol{t}_m \wedge T). \qquad (4.15)
\end{aligned}
$$

Remark 4.8. Assuming $t_1 \leq \cdots \leq t_m$, we get from (4.14) and (4.15)

$$
\begin{aligned}
&\Pr(X_{j:n} \wedge T \leq t_j, 1 \leq j \leq m) = \sum_{d=0}^{m} \Pr(X_{j:n} \wedge T \leq t_j, 1 \leq j \leq m, D_{\text{I}} = d)\\
&\quad= (1-F(T))^n \mathbb{1}_{[T,\infty)}(t_1) + \sum_{d=1}^{m-1} \mathbb{1}_{[T,\infty)}(t_{d+1})\binom{n}{d}(1-F(T))^{n-d} F_{1,\ldots,d:d}(\boldsymbol{t}_d \wedge T)\\
&\qquad + F_{1,\ldots,m:n}(\boldsymbol{t}_m \wedge T)\\
&\quad= \begin{cases}
1, & T \leq t_1\\
\binom{n}{j}(1-F(T))^{n-j} F_{1,\ldots,j:j}(\boldsymbol{t}_j) & \\
\quad + \sum_{d=j+1}^{m-1} \binom{n}{d}(1-F(T))^{n-d} F_{1,\ldots,d:d}(\boldsymbol{t}_j, T^{*d-j}) & t_j < T \leq t_{j+1},\\
\quad + F_{1,\ldots,m:n}(\boldsymbol{t}_j, T^{*m-j}), & j = 1, \ldots, m-1\\
F_{1,\ldots,m:n}(\boldsymbol{t}_m), & t_m < T
\end{cases}
\end{aligned}
$$

Obviously, this cumulative distribution function has jumps (for $F(T) < 1$) so that it is not absolutely continuous w.r.t. the Lebesgue measure. Moreover, it is a mixture of measures related to the marginal distribution of the first order statistics.

Although we have assumed that F has a density function f, the expression in (4.15) is valid for continuous distribution functions F, too. In fact, we get a quantile representation in terms of Type-I hybrid censored order statistics from a uniform distribution. This can be seen from (4.15) or from the construction process by using the monotonicity of the quantile function $F^{\leftarrow}$. Moreover, notice that we get from (4.2) and Theorem 2.36

$$D = \sum_{j=1}^{n} \mathbb{1}_{(-\infty,T]}(X_{j:n}) = \sum_{j=1}^{n} \mathbb{1}_{(-\infty,F(T)]}(F^{\leftarrow}(X_{j:n})) \stackrel{d}{=} \sum_{j=1}^{n} \mathbb{1}_{(-\infty,F(T)]}(U_{j:n}) = D_u.$$

Theorem 4.9. *Let $X_{j:n}$, $1 \le j \le D_I$, be Type-I hybrid censored order statistics from a continuous cumulative distribution function F with time threshold T. Then,*

$$X_{j:n}, 1 \le j \le D_I \quad \stackrel{d}{=} \quad F^{\leftarrow}\left(U_{j:n}\right), 1 \le j \le D_u \wedge m,$$

where $U_{j:n}$, $1 \le j \le n$, are order statistics from a uniform distribution and D_u denotes the number of uniform order statistics less than $F(T)$.

Remark 4.10. It should be mentioned that similar quantile representations can be established for the other (progressive) hybrid censoring schemes. For brevity, they will not be explicitly presented in the following.

The conditional cumulative distribution function $F^{X_{j:n}\wedge T, 1\le j\le m | D_I = d}$ can be directly deduced from (4.14) by dividing the expression by $\Pr(D_I = d)$. It follows that, for an absolutely continuous cumulative distribution function F, this measure has a density function w.r.t. the product measure $\lambda^d \otimes \bigotimes_{j=1}^{m-d} \delta_T$, where λ^d denotes the d-dimensional Lebesgue measure and δ_T is a one-point distribution in T. Therefore, conditionally on $D = D_I = 0$, we have

$$F^{X_{j:n}\wedge T, 1\le j\le m | D_I = 0}(\boldsymbol{t}_m) = \frac{F^{X_{j:n}\wedge T, 1\le j\le m, D=0}(\boldsymbol{t}_m)}{\Pr(D=0)} = \mathbb{1}_{[T,\infty)}(\min_{1\le j\le m} t_j)$$

which obviously is degenerated and independent of the cumulative distribution function F. The respective (conditional) probability mass function is given by $\Pr(X_{j:n} \wedge T = T, 1 \le j \le m \mid D = 0) = 1$. Conditionally on $D = d \in \{1, \ldots, m-1\}$, we have

$$f^{X_{j:n}, 1\le j\le D_I | D_I = d} = f^{X_{j:n}, 1\le j\le d | D = d}$$

so that the density function is given by

$$f^{X_{j:n}, 1\le j\le d | D = d}(\boldsymbol{t}_d) = \frac{1}{\Pr(D=d)} \binom{n}{d} (1 - F(T))^{n-d} f_{1,\ldots,d:d}(\boldsymbol{t}_d), \quad t_1 \le \cdots \le t_d \le T.$$

On the other hand, we can use (4.7) to get, for $\boldsymbol{t}_d \in \mathbb{R}^d_{\leq T}$ and $1 \leq d \leq m-1$,

$$\begin{aligned}
F^{X_{j:n},1\leq j\leq d|D=d}(\boldsymbol{t}_d) &= \frac{n!}{(n-d-1)!}\frac{f(T)(1-F(T))^{n-d-1}}{f_{d+1:n}(T)}\int_{-\infty}^{t_1}\cdots\int_{x_{d-1}}^{t_d}\prod_{j=1}^{d}f(x_j)d\boldsymbol{x}_d\\
&= \frac{1}{f_{d+1:n}(T)}\int_{-\infty}^{t_1}\cdots\int_{x_{d-1}}^{t_d} f_{1,\ldots,d+1:n}(\boldsymbol{x}_d, T)d\boldsymbol{x}_d\\
&= \int_{-\infty}^{t_1}\cdots\int_{x_{d-1}}^{t_d} f_{1,\ldots,d:n}(\boldsymbol{x}_d \mid X_{d+1:n}=T)d\boldsymbol{x}_d\\
&= F_{1,\ldots,d:n}(\boldsymbol{t}_d \mid X_{d+1:n}=T).
\end{aligned}$$

Hence, we obtain, for $t_1 \leq \cdots \leq t_d \leq T$,

$$F^{X_{j:n},1\leq j\leq d|D=d}(\boldsymbol{t}_d) = F_{1,\ldots,d:n}(\boldsymbol{t}_d \mid X_{d+1:n}=T).$$

For $D_{\text{I}} = m$, we obtain the same result interpreting $F_{1,\ldots,m:n}(\cdot \mid X_{m+1:n}=T)$ as the (right truncated) cumulative distribution function $F_{1,\ldots,m:n}/F_{m:n}(T)$, that is,

$$F^{X_{j:n},1\leq j\leq D_{\text{I}}|D_{\text{I}}=m} = F^{X_{j:n},1\leq j\leq m|D\geq m} = \frac{F_{1,\ldots,m:n}}{F_{m:n}(T)}.$$

Therefore, (4.15) can be written as

$$\begin{aligned}
&\Pr(X_{j:n}\wedge T\leq t_j, 1\leq j\leq m)\\
&\quad= \sum_{d=0}^{m}\Pr(D_{\text{I}}=d)\Pr(X_{j:n}\leq t_j, 1\leq j\leq d \mid D_{\text{I}}=d)\mathbb{1}_{[T,\infty)}(\min_{d+1\leq j\leq m} t_j)\\
&\quad= \sum_{d=0}^{m}\Pr(D_{\text{I}}=d)\mathbb{1}_{[T,\infty)}(\min_{d+1\leq j\leq m} t_j)F_{1,\ldots,d:n}(\boldsymbol{t}_d \mid X_{d+1:n}=T). \qquad (4.16)
\end{aligned}$$

Conditionally on $D_{\text{I}} = D = d \in \{1, \ldots, m-1\}$, we can interpret the d-th summand in (4.16) as the conditional joint cumulative distribution function of the first d order statistics, given $X_{d+1:n} = T$. Therefore, the corresponding Lebesgue density function is given by

$$f^{X_{j:n},1\leq j\leq d|D_{\text{I}}=d} = f_{1,\ldots,d:n}(\cdot \mid X_{d+1:n}=T) \qquad (4.17)$$

with support $\mathbb{R}^d_{\leq T}$, $1 \leq d \leq m-1$. For $D_{\text{I}} = m$, the joint density function is simply given by

$$f^{X_{j:n},1\leq j\leq m|D_{\text{I}}=m} = \frac{f_{1,\ldots,m:n}}{F_{m:n}(T)}$$

with support $\mathbb{R}^m_{\leq T}$. Notice that $\{D_{\text{I}} = m\} = \{D \geq m\}$.

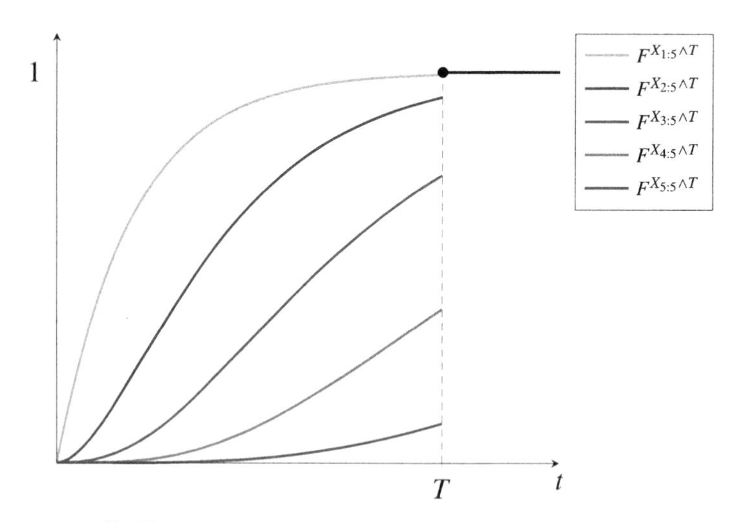

Figure 4.5 Plots of $F^{X_{j:n}\wedge T}$, $j \in \{1, \dots, 5\}$, for a standard exponential distribution with $n = 5$ and threshold $T = 1$.

These expressions will be quite useful in the derivation of the conditional density function of spacings for an exponential baseline distribution. Moreover, it obviously provides marginal cumulative distribution functions and density functions, given $D_\text{I} = d \in \{1, \dots, m\}$.

It should be noted that the (unconditional) one-dimensional marginals can be easily obtained directly from the definition. Namely, the cumulative distribution function of $X_{j:n} \wedge T$ is given by

$$F^{X_{j:n}\wedge T}(t) = \mathbb{1}_{[T,\infty)}(t)(1 - F_{j:n}(T)) + F_{j:n}(t \wedge T) = \begin{cases} F_{j:n}(t), & t < T \\ 1, & T \le t \end{cases}$$

which obviously has a point mass of $1 - F_{j:n}(T)$ in T, $j \in \{1, \dots, m\}$ (see Fig. 4.5).

Finally, for $d > 0$, the joint density function (w.r.t. an appropriate dominating measure) is given by

$$\begin{aligned} f^{X_{j:n},1\le j\le d,D_\text{I}}(\boldsymbol{t}_d, d) &= f^{X_{j:n},1\le j\le d,D_\text{I}}(\boldsymbol{t}_d, d) \\ &= f^{X_{j:n},1\le j\le d|D_\text{I}=d}(\boldsymbol{t}_d)\Pr(D_\text{I} = d) \\ &= \begin{cases} \frac{1-F(T)}{(n-d)f(T)} f_{1,\dots,d+1:n}(\boldsymbol{t}_d, T), & d \in \{1, \dots, m-1\} \\ f_{1,\dots,m:n}(\boldsymbol{t}_m), & d = m \end{cases}. \end{aligned} \tag{4.18}$$

The expressions in (4.18) yield the (conditional) likelihood function when the number of observed failures is positive. For (4.18), we find the following compact representation. This result will be very useful for inferential purposes.

Theorem 4.11. *For $d \in \{0, \dots, m\}$, the joint density function of $X_{1:n}, \dots, X_{D_I:n}, D_I$ can be written as*

$$f^{X_{j:n}, 1 \le j \le d, D_I}(\boldsymbol{t}_d, d) = \frac{n!}{(n-d)!} \Big(\prod_{j=1}^{d} f(t_j)\Big)(1 - F(w))^{n-d} \tag{4.19}$$

where $w = t_d \wedge T$.

Proof. First, we have two types of the density function in (4.18), that is,

$$\begin{aligned} f^{X_{j:n}, 1 \le j \le d, D_I}(\boldsymbol{t}_d, d) &= \frac{1 - F(T)}{(n-d)f(T)} f_{1,\dots,d+1:n}(\boldsymbol{t}_d, T), \quad d \in \{1, \dots, m-1\}, && \text{(type ⓘ)} \\ &= f_{1,\dots,m:n}(\boldsymbol{t}_m), \quad d = m. && \text{(type ⓘⓘ)} \end{aligned}$$

First, for type ⓘⓘ with $d = m$, the marginal density function $f_{1,\dots,d:n}$ has the form (see (2.3))

$$f_{1,\dots,d:n}(\boldsymbol{t}_d) = \frac{n!}{(n-d)!} \Big(\prod_{j=1}^{d} f(t_j)\Big)(1 - F(t_d))^{n-d}.$$

Similarly, using the representation of the marginal density function $f_{1,\dots,d+1:n}$, we get for type ⓘ

$$f_{1,\dots,d+1:n}(\boldsymbol{t}_d, T) = \frac{n!}{(n-d-1)!} \Big(\prod_{j=1}^{d} f(t_j)\Big) f(T)(1 - F(T))^{n-d-1},$$

so that, after some standard manipulation, the likelihood function for type ⓘ can be written as

$$f^{X_{j:n}, 1 \le j \le d, D_I}(\boldsymbol{t}_d, d) = \frac{n!}{(n-d)!} \Big(\prod_{j=1}^{d} f(t_j)\Big)(1 - F(T))^{n-d}.$$

Thus, the likelihood function can be written in both cases in the form (4.19) with appropriately chosen value w. □

Remark 4.12. **(a)** Recall that $W = X_{m:n} \wedge T$ denotes the test duration of the Type-I hybrid censored experiment.

(b) For $m = n$, we have $D_I = D$ and (4.18) equals the joint density function of Type-I censored data $X_{1:n}, \dots, X_{D:n}$ with sample size D.

4.3.3 Results for exponential distributions

In the following, we will provide some details about the distributions of Type-I hybrid censored order statistics when the population distribution is an exponential distribution with parameters $\vartheta > 0$ and $\mu \in \mathbb{R}$. Clearly, for $d > 0$ and $t_1 \leq \cdots \leq t_m$, the joint density function is given by

$$f^{X_{j:n},1\leq j\leq d,D_1}(\boldsymbol{t}_d, d) = \frac{n!}{(n-d)!\vartheta^d} \exp\Big(-\frac{1}{\vartheta}\sum_{j=1}^{d}(t_j-\mu) - \frac{n-d}{\vartheta}(w-\mu)\Big),$$

$$d \in \{1, \ldots, m\}, \quad (4.20)$$

where $w = t_m \wedge T$. Notice that, for $d \in \{0, \ldots, m\}$,

$$\Pr(D_1 = d) = \int_{\mathcal{W}_d^{\leq}(T)} f^{X_{j:n},1\leq j\leq d,D_1}(\boldsymbol{t}_d, d)\, d\boldsymbol{t}_d. \quad (4.21)$$

Using Lemma 4.4 for the exponential distribution, the above integral reads

$$\Pr(D_1 = d) = \begin{cases} \binom{n}{d}(1-e^{-(T-\mu)/\vartheta})^d e^{-(n-d)(T-\mu)/\vartheta}, & d \in \{0, \ldots, m-1\} \\ \sum_{j=m}^{n}\binom{n}{j}(1-e^{-(T-\mu)/\vartheta})^j e^{-(n-j)(T-\mu)/\vartheta}, & d = m \end{cases}. \quad (4.22)$$

In Cramer and Balakrishnan (2013), the (conditional) joint distribution of the normalized spacings

$$S_{j,n} = (n-j+1)(X_{j:n} - X_{j-1:n}), \quad j = 1, \ldots, D_1, \quad X_{0:n} = \mu,$$

has been obtained given $D_1 = d$, $d \in \{1, \ldots, m\}$. For $d \in \{1, \ldots, m-1\}$, it is given by

$$f^{S_{j,n},1\leq j\leq d|D_1=d}(\boldsymbol{x}_d) = \binom{n}{d}^{-1}\left(1-e^{-(T-\mu)/\vartheta}\right)^{-d}\vartheta^{-d}\exp\Big\{-\sum_{j=1}^{d}\frac{d-j+1}{n-j+1}\frac{x_j}{\vartheta}\Big\},$$

$$\boldsymbol{x}_d \in \mathcal{W}_d^{\leq}(T), \quad (4.23)$$

with support

$$\mathcal{W}_d^{\leq}(T) = \Big\{\boldsymbol{x}_d \mid x_j \geq 0, 1 \leq j \leq d, \sum_{j=1}^{d}\frac{x_j}{n-j+1} \leq T-\mu\Big\}. \quad (4.24)$$

For $D_1 = m$, we get

$$f^{S_{j,n},1\leq j\leq m|D_1=m}(\boldsymbol{x}_m) = f^{S_{j,n},1\leq j\leq m|D\geq m}(\boldsymbol{x}_m)$$

$$= \frac{\vartheta^{-m}}{F_{m:n}(T)}\exp\Big\{-\sum_{j=1}^{m}\frac{x_j}{\vartheta}\Big\}, \quad \boldsymbol{x}_m \in \mathcal{W}_m^{\leq}(T).$$

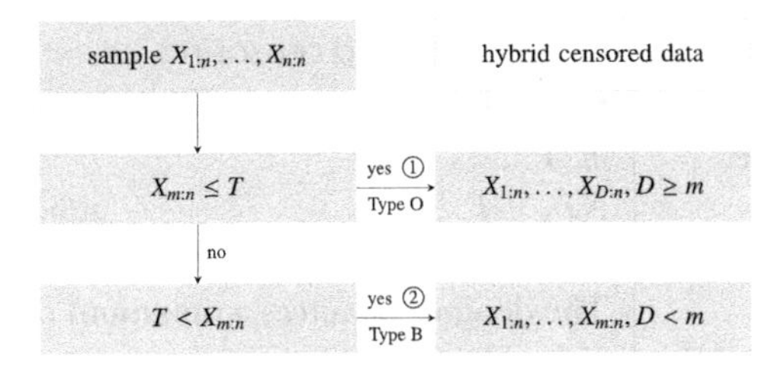

Figure 4.6 Sampling scenarios ①, ② in Type-II hybrid censoring.

Remark 4.13. Notice that the density functions have a product structure in the variables $x_1, x_2, \ldots$. However, the spacings are not independent since the support of the distribution is a simplex! Therefore, Sukhatme's independence result of spacings given in Theorem 2.38 is not true under time truncation. In fact, this illustrates the dependence structure of the spacings under Type-I hybrid censoring. Cramer and Balakrishnan (2013) also established expressions for the one-dimensional marginal density functions and cumulative distribution functions. They are connected to distributions of progressively Type-II censored statistics based on a one-step censoring plan.

4.4. Type-II hybrid censoring

4.4.1 Censoring design

Under the hybrid Type-II censoring scheme, the lifetime experiment is terminated when both $X_{m:n}$ has been observed and, if possible, the threshold T has been exceeded (see Fig. 4.6). Therefore, if the m-th failure time of the units is larger than the time T, then the experiment yields m observations (data scenario ②). If the m-th failure time does not exceed T, then the experiment will finally be stopped at time T. In the latter case, more than m failures may be observed (data scenario ①). The stopping time of the experiment is given by $T^* = X_{m:n} \vee T$. Notice that the experimental time is given by

$$W = X_{m:n} \vee (T \wedge X_{n:n}),$$

since the test will be terminated also when the last object in the life test will already fail before T. Characteristics of Type-II hybrid censoring are summarized in Table 4.4.

4.4.2 Derivation of the joint density function

We consider the complete (ordered) sample

$$X_{1:n}, \ldots, X_{m:n}, \; X_{m+1:n}, \ldots, X_{n:n}$$

Table 4.4 Characteristics of Type-II hybrid censoring.

Type-II hybrid censoring	Expression	Properties
Scheme parameters	m, T	
Termination time	$X_{m:n} \vee T$	unbounded
Sample size D_{II}	$D \vee m$	$D_{\text{II}} \in \{m, \dots, n\}$
	The design guarantees a minimum of m observations.	
Experimental time W	$X_{m:n} \vee (X_{n:n} \wedge T)$	

to analyze the Type-II hybrid censoring scheme. The (random) sample size of the Type-II hybrid censored data is given by the random counter

$$D_{\text{II}} = \Upsilon_{\text{II}}(D) = \max\{D, m\} = D \vee m = m + \sum_{j=m+1}^{n} \mathbb{1}_{(-\infty,T]}(X_{j:n}) \tag{4.25}$$

which has support $\{m, m+1, \dots, n\}$. Notice that

$$\begin{aligned}
\{D_{\text{II}} = m\} &= \{D = m\} \cup \{D < m\} = \{X_{m:n} \le T < X_{m+1:n}\} \cup \{X_{m:n} > T\}, \\
\{D_{\text{II}} = d\} &= \{D = d\} = \{X_{d:n} \le T < X_{d+1:n}\}, \quad d \in \{m+1, \dots, n-1\}, \\
\{D_{\text{II}} = n\} &= \{D = n\} = \{X_{n:n} \le T\}.
\end{aligned}$$

Thus, the event $\{D_{\text{II}} = m\}$ is composed of the disjoint events $\{D = m\}$ and $\{D < m\}$ which illustrates that the sample size $D_{\text{II}} = m$ may result from two different sampling situations. The Type-II hybrid censored sample is given by

$$X_{1:n}, \dots, X_{D_{\text{II}}:n}. \tag{4.26}$$

Since D has a binomial distribution with parameters n and $F(T)$, we get from (4.4) the probabilities $\Pr(D = d)$, $d \in \{m, \dots, n\}$, and

$$\Pr(D < m) = \Pr(X_{m:n} > T) = 1 - F_{m:n}(T).$$

Therefore, we get by analogy with Lemma 4.4 for the Type-I hybrid censoring design the following lemma. Notice that $F_{m:n}(T) = F_{m+1:n}(T) + \binom{n}{m} F^m(T)(1 - F(T))^{n-m}$.

Lemma 4.14. *The random counter D_{II} has a discrete distribution with support $\{m, \dots, n\}$ and probability mass function*

$$\Pr(D_{\text{II}} = d) = \begin{cases} 1 - F_{m+1:n}(T), & d = m \\ \binom{n}{d} F^d(T)(1 - F(T))^{n-d}, & d \in \{m+1, \dots, n\} \end{cases}.$$

As for Type-I hybrid censoring, one can consider the mean of the sample size as well as of the experimental time. They are given in the following lemma. The mean of D_{II} follows directly from Lemma 4.4.

Lemma 4.15. **(a)** $ED_{II} = E(D \vee m) = m + nF(T) - E(D \wedge m)$.
(b) $EW = EX_{m:n} - E(X_{m:n} \wedge T) + E(X_{n:n} \wedge T)$ *where* $E(X_{j:n} \wedge T)$, $j \in \{m, n\}$, *can be taken from Lemma 4.5.*

Proof. First, we get from Lemma 4.5 and from $D \sim bin(n, F(T))$:

$$\begin{aligned} E(D \vee m) &= \sum_{d=0}^{n} (d \vee m) \Pr(D = d) = m \sum_{d=0}^{m-1} \Pr(D = d) + \sum_{d=m}^{n} d \Pr(D = d) \\ &= m(1 - F_{m:n}(T)) + \sum_{d=0}^{n} d \Pr(D = d) - \sum_{d=0}^{m-1} d \Pr(D = d) \\ &= m + nF(T) - E(D \wedge m). \end{aligned}$$

Furthermore, the mean of the test duration $W = X_{m:n} \vee (T \wedge X_{n:n})$ results from the following calculation. First, notice that

$$X_{m:n} \vee (T \wedge X_{n:n}) = X_{m:n} \mathbb{1}_{(-\infty, X_{m:n}]}(T) + T \mathbb{1}_{(X_{m:n}, X_{n:n}]}(T) + X_{n:n} \mathbb{1}_{(X_{n:n}, \infty)}(T).$$

Then, we obtain

$$\begin{aligned} E[X_{m:n} \vee (T \wedge X_{n:n})] &= E\big(X_{m:n} \mathbb{1}_{(-\infty, X_{m:n}]}(T) + T \mathbb{1}_{(X_{m:n}, X_{n:n}]}(T) + X_{n:n} \mathbb{1}_{(X_{n:n}, \infty)}(T)\big) \\ &= \int_T^\infty t f^{X_{m:n}}(t)\, dt + T \Pr(X_{m:n} < T \le X_{n:n}) + \int_{-\infty}^T t f^{X_{n:n}}(t)\, dt \\ &= EX_{m:n} - \int_{-\infty}^T t f^{X_{m:n}}(t)\, dt + T(F_{m:n}(T) - F_{n:n}(T)) + \int_{-\infty}^T t f^{X_{n:n}}(t)\, dt \\ &= EX_{m:n} - E(X_{m:n} \wedge T) + E(X_{n:n} \wedge T). \end{aligned}$$

□

Remark 4.16. Using (4.6), one gets for an $Exp(\vartheta)$-distribution the identity

$$ED_{II} = m + nF(T/\vartheta) - \sum_{d=1}^{m-1} F_{d:n}(T/\vartheta) = m + \sum_{d=m+1}^{m} F_{d:n}(T/\vartheta). \tag{4.27}$$

The joint cumulative distribution function of $(X_{1:n}, \dots, X_{D_{II}:n})$ as in (4.26) can be derived using the results obtained for Type-I hybrid censoring (see also Cramer et al., 2016). As for Type-I hybrid censoring, we apply the law of total probability and get

$$\begin{aligned} \Pr(X_{j:n} \le t_j, 1 \le j \le D_{II}) = \sum_{d=m}^{n} \Pr(X_{j:n} \le t_j, 1 \le j \le d, D = d) \\ + \Pr(X_{j:n} \le t_j, 1 \le j \le m, D < m). \end{aligned} \tag{4.28}$$

Let $t_1, \ldots, t_n \in \mathbb{R}$. Thus, we have to calculate in the probabilities

$$\Pr(X_{j:n} \le t_j, 1 \le j \le m, D < m), \quad \Pr(X_{j:n} \le t_j, 1 \le j \le d, D = d), d \in \{m, \ldots, n\}.$$

For $d \in \{m, \ldots, n-1\}$, we get as in (4.13)

$$\Pr(X_{j:n} \le t_j, 1 \le j \le d, D = d) = F_{1,\ldots,d:n}(\boldsymbol{t}_d \wedge T) - F_{1,\ldots,d+1:n}(\boldsymbol{t}_d \wedge T, T).$$

For $d = n$, the expression reads

$$\Pr(X_{j:n} \le t_j, 1 \le j \le n, D = n) = F_{1,\ldots,n:n}(\boldsymbol{t}_n \wedge T).$$

Moreover, we have

$$\Pr(X_{j:n} \le t_j, 1 \le j \le m, D < m) = F_{1,\ldots,m:n}(\boldsymbol{t}_m) - F_{1,\ldots,m:n}(\boldsymbol{t}_m \wedge T).$$

Then, we get from (4.28) the desired representation of the joint cumulative distribution function

$$\begin{aligned}\Pr(X_{j:n} \le t_j, 1 \le j \le D_{\mathrm{II}}) &= F_{1,\ldots,m:n}(\boldsymbol{t}_m) - F_{1,\ldots,m:n}(\boldsymbol{t}_m \wedge T) \\ &+ \sum_{d=m}^{n-1} \Big(F_{1,\ldots,d:n}(\boldsymbol{t}_d \wedge T) - F_{1,\ldots,d+1:n}(\boldsymbol{t}_d \wedge T, T)\Big) + F_{1,\ldots,n:n}(\boldsymbol{t}_n \wedge T)\end{aligned}$$

By analogy with (4.14), this can be written as

$$= F_{1,\ldots,m:n}(\boldsymbol{t}_m) - F_{1,\ldots,m:n}(\boldsymbol{t}_m \wedge T) + \sum_{d=m}^{n} \binom{n}{d} (1 - F(T))^{n-d} F_{1,\ldots,d:d}(\boldsymbol{t}_d \wedge T).$$

Notice that, for $T \le t_1 \le \cdots \le t_m$, we have $F_{1,\ldots,d:d}(\boldsymbol{t}_d \wedge T) = F^d(T)$ and $F_{1,\ldots,m:n}(\boldsymbol{t}_m \wedge T) = F_{1,\ldots,m:n}(T^{*m}) = \sum_{d=m}^{n} \binom{n}{d} (1 - F(T))^{n-d} F^d(T)$, so that in this case

$$\Pr(X_{j:n} \le t_j, 1 \le j \le D_{\mathrm{II}}) = F_{1,\ldots,m:n}(\boldsymbol{t}_m).$$

If $t_1 \le \cdots \le t_m \le T$, we get

$$\Pr(X_{j:n} \le t_j, 1 \le j \le D_{\mathrm{II}}) = \sum_{d=m}^{n} \binom{n}{d} (1 - F(T))^{n-d} F_{1,\ldots,d:d}(\boldsymbol{t}_d \wedge T).$$

The conditional density function $f^{X_{1:n},\ldots,X_{d:n}|D=d}$, $d = m, \ldots, n$, can be directly obtained from Cramer and Balakrishnan (2013):

$$\begin{aligned} f^{X_{j:n}, 1 \le j \le d | D=d}(\boldsymbol{t}_d) &= f_{1,\ldots,d:n}(\boldsymbol{t}_d \mid X_{d+1:n} = T), \quad d = m, \ldots, n-1, \\ f^{X_{j:n}, 1 \le j \le n | D=n}(\boldsymbol{t}_n) &= \frac{f_{1,\ldots,n:n}(\boldsymbol{t}_n)}{F_{n:n}(T)} \mathbb{1}_{(-\infty, T]}(t_n), \end{aligned}$$

with $\boldsymbol{t}_n \in \mathbb{R}^n$. For $D < m$, we get the conditional density function

$$f^{X_{j:n},1\le j\le m|D<m}(\boldsymbol{t}_m) = \frac{f_{1,\ldots,m:n}(\boldsymbol{t}_m)}{1-F_{m:n}(T)} \mathbb{1}_{[T,\infty)}(t_m)$$

which belongs to a left truncated distribution.

For $D_{\text{II}} = m$, the joint density function of $X_{j:n}$, $1 \le j \le m$, $\{D_{\text{II}} = m\}$ can be written by using the same arguments as in the proof of Theorem 4.11 as

$$\begin{aligned} f^{X_{j:n},1\le j\le m,D_{\text{II}}}(\boldsymbol{t}_m, m) &= f^{X_{j:n},1\le j\le m,D<m}(\boldsymbol{t}_m)\mathbb{1}_{(T,\infty)}(t_m) + f^{X_{j:n},1\le j\le m,D}(\boldsymbol{t}_m, m)\mathbb{1}_{(-\infty,T]}(t_m) \\ &= \frac{n!}{(n-m)!}\Big(\prod_{j=1}^{m} f(t_j)\Big)(1-F(t_m \vee T))^{n-m} \end{aligned} \tag{4.29}$$

Furthermore, for $d \in \{m+1, \ldots, n\}$ and $t_1 \le \cdots \le t_n$, the joint density function is given by

$$\begin{aligned} f^{X_{j:n},1\le j\le d,D_{\text{II}}}(\boldsymbol{t}_d, d) &= f^{X_{j:n},1\le j\le d|D_{\text{II}}=d}(\boldsymbol{t}_d \mid d)\Pr(D_{\text{II}} = d) \\ &= \begin{cases} \frac{1-F(T)}{(n-d)f(T)} f_{1,\ldots,d+1:n}(\boldsymbol{t}_d, T), & d \in \{m+1, \ldots, n-1\} \\ f_{1,\ldots,n:n}(\boldsymbol{t}_n), & d = n \end{cases} \\ &= \frac{n!}{(n-d)!}\Big(\prod_{j=1}^{d} f(t_j)\Big)(1-F(w))^{n-d}, \quad d \in \{m+1, \ldots, n\}, \end{aligned} \tag{4.30}$$

where $w = t_d \vee (T \wedge t_n)$. The last equality follows as in the proof of Theorem 4.11. Therefore, combining (4.29) and (4.30), we find that (4.30) holds for $d \in \{m, \ldots, n\}$.

4.4.3 Results for exponential distributions

As for Type-I hybrid censoring, we consider $Exp(\mu, \vartheta)$-distributions. Then, for $d \in \{0, \ldots, n\}$ as realization of $D_{\text{II}} = D \vee m$, and $t_1 \le \cdots \le t_n$, the joint density function is given by

$$f^{X_{j:n},1\le j\le d,D_{\text{II}}}(\boldsymbol{t}_d, d) = \frac{n!}{(n-d)!\vartheta^d}\exp\Big(-\frac{1}{\vartheta}\sum_{j=1}^{d}(t_j-\mu) - \frac{n-d}{\vartheta}(w-\mu)\Big), \quad d \in \{m, \ldots, n\}. \tag{4.31}$$

In Cramer et al. (2016), the (conditional) joint distribution of the spacings

$$S_{j,n} = (n-j+1)(X_{j:n} - X_{j-1:n}),\ j = m, \ldots, D_{\text{II}}, \quad X_{0:n} = \mu,$$

has been obtained given $D = d$, $d \in \{m, \ldots, n\}$, and $D < m$, respectively.

First, we consider the case $d \in \{m, \ldots, n\}$. Then, the joint density function of $X_{1:n}, \ldots, X_{d:n}$ given $D = d$ can be directly taken from Cramer et al. (2016). Using the representations of the conditional joint density function in (4.31) and the density transformation theorem, we get directly the (conditional) joint density function of the normalized spacings (see also (4.23)), that is,

$$f^{S_{j,n},1\leq j\leq d|D=d}(\boldsymbol{x}_d)$$

$$= \binom{n}{d}^{-1} \left(1 - e^{-(T-\mu)/\vartheta}\right)^{-d} \vartheta^{-d} \exp\left\{-\sum_{j=1}^{d} \frac{d-j+1}{n-j+1} \frac{x_j}{\vartheta}\right\}, \quad \boldsymbol{x}_d \in \mathcal{W}_d^{\leq}(T),$$

with support $\mathcal{W}_d^{\leq}(T)$ as in (4.24). For $D < m$, we get

$$f^{S_{j,n},1\leq j\leq m|D<m}(\boldsymbol{x}_m) = \frac{\vartheta^{-m}}{1 - F_{m:n}(T)} \exp\left\{-\sum_{j=1}^{m} \frac{x_j}{\vartheta}\right\}, \quad \boldsymbol{x}_m \in \mathcal{W}_m^{>}(T),$$

where

$$\mathcal{W}_m^{>}(T) = \left\{\boldsymbol{x}_m \mid x_j \geq 0, 1 \leq j \leq m, \sum_{j=1}^{m} \frac{x_j}{n-j+1} > T - \mu\right\}.$$

Notice that $\mathcal{W}_m^{>}(T)$ is unbounded whereas $\mathcal{W}_m^{\leq}(T)$ is bounded.

From these expressions, marginal cumulative distribution functions of the spacings can be obtained, too (see Cramer et al., 2016).

4.5. Further hybrid censoring schemes

Based on the above basic hybrid censoring schemes, further hybrid censoring schemes have been proposed (see, e.g., Balakrishnan and Kundu, 2013; Górny and Cramer, 2018b). In the following, we present the most popular ones, the corresponding censoring mechanism as well as the resulting distributions. First, we illustrate a general approach called *modularization of hybrid censoring schemes* in order to derive conveniently the distributions of the hybrid censored samples. It provides a powerful tool to analyze the structure of hybrid censoring schemes so that we do not have to derive the distribution from scratch. Furthermore, it shows impressively that hybrid censoring schemes are composed of a few basic modules. Notice that we propose a slightly different decomposition than that established in Górny and Cramer (2018b).

4.5.1 Modularization of hybrid censoring schemes

Górny and Cramer (2018b) have developed a decomposition of hybrid censoring schemes in basic components which can be used to determine the distribution of various kinds of hybrid censored data. This idea will enable us to present the joint density functions of the observed data for the particular hybrid censoring schemes with little effort. Furthermore, it is also useful in estimation problems since, e.g., the likelihood function is directly related to the module type and the corresponding sampling scenario.

First, we illustrate this approach for Type-I and Type-II hybrid censoring (see density functions in (4.18) and (4.30)), respectively. Considering the random counter D introduced in (4.2), the joint density of the hybrid censored data has a different representation w.r.t. the values of D, that is,

(a) for Type-I hybrid censoring, we have either $D \geq m$ ① or $D = d$ with $d \in \mathfrak{D}_{\mathsf{I}} = \{1, \ldots, m-1\}$ in ② (see Fig. 4.4),

(b) for Type-II hybrid censoring, we have either $D = d$ with $d \in \mathfrak{D}_{\mathsf{II}} = \{m, \ldots, n\}$ in ①, or $D < m$ in ② (see Fig. 4.6).

Depending on the counter D, we have the module type O (with sample size $D = d$, d in some set $\mathfrak{D}_{\mathsf{HCS}}$) and, additionally, one of the module types A and B, respectively, where the sample size is fixed and equal to m. However, the random counter D has to satisfy either $D \geq m$ or $D < m$. Using the law of total probability (see (4.12)), the joint cumulative distribution functions can be written as[4]

$$\begin{aligned}
\text{Type-I HCS: } \Pr(X_{j:n} \wedge T \leq t_j, 1 \leq j \leq D_{\mathsf{I}}) &= \Pr(X_{j:n} \leq t_j, 1 \leq j \leq m, D \geq m) \\
&\quad + \sum_{d=1}^{m-1} \Pr(X_{j:n} \leq t_j, 1 \leq j \leq d, D = d), \\
\text{Type-II HCS: } \Pr(X_{j:n} \leq t_j, 1 \leq j \leq D_{\mathsf{II}}) &= \sum_{d=m}^{n} \Pr(X_{j:n} \leq t_j, 1 \leq j \leq d, D = d) \\
&\quad + \Pr(X_{j:n} \leq t_j, 1 \leq j \leq m, D < m).
\end{aligned}$$

Therefore, for Type-II hybrid censoring, the events in the probabilities with $d \in \mathfrak{D}_{\mathsf{II}} = \{m, \ldots, n\}$ have the same structure as those for $d \in \mathfrak{D}_{\mathsf{I}} = \{0, \ldots, m-1\}$ under Type-I hybrid censoring. For Type-I hybrid censoring, one has to study additionally the probability $\Pr(X_{j:n} \leq t_j, 1 \leq j \leq m, D \geq m)$ whereas, under Type-II censoring, we have to evaluate $\Pr(X_{j:n} \leq t_j, 1 \leq j \leq m, D < m)$.

Based on this observation, Górny and Cramer (2018b) have pointed out that the hybrid censoring schemes discussed so far in the literature are composed of different types of modules (or data scenarios). In fact, they identified three different types, but, since the random counter is defined slightly different in our presentation, we get four different module types leading to a simpler structure. In particular, these types are given by

[4] Under Type-I hybrid censoring, we exclude here the case of an empty sample, that is, the case $D_{\mathsf{I}} = 0$. Notice that inference is usually conducted under the condition that $D_{\mathsf{I}} \geq 1$ so that one is interested in the conditional joint cumulative distribution function.

Module type O: $X_{1:n}, \ldots, X_{D:n}$, $D = d$ with d in some set $\mathfrak{D}_{\mathsf{HCS}} \subseteq \{1, \ldots, n\}$,
Module type A: $X_{1:n}, \ldots, X_{m^\star:n}$, $D \geq m^\star$ for some $m^\star$,
Module type B: $X_{1:n}, \ldots, X_{m^\star:n}$, $D < m^\star$ for some $m^\star$,
and a fourth one which appears when at least two thresholds $T_i < T_{i+1}$ are involved in the construction of the hybrid censoring scheme (so-called **module type AB)**, see p. 145. Module type AB will occur for the first time in generalized Type-II hybrid censoring and will therefore only be introduced in Section 4.5.3. In Figs. 4.1, 4.2, 4.4 and 4.6–4.12, the module type is assigned to the respective sampling situation. Obviously, this illustrates the usefulness of the decomposition and yields easily the respective distributions.

4.5.2 Generalized Type-I hybrid censoring

Generalized hybrid censoring schemes have been proposed in Chandrasekar et al. (2004) in order to temper the drawbacks of the basic Type-I and Type-II hybrid censoring schemes. In particular, additional scheme parameters have been introduced. Extending the Type-I censoring scheme, the generalized Type-I hybrid censoring scheme guarantees the observation of a minimum quantity of k failures after stopping the experiment. In addition to the threshold-time $T > 0$ and the maximum number of observations m, a prefixed value $k \in \{1, \ldots, m-1\}$ is introduced. The *generalized Type-I hybrid censoring scheme* stops the life test at $X_{k:n}$ provided that the k-th failure $X_{k:n}$ occurs after time T. If the k-th failure will be observed before T, the experiment will be terminated at $\min\{T, X_{m:n}\} = X_{m:n} \wedge T$ as in Type-I hybrid censoring. Therefore, the stopping-time can be expressed as

$$T^* = (X_{k:n} \vee T) \wedge X_{m:n} \quad \text{with } k < m \text{ and } T > 0$$

showing that $T^* \leq X_{m:n}$. T^* equals the test duration W and can be written as

$$T^* = W = (X_{m:n} \wedge T) \vee X_{k:n}$$

which clearly reveals that $W = T^* \geq X_{k:n}$. As before, the random counter D denotes the number of observed failures till time T, that is, $D = \sum_{j=1}^{n} \mathbb{1}_{(-\infty,T]}(X_{j:n})$ as in (4.8). Then, the possibly occurring sampling situations are depicted in Fig. 4.7. Clearly, generalized Type-I hybrid censoring can be interpreted as a direct extension of Type-II hybrid censoring choosing $m = n$. Notice that this interpretation also explains the similar properties of these hybrid censoring schemes taken from Tables 4.4 and 4.5, respectively.

The observed sample size and the data are given by $D_{\mathsf{gI}} = (D \vee k) \wedge m$ and

$$X_{1:n}, \ldots, X_{D_{\mathsf{gI}}:n},$$

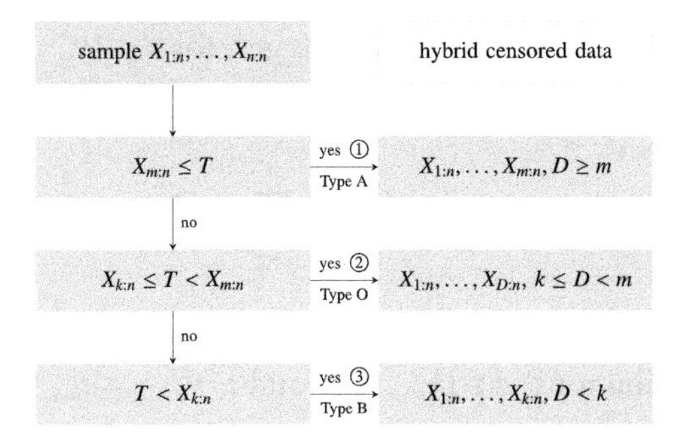

Figure 4.7 Sampling scenarios ①-③ in generalized Type-I hybrid censoring.

Table 4.5 Characteristics of generalized Type-I hybrid censoring.

generalized Type-I hybrid censoring	**Expression**	**Properties**
Scheme parameters	k, m with $k < m$, T	
Termination time	$(X_{k:n} \vee T) \wedge X_{m:n}$	unbounded
Sample size D_{gI}	$(k \vee D) \wedge m$	$D_{\mathrm{gI}} \in \{k, \dots, m\}$
	The design guarantees a minimum of k observations, the maximum number of observations equals m.	
Experimental time W	$X_{k:n} \vee (X_{m:n} \wedge T)$	

respectively. This shows that at least k failures will be observed. In the second case, that is, for $k \le D < m$, the life test is terminated at T (see Fig. 4.7). The other two scenarios correspond to Type-II censoring with k and m, respectively (conditionally on D).

Using the modularization approach, the joint cumulative distribution function can be written as

$$\begin{aligned}\Pr(X_{j:n} \le t_j, 1 \le j \le D_{\mathrm{gI}}) &= \Pr(X_{j:n} \le t_j, 1 \le j \le k, D < k)\\ &\quad + \sum_{d=k}^{m-1} \Pr(X_{j:n} \le t_j, 1 \le j \le d, D = d) + \Pr(X_{j:n} \le t_j, 1 \le j \le m, D \ge m).\end{aligned}$$

Using the expressions derived previously for the different probabilities, we get immediately the desired representation. Notice that, for module type O, the summation stops at $d = m - 1$. In particular, the joint density function reads ($D_{\mathrm{gI}} = d^* = (k \vee d) \wedge m, D = d$)

$$\begin{aligned} f^{X_{j:n}, 1 \le j \le d^*, D}(\boldsymbol{t}_{d^*}, d) &= f^{X_{j:n}, 1 \le j \le d^* | D = d}(\boldsymbol{t}_{d^*}) \Pr(D = d)\\ &= \begin{cases} f_{1,\dots,k:n}(\boldsymbol{t}_k), & d < k\\ \frac{1-F(T)}{(n-d)f(T)} f_{1,\dots,d+1:n}(\boldsymbol{t}_d, T), & d \in \{k, \dots, m-1\}\\ f_{1,\dots,m:n}(\boldsymbol{t}_m), & d \ge m \end{cases}. \end{aligned} \tag{4.32}$$

By analogy with Type-II hybrid censoring, we get, for $d \in \{k, \ldots, m\}$ and $t_1 \leq \cdots \leq t_m$, the joint density function as

$$f^{X_{j:n}, 1 \leq j \leq d, D_{\mathrm{gI}}}(\boldsymbol{t}_d, d) = \frac{n!}{(n-d)!} \Big(\prod_{j=1}^{d} f(t_j) \Big) (1 - F(w))^{n-d}, \tag{4.33}$$

where $w = t_d \vee (T \wedge t_m)$.

For $Exp(\mu, \vartheta)$-distributions, $d \in \{k, \ldots, m\}$ and $t_1 \leq \cdots \leq t_m$, we get (see Górny and Cramer, 2016)

$$f^{X_{j:n}, 1 \leq j \leq D_{\mathrm{gI}}, D_{\mathrm{gI}}}(\boldsymbol{t}_d, d) = \frac{n!}{(n-k)! \vartheta^k} \exp\Big(-\frac{1}{\vartheta} \sum_{j=1}^{k} (t_j - \mu) - \frac{n-k}{\vartheta} (w - \mu) \Big), \tag{4.34}$$

where $w = t_d \vee (T \wedge t_m)$.

4.5.3 Generalized Type-II hybrid censoring

For a previously fixed $m \in \{1, \ldots, n\}$, we consider instead of one threshold time two threshold times $0 < T_1 < T_2$. The *generalized Type-II hybrid censoring scheme* proceeds as follows: Assuming that the m-th failure has been observed before time T_1, the experiment will be stopped at T_1. If the m-th failure occurs between the threshold times T_1 and T_2, we terminate the experiment after the m-th failure. If less than m failures have been occurred until T_2, then the experiment will be terminated at T_2, which guarantees that the test duration does not exceed T_2.

After observing the m-th failure before time T_1, the experiment continues until T_1, which implies that in this particular scenario more than m failures can be observed. The following test duration W accomplishes the intended behavior of generalized Type-II hybrid censoring and is given by

$$W = (X_{m:n} \wedge T_2) \vee (T_1 \wedge X_{n:n}) = [X_{m:n} \vee (T_1 \wedge X_{n:n})] \wedge T_2 \quad \text{with } 0 < T_1 < T_2.$$

As in (4.2), two random counters D_1 and D_2 are defined as

$$D_i = D(T_i) = \sum_{j=1}^{n} \mathbb{1}_{(-\infty, T_i]}(X_{j:n}), \quad i = 1, 2, \tag{4.35}$$

which denote the number of observed failures until T_1 and T_2, respectively. The corresponding sampling situations are presented in Fig. 4.8. Characteristics are summarized in Table 4.6. Comparing this hybrid censoring scheme to Type-I hybrid censoring, it turns out that generalized Type-I hybrid censoring can be seen as an extension of Type-I hybrid censoring assuming $T_1 = 0$ (or, in general, as the left endpoint of support of F).

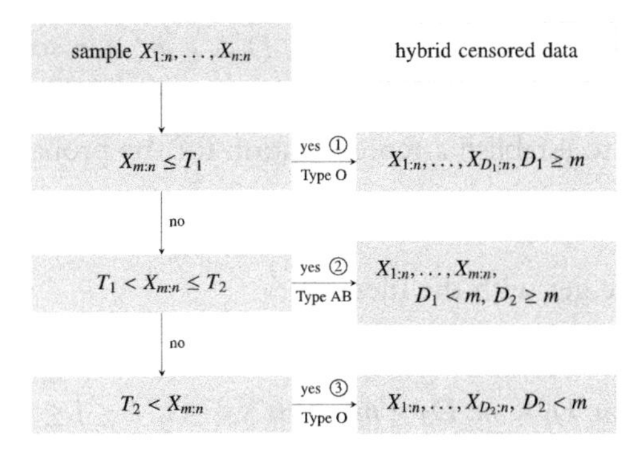

Figure 4.8 Sampling scenarios ①-③ in generalized Type-II hybrid censoring.

Table 4.6 Characteristics of generalized Type-II hybrid censoring.

generalized Type-II hybrid censoring	**Expression**	**Properties**
Scheme parameters	m, $T_1 < T_2$	
Termination time	$(X_{m:n} \vee T_1) \wedge T_2$	bounded by T_2
Sample size D_{gII}	$(D_1 \vee m) \wedge D_2$	$D_{\text{gII}} \in \{0, \ldots, n\}$
	The sample size may be zero with positive probability.	
Experimental time W	$[X_{m:n} \vee (T_1 \wedge X_{n:n})] \wedge T_2$	

The first and third setting correspond to censoring modules to module type O with random sample sizes D_1 and D_2, respectively. However, the second situation is different and involved both random counters D_1 and D_2. Therefore, the case $T_1 \leq X_{m:n} \leq T_2$ needs further analysis. However, the sample can be interpreted as a Type-II censored sample conditioned on $D_1 < m$ and $D_2 \geq m$ by taking into account the equivalence $D_1 \in \{0, \ldots, m-1\} \iff D_1 < m$.

Using the above mentioned modularization and the law of total probability, the joint cumulative distribution function can be written in the form

$$\begin{aligned}\Pr(X_{j:n} \leq t_j, 1 \leq j \leq D_{\text{gII}}) &= \sum_{d=m}^{n} \Pr(X_{j:n} \leq t_j, 1 \leq j \leq d, D_1 = d) \\ &+ \Pr(X_{j:n} \leq t_j, 1 \leq j \leq m, D_1 < m, D_2 \geq m) \\ &+ \sum_{d=0}^{m-1} \Pr(X_{j:n} \leq t_j, 1 \leq j \leq d, D_2 = d).\end{aligned}$$

The terms in the two sums correspond to module type O (with different counters of course) whereas the middle one is of module type AB, that is,

Module type AB: $X_{1:n}, \dots, X_{m^\star:n}$, $D_i < m^\star$, $D_{i+1} \geq m^\star$ for some $m^\star$ and thresholds $T_i < T_{i+1}$ with counters $D_j = D(T_j)$, $j \in \{i, i+1\}$, as in (4.35).

Therefore, we have to establish a representation for the probability

$$\Pr(X_{j:n} \leq t_j, 1 \leq j \leq m, D_1 < m, D_2 \geq m).$$

Proceeding as above, we get with the identity

$$\{D_1 < m\} \cap \{D_2 \geq m\} = \{T_1 \leq X_{m:n} \leq T_2\}:$$

$$\begin{aligned}
\Pr(X_{j:n} \leq t_j, 1 \leq j \leq m, D_1 < m, D_2 \geq m) &= \Pr(X_{j:n} \leq t_j, 1 \leq j \leq m, T_1 \leq X_{m:n} \leq T_2) \\
&= \Pr(X_{j:n} \leq t_j, 1 \leq j \leq m-1, T_1 \leq X_{m:n} \leq t_m \wedge T_2) \\
&= F_{1,\dots,m:n}(\boldsymbol{t}_m \wedge T_2) - F_{1,\dots,m:n}(\boldsymbol{t}_m \wedge T_1).
\end{aligned}$$

These results enable us to present the joint density function ($D_{\mathsf{gII}} = d^* = (d_1 \vee m) \wedge d_2, D_1 = d_1, D_2 = d_2$):

$$f^{X_{j:n}, 1 \leq j \leq d, D_1, D_2}(\boldsymbol{t}_d, d_1, d_2) = f^{X_{j:n}, 1 \leq j \leq d* | D_1 = d_1, D_2 = d_2}(\boldsymbol{t}_d \mid d) \Pr(D_1 = d_1, D_2 = d_2)$$

$$= \begin{cases} \frac{1-F(T_1)}{(n-d_1)f(T_1)} f_{1,\dots,d_1+1:n}(\boldsymbol{t}_{d_1}, T_1), & m \leq d_1 < n \\ f_{1,\dots,n:n}(\boldsymbol{t}_n), & d_1 = n \\ f_{1,\dots,m:n}(\boldsymbol{t}_m), & d_1 < m \leq d_2 \\ \frac{1-F(T_2)}{(n-d_2)f(T_2)} f_{1,\dots,d_2+1:n}(\boldsymbol{t}_{d_2}, T_2), & d_2 < m \end{cases}. \tag{4.36}$$

By analogy with Type-II hybrid censoring, we get, for $D_{\mathsf{gII}} = d \in \{0, \dots, n\}$ and $t_1 \leq \dots \leq t_n$, the joint density function as

$$f^{X_{j:n}, 1 \leq j \leq d, D_{\mathsf{gII}}}(\boldsymbol{t}_d, d) = \frac{n!}{(n-d)!} \Big(\prod_{j=1}^{d} f(t_j) \Big) (1 - F(w))^{n-d}, \tag{4.37}$$

where w is given in Table 4.7. Recalling the likelihood functions for the hybrid censoring schemes introduced in this chapter, we get the cases given in Table 4.7. Hence, the choice of w is completely determined by the value of the observed sample size d. A similar table can be established for the unified hybrid censoring schemes (see Table 6.5).

For $Exp(\mu, \vartheta)$-distributions, $d_2 > 0$ and $t_1 \leq \dots \leq t_m$, we get (see Górny and Cramer, 2016)

$$f^{X_{j:n}, 1 \leq j \leq d, D_{\mathsf{gII}}}(\boldsymbol{t}_d, d) = \frac{n!}{(n-d)! \vartheta^d} \exp\Big(-\frac{1}{\vartheta} \sum_{j=1}^{d} (t_j - \mu) - \frac{n-d}{\vartheta}(w - \mu) \Big),$$

$$d \in \{0, \dots, m\} \tag{4.38}$$

with w taken from Table 4.7.

Table 4.7 Types of likelihood functions for (hybrid) censoring schemes. d denotes the realization of the observed sample size D^*.

censoring scheme	joint density function	parameters	
		d	w
Type-I		$1, \dots, n-1$	T
		n	x_n
Type-II		m	x_m
Type-I hybrid	(4.18)	$1, \dots, m-1$	T
		m	x_m
Type-II hybrid	(4.30)	m	x_m
		$m+1, \dots, n-1$	T
		n	x_n
generalized Type-I hybrid	(4.32)	m	x_m
		$m+1, \dots, k-1$	T
		k	x_k
generalized Type-II hybrid	(4.36)	$m+1, \dots, n-1$	T_1
		n	x_n
		m	x_m
		$1, \dots, m-1$	T_2

4.5.4 Unified hybrid censoring schemes

Several authors have proposed extensions of the previous hybrid censoring schemes called *unified hybrid censoring*. In the following, we explain the details of these hybrid censoring schemes and provide some characteristics which are important in the statistical analysis. For details and proofs, we refer to the mentioned references as well as to the surveys provided by Górny and Cramer (2018b) and Górny (2017). The joint distribution of the sample can be obtained by analogy with the previous results by considering the module types O, A, B, and AB. This implies representations of the density functions as well as the respective likelihood functions (particularly in the case of exponential distributions). They can be assembled using the respective cases presented for the preceding hybrid censoring schemes (see (4.20), (4.31), (4.34), and (4.38)). This provides a convenient method to establish the likelihood functions for quite complicated hybrid censoring schemes. Therefore, we do not provide details for the following schemes. Expressions for the likelihood function can be directly obtained as special cases from those provided for the progressive censoring versions presented in (7.13)–(7.14) (see p. 218) by specifying the parameters appropriately.

Unified Type-I hybrid censoring

The unified Type-I hybrid censoring scheme has originally been proposed by Huang and Yang (2010) who called it *combined hybrid censoring sampling*. It has also been ad-

Table 4.8 Characteristics of unified Type-I hybrid censoring.

unified Type-I hybrid censoring	Expression	Properties
Scheme parameters	k, m with $k < m$, $T_1 < T_2$	
Termination time	$(X_{k:n} \wedge T_2) \vee (X_{m:n} \wedge T_1)$	bounded by T_2
Sample size D_{uI}	$(D_1 \wedge m) \vee (D_2 \wedge k)$	$D_{\text{uI}} \in \{0, \dots, m\}$
	The sample size may be zero with positive probability.	
Experimental time W	$(X_{k:n} \wedge T_2) \vee (X_{m:n} \wedge T_1)$	

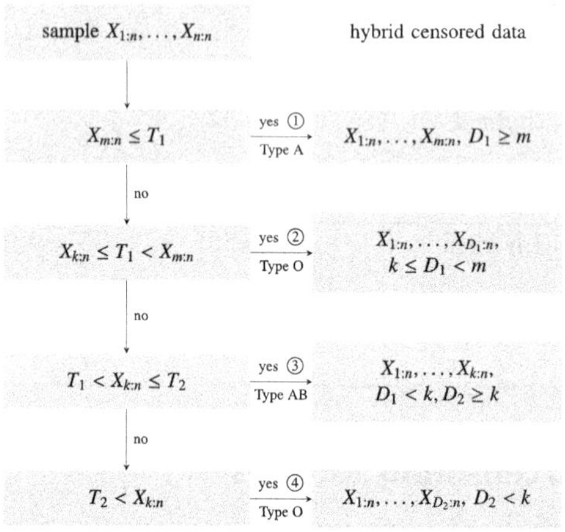

Figure 4.9 Sampling scenarios ①-④ in unified Type-I hybrid censoring.

dressed in Park and Balakrishnan (2012). The details of the hybrid censoring scheme are depicted in Table 4.8 and Fig. 4.9.

Unified Type-II hybrid censoring

The unified Type-II hybrid censoring scheme has originally been proposed by Balakrishnan et al. (2008d). The censoring scheme has also been discussed in Habibi Rad and Izanlo (2011) and Panahi and Sayyareh (2015). The details of the hybrid censoring scheme are depicted in Table 4.9 and Fig. 4.10.

Unified Type-III hybrid censoring

The unified Type-III hybrid censoring scheme has originally been proposed by Park and Balakrishnan (2012). The details of the hybrid censoring scheme are depicted in Table 4.10 and Fig. 4.11

Table 4.9 Characteristics of unified Type-II hybrid censoring.

unified Type-II hybrid censoring	Expression	Properties
Scheme parameters	k, m with $k < m$, $T_1 < T_2$	
Termination time	$(X_{k:n} \vee T_2) \wedge (X_{m:n} \vee T_1)$	unbounded
Sample size D_{uII}	$(D_1 \vee m) \wedge (D_2 \vee k)$	$D_{\text{uII}} \in \{k, \ldots, n\}$
	The design guarantees a minimum sample size of k observations.	
Experimental time W	$(X_{k:n} \vee T_2) \wedge [X_{m:n} \vee (T_1 \wedge X_{n:n})]$	

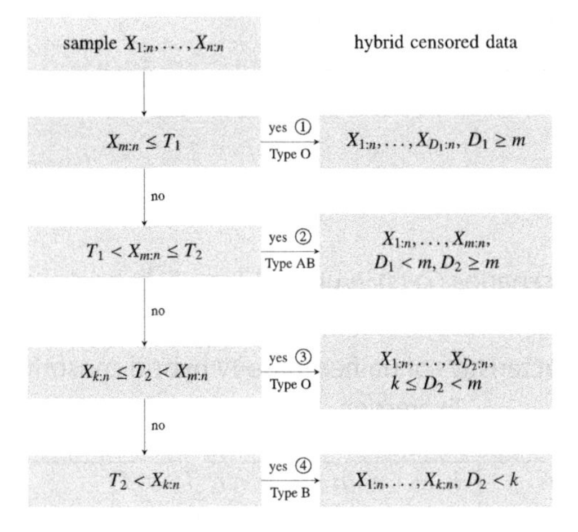

Figure 4.10 Sampling scenarios ①-④ in unified Type-II hybrid censoring.

Table 4.10 Characteristics of unified Type-III hybrid censoring.

unified Type-III hybrid censoring	Expression	Properties
Scheme parameters	k, m with $k < m$, $T_1 < T_2 < T_3$	
Termination time	$(X_{k:n} \wedge T_3) \vee [(X_{m:n} \vee T_1) \wedge T_2]$	bounded by T_3
Sample size D_{uIII}	$(k \wedge D_3) \vee ((m \vee D_1) \wedge D_2)$	$D_{\text{uIII}} \in \{0, \ldots, n\}$
	The sample size may be zero with positive probability.	
Experimental time W	$(X_{k:n} \wedge T_3) \vee [(X_{m:n} \vee [T_1 \wedge X_{n:n}]) \wedge T_2]$	

Unified Type-IV hybrid censoring

The unified Type-IV hybrid censoring scheme has originally been proposed by Park and Balakrishnan (2012) (see also Górny and Cramer, 2018b). The details of the hybrid censoring scheme are depicted in Table 4.11 and Fig. 4.12.

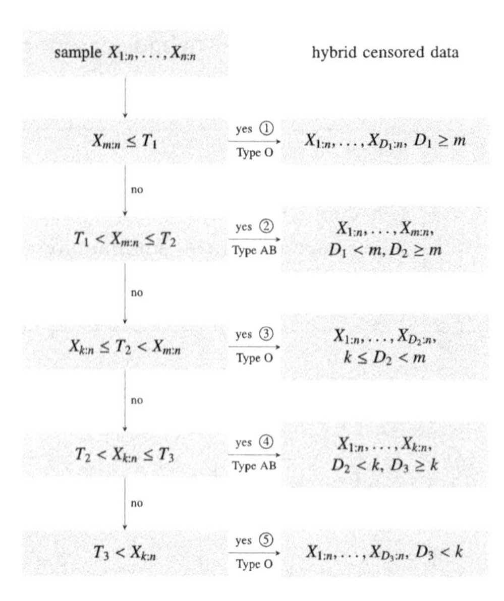

Figure 4.11 Sampling scenarios ①-⑤ in unified Type-III hybrid censoring.

Table 4.11 Characteristics of unified Type-IV hybrid censoring.

unified Type-IV hybrid censoring	Expression	Properties
Scheme parameters	k, m, r with $k < m < r$, $T_1 < T_2$	
Termination time	$(X_{r:n} \wedge T_1) \vee [(X_{k:n} \vee T_2) \wedge X_{m:n}]$	unbounded
Sample size D_{uIV}	$(r \wedge D_1) \vee ((k \vee D_2) \wedge m)$	$D_{\mathrm{uIV}} \in \{k, \dots, r\}$
	The design guarantees a minimum sample size of k observations.	
Experimental time W	$(X_{r:n} \wedge T_1) \vee [(X_{k:n} \vee T_2) \wedge X_{m:n}]$	

4.5.5 Designing new hybrid censoring schemes

As shown in the previous sections, hybrid censoring schemes can be generated by introducing requirements regarding certain order statistics, that is, considering a particular order statistic $X_{k:n}$ in the decision rule, as well as by introducing additional thresholds T_i. A very complex hybrid censoring scheme has been proposed in Górny and Cramer (2018a) in order to highlight this idea as well as to illustrate the power of the modularization technique. It takes into account three order statistics $X_{k:n} < X_{m:n} < X_{r:n}$ and three thresholds $T_1 < T_2 < T_3$ leading to six different sampling scenarios. The details and the decision rules of the hybrid censoring scheme are depicted in Fig. 4.13. However, although the censoring scheme looks quite complicated, a detailed analysis is possible using the modularization approach. The same comment applies to the unified

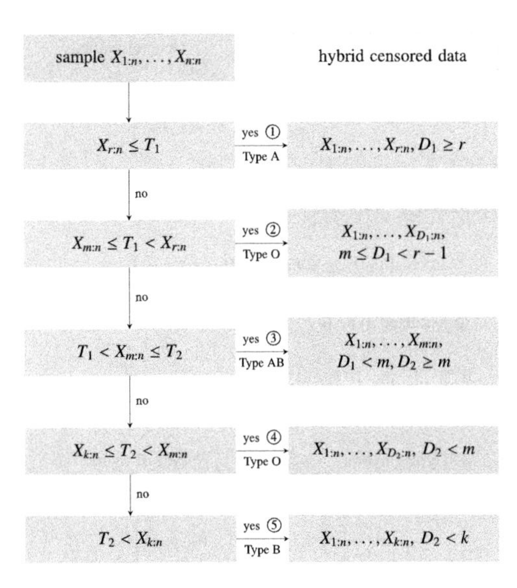

Figure 4.12 Sampling scenarios ①-⑤ in unified Type-IV hybrid censoring.

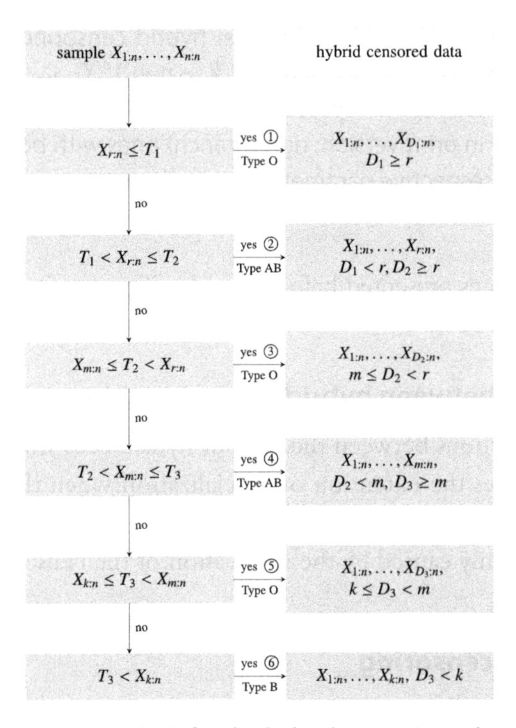

Figure 4.13 Sampling scenarios ①-⑥ for the hybrid censoring scheme proposed by Górny and Cramer (2018a).

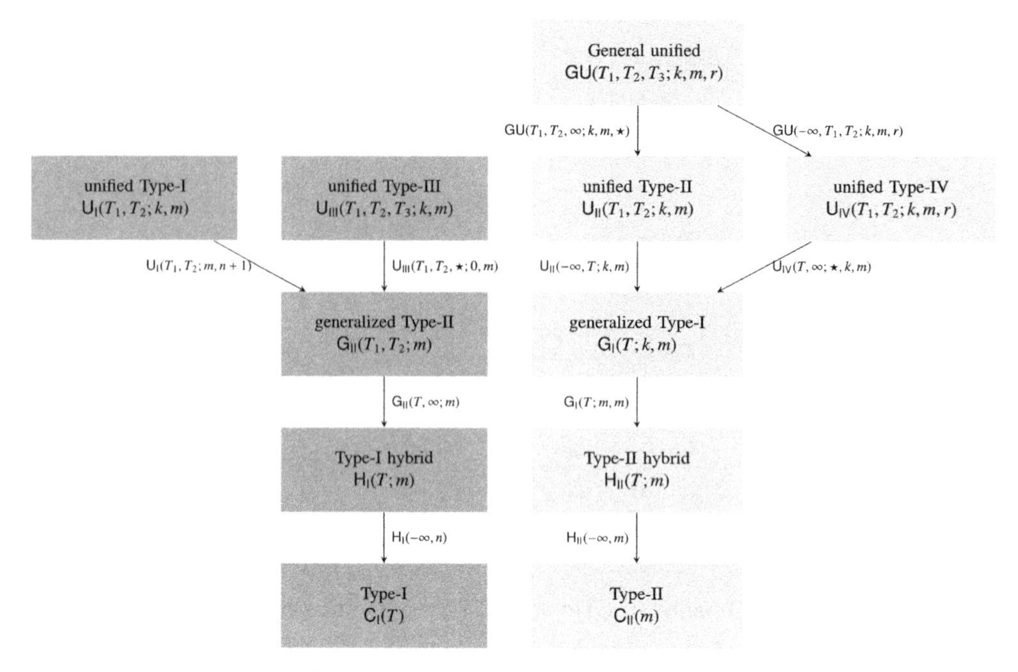

Figure 4.14 Connections between (progressive) hybrid censoring schemes. We use the convention that, for $k = 0$, $X_{0:n} = -\infty$ and, for $k = n + 1$, $X_{n+1:n} = \infty$. Green (dark gray in print version; left branch) boxes represent hybrid censoring schemes with unbounded test duration, blue (gray in print version; right branch) ones with bounded test duration. A star $\star$ indicates that the respective parameter is not relevant.

hybrid censoring schemes presented before. For details, we refer to Górny and Cramer (2018a,b) and Górny (2017).

4.5.6 Connections between hybrid censoring schemes

A survey of the connections between the various hybrid censoring schemes is provided in Fig. 4.14. It illustrates the extension or specialization when the involved parameters are appropriately chosen. Notice that the relations are independent of the underlying (ordered) sample and only caused by the application of the censoring mechanisms.

4.6. Joint (hybrid) censoring

In the preceding sections, it is assumed that the censored data is generated from an iid sample $X_1, \ldots, X_n$. One extension of this model is to assume a sample pooled from two independent iid subsamples with possibly different distribution. In joint Type-II censoring, the sample is based on two independent baseline samples

$$X_1, \ldots, X_{n_1} \overset{\text{iid}}{\sim} F_1 \quad \text{(Type-X)}$$
$$Y_1, \ldots, Y_{n_2} \overset{\text{iid}}{\sim} F_2 \quad \text{(Type-Y)}$$

of independent random variables. The censoring mechanism is applied to the pooled sample $X_1, \ldots, X_{n_1}, Y_1, \ldots, Y_{n_2}$ where it is assumed that the type of the failed unit is observable. Therefore, the data is given by $(\boldsymbol{C}, \boldsymbol{W})$, where

$$\boldsymbol{C} = (C_1, \ldots, C_m) \in \{0, 1\}^m, \qquad \boldsymbol{W} = (W_{1:n_1+n_2}, \ldots, W_{m:n_1+n_2}).$$

The indicators C_j have the value 1 if the failed unit is of Type X, and otherwise $C_j = 0$. $W_{j:n}$ denotes the j-th failure time in the censored experiment.

Results for such models have been presented in, e.g., Johnson and Mehrotra (1972), Bhattacharyya and Mehrotra (1981), Mehrotra and Bhattacharyya (1982), Bhattacharyya (1995, Section 7.5), Balakrishnan and Rasouli (2008), Su (2013), Shafay et al. (2014), and Zhu et al. (2020). An extension to k samples has been discussed in Balakrishnan and Su (2015) and Mondal and Kundu (2019b). Similar models under progressive Type-II censoring have been considered in, e.g., Rasouli and Balakrishnan (2010), Parsi and Bairamov (2009), Parsi et al. (2011), Doostparast et al. (2013), Balakrishnan et al. (2015), and Mondal and Kundu (2019c, 2020). Joint progressive Type-I censoring has been proposed in Ashour and Jones (2017). Another version of joint progressive censoring leading to more tractable results has been investigated in Mondal and Kundu (2019a, 2020) and Goel and Krishna (2022).

Joint hybrid censoring schemes are discussed in Su and Zhu (2016), Mao et al. (2017), Abo-Kasem et al. (2019), Abo-Kasem and Elshahhat (2022, 2021), and Shafay (2022).

CHAPTER 5

Inference for exponentially distributed lifetimes

Contents

5.1. Introduction

In this chapter, we assume that the life times are independent and exponentially distributed with mean ϑ, that is, $X_1, \ldots, X_n \stackrel{\text{iid}}{\sim} Exp(\vartheta)$. We present results for one- and two-parameter exponential distributions under various hybrid censoring schemes. In particular, the exact (conditional) distribution of the MLE of the distribution parameters are established in terms of moment generating functions and B-spline functions, respectively. We illustrate for Type-I and Type-II hybrid censoring in detail how these representations can be established. Furthermore, exact confidence intervals for the scale parameter are obtained using the method of pivoting the cumulative distribution function. Before starting with the presentation of results for exponentially distributed lifetimes, we establish a general form of the likelihood function which explains the structural similarities in the estimation process.

5.2. General expression for the likelihood function

Inference based on hybrid censored samples has been discussed for many lifetime distributions $F_{\boldsymbol{\theta}}$, $\boldsymbol{\theta} \in \Theta \subseteq \mathbb{R}^p$. In order to present a systematic approach to the topic, we

Hybrid Censoring Know-How
https://doi.org/10.1016/B978-0-12-398387-9.00013-1

illustrate the structure of the likelihood functions which is a fundamental tool in the analysis. We start with the likelihood function under Type-I hybrid censoring as given in (4.19), that is,

$$\mathscr{L}(\boldsymbol{\theta}; \boldsymbol{x}_d, d) = \frac{n!}{(n-d)!}\Big(\prod_{j=1}^{d} f_{\boldsymbol{\theta}}(x_j)\Big)(1 - F_{\boldsymbol{\theta}}(w))^{n-d} \tag{5.1}$$

with $w = x_d \wedge T$.[1] Comparing the likelihood functions under the various (hybrid) censoring schemes, it follows from the results presented in Chapter 4 that the likelihood has always the functional form in (5.1) where d is the observed value of the sample size D_{HCS} and w is the realization of the test duration W_{HCS} under the censoring scheme (see Table 5.1 for a summary of the expressions). In particular, we refer to the equations:

(1) (3.62): Type-I censoring
(2) (3.32): Type-II censoring
(3) (4.19): Type-I hybrid censoring
(4) (4.30): Type-II hybrid censoring
(5) (4.33): Generalized Type-I censoring
(6) (4.37): Generalized Type-II censoring

Respective results for the other hybrid censoring schemes can be taken from the results on progressive hybrid censoring schemes reported in Section 7.1.2.3. It should already be noted here that an analogous result is true for progressively Type-II censored underlying data (for details, see Chapter 7).

The expressions in (5.1) illustrate that, in order to obtain the MLE under a hybrid censoring scheme, one has to solve the same type of optimization problem. In particular, the log-likelihood function

$$\mathscr{L}^*(\boldsymbol{\theta}; \boldsymbol{x}_d, d) = c_d^* + \sum_{j=1}^{d} \log f_{\boldsymbol{\theta}}(x_j) + (n-d)\log(1 - F_{\boldsymbol{\theta}}(w)) \tag{5.2}$$

with $c_d^* = \log n! - \log(n-d)!$ needs to be addressed in the maximization process. In fact, this shows, that the likelihood equations are always of the same kind. In fact, the problem can be traced back to the optimization problems under Type-I and Type-II censoring, respectively. In particular, it has to be noted that type ⓘⓘ corresponds to common Type-II censoring whereas type ⓘ is related to Type-I censoring (see proof of Theorem 4.11). Thus, the respective results for these censoring schemes can be directly applied to hybrid censoring. This connection will simplify considerably the discussion of existence and uniqueness of MLEs for various lifetime distributions. A similar argument can be utilized in Bayesian inference choosing appropriate prior distributions.

[1] In order to simplify the notation, from now, we will denote the realizations $x_{1:n}, \ldots, x_{n:n}$ of the order statistics $X_{1:n}, \ldots, X_{n:n}$ by $\boldsymbol{x}_n = (x_1, \ldots x_n)$ (the same for the realizations of the progressively Type-II censored order statistics).

5.3. Type-I hybrid censoring

5.3.1 Likelihood inference and exact distribution of MLE

We consider likelihood inference based on Type-I hybrid censored data $X_{1:n}, \dots, X_{D_\text{I}:n}$ with sample size D_I, provided that $m \geq 2$. Therefore, the data is given by $\boldsymbol{x}_d$ with $x_1 \leq \dots \leq x_d < T$ and $D_\text{I} = d \in \{1, \dots, m\}$. Using (5.1) and assuming $d \geq 1$, the likelihood function in (5.1) reads for an $Exp(\vartheta)$-distribution

$$\mathscr{L}(\vartheta; \boldsymbol{x}_d, d) = \frac{n!}{(n-d)!\vartheta^d} \exp\Big\{ -\frac{1}{\vartheta}\Big[\sum_{j=1}^{d} x_j + (n-d)w\Big]\Big\} \tag{5.3}$$

$$= \frac{n!}{(n-d)!\vartheta^d} e^{-\mathsf{TTT}_d/\vartheta}, \quad 0 \leq x_1 \leq \dots \leq x_d \leq T, \tag{5.4}$$

where the total time on test TTT_d is defined by $(x_1 \leq \dots \leq x_m)$

$$\mathsf{TTT}_d = \sum_{j=1}^{d} x_j + (n-d)w, \quad d = \sum_{j=1}^{m} \mathbb{1}_{(-\infty, T]}(x_j). \tag{5.5}$$

It follows from the results of Childs et al. (2003) that the MLE of ϑ exists provided $D > 0$. It is given by

$$\widehat{\vartheta} = \frac{1}{D_\text{I}}\Big[\sum_{j=1}^{D_\text{I}} X_{j:n} + (n - D_\text{I})W\Big] = \frac{1}{D_\text{I}}\mathsf{TTT}_{D_\text{I}}. \tag{5.6}$$

Notice that $\mathsf{TTT}_0 = nT$ for $d = 0$, that is, $\mathsf{TTT}_{D_\text{I}}$ has point mass

$$\Pr_\vartheta(\mathsf{TTT}_{D_\text{I}} = nT) = \Pr_\vartheta(D_\text{I} = 0) = \Pr_\vartheta(X_{1:n} > T) = e^{-nT/\vartheta}$$

on the right endpoint of its support. Clearly, $0 \leq \mathsf{TTT}_{D_\text{I}} \leq nT$ due to the time truncation.

In the following, we present two approaches to derive the exact (conditional) density function of the MLE. Both methods can be applied to other hybrid censoring schemes, too.

Remark 5.1. **(a)** It should be mentioned that, for exponentially distributed lifetimes, the structure of the likelihood function under several hybrid censoring schemes has the form given in (5.4). Of course, the definition of the total time on test statistic TTT_d and the definition of the sample size $D_{\text{HCS}} = d$ depend on the particular (hybrid) censoring scheme. However, this structure explains that, e.g., Bayesian inference for ϑ assuming a gamma prior as in (5.18) leads to similar posterior distributions (see (5.19); see also comments in Section 3.2.4), Bayesian estimators (see (5.20)) and, thus, to similar statistical results. We will illustrate this property for the most common hybrid censoring schemes.

(b) In order to overcome the problem of a non-existing MLE for $D_\text{I} = 0$, Kundu and Koley (2017) considered inference for $\lambda = 1/\vartheta$. Notice that they assumed $m = n$. Since $(D_\text{I}, \text{TTT}_{D_\text{I}})$ is a sufficient statistic, they proposed the estimator

$$\widehat{\lambda} = \frac{D_\text{I}}{\text{TTT}_{D_\text{I}}} = \frac{D_\text{I}}{\sum_{j=1}^{D_\text{I}} X_{j:n} + (n - D_\text{I})W}$$

which equals the MLE when $D_\text{I} > 0$. However, it is well defined even if no failures have been observed, that is, $\widehat{\lambda} = 0$ in this case. As noted above, $\Pr_\lambda(\widehat{\lambda} = 0) = e^{-nT/\vartheta}$ so that the estimator has a point mass at 0.

5.3.1.1 Moment generating function

The most popular approach in deriving the (conditional) distribution of the MLE $\widehat{\vartheta}$ is based on the (conditional) moment generating function of $\widehat{\vartheta}$ (cf. Section 3.3, especially Theorem (3.45)). The method has been widely used for hybrid censored data. For Type-I censored, it has been applied first in Bartholomew (1963) (see Theorem 3.45; for applications to hybrid censored data, we refer to Chen and Bhattacharyya, 1988; Childs et al., 2003). In the following, we illustrate the procedure for Type-I hybrid censoring in detail since it can be applied similarly to the other hybrid censoring schemes, too. It can be seen as a paradigm in applying this method. However, our presentation is somewhat different to the approaches available in the literature in the following sense. In contrast to the proofs in Bartholomew (1963), Chen and Bhattacharyya (1988), and Childs et al. (2003) we avoid integration in the proof by using the expression in (4.20). We only need the probability mass function of D_I given in Lemma 4.4 and that the expressions in (4.20) are proper density functions for $d \in \{1, \ldots m\}$ and for any parameter value $\vartheta > 0$. For $m = n$, the representation in (5.7) reduces to that one established under Type-I censoring given in (3.65).

Theorem 5.2. *For $\vartheta > 0$, $t < 1/\vartheta$, and $d \in \{1, \ldots, m\}$, let $\vartheta_d = \frac{\vartheta}{1 - t\vartheta/d} > 0$. Then, the conditional moment generating function of the MLE $\widehat{\vartheta}$ under Type-I hybrid censoring is given by*

$$M_{\vartheta}^{D_\text{I} \geq 1}(t) = (1 - e^{-nT/\vartheta})^{-1} \Bigg[\sum_{d=1}^{m-1} \binom{n}{d} \frac{1}{(1 - t\vartheta/d)^d} (1 - e^{-T(1 - t\vartheta/d)/\vartheta})^d e^{-(n-d)T(1 - t\vartheta/d)/\vartheta}$$
$$+ \frac{1}{(1 - t\vartheta/m)^m} \sum_{j=m}^{n} \binom{n}{j} (1 - e^{-T(1 - t\vartheta/m)/\vartheta})^j e^{-(n-j)T(1 - t\vartheta/m)/\vartheta} \Bigg], \quad t < 1/\vartheta. \quad (5.7)$$

Proof. First, recall that $\widehat{\vartheta} = \frac{1}{d}\big[\sum_{j=1}^{d} x_j + (n - d)w\big] = \text{TTT}_d/d$ for data $x_1 \leq \cdots \leq x_d \leq T$, $D_\text{I} = d \in \{1, \ldots, m\}$. In order to emphasize the dependence on the parameter ϑ, we write $\Pr_\vartheta(D \geq 1), f_\vartheta^{X_{j:n}, 1 \leq j \leq D_\text{I}, D_\text{I}}$, etc. Furthermore, $\{D_\text{I} \geq 1\} = \{D \geq 1\}$. Then, given $D_\text{I} \geq 1$, we get from (4.20) for $t < 1/\vartheta$

$$M_\vartheta^{D_l \geq 1}(t) = E_\vartheta(e^{t\widehat{\vartheta}} \mid D_l \geq 1) = \frac{1}{\Pr_\vartheta(D_l \geq 1)} \sum_{d=1}^{m} \Pr_\vartheta(D_l = d) E_\vartheta(e^{-t\widehat{\vartheta}} \mid D_l = d)$$

$$= \frac{1}{\Pr_\vartheta(D \geq 1)} \sum_{d=1}^{m} \int_{\mathbb{R}^n_{\leq T}} e^{\frac{t}{d}\mathsf{TTT}_d} f_\vartheta^{X_{j:n}, 1 \leq j \leq d, D_l}(\boldsymbol{x}_d, d)\, d\boldsymbol{x}_d$$

$$= \frac{1}{\Pr_\vartheta(D \geq 1)} \sum_{d=1}^{m} \int_{\mathbb{R}^n_{>0, \leq T}} e^{\frac{t}{d}\mathsf{TTT}_d} \frac{n!}{(n-d)!\vartheta^d} \exp\Big(-\frac{1}{\vartheta}\mathsf{TTT}_d\Big)\, d\boldsymbol{x}_d$$

$$= \frac{1}{\Pr_\vartheta(D \geq 1)} \sum_{d=1}^{m} \int_{\mathbb{R}^n_{>0, \leq T}} \frac{n!}{(n-d)!\vartheta^d} \exp\Big(-\frac{d - t\vartheta}{d\vartheta}\mathsf{TTT}_d\Big)\, d\boldsymbol{x}_d$$

Writing $\vartheta_d = \frac{d\vartheta}{d - t\vartheta} = \frac{\vartheta}{1 - t\vartheta/d}$, we obtain from the density property as well as (4.21) and (4.22)

$$= \frac{1}{\Pr_\vartheta(D \geq 1)} \sum_{d=1}^{m} \frac{\vartheta_d^d}{\vartheta^d} \int_{\mathbb{R}^n_{\leq T}} f_{\vartheta_d}^{X_{j:n}, 1 \leq j \leq d, D_l}(\boldsymbol{x}_d, d)\, d\boldsymbol{x}_d$$

$$\overset{(4.21)}{=} \frac{1}{\Pr_\vartheta(D \geq 1)} \sum_{d=1}^{m} \frac{\vartheta_d^d}{\vartheta^d} \Pr_{\vartheta_d}(D_l = d)$$

$$\overset{(4.22)}{=} \frac{1}{\Pr_\vartheta(D \geq 1)} \Bigg[\sum_{d=1}^{m-1} \frac{1}{(1 - t\vartheta/d)^d} \binom{n}{d} (1 - e^{-T/\vartheta_d})^d e^{-(n-d)T/\vartheta_d}$$

$$+ \frac{1}{(1 - t\vartheta/m)^m} \sum_{j=m}^{n} \binom{n}{j} (1 - e^{-T/\vartheta_m})^j e^{-(n-j)T/\vartheta_m} \Bigg]$$

Using $\Pr_\vartheta(D \geq 1) = 1 - e^{-nT/\vartheta}$ and recalling the definition of ϑ_d, we get the desired representation. □

It should be noticed that the representation of the moment generating function in (5.7) is slightly different to that obtained by Childs et al. (2003). It can be shown by expanding the second term that these representations are identical.

The derivation of the (conditional) density function is based on the moment generating function of a (shifted) gamma distribution.

Lemma 5.3. *Let $X \sim \Gamma(\vartheta, d)$ with $d, \vartheta > 0$ and $c \in \mathbb{R}$. Then, the moment generating function M_{X+c} of $X + c$ is given by*

$$M_{d,\vartheta;c}(t) = M_{X+c}(t) = E(e^{t(X+c)}) = \frac{e^{ct}}{(1 - \vartheta t)^d}, \quad t < \frac{1}{\vartheta}.$$

In order to derive the (conditional) density function of the MLE, we expand the powers in the summation in (5.7) by the binomial theorem (see also proof of Theorem 3.45):

$$M_{\vartheta}^{D_I\geq 1}(t) = (1-e^{-nT/\vartheta})^{-1}\Bigg[\sum_{d=1}^{m-1}\sum_{k=0}^{d}\binom{n}{d}\binom{d}{k}(-1)^k e^{-(n-d+k)T/\vartheta}\frac{e^{(n-d+k)Tt/d}}{(1-t\vartheta/d)^d}$$
$$+\sum_{j=m}^{n}\sum_{k=0}^{j}\binom{n}{j}\binom{j}{k}(-1)^k e^{-(n-j+k)T/\vartheta}\frac{e^{(n-j+k)Tt/m}}{(1-t\vartheta/m)^m}\Bigg]$$

Using Lemma 5.3, we get

$$= (1-e^{-nT/\vartheta})^{-1}\Bigg[\sum_{d=1}^{m-1}\sum_{k=0}^{d}\binom{n}{d}\binom{d}{k}(-1)^k e^{-(n-d+k)T/\vartheta} M_{d,\vartheta/d;(n-d+k)T/d}(t)$$
$$+\sum_{j=m}^{n}\sum_{k=0}^{j}\binom{n}{j}\binom{j}{k}(-1)^k e^{-(n-j+k)T/\vartheta} M_{m,\vartheta/m;(n-j+k)T/m}(t)\Bigg].$$

Therefore, the conditional moment generating function of the MLE is a linear combination of moment generating functions of shifted gamma distributions. Inverting the moment generating functions yields directly the respective density function (see also proof of Theorem 3.46).

Theorem 5.4. *For $\vartheta > 0$, the conditional density function of the MLE $\widehat{\vartheta}$ under Type-I hybrid censoring is given by*

$$f_{\vartheta}^{\widehat{\vartheta}|D_I\geq 1}(t) = (1-e^{-nT/\vartheta})^{-1}\Bigg[\sum_{d=1}^{m-1}\sum_{k=0}^{d}\binom{n}{d}\binom{d}{k}(-1)^k e^{-(n-d+k)T/\vartheta} f_{\Gamma(\vartheta/d,d);(n-d+k)T/d}(t)$$
$$+\sum_{j=m}^{n}\sum_{k=0}^{j}\binom{n}{j}\binom{j}{k}(-1)^k e^{-(n-j+k)T/\vartheta} f_{\Gamma(\vartheta/m,m);(n-j+k)T/m}(t)\Bigg],\ t\in\mathbb{R}, \quad (5.8)$$

where $f_{\Gamma(\alpha,\beta);c}$ denotes the density function of a shifted gamma distribution as in Lemma 5.3.

Furthermore, replacing the density functions $f_{\Gamma(\alpha,\beta);c}$ by the corresponding cumulative distribution functions $F_{\Gamma(\alpha,\beta);c}$ or the survival function $\overline{F}_{\Gamma(\alpha,\beta);c}$, we get expressions for the cumulative distribution function and survival function of $\widehat{\vartheta}$ given $D_I\geq 1$, respectively.

Representation (5.8) in Theorem 5.4 is different to that presented in Childs et al. (2003) and Chen and Bhattacharyya (1988). It can also be found from the B-spline representation presented in the next section using the results of Górny and Cramer (2019b) who established a method that allows the gamma distribution representation to be converted to the B-spline representation.

Remark 5.5. The density function in Theorem 5.4 can be used to establish explicit expressions for (conditional) moments of the MLE as has been done in Childs et al. (2003). In particular, we get from the mean of a shifted gamma $\Gamma(\vartheta/d, d)$-distributed random variable $X+c$, that is, $E(X+c)=c+\vartheta$,

$$E_\vartheta(\widehat{\vartheta}\mid D_{\mathrm{I}}\geq 1)=\vartheta+T(1-e^{-nT/\vartheta})^{-1}\Bigg[\sum_{d=1}^{m-1}\sum_{k=0}^{d}\binom{n}{d}\binom{d}{k}(-1)^k e^{-(n-d+k)T/\vartheta}\frac{n-d+k}{d} + \sum_{j=m}^{n}\sum_{k=0}^{j}\binom{n}{j}\binom{j}{k}(-1)^k e^{-(n-j+k)T/\vartheta}\frac{n-j+k}{m}\Bigg].$$

5.3.1.2 Spacings, total time on test, and B-splines

In this section, we present a different way to calculate the density function of the MLE. It is based on the distribution of the spacings and due to Cramer and Balakrishnan (2013). It yields a representation of the (conditional) density function in terms of B-spline functions (see Appendix C). Such a (compact) representation will also be possible under other hybrid censoring schemes. In particular, we will illustrate that the B-spline representation will be very useful in studying properties of the distribution of the MLE.

As in the derivation of the moment generating function in Theorem 5.2, we make use of the law of total probability. Writing the MLE in terms of the total time on test (see (5.6)), we get

$$f_\vartheta^{\widehat{\vartheta}\mid D_{\mathrm{I}}\geq 1}(t)=\frac{1}{\Pr_\vartheta(D_{\mathrm{I}}\geq 1)}\sum_{d=1}^{m} f_\vartheta^{\mathrm{TTT}_d/d\mid D_{\mathrm{I}}=d}(t)\Pr_\vartheta(D_{\mathrm{I}}=d),\quad t\in\mathbb{R},\tag{5.9}$$

so that it is sufficient to determine the density function $f_\vartheta^{\mathrm{TTT}_d/d\mid D_{\mathrm{I}}=d}$, $d\in\{1,\ldots,m\}$. In order to find these density functions, we consider the spacings representation (5.10) of the MLE.

Representation (4.17) enables us to study the distribution of spacings in the case of an exponential distribution. Let $X_{1:n},\ldots,X_{D_{\mathrm{I}}:n}$ be a Type-I progressively hybrid censored sample from an $Exp(\vartheta)$-distribution, and suppose that $0<d\leq m-1$. Let

$$S_{j,n}=(n-j+1)(X_{j:n}-X_{j-1:n}),\quad 1\leq j\leq d,$$

be the normalized spacings of the first d random variables $X_{1:n},\ldots,X_{d:n}$, where $X_{0:n}=0$. Then, conditionally on $D_{\mathrm{I}}=D=d$, $S_{j,n}$, $1\leq j\leq d$, denote the normalized spacings of order statistics from an exponential distribution.

Given $D=d$, the total time on test statistic can be written as $(X_{0:n}=0)$ (cf. (5.6))

$$\mathrm{TTT}_d=\sum_{j=1}^{d}\Big(1-\frac{n-d}{n-j+1}\Big)S_{j,n}+(n-d)W,\quad 1\leq d\leq m.\tag{5.10}$$

The support of TTT_d is given by the interval $[(n-d)T, nT]$. As shown in Cramer and Balakrishnan (2013) this yields, for $1 \le d \le m-1$, the following representation of the density function

$$f_\vartheta^{S_{j,n},1\le j\le d-1,\text{TTT}_d|D_\text{I}=d}(\boldsymbol{w}_{d-1}, s) = \frac{(n-d)(n-d+1)}{\vartheta^{d+1} f_{\vartheta;d+1:n}(T)} e^{-s/\vartheta}. \tag{5.11}$$

For $d = m$, one gets

$$f_\vartheta^{S_{j,n},1\le j\le m-1,\text{TTT}_m|D_\text{I}=m}(\boldsymbol{w}_{m-1}, s) = \frac{1}{\vartheta^m F_{\vartheta;m:n}(T)} e^{-s/\vartheta}. \tag{5.12}$$

The supports of the density functions in (5.11) and (5.12) are given by the polyhedron

$$\mathcal{M}_{d-1}^{(s)} = \Big\{(w_1, \ldots, w_{d-1}) \,\Big|\, w_j \ge 0, 1 \le j \le d-1, s-(n-d+1)T \le \sum_{j=1}^{d-1} \beta_j w_j, \sum_{j=1}^{d-1} \alpha_j w_j \le s-(n-d)T\Big\},$$

where $s \in [(n-d)T, nT]$ and $\alpha_j = (d-j)/(n-j+1)$, $\beta_j = (d-j+1)/(n-j+1)$, $1 \le j \le d$. The preceding expressions enable us to establish the following properties of the density functions.

Remark 5.6. From expression (5.11), it is obvious that the (conditional) joint density function $f_\vartheta^{S_{j,n},1\le j\le d-1,\text{TTT}_d|D_\text{I}=d}$ is log-concave. Therefore, the marginals $f_\vartheta^{S_{j,n}|D_\text{I}=d}$, $1 \le j \le d-1$, and $f_\vartheta^{\text{TTT}_d|D_\text{I}=d}$ are log-concave since marginal density functions of log-concave distributions are also log-concave (see Brascamp and Lieb, 1975; Prékopa, 1973). Furthermore, any marginal distribution is (strongly) unimodal (see Cramer, 2004).

In the next step, we are interested in the (marginal) distribution of TTT_d (given $D_\text{I} = d$). Therefore, we have to integrate the density functions in (5.11) and (5.12) w.r.t. $\boldsymbol{w}_{d-1}$, $d \in \{2, \ldots, m\}$. Using results of Gerber (1981) and Cho and Cho (2001), the volume of $\mathcal{M}_{d-1}^{(s)}$ is given by

$$\text{Volume}(\mathcal{M}_{d-1}^{(s)}) = \frac{1}{d!}\binom{n}{d-1} \sum_{i=1}^{d+1} (-1)^{i-1} \binom{d}{i-1} [(n-i+1)T - s]_+^{d-1}, \quad 2 \le d \le m,$$

provided $(n-d)T \le s$.

The volume of $\mathcal{M}_{d-1}^{(s)}$ can be written in terms of a particular B-spline. Specifically, in our notation, the univariate B-Spline B_{d-1} of degree $d-1$ with knots $(n-d)T < \cdots < nT$ exhibits the representation (cf. (C.4))

$$B_{d-1}(s \mid (n-d)T, \ldots, nT) = \frac{1}{T^d(d-1)!} \cdot \sum_{i=1}^{d+1}(-1)^{i-1}\binom{d}{i-1}[(n-i+1)T - s]_+^{d-1},$$
$$s \in [(n-d)T, nT],$$

see de Boor (2001) and Dahmen and Micchelli (1986). Hence, this yields the identity

$$\text{Volume}(\mathcal{M}_{d-1}^{(s)}) = \frac{T^d}{d}\binom{n}{d-1}B_{d-1}(s \mid (n-d)T, \ldots, nT), \quad s \in [(n-d)T, nT].$$

Further information on the connection of B-splines to such volumes can be found in Curry and Schoenberg (1966), de Boor (1976), Dahmen and Micchelli (1986), Górny (2017), and Górny and Cramer (2019a).

This connection is quite useful as it shows that we do not have to take care about the support $[(n-d)T, nT]$ since $B_{d-1}(s \mid (n-d)T, \ldots, nT) = 0$ for $s \notin [(n-d)T, (n-d)T]$. In particular, the condition $0 \leq s \leq nT$ in Theorem 5.7 can be dropped and replaced by $s \in \mathbb{R}$. Moreover, we can use the properties of B-splines in all subsequent derivations.

Using the above representations, we obtain the following key result.

Theorem 5.7. *Let* $1 \leq d \leq m-1$. *The density function* $f^{\mathrm{TTT}_d | D_I = d}$ *is given by*

$$f_\vartheta^{\mathrm{TTT}_d | D_I = d}(s) = (d+1)\binom{n}{d+1}\frac{T^d}{\vartheta^{d+1} f_{d+1:n}(T)} B_{d-1}(s \mid (n-d)T, \ldots, nT)e^{-s/\vartheta},$$
$$0 \leq s \leq nT.$$

Furthermore, the density function $f^{\mathrm{TTT}_{dm} | D_I = m}$ *is given by*

$$f_\vartheta^{\mathrm{TTT}_m | D_I = m}(s) = \binom{n}{m}\frac{T^m}{\vartheta^m F_{m:n}(T)} B_{m-1}(s \mid 0, (n-m+1)T, \ldots, nT)e^{-s/\vartheta}, \quad 0 \leq s \leq nT.$$

Remark 5.8. B-spline functions have been connected to representations of density functions of L-statistics

$$\sum_{j=1}^{n} a_j U_{j:n}$$

based on uniform order statistics (see, e.g., Dahmen and Micchelli, 1986; Ignatov and Kaishev, 1985, 1989; Agarwal et al., 2002; Buonocore et al., 2009). In particular, one gets for the sum $L_n = \sum_{j=1}^{n} U_{j:n} = \sum_{j=1}^{n} U_j$ of $U_1, \ldots, U_n \overset{\text{iid}}{\sim} \mathit{Uniform}(0, b)$, $b > 0$, the density function

$$f^{L_n}(u) = B_{n-1}(u \mid 0, b, \ldots, nb), \quad u \in \mathbb{R},$$

(see Feller, 1968, p. 27–28; Buonocore et al., 2009). Furthermore, the distribution of the maximum spacing $Y_{n:n}$ in a sample $U_1, \ldots, U_{n-1} \overset{\text{iid}}{\sim} \mathit{Uniform}(0, 1)$, that is, $Y_i =$

$U_{i:n-1} - U_{i-1:n-1}$, $i = 1, \dots, n$, $U_{0:n-1} = 0$, $U_{n:n-1} = 1$, has the cumulative distribution function

$$F^{Y_{n:n}}(t) = (n-1)! t^n B_{n-1}(1 \mid 0, t, \dots, tb), \qquad t \in \mathbb{R}$$

(see David and Nagaraja, 2003, p. 135, (6.4.4); Barlevy and Nagaraja, 2015, Eq. (45); Górny, 2017, Example 3.1.8; Górny and Cramer, 2019a, Example 3.3)

Remark 5.9. **(a)** For $d \in \{1, \dots, m\}$, the density function $f_\vartheta^{\mathrm{TTT}_d|D_1=d}$ has support $[(n-d)T, nT]$. Therefore, $\lim_{T\to\infty} f_\vartheta^{\mathrm{TTT}_d|D_1=d}(s) = 0$ for any fixed $s > 0$ (choose $T > s/(n-d)$). For $d = m$, the situation is different. In this case, $n - m + 1 = 0$, so that we can choose T such that $s \in (0, T)$. Then, we get

$$\lim_{T\to\infty} f_\vartheta^{\mathrm{TTT}_m|D_1=m}(s) = \frac{1}{(m-1)!\vartheta^m} s^{m-1} e^{-s/\vartheta}, \qquad s > 0.$$

Hence, the distribution converges to a gamma distribution which is the distribution of TTT_m when Type-I censoring is not present in the model.

(b) From the preceding comment and (5.9), it is clear that the support of $\widehat{\vartheta}$ is a subset of $[0, nT]$. In particular, using properties of B-splines, we find that the support of TTT_d/d is given by $[(n/d - 1)T, nT/d]$, $1 \le d \le m-1$, and by $[0, nT/m]$, $d = m$. Thus, the support of $\widehat{\vartheta} = \mathrm{TTT}_{D_1}/D_1$ is given by union of possibly disjoint intervals, that is, by

$$\Big[0, \frac{n}{m}T\Big] \cup \bigcup_{d=1}^{m-1} \Big[\Big(\frac{n}{d} - 1\Big)T, \frac{n}{d}T\Big].$$

In particular, for $m = 1$, the support is always given by $[0, nT]$. For $m \ge 2$, we get for instance that the support has a gap at $(n/2, n-1)$ when $n \ge 3$. A second gap results when $n \ge 7$, that is, $(n/3, n/2 - 1)$, and $m \ge 3$. Notice that the threshold T can be seen as a scaling factor which does not affect the number of gaps, but its length.

A necessary and sufficient condition to hold for a gap to appear is given by the inequality

$$\frac{n}{d} < \frac{n}{d-1} - 1 \iff d(d-1) < n \qquad \text{for } d \in \{2, \dots, m\}.$$

Hence, an additional gap can appear only when n exceeds $2, 6, 12, 20, \dots$, i.e. n is an element of the set $\{\nu(\nu-1)+1 \mid \nu \in \mathbb{N}, \nu \ge 2\}$, and $d \le m$ satisfies the above inequality (see Fig. 5.1 and Figs. 5.3–5.6). Furthermore, for fixed n, the gaps remain at the same place once they have appeared which means that the support depends only on m and n specifying the set $\mathcal{G}_{n,m} = \{d \mid 2 \le d \le m, d(d-1) \le n\}$. In

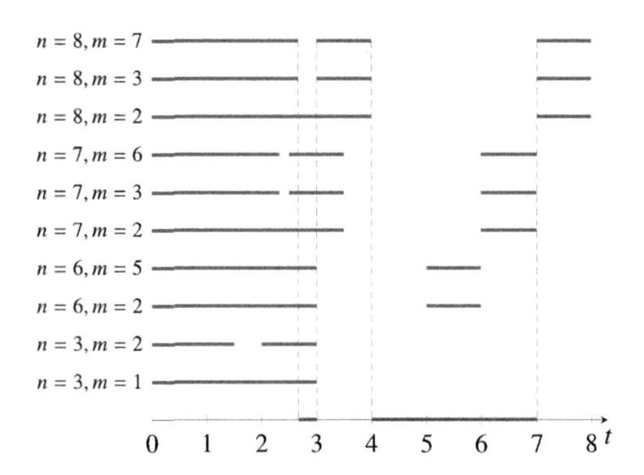

Figure 5.1 Support of $\widehat{\vartheta} = \mathrm{TTT}_{D_I}/D_I$ for selected values of n and m with corresponding gaps ($T = 1$). For $n = 8$ and $m \geq 3$, the gaps of the support are highlighted on the axis line.

fact, the gaps are given by the intervals

$$\Big(\frac{n}{d}T, \Big(\frac{n}{d-1} - 1\Big)T\Big), \qquad d \in \mathscr{G}_{n,m},$$

that is, for an appropriate choice of the maximal $d \in \mathscr{G}_{n,m}$

$$\Big(\frac{n}{2}T, (n-1)T\Big), \quad \Big(\frac{n}{3}T, \Big(\frac{n}{2} - 1\Big)T\Big), \quad \Big(\frac{n}{4}T, \Big(\frac{n}{3} - 1\Big)T\Big), \ldots .$$

(c) The cumulative distribution function of the total time on test is given by

$$\begin{aligned} F_{\vartheta}^{\mathrm{TTT}_{D_I}}(t) &= e^{-nT/\vartheta}\mathbb{1}_{[nT,\infty)}(t) + \sum_{j=1}^{m-1}\binom{n}{d}\frac{T^d}{\vartheta^d}\int_0^t B_{d-1}(s \mid (n-d)T, \ldots, nT)e^{-s/\vartheta}\,ds \\ &\quad + \binom{n}{m}\frac{T^m}{\vartheta^m}\int_0^t B_{m-1}(s \mid 0, (n-m+1)T, \ldots, nT)e^{-s/\vartheta}\,ds, \qquad t \geq 0. \end{aligned}$$

Plots for different values of ϑ are provided in Fig. 5.2. In particular, the plots illustrate the jump at the right endpoint of the support.

By using (5.9), we readily obtain the conditional density function of $\widehat{\vartheta}$.

Theorem 5.10. *For $\vartheta > 0$, the conditional density function of the MLE $\widehat{\vartheta}$ under Type-I hybrid censoring is given by*

$$\begin{aligned} f_{\vartheta}^{\widehat{\vartheta}|D_I \geq 1}(t) &= \frac{n}{1 - e^{-nT/\vartheta}}\Bigg[\sum_{d=1}^{m-1}\binom{n-1}{d-1}\frac{T^d}{\vartheta^d}B_{d-1}(dt \mid (n-d)T, \ldots, nT)e^{-dt/\vartheta} \\ &\quad + \binom{n-1}{m-1}\frac{T^m}{\vartheta^m}B_{m-1}(mt \mid 0, (n-m+1)T, \ldots, nT)e^{-mt/\vartheta}\Bigg], \qquad t \in \mathbb{R}. \quad (5.13) \end{aligned}$$

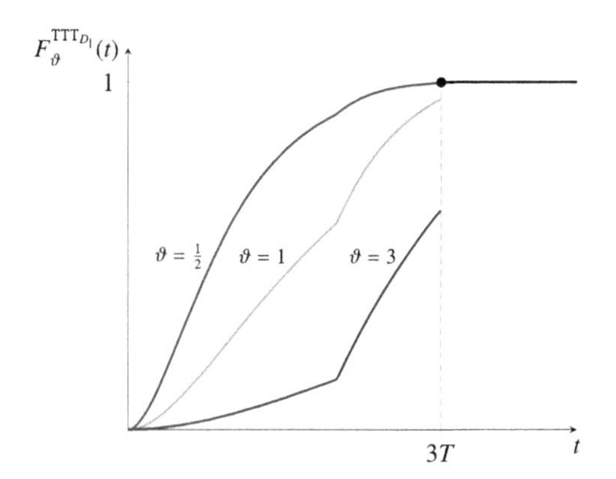

Figure 5.2 Plot of cumulative distribution function $F_{\vartheta}^{\mathrm{TTT}_{D_I}}$ for $n = 3$, $m = 2$, $T = 1$, $\vartheta \in \{.5, 1, 3\}$.

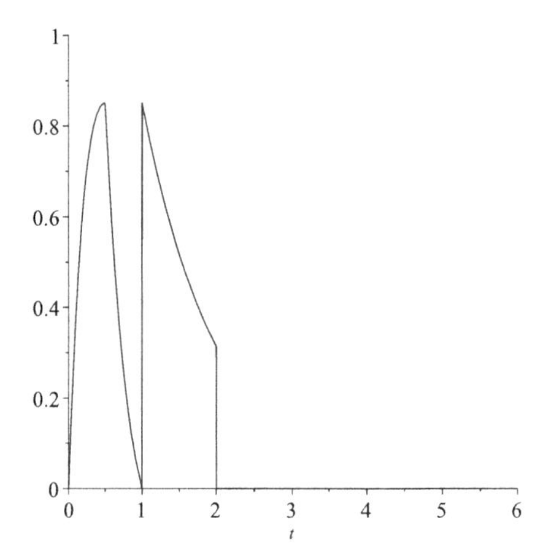

Figure 5.3 Plot of $f_{\vartheta}^{\widehat{\vartheta}|D_I \geq 1}$ with $n = m = 2$.

Remark 5.11. For $m = n$, the representation (5.13) simplifies to

$$f_{\vartheta}^{\widehat{\vartheta}|D_I\geq 1}(t) = \frac{n}{1 - e^{-nT/\vartheta}} \sum_{d=1}^{n} \binom{n-1}{d-1} \frac{T^d}{\vartheta^d} B_{d-1}(dt \mid (n-d)T, \ldots, nT) e^{-dt/\vartheta}, \quad t \in \mathbb{R}. \quad (5.14)$$

In this case, we apply Type-I censoring to the data (see also the expositions in Section 3.3). The stopping time is given by $T^* = X_{n:n} \wedge T$.

Figs. 5.3–5.6 provide plots of $f_{\vartheta}^{\widehat{\vartheta}|D_I\geq 1}$ for $T = 1$ and $\vartheta = 1$ as well as various values of $n \in \{2, 3, 4, 5\}$. They illustrate that the density function $f_{\vartheta}^{\widehat{\vartheta}|D_I\geq 1}$ is multi-modal and that its support may have gaps (see Remark 5.9).

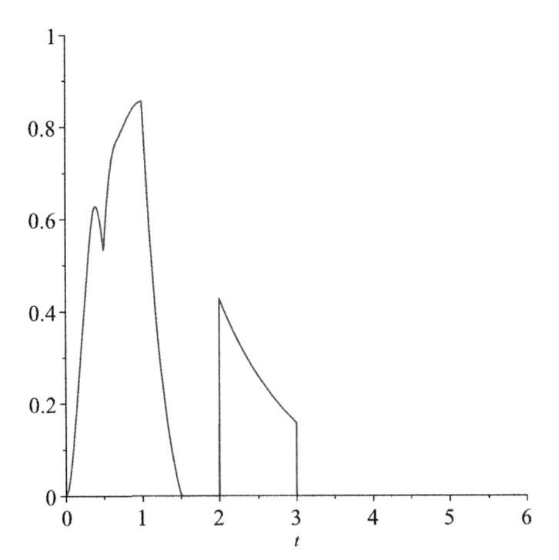

Figure 5.4 Plot of $f_{\vartheta}^{\widehat{\vartheta}|D_1\geq 1}$ with $n = m = 3$.

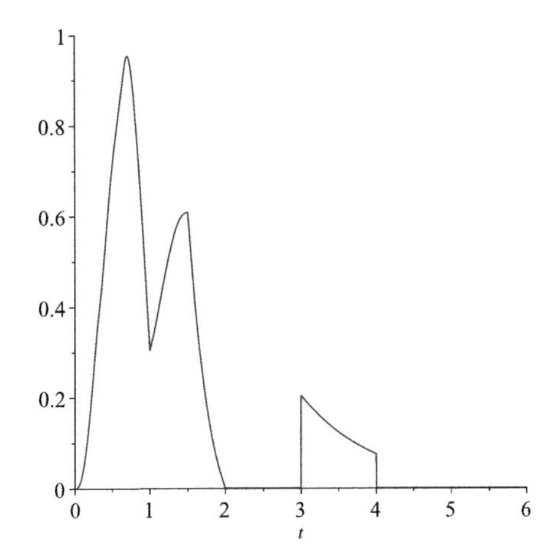

Figure 5.5 Plot of $f_{\vartheta}^{\widehat{\vartheta}|D_1\geq 1}$ with $n = m = 4$.

Remark 5.12. So far, the representation of the density functions and cumulative distribution functions in terms of B-spline functions is initially only an alternative, although more compact, representation than that in terms of shifted gamma distributions. However, we will see that especially for more complex hybrid censoring schemes, the representation has considerable advantages. The combination with the modularization approach of Górny and Cramer (2018b) provides clearly structured, concise representations of the (conditional) density functions of the maximum likelihood estimator.

Furthermore, it should be noted that the computation of the density functions and cumulative distribution functions can make direct use of B-spline function implemen-

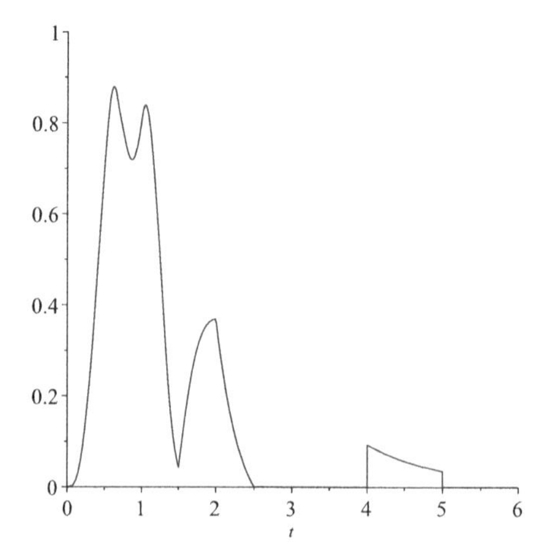

Figure 5.6 Plot of $f_{\vartheta}^{\widehat{\vartheta}|D_1\geq 1}$ with $n=m=5$.

tations in numerical software which generally should be numerically more stable than direct implementations of the shifted gamma expression. These representations have, for instance, alternating signs which support numerical cancellation and may lead to inaccurate results.

5.3.2 Confidence intervals

Several approaches have been proposed to construct (exact) confidence intervals for the parameter ϑ. We start with a widely used idea based on the method of pivoting the cumulative distribution function. This approach can also be applied to other hybrid censoring schemes.

5.3.2.1 Exact confidence intervals: pivoting the cumulative distribution function

Constructive exact confidence intervals for a parameter $\theta \in \Theta \subseteq \mathbb{R}$ by the method of pivoting a cumulative distribution function is a widely used approach in hybrid censoring (see, e.g., Casella and Berger, 2002, Theorem 9.2.12; Hahn et al., 2017, Section D6; Balakrishnan et al., 2014). For $\alpha_1, \alpha_2, \alpha \in (0,1)$ with $\alpha = \alpha_1 + \alpha_2$, a two-sided exact confidence interval for θ can be obtained, when, for all t in the support of $\widehat{\theta}$, the equations

$$\Pr_\theta(\widehat{\theta} > t) = 1 - \alpha_1 \quad \text{and} \quad \Pr_\theta(\widehat{\theta} > t) = \alpha_2 \tag{5.15}$$

have unique solutions w.r.t. $\theta \in \Theta$. Denoting these solutions by $\ell(t) < u(t)$, $[\ell(\widehat{\theta}), u(\widehat{\theta})]$ provides an exact $(1-\alpha)$-confidence interval for θ. To ensure the existence and uniqueness of the solutions, the following two conditions are generally imposed.

Condition 5.13. **(I)** *$\widehat{\theta}$ is stochastically monotone w.r.t. $\theta \in \Theta$, that is, the survival function $\Pr_\theta(\widehat{\theta} > t)$ is an increasing (decreasing) function of θ (for every fixed $t \in \mathbb{R}$);*
(II) *for any $\beta \in (0,1)$ and $t \in \operatorname{supp}(\widehat{\theta})$, the equation $\Pr_\theta(\widehat{\theta} > t) = \beta$ has a solution w.r.t. $\theta \in \Theta$.*

These two conditions are addressed in the following two subsections.

Stochastic monotonicity of the estimator

General approaches to verify Condition (I) have been proposed in Balakrishnan and Iliopoulos (2009, 2010) and van Bentum and Cramer (2019). The criteria are commonly called *Three Monotonicities Lemma (TML)*. We present the result in the more general form shown in van Bentum and Cramer (2019).

Lemma 5.14 (Generalized Three Monotonicities Lemma). *Let $X \sim F_\theta$ be a random variable so that its cumulative distribution function F_θ belongs to a parametric family of distributions with parameter $\theta \in \Theta \subseteq \mathbb{R}$. Furthermore, let $\boldsymbol{Y}$ be an n-dimensional random vector, so that, by the law of total probability, the survival function of X can be written as*

$$\Pr_\theta(X > x) = \int_{supp(\boldsymbol{Y})} \Pr_\theta(X > x \mid \boldsymbol{Y} = \boldsymbol{y})\, dP_\theta^{\boldsymbol{Y}}(\boldsymbol{y}), \quad x \in \mathbb{R}, \tag{5.16}$$

with $\theta \in \Theta$ and support $supp(\boldsymbol{Y}) \subseteq \mathbb{R}^n$ of $\boldsymbol{Y}$. Suppose the following monotonicities hold:

(M1) *For every $x \in \mathbb{R}$ and $\boldsymbol{y} \in supp(\boldsymbol{Y})$, the function $\Pr_\theta(X > x \mid \boldsymbol{Y} = \boldsymbol{y})$ is monotonically increasing in $\theta \in \Theta$, that is, for any $\boldsymbol{y} \in supp(\boldsymbol{Y})$, the distribution of X conditionally on $\boldsymbol{Y} = \boldsymbol{y}$ is monotonically increasing in $\theta \in \Theta$ w.r.t. the stochastic order.*

(M2) *For every $x \in \mathbb{R}$ and $\theta \in \Theta$, the function $\Pr_\theta(X > x \mid \boldsymbol{Y} = \boldsymbol{y})$ is monotonically decreasing (increasing) in $\boldsymbol{y} \in supp(\boldsymbol{Y})$, that is, the distribution of X conditionally on $\boldsymbol{Y} = \boldsymbol{y}$ is monotonically decreasing (increasing) in $\boldsymbol{y} \in supp(\boldsymbol{Y})$ w.r.t. the stochastic order.*

(M3) *For every increasing (decreasing) function $\Phi: \mathbb{R}^n \to \mathbb{R}$, the inequality $E_\theta \Phi(\boldsymbol{Y}) \le E_{\theta'} \Phi(\boldsymbol{Y})$ holds for all $\theta, \theta' \in \Theta$ with $\theta < \theta'$, that is, $\boldsymbol{Y}$ is monotonically decreasing (increasing) in $\theta \in \Theta$ w.r.t. the stochastic order.*

Then, $\Pr_\theta(X > x) \le \Pr_{\theta'}(X > x)$ for every $x \in \mathbb{R}$ and for all $\theta < \theta' \in \Theta$, i.e., X is increasing in $\theta \in \Theta$ w.r.t. the stochastic order.

Applying Lemma 5.14, we obtain the result in Theorem 5.16 ensuring the monotonicity of the (conditional) survival function. In the proof, we use the following result which has been proved in van Bentum and Cramer (2019) for generalized order statistics.

Lemma 5.15. *Let $X_{1:n}, \ldots, X_{n:n}$ be order statistics based on $Exp(\vartheta)$, $\vartheta > 0$. Then, for $m \in \{2, \ldots, n\}$, $[X_{1:n}, \ldots, X_{m-1:n} \mid X_{m:n} = s]$, $[X_{m+1:n}, \ldots, X_{n:n} \mid X_{m:n} = s]$ are*

(a) *independent,*

(b) *stochastically increasing in* $s > 0$,
(c) *stochastically increasing in* $\vartheta > 0$.

Furthermore, for random variables $V_1, \ldots, V_n \stackrel{iid}{\sim} Exp(1)$

(d) $[X_{m+1:n}, \ldots, X_{n:n} \mid X_{m:n} = s] \stackrel{d}{=} (s + \vartheta W_{m,i})_{i=m+1,\ldots,n}$ *where* $W_{m,i} = \sum_{j=m+1}^{i} V_j/(n-j+1)$, $m+1 \leq i \leq n$.

The stochastic monotonicity has first been established in Balakrishnan and Iliopoulos (2009). It can be deduced as a special case from results presented in Burkschat et al. (2016) and van Bentum and Cramer (2019), respectively.

Theorem 5.16. *The conditional survival function* $\overline{F}_{\vartheta}^{\widehat{\vartheta}|D_I \geq 1}(t) = \Pr_{\vartheta}(\widehat{\vartheta} > t \mid D_I \geq 1)$ *of the MLE* $\widehat{\vartheta}$ *under Type-I hybrid censoring given in Theorem 5.4 is decreasing in* $\vartheta > 0$ *for any fixed* $t > 0$.

Proof. Noticing that $\{D_I \geq 1\} = \{X_{1:n} \leq T\}$, we use (5.16) in the form

$$\begin{aligned}\Pr_{\vartheta}(\widehat{\vartheta} > t \mid D_I \geq 1) &= \Pr_{\vartheta}(\widehat{\vartheta} > t \mid X_{1:n} \leq T)\\ &= \int_0^T \Pr_{\vartheta}(\widehat{\vartheta} > t \mid X_{1:n} = s)\, dP_{\vartheta}^{X_{1:n}|X_{1:n} \leq T}(s), \qquad t \in \mathbb{R}.\end{aligned}$$

Let $D = \sum_{j=1}^{n} \mathbb{1}_{(-\infty, T]}(X_{j:n})$ be defined as in (4.2). Conditioning D on $X_{1:n} = s \in [0, T]$ yields

$$[D \mid X_{1:n} = s] \stackrel{d}{=} \mathbb{1}_{(0,T]}(s) + \sum_{i=2}^{n} \mathbb{1}_{(0,T]}(s + \vartheta W_{1,i}) = D^{(\vartheta,s)}, \text{ say}$$

where $W_{1,2}, \ldots, W_{1,n}$ are random variables as in Lemma 5.15. Then, $D^{(\vartheta,s)}$ is decreasing in both $\vartheta > 0$ and $s > 0$ w.r.t. the stochastic order. Conditionally on $X_{1:n} = s$, $s > 0$, the MLE has the representation

$$\left[\widehat{\vartheta} \mid X_{1:n} = s\right]_{\vartheta} \stackrel{d}{=} \frac{1}{D^{(\vartheta,s)}}\Big(s + \sum_{j=2}^{D^{(\vartheta,s)}} (s + \vartheta W_{1,j}) + (n - D^{(\vartheta,s)})(s + \vartheta W_{1,m}) \wedge T\Big), \qquad s \leq T,$$

for $\vartheta, s > 0$, where $\sum_a^b \ldots = 0$ whenever $a > b$.

Now, we prove the stochastic monotonicity of $\widehat{\vartheta}$. Let $0 < w_2 \leq \cdots \leq w_m$.

(M1) Let $0 < \vartheta < \vartheta'$. For $s \leq T$, we get

$$\begin{aligned}&\frac{1}{D^{(\vartheta,s)}}\Big(s + \sum_{j=2}^{D^{(\vartheta,s)}} (s + \vartheta w_j) + (n - D^{(\vartheta,s)})(s + \vartheta w_m) \wedge T\Big)\\ &\leq \frac{1}{D^{(\vartheta',s)}}\Big(s + \sum_{j=2}^{D^{(\vartheta,s)}} (s + \vartheta w_j) + (n - D^{(\vartheta,s)})(s + \vartheta w_m) \wedge T\Big).\end{aligned}$$

Noticing that $s+\vartheta w_j \leq (s+\vartheta w_m)\wedge T$ for $j\in\{2,\ldots,D^{(\vartheta,s)}\}$ and $D^{(\vartheta',s)}\leq D^{(\vartheta,s)}$, we have

$$\begin{aligned}
&\frac{1}{D^{(\vartheta',s)}}\Big(s+\sum_{j=2}^{D^{(\vartheta,s)}}(s+\vartheta w_j)+(n-D^{(\vartheta,s)})(s+\vartheta w_m)\wedge T\Big)\\
&\leq \frac{1}{D^{(\vartheta',s)}}\Big(s+\sum_{j=2}^{D^{(\vartheta',s)}}(s+\vartheta w_j)\\
&\qquad\qquad +\sum_{j=D^{(\vartheta',s)}+1}^{D^{(\vartheta,s)}}(s+\vartheta w_m)\wedge T+(n-D^{(\vartheta,s)})(s+\vartheta w_m)\wedge T\Big)\\
&=\frac{1}{D^{(\vartheta',s)}}\Big(s+\sum_{j=2}^{D^{(\vartheta',s)}}(s+\vartheta w_j)+(n-D^{(\vartheta',s)})(s+\vartheta w_m)\wedge T\Big)\\
&\leq \frac{1}{D^{(\vartheta',s)}}\Big(s+\sum_{j=2}^{D^{(\vartheta',s)}}(s+\vartheta' w_j)+(n-D^{(\vartheta',s)})(s+\vartheta' w_m)\wedge T\Big).
\end{aligned}$$

(M2) Let $0<s<s'<T$. Then, the property is established along the same lines as in the proof of **(M1)**.

(M3) It is sufficient to show that $\Pr_\vartheta(X_{1:n}\leq t\mid X_{1:n}\leq T)=\frac{1-e^{-nt/\vartheta}}{1-e^{-nT/\vartheta}}$, $0\leq t\leq T$, is a decreasing function of ϑ. Notice that this means that the right truncated exponential distribution with parameter ϑ and truncation point $T>0$ is stochastically increasing in ϑ. A formal proof can be found in Laumen (2017, p. 65/66).

Summing up, Lemma 5.15 yields that the MLE is stochastically increasing in $\vartheta\in(0,\infty)$. □

Limits of the survival function in ϑ

The second condition (II) in Condition 5.13 addresses the limits of the survival function w.r.t. the parameter ϑ. From Balakrishnan et al. (2014), we get the following result. Recall that the support of $F_\vartheta^{\widehat{\vartheta}|D_I\geq 1}$ is a subset of $[0,nT]$ (see Remark 5.9). The limit is plotted in Fig. 5.7. Alternatively, one may use the B-spline representation of the density function given in (5.13) and the respective representation of the cumulative distribution function as an integral of the density function (see Remark 5.24).

Lemma 5.17. *For $\vartheta>0$, let the conditional cumulative distribution function of the MLE $\widehat{\vartheta}$ under Type-I hybrid censoring be given by $F_\vartheta^{\widehat{\vartheta}|D_I\geq 1}$ (see Theorem 5.4). Then, for $0<t\leq nT$,*

$$\lim_{\vartheta\to 0}F_\vartheta^{\widehat{\vartheta}|D_I\geq 1}(t)=1,\qquad \lim_{\vartheta\to\infty}F_\vartheta^{\widehat{\vartheta}|D_I\geq 1}(t)=\begin{cases}0, & t\leq (n-1)T\\ \frac{(t-(n-1)T)_+}{T}, & (n-1)T<t\leq nT\end{cases}.$$

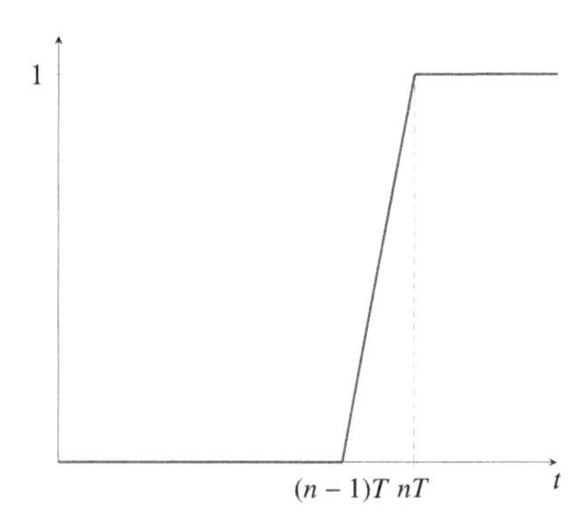

Figure 5.7 Limit $\lim_{\vartheta\to\infty} F_{\vartheta}^{\widehat{\vartheta}|D_1\geq 1}(t)$.

Proof. Let $0 \le t \le nT$. Using that $\lim_{\vartheta\to 0}(1-e^{-nT/\vartheta})=1$, $\lim_{\vartheta\to 0} F_{\Gamma(\vartheta/d,d);(n-d+k)T/d}(t)=1$, and (see Theorem 5.4)

$$F_{\vartheta}^{\widehat{\vartheta}|D_1\geq 1}(t)=(1-e^{-nT/\vartheta})^{-1}\Bigg[\sum_{d=1}^{m-1}\sum_{k=0}^{d}\binom{n}{d}\binom{d}{k}(-1)^k e^{-(n-d+k)T/\vartheta}F_{\Gamma(\vartheta/d,d);(n-d+k)T/d}(t)$$
$$+\sum_{j=m}^{n}\sum_{k=0}^{j}\binom{n}{j}\binom{j}{k}(-1)^k e^{-(n-j+k)T/\vartheta}F_{\Gamma(\vartheta/m,m);(n-j+k)T/m}(t)\Bigg],\ t\in\mathbb{R},$$

the first limit is obvious. For $\vartheta\to\infty$, we apply l'Hospital's rule to the ratio (see Balakrishnan et al., 2014)

$$\frac{F_{\Gamma(\vartheta/d,d);(n-d+k)T/d}(t)}{1-e^{-nT/\vartheta}}=\begin{cases}0, & d>1\\ \frac{(t-(n-1+k)T)_+}{nT}, & d=1\end{cases},\ k\in\{0,1\}.$$

Hence, we get

$$\lim_{\vartheta\to\infty} F_{\vartheta}^{\widehat{\vartheta}|D_1\geq 1}(t)=\begin{cases}0, & t\le(n-1)T\\ \frac{t}{T}-(n-1), & (n-1)T<t\le nT\\ 1, & nT\le t\end{cases}. \tag{5.17}$$

□

Lemma 5.17 tells us that the equations given in (5.15) may not have a solution. This problem is present only when $t\in((n-1)T,nT)$. According to Remark 5.9, this interval is relevant if and only if exactly one failure is observed, that is, $D=1$. Therefore, one may guess that a single observation is not enough to construct a confidence interval with finite endpoints and that the problem can be overcome by requiring more than one failure to be observed. However, as pointed out in Balakrishnan et al. (2014), this approach does not resolve the problem. They illustrated that the problem remains whatever number of observations is required. Notice that, for any $\vartheta > 0$, the probability $\Pr_\vartheta(D = 1 \mid D \geq 1) = n(e^{T/\vartheta} - 1)/(e^{nT/\vartheta} - 1)$ tends to zero for $n \to \infty$ quite fast. Therefore, although positive for any initial sample size, a larger sample size leads to a higher probability that the solutions exist.

5.3.2.2 *Alternative approaches*

Since the computation of exact (conditional) confidence interval is somewhat involved and poses a problem when n is large, Gupta and Kundu (1998) proposed an approximate confidence interval by using the asymptotic normality of the MLE $\widehat{\vartheta}$ and the observed Fisher information. Specifically, given $D_{\text{I}} = d$, the distribution of $\widehat{\vartheta}$ for large n is approximately normal with mean ϑ and variance $D_{\text{I}}/\widehat{\vartheta}^2$. Using this, an approximate $100(1-\alpha)\%$ confidence interval for ϑ can be easily constructed.

It should be mentioned that parametric and non-parametric bootstrap can also be used effectively for constructing approximate confidence intervals.

Another proposal to construct a confidence interval is due to Fairbanks et al. (1982). They proposed a $100(1-\alpha)\%$ confidence interval $\mathcal{C}_{D_{\text{I}}}$ for ϑ under Type-I hybrid censoring as follows.

$$
\begin{aligned}
\mathcal{C}_{D_{\text{I}}} &= \left[\frac{2\mathsf{TTT}_{D_{\text{I}}}}{\chi^2_{2,\alpha/2}}, \infty\right), && \text{if } D_{\text{I}} = 0,\\
\mathcal{C}_{D_{\text{I}}} &= \left[\frac{2\mathsf{TTT}_{D_{\text{I}}}}{\chi^2_{2D_{\text{I}}+2,\alpha/2}}, \frac{2\mathsf{TTT}_{D_{\text{I}}}}{\chi^2_{2D_{\text{I}},1-\alpha/2}}\right], && \text{if } 1 \leq D_{\text{I}} \leq m-1,\\
\mathcal{C}_{D_{\text{I}}} &= \left[\frac{2\mathsf{TTT}_{D_{\text{I}}}}{\chi^2_{2D_{\text{I}},\alpha/2}}, \frac{2\mathsf{TTT}_{D_{\text{I}}}}{\chi^2_{2D_{\text{I}},1-\alpha/2}}\right], && \text{if } D_{\text{I}} = m,
\end{aligned}
$$

where $\mathsf{TTT}_{D_{\text{I}}}$ is the total time on test given in (5.6) and $\chi^2_{k,\beta}$ denotes the upper β percentage point of a χ^2-distribution with k degrees of freedom. Motivated by the *replacement setting*, Fairbanks et al. (1982) argued that this interval could be used as a confidence interval but failed to provide a formal proof. However, simulations provided in Gupta and Kundu (1998) illustrate that this approach provides reasonable coverage probabilities.

5.3.3 Bayesian inference

Draper and Guttman (1987) considered Bayesian inference for the parameter ϑ based on Type-I hybrid censored data. For this purpose, they assumed that ϑ has an inverse gamma prior $\mathsf{I}\Gamma(\lambda, \beta)$ with density function

$$
\pi(\vartheta) = \frac{\lambda^\beta}{\Gamma(\beta)} \vartheta^{-(\beta+1)} e^{-\lambda/\vartheta}, \quad \vartheta > 0, \tag{5.18}
$$

where $\beta > 0$ and $\lambda > 0$ are the hyper-parameters (see Section 3.2.4 and Section 3.3 for Type-II and Type-I censoring, respectively). For the choice $\beta = \lambda = 0$, the prior in (5.18) becomes a non-informative prior. Based on the inverse gamma prior in (5.18) and the likelihood function in (5.3), the posterior density function $p(\cdot \mid \text{data})$ of ϑ becomes

$$p(\vartheta \mid \text{data}) = \frac{(\mathsf{TTT}_d + \lambda)^{d+\beta}}{\Gamma(d+\beta)} \vartheta^{-(d+\beta+1)} e^{-(\mathsf{TTT}_d+\lambda)/\vartheta}, \quad \vartheta > 0, \tag{5.19}$$

where the total time on test TTT_d is defined in (5.5). Since the posterior density function is the density function of an inverse gamma distribution (see Definition B.8) with parameters $d + \beta > 2$ and $\mathsf{TTT}_d + \lambda$, the Bayesian estimator of ϑ under the squared-error loss function, being the posterior mean, is simply obtained as

$$\widehat{\vartheta}_{\mathsf{B}} = \frac{\mathsf{TTT}_{D_{\mathsf{I}}} + \lambda}{D_{\mathsf{I}} + \beta - 1} \tag{5.20}$$

provided $D_{\mathsf{I}} + \beta > 1$. It is clear that the MLE of ϑ presented in (5.6) coincides with the Bayesian estimator in (5.20) when $\beta = 1$ and $\lambda = 0$. Moreover, a $100(1-\alpha)\%$ credible interval for ϑ is easily obtained as

$$\left(\frac{2(\mathsf{TTT}_{D_{\mathsf{I}}} + \lambda)}{\chi^2_{2(D_{\mathsf{I}}+\beta),\alpha/2}}, \frac{2(\mathsf{TTT}_{D_{\mathsf{I}}} + \lambda)}{\chi^2_{2(D_{\mathsf{I}}+\beta),1-\alpha/2}} \right)$$

provided $D_{\mathsf{I}} + \beta > 0$ with an integer β, where $\chi^2_{k,\alpha}$ denotes as before the upper α percentage point of the χ^2-distribution with k degrees of freedom.

5.3.4 Two-parameter exponential distribution

Childs et al. (2012) discussed likelihood inference for the parameters of a two-parameter exponential distribution $Exp(\mu, \vartheta)$ when the data are Type-I hybrid censored. It has been observed by Childs et al. (2012) that, for any $m \in \{1, \dots, n\}$, the MLEs of ϑ and μ do not exist when $d = 0$. Based on the observed sample as above, the likelihood function is given by

$$\begin{aligned}
\mathscr{L}(\mu, \vartheta; \boldsymbol{x}_d, d) &= \frac{n!}{(n-d)!\vartheta^d} \exp\left\{ -\frac{1}{\vartheta}\left[\sum_{j=1}^{d} (x_j - \mu) + (n-d)(w-\mu) \right] \right\} \\
&= \frac{n!}{(n-d)!\vartheta^d} \exp\left\{ -\frac{1}{\vartheta}\left[\sum_{j=2}^{d} (n-j+1)(x_j - x_{j-1}) + (n-d)(w - x_d) \right] \right\} \\
&\quad \times e^{-n(x_1-\mu)/\vartheta} \mathbb{1}_{(-\infty, x_1]}(\mu),
\end{aligned} \tag{5.21}$$

for $x_1 \le \cdots \le x_d \le T$. The likelihood function in (5.21) is an increasing function in μ for any $\vartheta > 0$. Therefore, $\widehat{\mu} = x_1$ is the MLE of μ leading to the upper bound

$$\mathscr{L}(\widehat{\mu}, \vartheta; \boldsymbol{x}_d, d) = \frac{n!}{(n-d)!\vartheta^d} \exp\left\{ -\frac{\mathsf{TTT}_d^*}{\vartheta} \right\}$$

with

$$\mathsf{TTT}_d^* = \sum_{j=2}^{d} (n-j+1)(x_j - x_{j-1}) + (n-d)(w - x_d). \tag{5.22}$$

When $m = 1$ (and $d = 1$), $\mathsf{TTT}_1^* = 0$ so that $\mathscr{L}(\widehat{\mu}, \vartheta; \boldsymbol{x}_1, 1) = \frac{n}{\vartheta}$ is a decreasing function of ϑ. Thus, the MLE of ϑ does not exist although the MLE of μ is given by $\widehat{\mu} = X_{1:n}$. Hence, a single observation is not sufficient to estimate both parameters in this case.

For $m > 1$, the situation is different. Using standard arguments, the MLEs of μ and ϑ are given by

$$\widehat{\mu} = X_{1:n} \quad \text{and} \quad \widehat{\vartheta} = \frac{1}{D_{\mathrm{I}}}\Big[\sum_{j=1}^{D_{\mathrm{I}}} (X_{j:n} - X_{j-1:n}) + (n - D_{\mathrm{I}})(W - X_{1:n})\Big].$$

Notice that in the case of one observation, i.e., $D_{\mathrm{I}} = 1$, ϑ is estimated by $\widehat{\vartheta} = T - X_{1:n}$. Hence, the knowledge that the experiment is terminated at time T yields an estimator of the scale parameter.

Notice that $\widehat{\vartheta}$ is bounded and that $\widehat{\vartheta} = \frac{1}{D_{\mathrm{I}}}\mathsf{TTT}_{D_{\mathrm{I}}}^* d$, with $\mathsf{TTT}_{D_{\mathrm{I}}}^*$ as in (5.22). Thus, we get from $x_j \le T$, $1 \le j \le d$, the bounds

$$\mu \le \widehat{\mu} \le T, \quad 0 \le \widehat{\vartheta} \le (n-1)(T-\mu), \qquad d \ge 1,$$

which shows that both $\widehat{\mu}$ and $\widehat{\vartheta}$ have finite supports provided $D_{\mathrm{I}} = d \ge 1$. Distributional properties of the MLEs have been studied in Cramer and Balakrishnan (2013).

Theorem 5.18. *The distribution of $\widehat{\mu}$ is a right truncated Exp$(\mu, \vartheta/n)$-distribution with truncation point $T \ge \mu$. Specifically,*

$$\Pr_{\mu,\vartheta}\left(\frac{\widehat{\mu} - \mu}{\vartheta} > t \,\Big|\, D_I \ge 1\right) = \frac{e^{-nt} - e^{-n(T-\mu)/\vartheta}}{1 - e^{-n(T-\mu)/\vartheta}}, \quad 0 \le t \le (T-\mu)/\vartheta.$$

Furthermore, for any $\vartheta > 0$, $\Pr_{\mu,\vartheta}\big(\widehat{\mu} > t \mid D_I \ge 1\big)$ is an increasing function of μ.

The (conditional) distribution of $\mathsf{TTT}_{D_{\mathrm{I}}}^*$ has been established in Cramer and Balakrishnan (2013) (see also Górny, 2017, Section 4.2.2) and can be used to find the bivariate density function of $\widehat{\mu}$ and $\widehat{\vartheta}$. As pointed out in Górny (2017), the conditional density function does only exist for $D_{\mathrm{I}} \ge 2$ (which implies $m > 1$). In this case, introducing the notation $T_x = T - x$, we get under Type-I hybrid censoring

$$f_{\mu,\vartheta}^{\widehat{\mu},\widehat{\vartheta}|D_{\mathrm{I}}\geq 2}(x,t) = \frac{n}{\Pr(D_{\mathrm{I}}\geq 2)} e^{-n(x-\mu)/\vartheta}$$
$$\times \Bigg[\sum_{d=2}^{m-1} d \binom{n-1}{d-1} \frac{T_x^{d-1}}{\vartheta^d} B_{d-2}\Big(dt \mid (n-d)T_x, \ldots, (n-1)T_x\Big) e^{-dt/\vartheta}$$
$$+ m \binom{n-1}{m-1} \frac{T_x^{m-1}}{\vartheta^m} B_{m-2}\Big(dt \mid 0, (n-m+1)T_x, \ldots, (n-1)T_x\Big) e^{-mt/\vartheta} \Bigg],$$
$$\mu \leq x \leq T, 0 \leq t \leq \frac{(n-1)}{d}(T-\mu),$$

where

$$\Pr(D_{\mathrm{I}} \geq 2) = 1 - \Pr(D=0) - \Pr(D=1) = 1 + (n-1)e^{-n(T-\mu)/\vartheta} - ne^{-(n-1)(T-\mu)/\vartheta}.$$

Expressions for $f_{\mu,\vartheta}^{\widehat{\vartheta}|D_{\mathrm{I}}\geq 1}$ can be obtained from this expression by integration w.r.t. x and proceeding as in the case of a known location parameter. Further discussions in this direction can be found Section 7.2.2.2 for progressive hybrid censoring where the connection to the one-parameter exponential distribution is illustrated in detail (see also Górny, 2017).

Remark 5.19. Bayesian inference has also been discussed for the two-parameter exponential distribution (see Kundu et al., 2013). They assumed a uniform prior for the location parameter and an (independent) gamma prior for the scale parameter $\lambda = 1/\vartheta$ (for Type-II hybrid censoring, see Bayoud, 2014).

5.4. Type-II hybrid censoring

Inference for Type-II hybrid censored data can be conducted similarly to Type-I hybrid censoring. In particular, the methods applied under Type-I hybrid censoring can also be used by adapting them properly to the respective data. Similar comments apply to other hybrid censoring schemes. Therefore, we illustrate how these approaches can be utilized under Type-II hybrid censoring. For the other hybrid censoring schemes, we will only provide results and references where details can be found. We assume in the following Type-II hybrid censored data $X_{1:n}, \ldots, X_{D_{\mathrm{II}}:n}$ with $D_{\mathrm{II}} = D \vee m$ as described in Fig. 4.6 (see also (4.26)) and exponentially distributed lifetimes.

5.4.1 Likelihood inference and exact distribution of MLE

Taking into account the random variable $D_{\mathrm{II}} = D \vee m$ in (4.25), we have the data $\boldsymbol{x}_d$ with $x_1 \leq \cdots \leq x_d$ and $D_{\mathrm{II}} = d \in \{m, \ldots, n\}$. Using (4.31), the likelihood function reads

$$\mathscr{L}(\vartheta; \boldsymbol{x}_d, d) = f_\vartheta^{X_{j:n}, 1\leq j\leq D_{\mathrm{II}}, D}(\boldsymbol{x}_d, d)$$

$$= \begin{cases} f_{\vartheta;1,\dots,m:n}(\boldsymbol{t}_m), & d < m \\ \frac{F_\vartheta(T)}{(n-d)f_\vartheta(T)} f_{1,\dots,d+1:n}(\boldsymbol{x}_d, T), & d \in \{m, \dots, n-1\} \\ f_{\vartheta;1,\dots,n:n}(\boldsymbol{x}_n), & d = n \end{cases}.$$

For an $Exp(\mu, \vartheta)$-distribution, this yields the expression

$$\mathscr{L}(\mu, \vartheta; \boldsymbol{x}_d, d) = \frac{n!}{(n-d)!\vartheta^d} \exp\Big(-\frac{1}{\vartheta}\sum_{j=1}^{d}(x_j - \mu) - \frac{n-d}{\vartheta}(w-\mu)\Big), \tag{5.23}$$

where

$$w = T + \mathbb{1}_{\{m\}}(d)(x_m - T) + \mathbb{1}_{\{n\}}(d)(x_n - T).$$

Assuming $\mu = 0$ and introducing $\mathsf{TTT}_d = \sum_{j=1}^{d} x_j - (n-d)w$, we get

$$\mathscr{L}(\vartheta; \boldsymbol{x}_d, d) = \frac{n!}{(n-d)!\vartheta^d} \exp\Big(-\frac{1}{\vartheta}\mathsf{TTT}_d\Big),$$

where the total time on test TTT_d is defined by $(x_1 \le \dots \le x_m)$

$$\mathsf{TTT}_d = \sum_{j=1}^{d} x_j + (n-d)w, \quad d = m \vee \sum_{j=1}^{n} \mathbb{1}_{(-\infty,T]}(x_j). \tag{5.24}$$

Thus, the MLE of ϑ is given by (see Childs et al., 2003; Cramer et al., 2016)

$$\widehat{\vartheta} = \frac{1}{D_{\text{II}}}\Big[\sum_{j=1}^{D_{\text{II}}} X_{j:n} + (n - D_{\text{II}})W\Big] = \frac{1}{D_{\text{II}}}\mathsf{TTT}_{D_{\text{II}}}.$$

As for Type-I hybrid censoring, the exact distribution can be studied. It can be derived using either the moment generating function approach or the spacings approach. For illustration, we show the similarity to the case of Type-I hybrid censoring. Notice that we do not have to condition on a minimum sample size since Type-II hybrid censoring ensures a minimum of m observations. The following representation of the moment generating function is due to Childs et al. (2003).

Theorem 5.20. *For $\vartheta > 0$, $t < m/\vartheta$, and $d \in \{m, \dots, n\}$, let $\vartheta_d = \frac{\vartheta}{1-t\vartheta/d} > 0$. Then, the moment generating function of the MLE $\widehat{\vartheta}$ under Type-II hybrid censoring is given by*

$$M_\vartheta(t) = \frac{1}{(1-t\vartheta/m)^m}\sum_{j=0}^{m-1}\binom{n}{j}(1 - e^{-T(1-t\vartheta/m)/\vartheta})^j e^{-(n-j)T(1-t\vartheta/m)/\vartheta}$$

$$+\sum_{d=m}^{n}\binom{n}{d}\frac{1}{(1-t\vartheta/d)^{d}}(1-e^{-T(1-t\vartheta/d)/\vartheta})^{d}e^{-(n-d)T(1-t\vartheta/d)/\vartheta}, \qquad t<m/\vartheta.$$

Proof. First, recall that $\widehat{\vartheta}=\mathrm{TTT}_d/d$ for data $x_1\le\cdots\le x_d\le T$, $D_{\mathrm{II}}=d\in\{m,\ldots,n\}$. Moreover, $\{D_{\mathrm{II}}=m\}=\{D<m\}\cup\{D=m\}$ and $\{D_{\mathrm{II}}=d\}=\{D=d\}$, $d\in\{m+1,\ldots,n\}$. Then, we get from (4.20) for $t<1/\vartheta$

$$\begin{aligned}
M_\vartheta(t)&=E_\vartheta(e^{t\widehat{\vartheta}})=\Pr_\vartheta(D<m)E_\vartheta(e^{-t\widehat{\vartheta}}\mid D<m)+\sum_{d=m}^{n}\Pr_\vartheta(D=d)E_\vartheta(e^{-t\widehat{\vartheta}}\mid D=d)\\
&=\int_{\mathbb{R}^m_{\le}} e^{t\cdot\mathrm{TTT}_m/m}f_\vartheta^{X_{j:n},1\le j\le m,D<m}(\boldsymbol{x}_m)\mathbb{1}_{[T,\infty)}(x_m)\,d\boldsymbol{x}_m\\
&\qquad+\sum_{d=m}^{n}\int_{\mathbb{R}^d_{\le T}} e^{t\cdot\mathrm{TTT}_d/d}f_\vartheta^{X_{j:n},1\le j\le d,D}(\boldsymbol{x}_d,d)\,d\boldsymbol{x}_d\\
&=\frac{\vartheta_m^m}{\vartheta^m}\int_{\mathbb{R}^m_{\le}} e^{t\cdot\mathrm{TTT}_m/m}f_\vartheta^{X_{j:n},1\le j\le m,D<m}(\boldsymbol{x}_m)\mathbb{1}_{[T,\infty)}(x_m)\,d\boldsymbol{x}_m\\
&\quad+\sum_{d=m}^{n}\frac{\vartheta_d^d}{\vartheta^d}\int_{\mathbb{R}^d_{\le T}} e^{t\cdot\mathrm{TTT}_d/d}f_\vartheta^{X_{j:n},1\le j\le d,D}(\boldsymbol{x}_d,d)\,d\boldsymbol{x}_d
\end{aligned}$$

with $\vartheta_d=\frac{d\vartheta}{d-t\vartheta}=\frac{\vartheta}{1-t\vartheta/d}$. Proceeding as in the proof of Theorem 5.2, we obtain from the density property as well as (4.21) and (4.22)

$$\begin{aligned}
&=\frac{\vartheta_m^m}{\vartheta^m}\Pr_{\vartheta_d}(D<m)+\sum_{d=1}^{m}\frac{\vartheta_d^d}{\vartheta^d}\Pr_{\vartheta_d}(D=d)\\
&=\frac{1}{(1-t\vartheta/m)^m}\sum_{j=0}^{m-1}\binom{n}{j}(1-e^{-T/\vartheta_m})^j e^{-(n-j)T/\vartheta_m}\\
&\quad+\sum_{d=m}^{n}\frac{1}{(1-t\vartheta/d)^d}\binom{n}{d}(1-e^{-T/\vartheta_d})^d e^{-(n-d)T/\vartheta_d}.
\end{aligned}$$

Recalling the definition of ϑ_d, we get the desired representation. □

A comparison of the proofs of Theorem 5.2 and 5.20 shows obvious similarities. In fact, this approach can be utilized in the same manner for other hybrid censoring schemes. Inverting the moment generating function yields then expressions for the density function, the cumulative distribution function, and the survival function. Alternatively, one can also use the spacings approach as has been done in Cramer et al. (2016) which leads to a compact representation in terms of B-splines. In addition to the representations of $f_\vartheta^{\mathrm{TTT}_d|D=d}$, $d=m,\ldots,n$, given in Theorem 5.7, we need the following result for $f_\vartheta^{\mathrm{TTT}_m|D<m}$ due to Cramer et al. (2016). A simple argument to derive the conditional

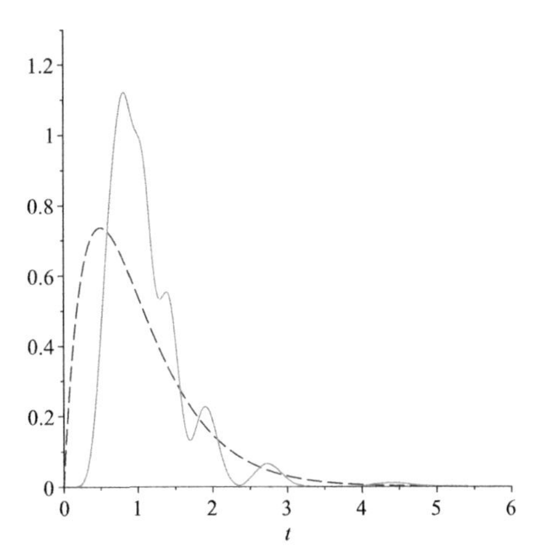

Figure 5.8 Plots of $f_{\vartheta}^{\widehat{\vartheta}}$ and density function of $\Gamma(\vartheta/m, m)$-distribution with $n = 10, m = 2, \vartheta = 1, T = 1$.

density function given in Theorem 5.21 is provided in the derivation of the same result for progressively Type-II censored data (see Theorem 7.8).

Theorem 5.21. *The conditional density function of* TTT_m *given* $D < m$ *is given by*

$$f_{\vartheta}^{\mathrm{TTT}_m|D<m}(s) = \frac{s^{m-1}e^{-s/\vartheta}}{(m-1)!\vartheta^m(1-F_{m:n}(T))} - \binom{n}{m}\frac{T^m}{\vartheta^m(1-F_{m:n}(T))}B_{m-1}(s\,|\,0,(n-m+1)T,\ldots,nT)e^{-s/\vartheta}, \quad s \geq 0.$$

Summing up, we arrive at the following representation of the density function of the MLE.

Theorem 5.22. *The density function of the MLE* $\widehat{\vartheta}$ *under Type-II hybrid censoring is given by*

$$f_{\vartheta}^{\widehat{\vartheta}}(s) = \frac{m^m s^{m-1}e^{-ms/\vartheta}}{(m-1)!\vartheta^m} - m\binom{n}{m}\frac{T^m}{\vartheta^m}B_{m-1}(ms\,|\,0,(n-m+1)T,\ldots,nT)e^{-ms/\vartheta} + \sum_{d=m}^{n} d\binom{n}{d}\frac{T^d}{\vartheta^d}B_{d-1}(ds\,|\,(n-d)T,\ldots,nT)e^{-ds/\vartheta}, \quad s \geq 0.$$

Some plots of density functions are provided in Figs. 5.8–5.11. The figures illustrate that the density function of $\widehat{\vartheta}$ tends to the density function of a $\Gamma(\vartheta/m, m)$-distribution

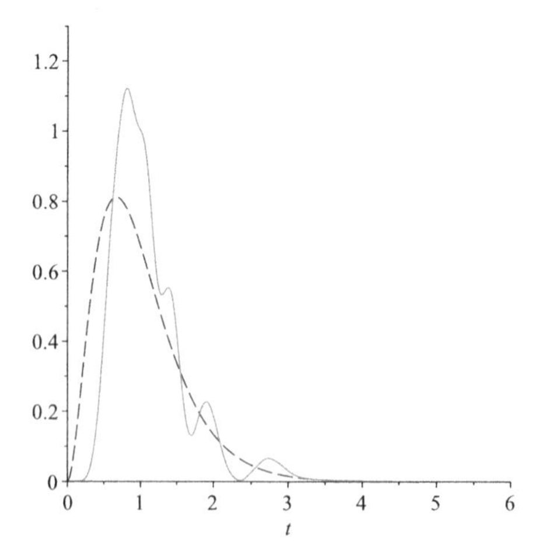

Figure 5.9 Plots of $f_{\vartheta}^{\widehat{\vartheta}}$ and density function of $\Gamma(\vartheta/m, m)$-distribution with $n = 10, m = 3$, $\vartheta = 1, T = 1$.

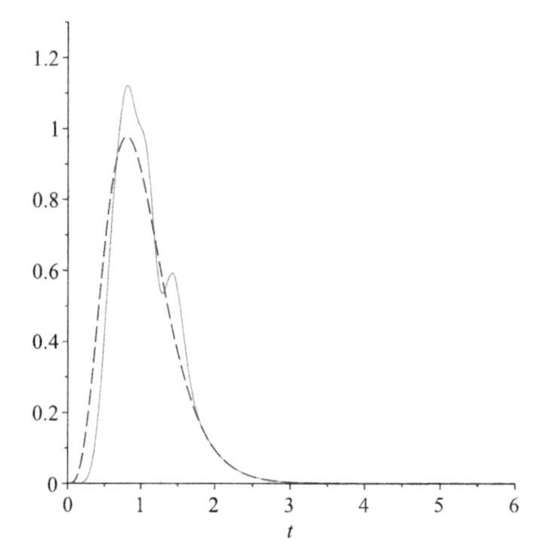

Figure 5.10 Plots of $f_{\vartheta}^{\widehat{\vartheta}}$ and density function of $\Gamma(\vartheta/m, m)$-distribution with $n = 10, m = 5$, $\vartheta = 1, T = 1$.

for larger values of m which is given by

$$f_{\Gamma(\vartheta/m,m)}(s) = \frac{m^m s^{m-1} e^{-ms/\vartheta}}{(m-1)!\vartheta^m}, \quad s \geq 0.$$

The expression of the density function in Theorem 5.22 can directly be applied to calculate expressions for the cumulative distribution function, the survival function as well as for moments. Expressions can be easily derived from Cramer et al. (2016). It

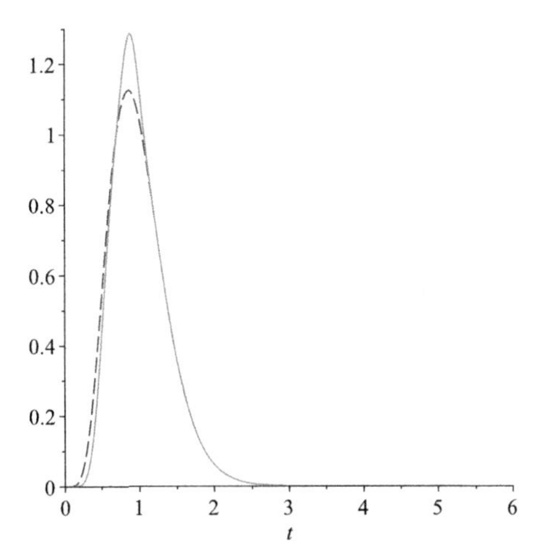

Figure 5.11 Plots of $f_{\widehat{\vartheta}}^{\vartheta}$ and density function of $\Gamma(\vartheta/m, m)$-distribution with $n = 10, m = 7, \vartheta = 1, T = 1$.

should be noted that the first term in the expression of the density function corresponds to a $\Gamma(\vartheta/m, m)$-distribution. Notice that the choice $m = n$ yields exactly this distribution, that is, for $m = n$, we have $2n\widehat{\vartheta}/\vartheta \sim \chi^2_{2n}$. In this case, no censoring occurs so that we end up with the complete sample. Thus, Theorem 5.22 includes this well-known property as a special case (see, e.g., Arnold et al., 2008; Balakrishnan et al., 2001).

Confidence intervals

As for Type-I hybrid censoring, exact confidence intervals can be constructed using the pivoting method described in Section 5.3.2.1. In order to apply it, one has to check Condition 5.13. The stochastic monotonicity of the MLE $\widehat{\vartheta}$ has been established in Balakrishnan and Iliopoulos (2009) (see also van Bentum and Cramer, 2019, Example 1, for Type-II hybrid censored generalized order statistics). The condition on the limits of the cumulative distribution function of $\widehat{\vartheta}$ can be easily checked using the representation of the density function in Theorem 5.22. In particular, Lemma 5.23 is useful in the derivation. Notice that the restriction on the parameters $\gamma_1, \ldots, \gamma_{d+1}$ is not necessary but satisfied for the discussed hybrid censoring schemes.

Lemma 5.23. *Let $d \in \mathbb{N}$, d, and $B_{d-1}(\cdot \mid \gamma_{d+1}, \ldots, \gamma_1)$ be a B-spline with knots $0 \leq \gamma_{d+1} < \cdots < \gamma_1$. Then, for any $t > 0$,*

$$\lim_{\vartheta \to 0} \frac{1}{\vartheta^d} \int_0^t B_{d-1}(s \mid \gamma_{d+1}, \ldots, \gamma_1) e^{-s/\vartheta} \, ds = 0,$$

$$\lim_{\vartheta \to \infty} \frac{1}{\vartheta^d} \int_0^t B_{d-1}(s \mid \gamma_{d+1}, \ldots, \gamma_1) e^{-s/\vartheta} \, ds = 0.$$

Proof. First, for any $d > 0$ and $s > 0$, we have $\lim_{\vartheta\to 0} e^{-ds/\vartheta}/\vartheta^d = \lim_{\vartheta\to\infty} e^{-ds/\vartheta}/\vartheta^d = 0$. Furthermore, $B_{d-1}(s \mid \gamma_{d+1}, \ldots, \gamma_1)$ is bounded for any $s > 0$ and $\vartheta > 0$ (B-splines are continuous and have a bounded support). Using Lebesgue's dominated convergence theorem (see, e.g., Klenke, 2014), we get, e.g.,

$$\lim_{\vartheta\to 0} \frac{1}{\vartheta^d} \int_0^t B_{d-1}(s \mid \gamma_{d+1}, \ldots, \gamma_1) e^{-ds/\vartheta}\, ds$$
$$= \int_0^t B_{d-1}(s \mid \gamma_{d+1}, \ldots, \gamma_1) \Big(\lim_{\vartheta\to 0} \frac{1}{\vartheta^d} e^{-ds/\vartheta} \Big)\, ds = 0.$$

□

Remark 5.24. A similar result can be used to prove the limits under Type-I hybrid censoring. In this case, we can use (5.13). Here, for $n, d, s, T > 0$, we have to consider the limits

$$\lim_{\vartheta\to 0} \frac{e^{-ds/\vartheta}}{(1 - e^{-nT/\vartheta})\vartheta^d} = 0, \quad \lim_{\vartheta\to\infty} \frac{e^{-ds/\vartheta}}{(1 - e^{-nT/\vartheta})\vartheta^d} = \begin{cases} \frac{1}{nT}, & d = 1 \\ 0, & d > 1 \end{cases}.$$

Then, integrating the density function in (5.13) and using Lebesgue's dominated convergence theorem, we get with (C.5)

$$\lim_{\vartheta\to\infty} F_\vartheta^{\widehat{\vartheta}|D\geq 1}(t) = \int_0^t B_0(s \mid (n-1)T, nT)\, ds = \frac{(t - (n-1)T)_+ - (t - nT)_+}{T}, \quad t \in \mathbb{R}$$

which equals (5.17).

From Lemma 5.23 and Theorem 5.22, we get directly the following result.

Theorem 5.25. *For any $\alpha \in (0, 1)$, an exact $100(1-\alpha)\%$ confidence interval can be obtained under Type-II hybrid censoring using the method of pivoting the cumulative distribution function.*

Proof. As pointed out above, the MLE $\widehat{\vartheta}$ is stochastically increasing in $\vartheta > 0$. Furthermore, Lemma 5.23 shows that, for any $t > 0$,

$$\lim_{\vartheta\to\infty} F_\vartheta^{\widehat{\vartheta}}(s) = \lim_{\vartheta\to\infty} \int_0^t \frac{m^m s^{m-1} e^{-ms/\vartheta}}{(m-1)!\vartheta^m}\, ds = \lim_{\vartheta\to\infty} F_{\Gamma(1,m)}\Big(\frac{mt}{\vartheta}\Big) = 0.$$

Notice that we must not exchange integration and limit in this case (as in Lemma 5.23) since the function h given by $h(s) = \frac{m^m s^{m-1}}{(m-1)!}$, $s > 0$, is not bounded on $[0, \infty)$. Similarly, we get $\lim_{\vartheta\to 0} F_\vartheta^{\widehat{\vartheta}}(t) = \lim_{\vartheta\to 0} F_{\Gamma(1,m)}\big(\frac{mt}{\vartheta}\big) = 1$ so that Condition 5.13 is satisfied. This proves the assertion. □

As for Type-I hybrid censoring, one may consider asymptotic confidence intervals for ϑ using a normal approximation and the observed Fisher information.

5.4.2 Bayesian inference

In this case as well, the Bayesian estimate and the associated credible intervals can be constructed very much along the lines of Type-I hybrid censoring (see Bayoud, 2014). Upon using the same prior as in (5.18), the posterior density function of ϑ becomes

$$p(\vartheta \mid \text{data}) = \frac{(\mathrm{TTT}_d + \lambda)^{d+\beta}}{\Gamma(d+\beta)} \vartheta^{-(d+\beta+1)} e^{-(\mathrm{TTT}_d+\lambda)\vartheta},$$

where TTT_d is given in (5.24).

Since $2(\mathrm{TTT}_d + \lambda)/\theta$ is distributed as $\chi^2_{2(\beta+\lambda)}$ a posteriori, the Bayesian estimate of ϑ under the squared-error loss function is given by

$$\widehat{\vartheta}_{\mathrm{B}} = \frac{\mathrm{TTT}_{D_{\mathrm{II}}} + \lambda}{D_{\mathrm{II}} + \beta - 1}.$$

Since $D_{\mathrm{II}} \geq m$, the condition $D_{\mathrm{II}} + \beta - 1 > 0$ is always satisfied. As expected, the Bayesian estimate under the non-informative prior coincides with the MLE of ϑ in this case as well. Moreover, a $100(1-\alpha)\%$ credible interval of ϑ can be obtained as

$$\left(\frac{2(\mathrm{TTT}_{D_{\mathrm{II}}} + \lambda)}{\chi^2_{2(D_{\mathrm{II}}+\beta),\alpha/2}}, \frac{2(\mathrm{TTT}_{D_{\mathrm{II}}} + \lambda)}{\chi^2_{2(D_{\mathrm{II}}+\beta),1-\alpha/2}} \right).$$

5.4.3 Two-parameter exponential distribution

Results for the two-parameter exponential distribution in presence of Type-II hybrid censoring can be taken from Burkschat et al. (2016). First, it should be noted that the MLE of ϑ does not exist when $m = 1$ and $d = 0$ (see (5.25)). Therefore, we have to assume $m \geq 2$ by design or we have to assume that at least one failure is observed before T, that is, $D \geq 1$. In this case, inference must be conditional on that event (as it is for Type-I hybrid censoring). In order to exclude this case, let $m \geq 2$.

To find the bivariate density function of the MLE, we first state that the MLE of μ is given by $\widehat{\mu} = x_1$ for $D < m$ and $D = d \in \{m, \ldots, n\}$, which follows from the likelihood function (5.23) which is easily seen to be increasing in μ for any $\vartheta > 0$. This yields the upper bound

$$\begin{aligned} \mathscr{L}(\widehat{\mu}, \vartheta; \boldsymbol{x}_d, d) &= \frac{n!}{(n-d)!\vartheta^d} \exp\Big(-\frac{1}{\vartheta} \sum_{j=1}^{d} (x_j - x_1) - \frac{n-d}{\vartheta}(w - x_1) \Big) \\ &= \frac{n!}{(n-d)!\vartheta^d} \exp\Big(-\frac{1}{\vartheta} \mathrm{TTT}_d^* \Big) \end{aligned} \tag{5.25}$$

where

$$\mathsf{TTT}_d^* = \sum_{j=2}^{d} (x_j - x_1) + (n - d)(w - x_1). \tag{5.26}$$

Then, the MLE of ϑ for the two-parameter case is given by $\hat{\vartheta} = \frac{1}{D_{\text{II}}}\mathsf{TTT}_{D_{\text{II}}}$, where $\mathsf{TTT}_{D_{\text{II}}}^*$ denotes the modified total time on test statistic introduced in (5.26).

The bivariate density function of the MLE $(\widehat{\mu}, \widehat{\vartheta})$ can now be established, that is, with $T_x = T - x$

$$\begin{aligned}
f_{\mu,\vartheta}^{\widehat{\mu},\widehat{\vartheta}}(x,t) &= \frac{n}{\vartheta^m} \exp\left\{-\frac{n(x-\mu)+mt}{\vartheta}\right\} \\
&\times \left[\frac{m^{m-1}t^{m-2}}{(m-2)!} - T_x^{m-1}\binom{n-1}{m-1} B_{m-2}\Big(mt \mid 0, (n-m+1)T_x, \ldots, (n-1)T_x\Big)\right] \\
&+ n\sum_{d=m}^{n} \frac{dT_x^{d-1}}{\vartheta^d}\binom{n-1}{d-1} B_{d-2}\Big(dt \mid (n-d)T_x, \ldots, (n-1)T_x\Big) \exp\left\{-\frac{n(x-\mu)+dt}{\vartheta}\right\}, \\
& \qquad x \geq \mu, \quad t \geq 0.
\end{aligned}$$

Results for the marginal distribution of $\hat{\vartheta}$ can be obtained by integrating $f^{\hat{\mu},\hat{\vartheta}}$ w.r.t. the variable x. Further details on the derivation of this density function are provided in Section 7.2.2.2.

The distribution of $\hat{\mu}$ follows immediately from the distribution of $X_{1:n}$, i.e., the survival function is given by

$$\Pr_{\mu,\vartheta}(\widehat{\mu} > t) = e^{-n(t-\mu)/\vartheta}, \quad t \geq \mu.$$

From this expression we further find, that $\Pr_{\mu,\vartheta}(\widehat{\mu} > t)$ is an increasing function in μ for $t \geq \mu$ and any $\vartheta > 0$.

5.5. Further hybrid censoring schemes

The derivations of the MLEs, their existence conditions, as well as respective distributional results can be established by similarity with those obtained under Type-I/II hybrid censoring. A similar comment applies to the construction of confidence intervals and Bayesian inference, respectively. Further details can also be found for the more general data situation of progressively Type-II censored data discussed in Chapter 7. For brevity, we sketch only the results and do not present details on the derivations.

5.5.1 Generalized hybrid censoring

Exact statistical inference for generalized Type-I/II hybrid censoring schemes has been first discussed in Chandrasekar et al. (2004). As for Type-I and Type-II hybrid censoring, the moment generating function approach can be utilized.

From the joint density function in (4.34), the likelihood function for an $Exp(\vartheta)$-distribution under generalized Type-I hybrid censoring is given by

$$\begin{aligned}\mathscr{L}(\vartheta; \boldsymbol{x}_d, d) &= \frac{n!}{(n-d)!\vartheta^d} \exp\Big(-\frac{1}{\vartheta}\sum_{j=1}^{d} x_j - \frac{n-d}{\vartheta}w\Big)\\ &= \frac{n!}{(n-d)!\vartheta^d} \exp\big(-\mathsf{TTT}_d/\vartheta\big),\end{aligned}$$

where d is the realization of $D_{\mathsf{gI}} = (D \wedge m) \vee k$, $w = (x_m \wedge T) \vee x_k$, and $\mathsf{TTT}_d = \sum_{j=1}^{d} x_j + (n-d)w$ represents the total time on test statistic. Therefore, the MLE is obviously be given by

$$\widehat{\vartheta} = \frac{1}{D_{\mathsf{gI}}}\mathsf{TTT}_{D_{\mathsf{gI}}} = \frac{1}{D_{\mathsf{gI}}}\Big(\sum_{j=1}^{D_{\mathsf{gI}}} X_{j:n} + (n - D_{\mathsf{gI}})W\Big).$$

The distribution of the MLE has been obtained in Chandrasekar et al. (2004) applying the moment generating function approach. A B-spline representation of the density function can be found in Górny and Cramer (2016, Theorem 3.2), that is,

$$\begin{aligned}f_{\vartheta}^{\widehat{\vartheta}}(t) =& f_{\Gamma(\vartheta/k,k)}(t) - \frac{kT^k}{\vartheta^k}\binom{n}{k} B_{k-1}(kt \mid 0, (n-k+1)T, \ldots, nT)e^{-kt/\vartheta}\\ &+ \sum_{d=k}^{m-1} \frac{dT^d}{\vartheta^d}\binom{n}{d} B_{d-1}(dt \mid (n-d)T, \ldots, nT)e^{-dt/\vartheta}\\ &+ \frac{mT^m}{\vartheta^m}\binom{n}{m} B_{m-1}(mt \mid 0, (n-m+1)T, \ldots, nT)e^{-mt/\vartheta}, \quad t \in \mathbb{R}.\end{aligned}$$

The stochastic monotonicity of the MLE has been established in Balakrishnan and Iliopoulos (2009) (for an alternative proof, see van Bentum and Cramer, 2019).

Clearly, Bayesian inference for ϑ with a gamma prior as in (5.18) can be applied leading to the posterior density function (5.19) taking into account the appropriate choice of d^* and the total time on test statistic TTT_{d^*} given above.

For generalized Type-II hybrid censoring, the likelihood function for $Exp(\vartheta)$-distributed lifetimes reads (see (4.38); $D_{\mathsf{gII}} = (D_1 \vee m) \wedge D_2$)

$$\mathscr{L}(\vartheta; \boldsymbol{x}_d, d) = \frac{n!}{(n-d)!\vartheta^d} \exp\Big(-\frac{1}{\vartheta}\sum_{j=1}^{d} x_j - \frac{n-d}{\vartheta}w\Big),$$

Table 5.1 Expressions of the observed sample size D_{HCS} and the test duration W_{HCS} under various hybrid censoring schemes. The counters D, D_1, D_2 are defined as in (4.2) and (7.3), respectively.

censoring scheme	notation	observed sample size D_{HCS}	test duration W_{HCS} (as in (5.27))
Type-II	D	m	$X_{m:n}$
Type-I	D	D	$X_{n:n} \wedge T$
Type-I hybrid	D_{I}	$D \wedge m$	$X_{m:n} \wedge T$
Type-II hybrid	D_{II}	$D \vee m$	$X_{m:n} \vee (T \wedge X_{n:n})$
Generalized Type-I hybrid	D_{gI}	$(k \vee D) \wedge m$	$(X_{k:n} \vee T) \wedge X_{m:n}$
Generalized Type-II hybrid	D_{gII}	$(D_1 \vee m) \wedge D_2$	$[(T_1 \wedge X_{n:n}) \vee X_{m:n}] \wedge T_2$
Unified Type-I hybrid	D_{uI}	$(D_1 \wedge m) \vee (D_2 \wedge k)$	$(X_{k:n} \wedge T_2) \vee (X_{m:n} \wedge T_1)$
Unified Type-II hybrid	D_{uII}	$(D_1 \vee m) \wedge (D_2 \vee k)$	$(X_{k:n} \vee T_2) \wedge [X_{m:n} \vee (T_1 \wedge X_{n:n})]$
Unified Type-III hybrid	D_{uIII}	$(k \wedge D_3) \vee ((m \vee D_1) \wedge D_2)$	$(X_{k:n} \wedge T_3) \vee [(X_{m:n} \vee [T_1 \wedge X_{n:n}]) \wedge T_2]$
Unified Type-IV hybrid	D_{uIV}	$(r \wedge D_1) \vee ((k \vee D_2) \wedge m)$	$(X_{r:n} \wedge T_1) \vee [(X_{k:n} \vee T_2) \wedge X_{m:n}]$

so that the MLE is given by $\widehat{\vartheta} = \mathrm{TTT}_{D_{\mathrm{gII}}}/D_{\mathrm{gII}}$ with total time on test statistic

$$\mathrm{TTT}_{D_{\mathrm{gII}}} = \sum_{j=1}^{D_{\mathrm{gII}}} X_{j:n} + (n - D_{\mathrm{gII}})W$$

and $D_{\mathrm{gII}} = (D_1 \vee m) \wedge D_2$. Notice that $\{D_{\mathrm{gII}} \geq 1\} = \{D_2 \geq 1\}$. Further details are provided in Section 7.2.6. In particular, the conditional density function of $\widehat{\vartheta}$ given $D_{\mathrm{gII}} \geq 1$, that is, for $t \geq 0$,

$$\begin{aligned} f_{\vartheta}^{\widehat{\vartheta}|D_{\mathrm{gII}}\geq 1}(t) = {} & \frac{1}{1-e^{-nT_2/\vartheta}} \Bigg(\sum_{d_1=m}^{n} \frac{d_1 T_1^{d_1}}{\vartheta^{d_1}} \binom{n}{d_1} B_{d_1-1}(d_1 t \mid (n-d_1)T_1, \ldots, nT_1) e^{-d_1 t/\vartheta} \\ & + \frac{mT_2^m}{\vartheta^m} \binom{n}{m} B_{m-1}(mt \mid 0, (n-m+1)T_2, \ldots, nT_2) e^{-mt/\vartheta} \\ & - \frac{mT_1^m}{(m-1)!\vartheta^m} \binom{n}{m} B_{m-1}(mt \mid 0, (n-m+1)T_1, \ldots, nT_1) e^{-mt/\vartheta} \\ & + \sum_{d_2=1}^{m-1} \frac{d_2 T_2^{d_2}}{\vartheta^{d_2}} \binom{n}{d_2} B_{d_2-1}(d_2 t \mid (n-d_2)T_2, \ldots, \gamma_1 T_2) e^{-d_2 t/\vartheta} \Bigg), \quad t \in \mathbb{R}. \end{aligned}$$

5.5.2 Unified hybrid censoring

Inference for exponential lifetimes under a unified hybrid censoring scheme can be conduced as for the previously discussed hybrid censoring schemes. Therefore, we not provide details here and restrict ourselves to the presentation of the total time on test statistic – which directly yields the MLE for ϑ – and some references to the original contributions (see also the comment in Section 7.2.7). In fact, the total time on test

Table 5.2 Hybrid censoring schemes applied to Lawless's electrical insulation data. Notice that the MLEs are given by the total time on test $\mathrm{TTT}_{D_{\mathrm{HCS}}} = \sum_{j=1}^{D_{\mathrm{HCS}}} X_{j:n} + (n - D_{\mathrm{HCS}})W_{\mathrm{HCS}}$ with observed sample size D_{HCS} and the test duration W_{HCS} as in Table 5.1. Furthermore, we get the values $D = D_1 = 8$, $D_2 = 10$, and $D_3 = 10$. The columns of the involved order statistics are highlighted in blue (gray in print version) color.

scheme	parameters						data												D_{HCS}	W_{HCS}	MLE
	m	k	r	$T(T_1)$	T_2	T_3	1	2	3	4	5	6	7	8	9	10	11	12			
complete							12.3	21.8	24.4	28.6	43.2	46.9	70.7	75.3	95.5	98.1	138.6	151.9	12	151.9	67.275
Type-I				90			12.3	21.8	24.4	28.6	43.2	46.9	70.7	75.3					8	90	85.400
Type-II	11						12.3	21.8	24.4	28.6	43.2	46.9	70.7	75.3	95.5	98.1	138.6		11	138.6	72.182
Type-I H	11			90			12.3	21.8	24.4	28.6	43.2	46.9	70.7	75.3					8	90	85.400
Type-II H	11			90			12.3	21.8	24.4	28.6	43.2	46.9	70.7	75.3	95.5	98.1	138.6		11	138.6	72.182
G Type-I H	11	7		90			12.3	21.8	24.4	28.6	43.2	46.9	70.7	75.3	95.5	98.1	138.6		8	90	85.400
G Type-II H	11			90	100		12.3	21.8	24.4	28.6	43.2	46.9	70.7	75.3	95.5	98.1	138.6		10	100	71.680
U Type-I H	11	7		90	100		12.3	21.8	24.4	28.6	43.2	46.9	70.7	75.3	95.5	98.1	138.6		8	90	85.400
U Type-II U	11	7		90	100		12.3	21.8	24.4	28.6	43.2	46.9	70.7	75.3	95.5	98.1	138.6		10	100	71.680
U Type-III H	11	7		90	100	120	12.3	21.8	24.4	28.6	43.2	46.9	70.7	75.3	95.5	98.1	138.6		10	100	71.680
U Type-IV U	11	7	5	90	100		12.3	21.8	24.4	28.6	43.2	46.9	70.7	75.3	95.5	98.1	138.6		10	100	71.680

statistic has the form

$$\mathrm{TTT}_{D_{\mathrm{HCS}}} = \sum_{j=1}^{D_{\mathrm{HCS}}} X_{j:n} + (n - D_{\mathrm{HCS}}) W_{\mathrm{HCS}} \tag{5.27}$$

where D_{HCS} denotes the observed sample size and W_{HCS} the test duration under the hybrid censoring scheme applied (see Table 5.1). In particular, the total time on test statistic is constructed as

$$\begin{aligned}&\text{total time on test}\\ &\quad = \text{sum of observed failure times} + \text{no. of censored items} \times \text{test duration}.\end{aligned} \tag{5.28}$$

Notice that a similar representation holds under progressive censoring where the order statistics have to be replaced by the corresponding progressively Type-II censored order statistics.

5.5.3 Illustrative example

Example 5.26. In order to illustrate the results for the various hybrid censoring schemes, we consider the complete electrical insulation data

$$12.3, 21.8, 24.4, 28.6, 43.2, 46.9, 70.7, 75.3, 95.5, 98.1, 138.6, 151.9,$$

already discussed in Example 3.6. It is subject to the various hybrid censoring schemes introduced so far. We report the respective hybrid censored samples for the given parameters as well as the corresponding MLEs in Table 5.2.

CHAPTER 6

Inference for other lifetime distributions

Contents

6.1. Introduction

In this chapter, we present results for data from various lifetime distributions under hybrid censoring. We start with two-parameter Weibull distributions, where the presentation is more detailed than in the other cases. This will illustrate the application of the ideas presented in Section 5.2 showing that the derivations simplify considerably. As a matter of fact, the number of measurements d and the observed measurements $\boldsymbol{x}_d$ determine completely the case which has to be considered.

6.2. Weibull distributions

6.2.1 Type-I hybrid censoring

Type-I hybrid censoring for Weibull-distributed data has firstly been discussed in Kundu (2007) who used the parametrization *Weibull*(λ^β, β) of the Weibull distribution. From (5.2), the log-likelihood function reads for *Weibull*(ϑ, β)-distributions,

$$\mathscr{L}^*(\vartheta, \beta; \boldsymbol{x}_d, d) = c_d^* + d\log(\beta/\vartheta) + (\beta - 1)\sum_{j=1}^{d}\log x_j - \frac{1}{\vartheta}\sum_{j=1}^{d}x_j^\beta - \frac{n-d}{\vartheta}w^\beta.$$

Hybrid Censoring Know-How
https://doi.org/10.1016/B978-0-12-398387-9.00014-3

Table 6.1 Types of likelihood functions for Type-I hybrid censoring schemes.

type	d	w
ⓘ	$1, \ldots, m-1$	T
ⓘⓘ	m	x_m

Proceeding as in Balakrishnan and Kateri (2008), it can be shown that the log-likelihood has a unique maximum which is attained by $\widehat{\vartheta}$ and $\widehat{\beta}$ where

$$\widehat{\vartheta} = \widehat{\vartheta}(\widehat{\beta}) = \frac{1}{d}\Big(\sum_{j=1}^{d} x_j^{\widehat{\beta}} + (n-d)w^{\widehat{\beta}}\Big) \tag{6.1}$$

and $\widehat{\beta}$ is the unique solution of the equation

$$0 = \frac{d}{\beta} - \frac{\sum_{j=1}^{d} x_j^{\beta} \log x_j + (n-d)w^{\beta} \log w}{\widehat{\vartheta}(\beta)} + \sum_{j=1}^{d} \log x_j.$$

Notice that the crucial condition enabling this result is the fact that $x_1 \leq \cdots \leq x_d \leq w$ are increasingly ordered which is obviously satisfied in the models under study (for details, see Section 3.2.3.4).

In view of the comments in Section 5.2, we see that the problem is traced back to the same problems under Type-I and Type-II censoring, respectively. We have only to identify the effectively applied censoring mechanism and the values of the parameters as given in Table 4.7. Hence, we have to take into account in the present situation the scenarios given in Table 6.1.

As has been proposed in Kundu (2007), iterative procedures like the Newton-Raphson method or fixpoint procedures may be used to determine the estimates. In order to find appropriate starting values, Kundu (2007) proposed approximate MLEs (see Balakrishnan and Varadan, 1991 as well as the exposition in Section 3.2.3.1). If one takes into account the comments in Section 5.2, then its clear that we can directly apply results already obtained for Type-I and Type-II censoring by discussing the likelihood of types ⓘ and ⓘⓘ, respectively. For Weibull distributed data, this estimation concept is based on the transformation of the data to a location-scale family of extreme value distributions which is appropriately linearized afterwards. The resulting linear equations can be explicitly solved for the parameters ϑ and β yielding the so-called approximate

MLEs. In the present setting, the respective linearized equations read

$$-\sum_{j=1}^{d}(a_j - b_j(\beta \log x_j - \log \vartheta)) - (n-d)(a_d - b_d(\beta \log w - \log \vartheta)) + n - d = 0$$

$$\begin{aligned} -d - \sum_{j=1}^{d}(a_j - b_j(\beta \log x_j - \log \vartheta))(\beta \log x_j - \log \vartheta) \\ -(n-d)(1 - a_d + b_d(\beta \log w - \log \vartheta))(\beta \log w - \log \vartheta) = 0 \end{aligned} \tag{6.2}$$

where

$$a_j = 1 + \log(q_j)\big[-\log(-\log q_j)\big], \quad b_j = -\log q_j, \qquad 1 \le j \le d$$

(for further details, see Kundu, 2007). The values $q_1 \ldots, q_d$ are defined based on the particular type ⓘ and ⓘⓘ, respectively, induced by the observed measurements. For Type-II censored data, results of Balakrishnan and Varadan (1991) are directly applied who had discussed Type-II double censored data (see also Balasooriya and Balakrishnan, 2000). In this case, the proportions are defined by

$$q_j = 1 - \frac{j}{n+1}, \quad 1 \le j \le d,$$

which result from means of order statistics from a standard uniform distribution. For Type-I censoring and inspired by Balakrishnan and Varadan (1991), Joarder et al. (2011) have proposed the modified proportion

$$q_d = 1 - \frac{d + 1/2}{n+1} \tag{6.3}$$

for the maximum index $j = d$. Therefore, these values will be defined in the following as

$$\begin{aligned} &\text{type ⓘ}: \quad q_j = 1 - \frac{j}{n+1}, 1 \le j \le d-1, \quad q_d = 1 - \frac{d+1/2}{n+1}, \\ &\text{type ⓘⓘ}: \quad q_j = 1 - \frac{j}{n+1}, 1 \le j \le d. \end{aligned}$$

Using the notation in Table 6.2, the equations in (6.2) can be written as

$$\log \vartheta = \zeta\beta - \tau, \qquad \beta^2 - \frac{\delta}{\eta}\beta - \frac{d}{\eta} = 0, \tag{6.4}$$

where $\eta > 0$. Solving the quadratic equation (6.4) for the parameters yields the estimators

$$\widehat{\vartheta}^* = e^{\zeta\widehat{\beta}^* - \tau}, \qquad \widehat{\beta}^* = \frac{\delta + \sqrt{\delta^2 + 4d\eta}}{2\eta} = \frac{2d}{-\delta + \sqrt{\delta^2 + 4d\eta}}$$

Table 6.2 Quantities in the derivation of approximate MLEs for the Weibull parameters under Type-I hybrid censoring.

quantity	Type-I hybrid
w	$x_m \wedge T$
d	$D \wedge m$
ζ	$\dfrac{\sum_{j=1}^{d} b_j \log x_j + b_d(n-d)\log w}{\sum_{j=1}^{d} b_j + b_d(n-d)}$
τ	$\dfrac{\sum_{j=1}^{d} a_j - (n-d)(1-a_d)}{\sum_{j=1}^{d} b_j + b_d(n-d)}$
δ	$\sum_{j=1}^{d} a_j(\log x_j - \zeta) - (n-d)(1-a_d)(\log w - \zeta)$ $-2\tau\sum_{j=1}^{d} b_j(\log x_j - \zeta) - 2(n-d)\tau b_d(\log w - \zeta)$
η	$\sum_{j=1}^{d} b_j(\log x_j - \zeta)^2 + (n-d)b_d(\log w - \zeta)^2$

with values ζ, τ, δ, and η given in Table 6.2. Notice that we use only the positive root of the quadratic equation since β is assumed to be positive.

In Bayesian inference, we get similar results. Similar as in Kundu (2007), we obtain with an inverse gamma prior $\mathsf{I\Gamma}(b, a)$ and density function (see Section 3.2.4)

$$\pi_1(\vartheta) \propto \vartheta^{-a-1} e^{-b/\vartheta}, \quad \vartheta > 0, \tag{6.5}$$

for ϑ and an arbitrary prior π_2 for β with a support in $(0, \infty)$ (independent of ϑ) the joint density function

$$\begin{aligned} f(\text{data}, \beta, \vartheta) &= f(\mathbf{x}_d, d, \beta, \vartheta) \\ &= e^{c_d^*}\beta^d \vartheta^{-(a+d+1)}\Big(\prod_{j=1}^{d} x_j\Big)^{\beta-1} \exp\Big(-\frac{1}{\vartheta}\Big(\sum_{j=1}^{d} x_j^\beta + (n-l)w^\beta + b\Big)\Big)\pi_2(\beta). \end{aligned}$$

Notice that Kundu (2007) used a gamma prior due to a different parametrization of the Weibull distribution. Then, the (conditional) distribution of ϑ given β and the data is given by an inverse gamma distribution

$$\mathsf{I\Gamma}\Big(\sum_{j=1}^{d} x_j^\beta + (n-l)w^\beta + b, a+d\Big).$$

For $d = 0$, we get $\mathsf{I\Gamma}(nT^\beta + b, a)$.

This result can be used to compute the joint posterior density function given by

$$\mathscr{L}(\beta, \vartheta \mid \text{data}) = \frac{f(\text{data}, \beta, \vartheta)}{\int_0^\infty \int_0^\infty f(\text{data}, z, t)\,dz\,dt},$$

Table 6.3 Types of likelihood functions for Type-II hybrid censoring schemes. The case $d = n$ can be included in the case $d = d$ since in that case $X_{n:n} \leq T$. Thus, we have a complete sample and no censoring occurs.

type	d	w
(ii)	m	x_m
(i)	$m+1, \ldots, n-1$	T
(ii)	n	x_n

which cannot be computed analytically. Kundu (2007) proposed Gibbs sampling procedures to compute Bayes estimates of β and ϑ. In order to perform the Gibbs sampling procedures, the prior π_2 of β is supposed log-concave (which is satisfied for Weibull and gamma density functions when the shape parameter is at least equal to one; normal and log-normal priors satisfy the assumption always). This ensures that the conditional density function of β given ϑ and the data is also log-concave. A simulation algorithm based on an idea of Geman and Geman (1984) can be found in Kundu (2007). Samples from a log-concave density function are generated using an approach of Devroye (1984). Furthermore, the results are utilized to generate HPD credible intervals using a method proposed by Chen and Shao (1999).

6.2.2 Type-II hybrid censoring

Type-II hybrid censored data from a Weibull distribution has been addressed in Banerjee and Kundu (2008) paralleling the results of Kundu (2007) for Type-I hybrid censoring. Applying the results presented in Section 5.2 and the classification of types in Table 4.7, the resulting formulas are similar to those obtained under Type-I hybrid censoring given in Section 6.2.1. This comment applies to the MLEs, the AMLEs, and the Bayesian estimates. The respective results are directly obtained using the type of the likelihood in Table 4.7 and the respective results in Section 6.2.1. The respective values for the three scenarios taken from Table 4.7 are presented in Table 6.3.

Assuming gamma priors for both the scale and the shape parameter of the Weibull distribution, Banerjee and Kundu (2008) additionally considered Lindley's approximation to derive approximate Bayesian estimates (see Lindley, 1980).

6.2.3 Further hybrid censoring schemes

Results for Weibull lifetimes under other hybrid censoring schemes can directly be established using the approach illustrated in Section 5.2. For generalized Type-I and

Table 6.4 Types of likelihood functions for generalized hybrid censoring schemes. The case $d = n$ can be included in the case $d = d_1$ since in that case $X_{n:n} \leq T_1$. Thus, we have a complete sample and no censoring occurs.

censoring scheme	type	parameters	
		d	w
generalized Type-I hybrid	(ii)	m	x_m
	(i)	d	T
	(ii)	k	x_k
generalized Type-II hybrid	(i)	d_1	T_1
	(ii)	n	x_n
	(ii)	m	x_m
	(i)	d_2	T_2

Type-II hybrid censoring, such results have been presented in Lee (2020, 2021), respectively. In this case, the corresponding scenarios given in Table 6.4 need to be taken into account.

Habibi Rad and Yousefzadeh (2014) had discussed the above mentioned inferential problems for the unified Type-II hybrid censoring scheme. However, Table 6.5 illustrates how the corresponding inferential results can be established for all unified hybrid censoring schemes presented in Section 4.5.4.

6.3. Further distributions

Hybrid censoring has also been discussed for various other lifetime distributions. In the following sections, we summarize some references without giving details about the results. As illustrated for the Weibull distribution, the respective results can be directly obtained from the results under Type-I and Type-II censoring, respectively, by taking into account the types of the likelihood given in Tables 4.7 and 6.5.

6.3.1 Log-normal distributions

6.3.1.1 Type-I hybrid censoring

Habibi Rad and Yousefzadeh (2010) and Dube et al. (2011) discussed log-normal lifetime distributions *log*-$N(\mu, \sigma^2)$ with parameters $\sigma > 0$, $\mu \in \mathbb{R}$. For the two types, we get from (5.1) the likelihood function as

$$\mathscr{L}(\mu, \sigma; \boldsymbol{x}_d, d) = \frac{e^{c_d^*}}{\sigma^d} \exp\Big\{ -\frac{1}{2\sigma^2} \sum_{j=1}^{d} (\log x_j - \mu)^2 \Big\} \Big\{ 1 - \Phi\Big(\frac{\log w - \mu}{\sigma}\Big) \Big\}^{n-d}$$

Table 6.5 Types of likelihood functions for unified hybrid censoring schemes.

censoring scheme	type	parameters	
		d	w
unified Type-I hybrid	(ii)	m	x_m
	(i)	d_1	T_1
	(ii)	k	x_k
	(i)	d_2	T_2
unified Type-II hybrid	(i)	d_1	T_1
	(ii)	n	x_n
	(ii)	m	x_m
	(i)	d_2	T_2
	(ii)	k	x_k
unified Type-III hybrid	(i)	d_1	T_1
	(ii)	n	x_n
	(ii)	m	x_m
	(i)	d_2	T_2
	(ii)	k	x_k
	(i)	d_3	T_3
unified Type-IV hybrid	(ii)	r	x_r
	(i)	d_1	T_1
	(ii)	m	x_m
	(i)	d_2	T_2
	(ii)	k	x_k

provided $d > 0$. This expression can already be found in Gupta (1952) for the normal distribution under Type-II censoring. Type-I and Type-II censoring are discussed in Cohen (1991, Section 2.5) (see also Cohen (1959)). Existence and uniqueness of the MLEs can be deduced from the results presented in Balakrishnan and Mi (2003) for general progressively Type-II censored samples.

For $d = 0$, we have

$$\mathscr{L}(\mu, \sigma; 0) = \Big\{1 - \Phi\Big(\frac{\log w - \mu}{\sigma}\Big)\Big\}^n,$$

so that the MLE of μ and σ does not exist for $d = 0$ since the likelihood increases with both μ and σ.

For $d > 0$, the log-likelihood function is given by

$$\mathscr{L}^*(\mu, \sigma; \boldsymbol{x}_d, d) = c_d^* \\ - d\log(\sigma) - \frac{1}{2\sigma^2}\sum_{j=1}^{d}(\log x_j - \mu)^2 - (n-d)\log\Big\{1 - \Phi\Big(\frac{\log w - \mu}{\sigma}\Big)\Big\}$$

leading to the likelihood equations

$$\frac{\partial}{\partial\mu}\mathcal{L}^*(\mu,\sigma;\boldsymbol{x}_d,d)=\frac{1}{\sigma^2}\sum_{j=1}^{d}(\log x_{j:n}-\mu)-\frac{n-d}{\sigma}\frac{\varphi\left(\frac{\log w-\mu}{\sigma}\right)}{1-\Phi\left(\frac{\log w-\mu}{\sigma}\right)}$$

$$\frac{\partial}{\partial\sigma}\mathcal{L}^*(\mu,\sigma;\boldsymbol{x}_d,d)=-\frac{d}{\sigma}+\frac{1}{\sigma^3}\sum_{j=1}^{d}(\log x_{j:n}-\mu)^2+\frac{n-d}{\sigma}\frac{\log w-\mu}{\sigma}\frac{\varphi\left(\frac{\log w-\mu}{\sigma}\right)}{1-\Phi\left(\frac{\log w-\mu}{\sigma}\right)}. \tag{6.6}$$

The likelihood equations have to be solved numerically, e.g., by a Newton-Raphson procedure or direct maximization. Manipulating the likelihood equations, Habibi Rad and Yousefzadeh (2010) showed that σ can be written as a function of μ, i.e.,

$$\sigma=\sqrt{\frac{1}{d}}\sqrt{(\log w-\mu)\sum_{j=1}^{d}(\log x_j-\mu)-\sum_{j=1}^{d}(\log x_j-\mu)^2}=\sigma(\mu)$$

leading to a fixed point equation in μ, that is,

$$\mu=\nu(\mu)=\frac{1}{d}\left((n-d)\sigma(\mu)\frac{\varphi\left(\frac{\log w-\mu}{\sigma}\right)}{1-\Phi\left(\frac{\log w-\mu}{\sigma}\right)}+\sum_{j=1}^{d}\log x_j\right).$$

They computed the MLE using the resulting fixed point procedure for μ. However, Dube et al. (2011) reported that they faced convergence problems for the Newton-Raphson procedure so that they proposed an EM algorithm type approach inspired by results obtained in Ng et al. (2002) (cf. Section 3.2.3.5). The observed data $(d,\boldsymbol{x}_d)$ is considered as an incomplete sample of $(x_1,\ldots,x_n)$. Here, $(x_{d+1},\ldots,x_n)=(u_{d+1},\ldots,u_n)$ are interpreted as censored/missing data where the number of measurements d is supposed known. Denoting the corresponding random vectors by $\boldsymbol{X}$ and $\boldsymbol{U}$, $\boldsymbol{Z}=(\boldsymbol{X},\boldsymbol{U})$ forms the complete data. Then, the corresponding pseudo-log-likelihood function for the complete data reads

$$\begin{aligned}\mathcal{L}_c^*(\mu,\sigma;\boldsymbol{z})&=-n\log(\sigma)-\frac{1}{2\sigma^2}\sum_{j=1}^{n}(\log z_j-\mu)^2\\&=-n\log(\sigma)-\frac{1}{2\sigma^2}\sum_{j=1}^{d}(\log x_j-\mu)^2-\frac{1}{2\sigma^2}\sum_{j=d+1}^{n}(\log u_j-\mu)^2\end{aligned}$$

In the E-step of the algorithm, the terms depending on u_j are replaced by conditional expectations of random variables $U_{d+1},\ldots,U_n$ which are considered as iid normally distributed random variables with mean μ_0 and variance σ_0^2. The conditional expectations are computed assuming that the true parameter values are μ_0 and σ_0 so that we get

$$\begin{aligned}
&\mathcal{L}_e^*(\mu, \sigma; \boldsymbol{x}_d, d) \\
&= -n\log(\sigma) - \frac{1}{2\sigma^2}\sum_{j=1}^{d}(\log x_j - \mu)^2 - \frac{1}{2\sigma^2}\sum_{j=d+1}^{n} E_{\mu_0,\sigma_0}\big((\log U_j - \mu)^2 \mid U_j > w\big) \\
&= -n\log(\sigma) - \frac{1}{2\sigma^2}\sum_{j=1}^{d}(\log x_j - \mu)^2 - \frac{n-d}{2\sigma^2} E_{\mu_0,\sigma_0}\big((\log U_n - \mu)^2 \mid U_n > w\big) \\
&= -n\log(\sigma) - \frac{1}{2\sigma^2}\sum_{j=1}^{d}(\log x_j - \mu)^2 - \frac{n-d}{2\sigma^2} E_{\mu_0,\sigma_0}\big((\log U_n)^2 \mid U_n > w\big) \\
&\quad + \frac{n-d}{\sigma^2}\mu E_{\mu_0,\sigma_0}\big(\log U_n \mid U_n > w\big) - \frac{n-d}{2\sigma^2}\mu^2 \\
&= -n\log(\sigma) - \frac{1}{2\sigma^2}\sum_{j=1}^{n}(z_j^* - \mu)^2 \\
&\quad - \frac{n-d}{2\sigma^2}\Big(E_{\mu_0,\sigma_0}\big((\log U_n)^2 \mid U_n > w\big) - E^2_{\mu_0,\sigma_0}\big(\log U_n \mid U_n > w\big)\Big) \qquad (6.7)
\end{aligned}$$

with $z_j^* = \log x_j$, $1 \le j \le d$, $z_j^* = E_{\mu_0,\sigma_0}\big(\log U_n \mid U_n > w\big)$, $d+1 \le j \le n$.

Furthermore, as pointed out in Dube et al. (2011) (see also Ng et al., 2002), the conditional expectations can be expressed as

$$\begin{aligned}
E_{\mu_0,\sigma_0}\big(\log U_j \mid U_j > w\big) &= \mu_0 + \sigma_0 Q = M(\mu_0, \sigma_0), \\
E_{\mu_0,\sigma_0}\big((\log U_j)^2 \mid U_j > w\big) &= \sigma_0^2(1 + \xi Q) + 2\sigma_0\mu_0 Q + \mu_0^2 = V(\mu_0, \sigma_0)
\end{aligned}$$

with $Q = \frac{\varphi(\xi)}{1-\Phi(\xi)}$, $\xi = \frac{\log w - \mu}{\sigma}$.

In the M-step, one has to maximize $\mathcal{L}_e^*(\mu, \sigma; \boldsymbol{x}_d, d)$ as in (6.7) w.r.t. μ and σ. It is easy to see that the maximum of $\mathcal{L}_e^*(\mu, \sigma; \boldsymbol{x}_d, d)$ is attained for the quantities

$$\begin{aligned}
\widehat{\mu} &= \frac{1}{n}\sum_{j=1}^{n} z_j^* = \frac{1}{n}\Big(\sum_{j=1}^{d}\log x_j + (n-d)M(\mu_0, \sigma_0)\Big) \\
\widehat{\sigma}^2 &= \frac{1}{n}\Big(\sum_{j=1}^{d}(\log x_j)^2 + (n-d)V(\mu_0, \sigma_0)\Big) - \widehat{\mu}^2
\end{aligned} \qquad (6.8)$$

The expressions in (6.8) are now used to define update equations for the EM-algorithm, that is,

$$\mu^{(k+1)} = \frac{1}{n}\sum_{j=1}^{d} z_j^* = \frac{1}{n}\Big(\sum_{j=1}^{d}\log x_j + (n-d)M(\mu^{(k)},\sigma^{(k)})\Big)$$
$$\sigma^{(k+1)} = \sqrt{\frac{1}{n}\Big(\sum_{j=1}^{n}(\log x_j)^2 + (n-d)V(\mu^{(k)},\sigma^{(k)})\Big) - \big(\mu^{(k+1)}\big)^2}, \qquad k = 1, 2, \ldots$$

with appropriately chosen initial values $\mu^{(0)}$, $\sigma^{(0)}$.

In order to find appropriate initial values, Dube et al. (2011) proposed approximate MLEs as introduced in Section 3.2.3.1. Those have been obtained contemporaneously in Habibi Rad and Yousefzadeh (2010). Expanding the hazard rate

$$h(t) = \frac{\varphi(t)}{1-\Phi(t)} = \frac{\varphi(t)}{\Phi(-t)}$$

at $q_d = \Phi^{-1}(d/(n+1))$ (type ②) and $q_d = \Phi^{-1}((d+1/2)/(n+1))$ (type ①; Habibi Rad and Yousefzadeh, 2010 used the same value as for type ②) in a Taylor polynomial of order 1, we get

$$h\Big(\frac{\log w-\mu}{\sigma}\Big) \doteq h(q_d) + h'(q_d)\Big(\frac{\log w-\mu}{\sigma} - q_d\Big) = a_d + b_d\frac{\log w-\mu}{\sigma}$$

where $a_d = h(q_d) - q_d h'(q_d)$ and $b_d = h'(q_d) \geq 0$. Then, we find from (6.6) the linearized likelihood equations

$$\frac{\partial}{\partial\mu}\mathscr{L}^*(\mu,\sigma;\boldsymbol{x}_d,d) \doteq \frac{1}{\sigma^2}\sum_{j=1}^{d}(\log x_j-\mu) - \frac{n-d}{\sigma}\Big(a_d + b_d\frac{\log w-\mu}{\sigma}\Big)$$
$$\frac{\partial}{\partial\sigma}\mathscr{L}^*(\mu,\sigma;\boldsymbol{x}_d,d) \doteq -\frac{d}{\sigma} + \frac{1}{\sigma^3}\sum_{j=1}^{d}(\log x_j-\mu)^2 + \frac{n-d}{\sigma}\frac{\log w-\mu}{\sigma}\Big(a_d + b_d\frac{\log w-\mu}{\sigma}\Big).$$

Using the notation from Table 6.6, we get

$$\mu = \zeta + \tau\sigma, \qquad \sigma^2 - \frac{\delta}{d}\sigma - \frac{\eta}{d} = 0. \tag{6.9}$$

This yields the approximate MLEs

$$\widehat{\mu} = \zeta + \tau\widehat{\sigma}, \qquad \widehat{\sigma} = \frac{\delta + \sqrt{\delta^2 + 4d\eta}}{2d},$$

where $\widehat{\sigma}$ is the only positive root of the quadratic equation in (6.9).

Singh and Tripathi (2016) discussed Bayesian inference for log-normal data under Type-I hybrid censoring using a non-informative prior as well as a bivariate prior

$$\pi(\mu,\sigma) = \pi_1(\sigma)\,\pi_2(\mu \mid \sigma)$$

Table 6.6 Quantities in the derivation of approximate MLEs for the log-normal parameters under Type-I hybrid censoring.

quantity	Type-I hybrid
w	$x_m \wedge T$
d	$D \wedge m$
ζ	$\dfrac{\sum_{j=1}^{d} \log x_j + (n-d)b_d \log w}{d+(n-d)b_d}$
τ	$\dfrac{(n-d)a_d}{d+b_d(n-d)}$
δ	$(n-d)a_d(\log w - \zeta)$
η	$\sum_{j=1}^{d} (\log x_j - \zeta)^2 + (n-d)b_d(\log w - \zeta)^2$

where π_1 denotes the density function of an $\mathsf{I\Gamma}(p_1, p_2)$-distribution with parameters $p_1, p_2 > 0$ and $\pi_2(\cdot \mid \sigma)$ denotes the (conditional) density function of a normal distribution with parameters $a \in \mathbb{R}$ and $\sigma/b > 0$ (given the value of σ). They considered Lindley's approximation method, importance sampling, and the software OpenBUGS (see Kelly and Smith, 2011) to compute estimates as well as credible intervals for the parameters of interest.

6.3.1.2 *Further hybrid censoring schemes*

As has been presented for the Weibull distribution in Sections 6.2.2 and 6.2.3, results for other censoring schemes can be directly obtained from the Type-I hybrid censoring case by taking into account the values and type specifications given in Tables 6.3–6.5.

6.3.2 Generalized exponential distributions

Generalized exponential distributions (also called exponentiated exponential distributions) with density function f and cumulative distribution function F defined by

$$f(t) = \alpha\lambda(1-e^{-\lambda t})^{\alpha-1}, \quad F(t) = (1-e^{-\lambda t})^{\alpha}, \quad t > 0,$$

have been proposed in Gupta and Kundu (1999) and extensively studied for various censoring schemes (for reviews, see Gupta and Kundu, 2007; Nadarajah, 2011). Type-I hybrid censoring has been considered in Kundu and Pradhan (2009). In particular, using the data notation as above, the log-likelihood function is given by

$$\begin{aligned}\mathcal{L}^*(\alpha, \lambda; \boldsymbol{x}_d, d) &= c_d^* + d\log(\alpha) + d\log(\lambda)\\ &\quad - \lambda\sum_{j=1}^{d} x_j + (\alpha-1)\sum_{j=1}^{d}\log(1-e^{-\lambda x_j}) + (n-d)\log\left(1-(1-e^{-\lambda w})^{\alpha}\right)\end{aligned}$$

when $d > 0$. For $d = 0$, $\mathscr{L}^*(\alpha, \lambda; 0) = n\big(1 - (1 - e^{-\lambda w})^\alpha\big)$ which is easily seen to be increasing in both parameters so that the MLE does not exist. Following the ideas presented in Gupta and Kundu (2002) for Type-I and Type-II censored data, the corresponding likelihood equations read

$$\frac{\partial}{\partial \alpha}\mathscr{L}^*(\alpha, \lambda; \boldsymbol{x}_d, d) = \frac{d}{\alpha} + \sum_{j=1}^{d} \log(1 - e^{-\lambda x_j}) - (n-d)\frac{\log(1 - e^{-\lambda w})(1 - e^{-\lambda w})^\alpha}{1 - (1 - e^{-\lambda w})^\alpha} = 0,$$

$$\frac{\partial}{\partial \lambda}\mathscr{L}^*(\alpha, \lambda; \boldsymbol{x}_d, d) = \frac{d}{\lambda} + (\alpha - 1)\sum_{j=1}^{d} \frac{x_j e^{-\lambda x_j}}{1 - e^{-\lambda x_j}} - \sum_{j=1}^{d} x_j$$

$$- (n-d)\frac{\alpha w e^{-\lambda w}(1 - e^{-\lambda w})^{\alpha - 1}}{1 - (1 - e^{-\lambda w})^\alpha} = 0.$$

As mentioned in Gupta and Kundu (2002), these equations can obviously be written as a fixed point equation $(\alpha, \lambda) = h(\alpha, \lambda)$ with an appropriate function h leading to an iterative procedure. Alternatively, Gupta and Kundu (2002) proposed a Newton-Raphson procedure to solve the equations (notice that Habibi Rad and Izanlo, 2011 reported convergence problems). An EM-algorithm approach can be found in Kundu and Pradhan (2009) which follows the ideas presented in detail for the log-normal distribution. Therefore, we do not present further details here. It should be noted that, in the maximization step, a fixed point equation in λ has to be solved by an iterative procedure (see Gupta and Kundu, 2001). Clearly, expressions for the approximate MLEs (to be used as initial values in the computation) can be established along the lines of Asgharzadeh (2009).

Likelihood inference under Type-II hybrid censoring has been touched upon in Sen et al. (2018d) and Valiollahi et al. (2017) who focused on optimum life testing plans and prediction problems, respectively. Habibi Rad and Izanlo (2011) adapted the EM-algorithm approach to unified Type-I hybrid censoring as introduced by Balakrishnan et al. (2008d). As shown in Section 5.2, these results as well as paralleling ones for the other hybrid censoring schemes can be obtained using the previously presented results and approaches for Type-I hybrid censored data.

6.3.3 Birnbaum-Saunders distribution

Likelihood inference for Birnbaum-Saunders distributions with parameters $\alpha, \beta > 0$ and density function and cumulative distribution function

$$f(t) = \frac{1}{2\sqrt{2\pi}\alpha\beta}\left[\left(\frac{\beta}{t}\right)^{1/2} + \left(\frac{\beta}{t}\right)^{3/2}\right]\exp\left[-\frac{1}{2\alpha^2}\left(\frac{\beta}{t} - \frac{t}{\beta} - 2\right)\right], \quad t > 0$$

$$F(t) = \Phi\left(\frac{1}{\alpha}\left(\sqrt{\frac{\beta}{t}} - \sqrt{\frac{t}{\beta}}\right)\right), \quad t > 0,$$

under Type-I and Type-II hybrid censoring has been discussed in Balakrishnan and Zhu (2014). In particular, they have studied existence and uniqueness of the MLEs where the results are obtained from the analogous ones established under Type-I and Type-II censoring. In view of the preliminary remarks in this chapter, this is obviously possible. They showed that, using our notation, the MLE of α can be written in terms of the MLE of β, that is,

$$\widehat{\alpha} = \alpha(\widehat{\beta}) = \sqrt{\frac{\sum_{j=1}^{d}(w - x_j)(x_j^{-1} - \widehat{\beta}^{-1})}{\sum_{j=1}^{d}(w + x_j)/(x_j + \widehat{\beta})}}.$$

The MLEs of the parameters α and β exist when $d \geq 2$ and the inequality

$$(n-d)\frac{\varphi\left(\sqrt{\frac{\sum_{j=1}^{d}(w+x_j)/w}{\sum_{j=1}^{d}(w-x_j)/x_j}}\right)}{\Phi\left(\sqrt{\frac{\sum_{j=1}^{d}(w+x_j)/w}{\sum_{j=1}^{d}(w-x_j)/x_j}}\right)} < \frac{1}{\sqrt{w}}\left[\sum_{j=1}^{d} x_j\sqrt{\frac{\sum_{j=1}^{d}(w-x_j)/x_j}{\sum_{j=1}^{d}(w+x_j)}} + d\sqrt{\frac{\sum_{j=1}^{d}(w+x_j)}{\sum_{j=1}^{d}(w-x_j)/x_j}}\right]$$

is satisfied. In this case, the MLEs are unique. The respective equation to determine $\widehat{\beta}$ can be found in Ng et al. (2006) and Balakrishnan and Kundu (2019, p. 19). For $d \leq 1$, the MLEs do not exist. The same problem has been studied for Laplace Birnbaum-Saunders distributions in Zhu et al. (2019).

6.3.4 Laplace distributions

Su et al. (2018) presented MLEs for a location-scale family of Laplace distributions *Laplace*(μ, σ) with location parameter $\mu \in \mathbb{R}$ and scale parameter $\sigma > 0$ (see Definition B.15).

The likelihood function in this case is given by

$$\mathscr{L}(\mu, \sigma; \boldsymbol{x}_d, d) = e^{c_d^*}\prod_{j=1}^{d} f_{\mu,\sigma}(x_j)\left[1 - F_{\mu,\sigma}(w)\right]^{n-d}$$

with $f_{\mu,\sigma}$ as in (B.2).

In this presentation, we use the results of Balakrishnan and Cutler (1996) (see p. 90 as well Childs and Balakrishnan, 1997; Kotz et al., 2001; Iliopoulos and Balakrishnan, 2011; Zhu and Balakrishnan, 2016) obtained for Type-II censored data. Using the decomposition of the problem as detailed in Section 5.2, we find for the Type-II censoring case ⓘⓘ the MLEs of μ and σ as follows ($X_{m:n} \leq T$):

(a) Case $m < n/2$:

$$\widehat{\mu} = X_{m:n} + \widehat{\sigma}\frac{n}{2m}$$

$$\widehat{\sigma} = \frac{1}{m+1}\left[X_{m:n} - \sum_{j=1}^{m} X_{j:n}\right]$$

(b) Case $m \geq n/2$:

$$\widehat{\mu} = \begin{cases} X^* \in [X_{k:2k}, X_{k+1:2k}], & n = 2k, \\ X_{k+1:2k+1}, & n = 2k+1, \end{cases}$$

$$\widehat{\sigma} = \begin{cases} \frac{1}{m+1}\left[(n-m)X_{m:n} + \sum_{j=k+1}^{m} X_{j:n} - \sum_{j=1}^{m} X_{j:n}\right] & n = 2k \\ \frac{1}{m+1}\left[(n-m)X_{m:n} + \sum_{j=k+2}^{m} X_{j:n} - \sum_{j=1}^{m} X_{j:n}\right] & n = 2k+1 \end{cases}$$

Under Type-I censoring, that is, hybrid censoring of type ①, the likelihood function has the same structure with x_m replaced by T, m by D_I (which must be positive in order to ensure existence of the MLEs). Following Zhu and Balakrishnan (2017) (see also Zhu, 2015), the MLEs are given by similar expressions (provided $D_\text{I} \geq 1$), that is,

(a) Case $D_\text{I} < n/2$:

$$\widehat{\mu} = T + \widehat{\sigma}\frac{n}{2D_\text{I}}$$

$$\widehat{\sigma} = \frac{1}{D_\text{I}}\left[T - \sum_{j=1}^{D} X_{j:n}\right]$$

(b) Case $D_\text{I} \geq n/2$:

$$\widehat{\mu} = \begin{cases} X^* \in [X_{k:2k}, X_{k+1:2k} \wedge T], & n = 2k \\ X_{k+1:2k+1}, & n = 2k+1 \end{cases}$$

$$\widehat{\sigma} = \begin{cases} \frac{1}{D_\text{I}}\left[(n-D_\text{I})T + \sum_{j=k+1}^{D_\text{I}} X_{j:n} - \sum_{j=1}^{m} X_{j:n}\right] & n = 2k \\ \frac{1}{D_\text{I}}\left[(n-D_\text{I})T + \sum_{j=k+2}^{D_\text{I}} X_{j:n} - \sum_{j=1}^{m} X_{j:n}\right] & n = 2k+1 \end{cases}$$

Using the results of Section 5.2, these expressions can be utilized to find the MLEs under the other hybrid censoring schemes.

Applying the (conditional) moment generating function approach, the exact (conditional) distribution of the MLEs can be established. For details, we refer to Su et al. (2018) which is based on previous results of Iliopoulos and Balakrishnan (2011), Zhu (2015), and Zhu and Balakrishnan (2016, 2017).

6.3.5 Uniform distributions

Górny and Cramer (2019c) discussed Type-I hybrid censored data from a *Uniform*(a, b)-distributions. They found that the MLE of b does not exist when $D_\text{I} = 0$. For $D_\text{I} > 0$, the MLEs are as follows:

(a) If the left endpoint of support $a < T$ is known, then the MLE of b is given by

$$\widehat{b} = \frac{nX_{m:n} \wedge T - (n - D_\mathrm{I})a}{m}$$

(b) For an unknown parameter a, the MLEs $\widehat{a}$ and $\widehat{b}$ are given by

$$\widehat{a} = X_{1:n} \quad \text{and} \quad \widehat{b} = \frac{nX_{m:n} \wedge T - (n - D_\mathrm{I})\widehat{a}}{m}.$$

Notice that $D_\mathrm{I} > 1$ implies that $\widehat{a} = X_{1:n} \leq T$. Moreover, the representations of the MLE $\widehat{b}$ show that it exceeds a and $\widehat{a}$, respectively.

As Górny and Cramer (2019c) noted, the distributions of the MLEs will have jumps. For a known left endpoint of support a, $\widehat{b}$ has a discrete distribution when $d \in \{1, \ldots, m-1\}$ with probabilities

$$\Pr\left(\widehat{b} = \frac{nT - (n-d)a}{d}\right) = \Pr(D = d), \quad d \in \{1, \ldots, m-1\}.$$

Therefore, for $a < T < b$ and $t \geq 0$, the conditional cumulative distribution function of the MLE $\widehat{b}$ under Type-I hybrid censoring is given by

$$F^{\widehat{b}|D\geq 1}(t) = \frac{1}{1 - \left(\frac{b-T}{b-a}\right)^n}\left[\sum_{d=1}^{m-1} \mathbb{1}_{[n(T-a),\infty)}\big(d(t-a)\big)\,\frac{b-T}{n-d} f_{d+1:n}(T) + F_{m:n}\Big(a + \min\left\{\tfrac{m}{n}(t-a),\, T-a\right\}\Big)\right].$$

An exemplary plot is provided in Fig. 6.1. Notice that $F^{\widehat{b}|D\geq 1}$ is continuous in the interval $[a, a + \frac{n}{m-1}(T-a))$ and has jumps at

$$a + \frac{n}{m-1}(T-a), \ldots, a + \frac{n}{2}(T-a), a + n(T-a), \quad d \in \{1, \ldots, m-1\}. \tag{6.10}$$

Furthermore,

$$\lim_{T \to b-} F^{\widehat{b}|D\geq 1}(t) = F_{m:n}\Big(a + \min\{\tfrac{m}{n}(t-a), b-a\}\Big), \quad a \leq t,$$

which equals the cumulative distribution function of the MLE under Type-II censoring.

For $m = 1$ and an unknown parameter a, the bivariate MLE $(\widehat{a}, \widehat{b})$ is given by

$$(\widehat{a}, \widehat{b}) = \big(U_{1:n}, U_{1:n}\big),$$

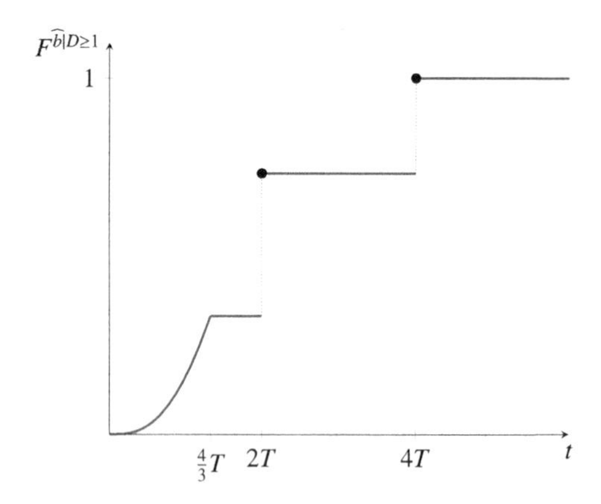

Figure 6.1 Plot of (conditional) cumulative distribution function $F^{\widehat{b}|D\geq 1}$ of $\widehat{b}$ with $a=0$ known and $b=2$ unknown ($T=1, n=4, m=4$).

with cumulative distribution function

$$F^{\widehat{a},\widehat{b}|D\geq 1}(s,t)=\frac{F_{1:n}\big(\min\{s,t,T\}\big)}{F_{1:n}(T)},\qquad s,t\geq a.$$

For $m\geq 2$, the conditional (joint) cumulative distribution function $F^{\widehat{a},\widehat{b}|D\geq 1}$ is given by

$$\begin{aligned}F^{\widehat{a},\widehat{b}|D\geq 1}(s,t)&=\frac{1}{1-\left(\frac{b-T}{b-a}\right)^n}\\&\times\Bigg[\mathbb{1}_{[nT-(n-1)\min\{s,T\},\infty)}(t)\\&\quad\times\Big[F_{1:n}(\min\{s,T\})-F_{1:n}\Big(\tfrac{nT-t}{n-1}\Big)+F_{1,2:n}\Big(\tfrac{nT-t}{n-1},T\Big)-F_{1,2:n}(\min\{s,T\},T)\Big]\\&\quad+\sum_{d=2}^{m-1}\mathbb{1}_{[nT-(n-d)\min\{s,T\},\infty)}(dt)\\&\quad\times\Big[F_{1,d:n}(s,T)-F_{1,d:n}\Big(\tfrac{nT-dt}{n-d},T\Big)+F_{1,d+1:n}\Big(\tfrac{nT-dt}{n-d},T\Big)-F_{1,d+1:n}(s,T)\Big]\\&\quad+\tfrac{m}{n}\int_a^{s\wedge T}\int_a^t f_{1,m:n}\Big(u,\tfrac{mx+(n-m)u}{n}\Big)\,dx\,du\Bigg],\qquad s,t\geq a.\end{aligned}$$

The cumulative distribution function of $\widehat{a}$ conditionally on $D\geq 1$ is given by

$$F^{\widehat{a}|D\geq 1}(s)=\frac{1-\left(\frac{b-\min\{T,s\}}{b-a}\right)^n}{1-\left(\frac{b-T}{b-a}\right)^n},\qquad s\geq a.$$

The cumulative distribution function of $\widehat{b}$ conditionally on $D \geq 1$ is given by

$$F^{\widehat{b}|D\geq 1}(t) = \frac{1}{1-\left(\frac{b-T}{b-a}\right)^n}$$
$$\times \left[\mathbb{1}_{[T,\infty)}(t) \sum_{d=1}^{m-1} \left[F_{d:n}(T) - F_{1,d:n}\left(\tfrac{nT-dt}{n-d}, T\right) + F_{1,d+1:n}\left(\tfrac{nT-dt}{n-d}, T\right) - F_{d+1:n}(T) \right] \right.$$
$$\left. + \frac{m}{n} \int_a^T \int_a^t f_{1,m:n}\left(u, \tfrac{mx+(n-m)u}{n}\right) dx du \right], \quad t \geq a.$$

Notice that the conditional cumulative distribution function of $\widehat{b}$ has jumps at the same points given in (6.10) for the case of a known left endpoint of support a.

6.3.6 ...and even more distributions

Various hybrid censoring schemes have also been discussed for other distributions. Particularly, several estimation concepts have been applied, (approximate) likelihood inference, Bayesian inference, product spacings estimation. Subsequently, we provide an incomplete list with respective references to find further information. It should be noticed that many results can be directly obtained from the Type-I and Type-II censoring cases as illustrated in Section 5.2. Therefore, the results can easily be extended to more complicated hybrid censoring schemes.

Burr XII distribution: Rastogi and Tripathi (2011), Rastogi and Tripathi (2013b) (Type-I hybrid), Panahi and Sayyareh (2015) (unified hybrid)

Dagum distribution: Emam and Sultan (2021) (unified Type-I/Type-II hybrid)

Chen distribution: Pundir and Gupta (2018) (Type-I hybrid)

Exponential-type and Weibull-type distributions: Dey and Pradhan (2014), Kayal et al. (2018), Monfared et al. (2022), Khan and Mitra (2021) (Type-I hybrid), Dey and Pradhan (2014), Kohansal et al. (2015), Singh and Goel (2018), Salah et al. (2021) (Type-II hybrid), Jeon and Kang (2021) (unified hybrid)

Gamma type distributions: Yadav et al. (2019a) (Type-II hybrid)

Gumbel distribution: Mahto et al. (2022b) (Type-I hybrid)

Kumaraswamy distribution: Sultana et al. (2018) (Type-I hybrid)

Lindley type distributions: Gupta and Singh (2013), Al-Zahrani and Gindwan (2014), Singh et al. (2016), Basu et al. (2019), Du et al. (2021) (Type-I hybrid), Singh et al. (2014), Al-Zahrani and Gindwan (2014) (Type-II hybrid)

(Log)-logistic distributions: Asgharzadeh et al. (2013), Hyun et al. (2016), Raqab et al. (2021) (Type-I hybrid), Hyun et al. (2016) (Type-II hybrid)

Normal-type distributions: Sultana and Tripathi (2018) (Type-I hybrid)

Pareto/Lomax distribution: Cheng and Zhao (2016), Prakash (2020) (Type-I hybrid), Yadav et al. (2019b), Çetinkaya (2020) (Type-II hybrid), Sultan and Emam (2021)

Bathtub-shaped distribution: Rastogi and Tripathi (2013a) (Type-I hybrid)

Further distributions: Data from the family of distributions with cumulative distribution function given by

$$F(t) = 1 - \exp(-\psi_\theta(t)), \quad t \in \mathbb{R},$$

where ψ_θ is an increasing, continuous and differentiable function, with $\psi_\theta \to 0$ for $t \to -\infty$ and $\psi_\theta(t) \to \infty$ as $t \to \infty$, has been extensively studied under hybrid censoring schemes in terms of Bayesian inference. Notice that this family includes, e.g., exponential, (compound) Weibull, Gompertz, Burr XII, and Pareto distributions (see AL-Hussaini, 1999). Shafay and Balakrishnan (2012) and Balakrishnan and Shafay (2012) considered generalized Type-I and Type-II hybrid censoring schemes, respectively. Generalized Type-II hybrid censored samples have been investigated by Shafay (2016a). Mohie El-Din et al. (2017) discussed unified Type-II hybrid censoring.

CHAPTER 7

Progressive hybrid censored data

Contents

7.1. Progressive (hybrid) censoring schemes

So far, the idea of hybrid censoring has been applied to order statistics $X_{1:n} \leq \cdots \leq X_{n:n}$ resulting in the various hybrid censoring schemes presented in Chapter 4. The basic ingredients in the definition are the ordering of the data as well as a combination of Type-I and Type-II censoring operated on the data. In the following, we illustrate these procedures for progressively Type-II censored data $X_{1:m:n} \leq \cdots \leq X_{m:m:n}$ (see Fig. 2.1). As a matter of fact, an appropriate name for the model would therefore be *hybrid censored progressively Type-II censored data*. However, since the term *progressive hybrid censoring* has been generally accepted, we will use it in the following.

7.1.1 Progressive Type-I censoring as (Type-I) hybrid censoring scheme

As mentioned above, the baseline data will be progressively Type-II censored order statistics in the ensuing discussion. However, it should be mentioned that any hybrid censoring scheme may be applied to other kinds of (progressively) censored data. For instance, progressive Type-I censored data as introduced in Cohen (1963) may also been discussed in this framework by interpreting the final censoring time τ_m as a Type-I

Hybrid Censoring Know-How
https://doi.org/10.1016/B978-0-12-398387-9.00015-5

censoring threshold. It should be mentioned that the terminology introduced by Cohen (1963) is somewhat misleading since the measurements in the originally proposed sampling design are not really Type-I censored as it allows observation after the final withdrawal time τ_m. For this reason, Laumen and Cramer (2019) proposed the term *progressive censoring with fixed censoring times* for Cohen's progressive Type-I censoring scheme (see also the comments in Cramer, 2021; Balakrishnan and Cramer, 2023). Understanding $\tau_m = T$ as the termination time of the experiment, the progressive Type-I censoring scheme (as it has been understood after publication of the monograph Balakrishnan and Aggarwala, 2000) can be seen as a Type-I hybrid censored version of progressively censored data with fixed withdrawal (censoring) times $\tau_1 < \cdots < \tau_{m-1}$. For reviews of respective results, we refer to Balakrishnan and Cramer (2014, 2023). Notice that all the hybrid censoring schemes mentioned in Chapter 4 can also be applied to that kind of data. However, apparently, such models seem not have been studied in the literature so far.

7.1.2 Hybrid censoring schemes based on progressively Type-II censored data

In the literature, progressive hybrid censoring schemes are commonly based on progressively Type-II censored samples. Consider a progressively Type-II right censored sample $X_{1:m:n} \leq \cdots \leq X_{m:m:n}$ with initial censoring plan $\mathscr{R} = (R_1, \ldots, R_m)$ prespecified at the start of the life test. Notice that the final censoring time $X_{m:m:n}$ can be interpreted in terms of Type-II right censoring where we intend to observe m failures and R_j items are withdrawn at $X_{j:m:n}$, $1 \leq j \leq m-1$. In this sense, the (extended) censoring plan

$$\mathscr{R}_* = (R_1, \ldots, R_{m-1}, 0^{*R_m+1}) \tag{7.1}$$

and the *(complete) progressively Type-II censored sample*

$$X_{1:m+R_m:n} \leq \cdots \leq X_{m:m+R_m:n} \leq X_{m+1:m+R_m:n} \leq \cdots \leq X_{m+R_m:m+R_m:n}$$

have been introduced by Childs et al. (2008) to allow the definition of Type-II progressive hybrid censoring (see also Cramer et al., 2016; Górny and Cramer, 2016). Notice that the part 0^{*R_m+1} of the censoring plan $\mathscr{R}_*$ means that no withdrawals occur after $X_{m:m:n}$, that is, all failures at $X_{m:m+R_m:n}, \ldots, X_{m+R_m:m+R_m:n}$ are possibly observed. In fact, right censoring is dropped in the life test in order to have possible measurements after $X_{m:m:n} = X_{m:m+R_m:n}$. In this regard, we can view $n^* = m + R_m$ as the sample size of the complete progressively Type-II censored sample (see Fig. 7.1) which yields the notation

$$X_{1:n^*:n}, \ldots, X_{n^*:n^*:n}.$$

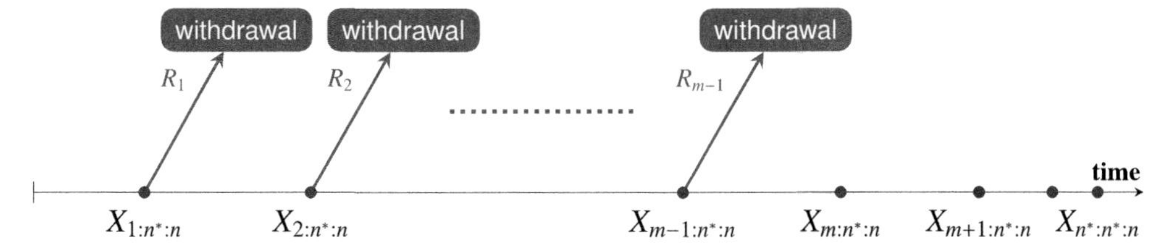

Figure 7.1 Complete (or extended) progressively Type-II censored data with censoring plan $\mathscr{R}_* = (R_1, \ldots, R_{m-1}, 0^{*R_m+1})$ as introduced in Childs et al. (2008).

Applying Type-II right censoring to these data with m intended observations, we get the *progressively Type-II right censored sample*

$$(X_{1:m:n}, \ldots, X_{m:m:n}) = (X_{1:n^*:n}, \ldots, X_{m:n^*:n})$$

as illustrated in Fig. 2.1. Furthermore,

$$\gamma_j^* = \begin{cases} \sum_{i=j}^{m}(R_i + 1), & 1 \leq j \leq m \\ n^* - j + 1 = m + R_m - j + 1, & m + 1 \leq j \leq n^* \end{cases}$$

with $\gamma_m = R_m + 1$.

However, although this view of the model easily allows to define all the hybrid censored schemes presented in Chapter 4 for progressively Type-II censored data, we will proceed with a more general model which is in accordance with both the hybrid censoring mechanism for order statistics and allows a comparable notation. Furthermore, the extended sample introduced by Childs et al. (2008) is included by considering the particular censoring plan $\mathscr{R}_*$ as in (7.1).

Therefore, we consider in the following a more general situation, that is, the (complete) progressively Type-II censored sample

$$X_{1:n^*:n}, \ldots, X_{n^*:n^*:n} \tag{7.2}$$

with censoring plan $\mathscr{R} = (R_1, \ldots, R_{n^*})$ and baseline sample size $n = \sum_{j=1}^{n^*}(R_j + 1)$. This approach underlines the similarities to the hybrid censoring schemes (based on order statistics' data). Clearly, we have $\gamma_j = \sum_{i=j}^{n^*}(R_i + 1)$, $1 \leq j \leq n^*$. For convenience, we define $\gamma_{n^*+1} = 0$. The parameter m is reserved for the Type-II right censored data in (7.2) with $m \leq n^*$ intended observations leading to the marginal sample (or Type-II right censored progressively Type-II censored sample)

$$(X_{1:m:n}, \ldots, X_{m:m:n}) = (X_{1:n^*:n}, \ldots, X_{m:n^*:n})$$

where a total number of $\gamma_m - 1 = \sum_{j=m}^{n^*}(R_j + 1) - 1$ items are withdrawn at time $X_{m:n^*:n}$ from the life test (see Fig. 7.2).

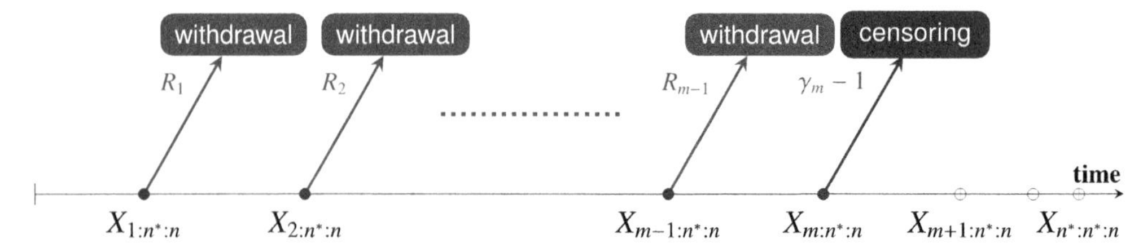

Figure 7.2 Progressively Type-II right censored data with m observations and censoring plan $\mathscr{R}_* = (R_1, \ldots, R_{n^*})$. The measurements $X_{m+1:n^*:n}, \ldots, X_{n^*:n^*:n}$ are censored which is indicated by the circle $\circ$.

In order to illustrate the construction and properties of progressive hybrid censoring schemes, we present the basic schemes of Type-I and Type-II progressive hybrid censoring originally introduced in Childs et al. (2008) and Kundu and Joarder (2006b) in detail. Comparing Figs. 4.4, 7.4 and Tables 4.3, 7.1 as well as Figs. 4.6, 7.5 and Tables 4.4, 7.2, the structural similarities to hybrid censoring are striking. In particular, the progressive hybrid censoring schemes can be decomposed into the same basic module types O, A, B, AB as given in Sections 4.5.1 and 4.5.3 (see also Górny and Cramer, 2018b). Therefore, the remaining progressive hybrid censoring schemes are only sketched.

By similarity with hybrid censoring schemes for order statistics, we first introduce random counters which represent the sample size of the progressive hybrid censored sample. We proceed as in Section 4.2. Given the threshold T, we introduce the random counter $D = D(T)$ as in (4.2) by (see (2.55))

$$D = \sum_{j=1}^{n^*} \mathbb{1}_{(-\infty, T]}(X_{j:n^*:n}). \tag{7.3}$$

As above, we write $D_i = D(T_i)$ for a threshold T_i. Then, for $d \in \{1, \ldots, n^*\}$, we have

$$D \geq d \iff X_{d:n^*:n} \leq T$$

so that the probability mass function of D is given by

$$\Pr(D = d) = \begin{cases} 1 - F_{1:n}(T), & d = 0 \\ F_{d:n^*:n}(T) - F_{d+1:n^*:n}(T), & d \in \{1, \ldots, n^* - 1\} \\ F_{n^*:n^*:n}(T), & d = n^* \end{cases} \tag{7.4}$$

with support $\{0, \ldots, n^*\}$.

7.1.2.1 Type-I progressive hybrid censoring

Defining the random counter $D_\mathrm{I} = \sum_{j=1}^{m} \mathbb{1}_{(-\infty,T]}(X_{j:n^*:n}) = D \wedge m$ as in (4.5) as the sample size of the Type-I progressive hybrid censored data (see below), the respective probability mass function is given in Lemma 7.1 (see also Balakrishnan and Cramer, 2014, Corollary 2.4.7, Lemma 2.5.4).

Lemma 7.1. *The probability mass function of D_I is given by the probabilities*

$$\begin{aligned}
\Pr(D_I = 0) &= \Pr(X_{1:n} > T) = (1 - F(T))^n \\
\Pr(D_I = m) &= \Pr(D \geq m) = \Pr(X_{m:n^*:n} \leq T) = F_{m:n^*:n}(T),
\end{aligned}$$

with

$$F_{m:n^*:n}(T) = F_{m:m:n}(T) = 1 - \Big(\prod_{i=1}^{m} \gamma_i\Big) \sum_{j=1}^{m} \frac{1}{\gamma_j} a_{j,m} (1 - F(t))^{\gamma_j}, \quad t \in \mathbb{R}.$$

For $d \in \{1, \dots, m-1\}$, we obtain the identity

$$\Pr(D_I = d) = \Pr(D = d) = \frac{1 - F(T)}{\gamma_{d+1} f(T)} f_{d+1:n^*:n}(T)$$

in terms of the density function $f_{d+1:n^*:n} = f_{d+1:m:n}$ (provided it exists and $f(T) > 0$).

Notice that, from (2.50),

$$f_{d+1:n^*:n}(t) = \Big(\prod_{j=1}^{d+1} \gamma_j\Big) f(t) \sum_{j=1}^{d+1} a_{j,d+1} (1 - F(t))^{\gamma_j - 1}, \quad t \in \mathbb{R}.$$

By analogy with Lemma 4.5, we get the following result from Lemma 7.1.

Lemma 7.2. **(a)** $ED_I = E(D \wedge m) = \frac{1-F(T)}{f(T)} \sum_{d=0}^{m-1} \frac{d}{\gamma_{d+1}} f_{d+1:n^*:n}(T) + mF_{m:n^*n}(T) = \sum_{d=1}^{m} F_{d:n^*n}(T)$
(b) $EW = T(1 - F_{m:n^*n}(T)) + \int_{-\infty}^{T} t f_{m:n^*:n}(t)\, dt$

Type-I progressive hybrid censoring has been introduced in Kundu and Joarder (2006b) and Balakrishnan et al. (2008d). This scheme can be considered as the first hybrid censoring scheme based on progressively Type-II censored data. By similarity with (4.8), the Type-I progressive hybrid censored data is given by the random variables

$$\min(X_{j:n^*:n}, T) = X_{j:n^*:n} \wedge T, \quad 1 \leq j \leq m. \tag{7.5}$$

An illustration of the Type-I censoring case is shown in Fig. 7.3.

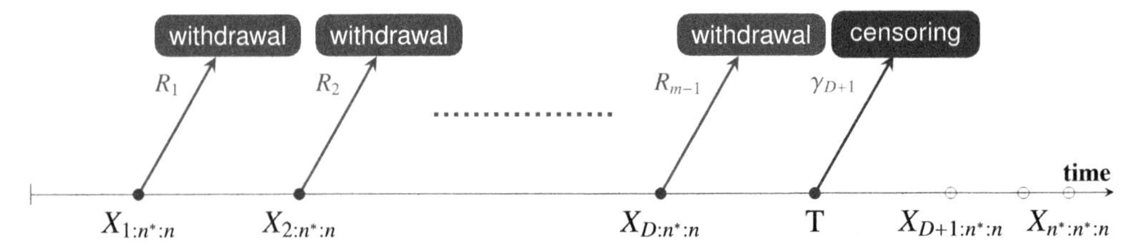

Figure 7.3 Type-I censoring of progressively Type-II censored data with censoring plan $\mathscr{R} = (R_1, \ldots, R_m)$ and time threshold T.

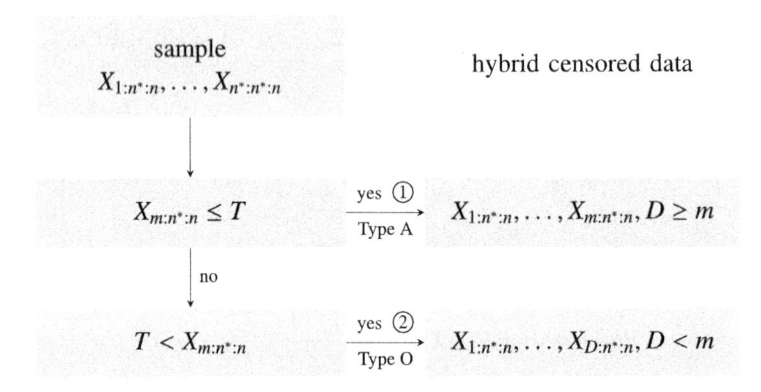

Figure 7.4 Sampling scenarios ①, ② in Type-I progressive hybrid censoring.

In particular, we get two data scenarios ① and ② as depicted in Fig. 7.4 and the effective sample size is given by $D_{\text{I}} = \min\{D, m\} = D \wedge m$, so that, from (7.5), the Type-I progressive hybrid censored sample is given by

$$X_{1:n^*:n}, \ldots, X_{D_{\text{I}}:n^*:n}. \tag{7.6}$$

As in standard Type-I hybrid censoring, the experiment may terminate without observing a failure, that is, $D_{\text{I}} = 0$ with probability $\Pr(D_{\text{I}} = 0) = \Pr(D = 0) = (1 - F(T))^n > 0$ since $F(T) < 1$ (see (7.4)). The termination time of the experiment is given by $X_{D_{\text{I}}:n^*:n} \wedge T$. Characteristics of Type-I progressive hybrid censoring are summarized in Table 7.1.

As shown in Cramer and Balakrishnan (2013), the derivation of the distribution is along the lines of the Type-I hybrid censoring scheme taking into account the distribution of the underlying sample given in (7.6) (see also (4.15)). In particular, the joint cumulative distribution function of $X_{1:n^*:n} \wedge T, \ldots, X_{m:n^*:n} \wedge T$ is given by the expression

$$\Pr(X_{j:n^*:n} \wedge T \leq t_j, 1 \leq j \leq m) = \sum_{d=0}^{m} \Pr(X_{j:n^*:n} \wedge T \leq t_j, 1 \leq j \leq m, D = d)$$

Table 7.1 Characteristics of Type-I progressive hybrid censoring.

Type-I progressive hybrid censoring	Expression	Properties
Scheme parameters	m, T	
Stopping time T^*	$X_{m:n^*:n} \wedge T$	bounded by T
Sample size D_I	$D \wedge m$	$D_I \in \{0, \ldots, m\}$
	Sample size may be zero with positive probability.	
Experimental time W	$X_{m:n^*:n} \wedge T$	

$$= \sum_{d=0}^{m-1} \mathbb{1}_{[T,\infty)}\Big(\min_{d+1 \le j \le m} t_j\Big) C_d (1 - F(T))^{\gamma_{d+1}} F_{1,\ldots,d:d:n-\gamma_{d+1}}(\boldsymbol{t}_d \wedge T)$$
$$+ F_{1,\ldots,m:m:n}(\boldsymbol{t}_m \wedge T).$$

Notice that $n - \gamma_{d+1} = d + \sum_{i=1}^{d} R_j$ equals the sum of both failed and progressively censored items until the d-th failure $X_{d:n^*:n}$. Furthermore, we get as in Theorem 4.9 a quantile representation, that is,

$$X_{j:n^*:n} \wedge T, 1 \le j \le m \stackrel{d}{=} F^{\leftarrow}\big(U_{j:n^*:n} \wedge F(T)\big), 1 \le j \le m,$$

where $U_{j:n^*:n} \wedge F(T)$, $1 \le j \le m$, are Type-I progressive hybrid censored order statistics from a uniform distribution with time threshold $F(T)$.

Finally, let $D_I = D = d \in \{1, \ldots, m-1\}$. Then, we get the (corresponding) conditional density function

$$f^{X_{j:n^*:n}, 1 \le j \le d | D_I = d} = f_{1,\ldots,d:n^*:n}(\cdot \mid X_{d+1:n^*:n} = T)$$

with support $\mathbb{R}^d_{\le T}$, $1 \le d \le m-1$. For $D_I = m$, the joint density function reads

$$f^{X_{j:n^*:n}, 1 \le j \le m | D_I = m} = \frac{f_{1,\ldots,m:m:n}}{F_{m:m:n}(T)}$$

with support $\mathbb{R}^m_{\le T}$. Notice that $\{D_I = m\} = \{D \ge m\}$.

Finally, for $d > 0$, the joint density function is given by

$$f^{X_{j:n^*:n}, 1 \le j \le d, D_I}(\boldsymbol{t}_d, d) = f^{X_{j:n^*:n}, 1 \le j \le d | D_I = d}(\boldsymbol{t}_d \mid d) \Pr(D_I = d)$$
$$= \begin{cases} \frac{1-F(T)}{\gamma_{d+1} f(T)} f_{1,\ldots,d+1:m:n}(\boldsymbol{t}_d, T), & d \in \{1, \ldots, m-1\} \\ f_{1,\ldots,m:m:n}(\boldsymbol{t}_m), & d = m \end{cases}. \tag{7.7}$$

This expression will be of great interest in statistical inference since (7.7) defines the likelihood function. Proceeding as in Section 5.2 with an absolutely continuous lifetime cumulative distribution function $F_{\boldsymbol{\theta}}$ and density function $f_{\boldsymbol{\theta}}$, $\boldsymbol{\theta} \in \Theta$, we find using the representations in (7.7)

$$\begin{aligned}
&\mathscr{L}(\boldsymbol{\theta}; \boldsymbol{x}_d, d) \\
&= \frac{1-F_{\boldsymbol{\theta}}(T)}{(n-d)f_{\boldsymbol{\theta}}(T)} f_{\boldsymbol{\theta},1,\ldots,d+1:m:n}(\boldsymbol{x}_d, T), \qquad d \in \{1,\ldots,m-1\}, \quad (\text{type } ⓘ) \qquad (7.8)\\
&= f_{\boldsymbol{\theta},1,\ldots,m:m:n}(\boldsymbol{x}_m), \qquad d = m. \quad (\text{type } ⓘⓘ)
\end{aligned}$$

Then, for type ⓘⓘ with $d = m$, the marginal density function $f_{\boldsymbol{\theta},1,\ldots,d:m:n}$ has the form

$$f_{\boldsymbol{\theta},1,\ldots,d:n}(\boldsymbol{x}_d) = \prod_{j=1}^{d} \gamma_j \Big(\prod_{j=1}^{d} f_{\boldsymbol{\theta}}(x_j)(1-F_{\boldsymbol{\theta}}(x_j))^{R_j} \Big)(1-F_{\boldsymbol{\theta}}(x_d))^{\gamma_{d+1}},$$

where the identity $\gamma_{d+1} = \gamma_d - R_d - 1$ has been utilized. Using the representation of the marginal density function $f_{\boldsymbol{\theta},1,\ldots,d+1:m:n}$, we get for the type ⓘ–scenario the following expression for the density function

$$f_{\boldsymbol{\theta},1,\ldots,d+1:m:n}(\boldsymbol{x}_d, T) = \prod_{j=1}^{d+1} \gamma_j \Big(\prod_{j=1}^{d} f_{\boldsymbol{\theta}}(x_j)(1-F_{\boldsymbol{\theta}}(x_j))^{R_j} \Big) f_{\boldsymbol{\theta}}(T)(1-F_{\boldsymbol{\theta}}(T))^{\gamma_{d+1}-1}$$

with $d \in \{1,\ldots,m-1\}$. Therefore, the likelihood function for type ⓘ can be written as

$$\mathscr{L}(\boldsymbol{\theta}; \boldsymbol{x}_d, d) = \prod_{j=1}^{d} \gamma_j \Big(\prod_{j=1}^{d} f_{\boldsymbol{\theta}}(x_j)(1-F_{\boldsymbol{\theta}}(x_j))^{R_j} \Big)(1-F_{\boldsymbol{\theta}}(T))^{\gamma_{d+1}}.$$

Thus, by similarity with hybrid censoring, introducing a real value w, the likelihood function can be written in both cases in the form

$$\mathscr{L}(\boldsymbol{\theta}; \boldsymbol{x}_d, d) = \prod_{j=1}^{d} \gamma_j \Big(\prod_{j=1}^{d} f_{\boldsymbol{\theta}}(x_j)(1-F_{\boldsymbol{\theta}}(x_j))^{R_j} \Big)(1-F_{\boldsymbol{\theta}}(w))^{\gamma_{d+1}} \qquad (7.9)$$

with appropriately chosen value w, respectively (see Table 7.3). Thus, we arrive at a similar representation for the likelihood as for Type-I hybrid censoring (see (5.1)). As a consequence, we can solve the respective inferential problems in the same manner as for hybrid censoring schemes taking into account respective results for progressively Type-II censored data. In order to illustrate this, we will present the respective likelihood functions for the remaining progressive hybrid censoring schemes.

7.1.2.2 Type-II progressive hybrid censoring

This hybrid censoring scheme has been proposed by Balakrishnan et al. (2008d). Paralleling the distributional results for Type-I progressive hybrid censoring, Cramer et al.

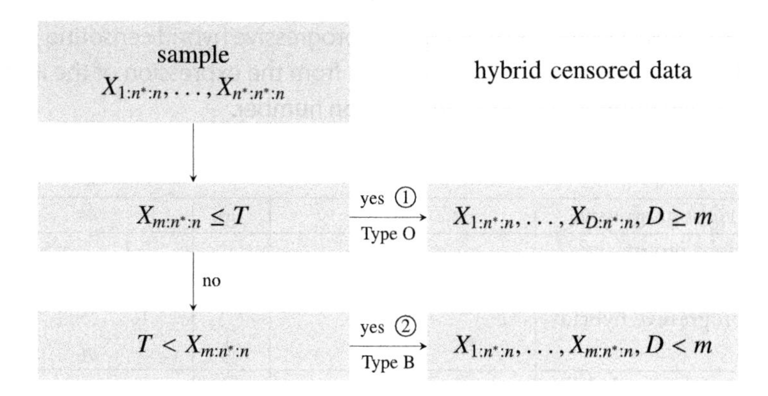

Figure 7.5 Sampling scenarios ①, ② in Type-II progressive hybrid censoring.

Table 7.2 Characteristics of Type-II progressive hybrid censoring.

Type-II progressive hybrid censoring	Expression	Properties
Scheme parameters	m, T	
Stopping time T^*	$X_{m:n^*:n} \vee T$	unbounded
Sample size D_{II}	$D \vee m$	$D_{\text{II}} \in \{m, \ldots, n^*\}$
	The design guarantees a minimum of m observations.	
Experimental time W	$X_{m:n^*:n} \vee (X_{n^*:n^*:n} \wedge T)$	

(2016) established expressions for the joint cumulative distribution function and density function. As indicated by the sampling outcomes shown in Fig. 7.5, the sample size is given by $D_{\text{II}} = D \vee m$ (see Table 7.2). According to Cramer et al. (2016), the corresponding probability mass function of D_{II} can be obtained from the probabilities

$$\begin{aligned}
\Pr(D < m) &= \Pr(X_{m:n^*:n} > T) = 1 - F_{m:n^*:n}(T), \\
\Pr(D = d) &= F_{d:n^*:n}(T) - F_{d+1:n^*:n}(T) \\
&= \frac{1 - F(T)}{\gamma_{d+1} f(T)} f_{d+1:n^*:n}(T), \qquad d \in \{m, \ldots, n^* - 1\}, \\
\Pr(D = n^*) &= P(X_{n^*:n^*:n} \leq T) = F_{n^*:n^*:n}(T).
\end{aligned}$$

Performing similar steps as in the derivation of (7.7), the joint density function can be written as

$$\begin{aligned}
f^{X_{j:n^*:n}, 1 \leq j \leq m, D < m}(\boldsymbol{t}_m) &= f_{1,\ldots,m:n^*:n}(\boldsymbol{t}_m), \\
f^{X_{j:n^*:n}, 1 \leq j \leq d, D}(\boldsymbol{t}_d, d) &= \begin{cases} \frac{1-F(T)}{\gamma_{d+1} f(T)} f_{1,\ldots,d+1:m:n}(\boldsymbol{t}_d, T), & d \in \{m, \ldots, n^* - 1\} \\ f_{1,\ldots,n^*:n^*:n}(\boldsymbol{t}_{n^*}), & d = n^* \end{cases}.
\end{aligned} \tag{7.10}$$

Table 7.3 Types of likelihood functions for progressive hybrid censoring schemes. The appropriate range for *d* has to be taken from the expression of the likelihood function given under the respective equation number.

censoring scheme	joint density function	type	parameters	
			d	w
Type-II right progressive		(ii)	m	x_m
Type-I right progressive		(i)	$1, \dots, n^*$	T
		(ii)	n^*	x_{n^*}
Type-I progressive hybrid	(7.7)	(i)	$1, \dots, m-1$	T
		(ii)	m	x_m
Type-II progressive hybrid	(7.10)	(ii)	m	x_m
		(i)	$m+1, \dots, n^*-1$	T
		(ii)	n^*	x_{n^*}
generalized Type-I progressive hybrid	(7.11)	(ii)	m	x_m
		(i)	$m+1, \dots, k-1$	T
		(ii)	k	x_k
generalized Type-II progressive hybrid	(7.12)	(i)	$m+1, \dots, n^*-1$	T_1
		(ii)	n^*	x_{n^*}
		(ii)	m	x_m
		(i)	$1, \dots, m-1$	T_2

As a consequence, the likelihood function can be written in the form (7.9) with appropriately chosen values for w (see Table 7.3).

7.1.2.3 *Joint density function and likelihood function for other progressive hybrid censoring schemes*

In this section, we present only expressions for the joint density function of the respective progressive hybrid censoring schemes. For details, we refer to the references mentioned as well as to the presentation in Section 4.5 for standard hybrid censoring schemes. This particularly applies to the sampling situation depicted in Figs. 4.7–4.13 where the order statistics have to be replaced by the corresponding progressively Type-II censored order statistics. The representations of the joint density function yield directly an expression of the form (7.9) for the likelihood function choosing the values of d and w as given in Table 7.3.

Furthermore, the joint density function (and thus, the likelihood function) under the corresponding hybrid censoring scheme results by choosing

$$n^* = n, \quad R_1 = \cdots = R_n = 0,$$

so that $\gamma_j = n - j + 1$, $1 \le j \le n$. Note also that the function Υ introduced in Remark 4.2 is identical leading to the test durations given in Table 5.1 (with the order statistics replaced by the corresponding progressively Type-II censored order statistics).

Generalized Type-I progressive hybrid censoring

This progressive hybrid censoring scheme has independently been proposed in Cho et al. (2015b) and Górny and Cramer (2016) with scheme parameters $1 \le m < k \le n^*$ and threshold T (for $k = n^*$ it is identical to Type-II progressive hybrid censoring). The corresponding joint density function is given by

$$f^{X_{j:n^*:n}, 1\le j\le d\vee m, D}(\boldsymbol{t}_{d\vee m}, d) = \begin{cases} f_{1,\dots,m:m:n}(\boldsymbol{t}_m), & d < m \\ \frac{1-F(T)}{\gamma_{d+1} f(T)} f_{1,\dots,d+1:n^*:n}(\boldsymbol{t}_d, T), & d \in \{m, \dots, k-1\} \\ f_{1,\dots,k:k:n}(\boldsymbol{t}_k), & d = k \end{cases} \tag{7.11}$$

The parameters of the likelihood function with representation as in (7.9) and the respective types are given in Table 7.3. Further details are given in Cho et al. (2015b), Górny and Cramer (2016), and Górny (2017) (see also Górny and Cramer, 2018b).

Generalized Type-II progressive hybrid censoring

This progressive hybrid censoring scheme has independently been proposed in Górny and Cramer (2016) and Lee et al. (2016a) with scheme parameters $1 \le m \le n^*$ and thresholds $T_1 < T_2$. The corresponding joint density function is given by

$$f^{X_{j:n^*:n}, 1\le j\le d^*, D_1, D_2}(\boldsymbol{t}_{d^*}, d_1, d_2) = \begin{cases} \frac{1-F(T_1)}{\gamma_{d_1+1} f(T_1)} f_{1,\dots,d_1+1:n^*:n}(\boldsymbol{t}_{d_1}, T_1), & m \le d_1 < n^* \\ f_{1,\dots,n^*:n^*:n}(\boldsymbol{t}_n), & d_1 = n^* \\ f_{1,\dots,m:n^*:n}(\boldsymbol{t}_m), & d_1 < m \le d_2 \\ \frac{1-F(T_2)}{\gamma_{d_2+1} f(T_2)} f_{1,\dots,d_2+1:n^*:n}(\boldsymbol{t}_{d_2}, T_2), & d_2 < m \end{cases} \tag{7.12}$$

with $d^* = \min\{\max\{d_1, m\}, d_2\}$. The parameters of the likelihood function with form as in (7.9) and the respective types are given in Table 7.3. Further details are given in Górny and Cramer (2016), Lee et al. (2016a), and Górny (2017) (see also Górny and Cramer, 2018b).

Unified progressive hybrid censoring

Inspired by the unified hybrid censoring schemes presented in Section 4.5.4, Górny (2017) and Górny and Cramer (2018b) proposed corresponding versions based on progressively Type-II censored data. The scheme parameters as well as the thresholds satisfy the same conditions as those for the corresponding hybrid censoring schemes. For completeness, we present the respective joint density functions. The expression for w in the likelihood function with form as in (7.9) as well as the respective types can be taken from Table 6.5 where n has to be replaced by n^*.

(a) Unified Type-I progressive hybrid censoring

$$f^{X_{j:n^*:n},1\le j\le d^*,D_1,D_2}(\boldsymbol{t}_{d^*},d_1,d_2)=\begin{cases}f_{1,\ldots,m:n^*:n}(\boldsymbol{t}_m), & d_1=m\\ \frac{1-F(T_1)}{\gamma_{d_1+1}f(T_1)}f_{1,\ldots,d_1+1:n^*:n}(\boldsymbol{t}_{d_1},T_1), & k\le d_1\le m-1\\ f_{1,\ldots,k:n^*:n}(\boldsymbol{t}_k), & d_1<k=d_2\\ \frac{1-F(T_2)}{\gamma_{d_2+1}f(T_2)}f_{1,\ldots,d_2+1:k:n}(\boldsymbol{t}_{d_2},T_2), & d_2<k\end{cases} \tag{7.13}$$

(b) Unified Type-II progressive hybrid censoring

$$f^{X_{j:n^*:n},1\le j\le d^*,D_1,D_2}(\boldsymbol{t}_{d^*},d_1,d_2)=\begin{cases}\frac{1-F(T_1)}{\gamma_{d_1+1}f(T_1)}f_{1,\ldots,d_1+1:n^*:n}(\boldsymbol{t}_{d_1},T_1), & m\le d_1<n^*\\ f_{1,\ldots,n^*:n^*:n}(\boldsymbol{t}_{n^*}), & d_1=n^*\\ f_{1,\ldots,m:n^*:n}(\boldsymbol{t}_m), & d_1<m=d_2\\ \frac{1-F(T_2)}{\gamma_{d_2+1}f(T_2)}f_{1,\ldots,d_2+1:k:n}(\boldsymbol{t}_{d_2},T_2), & k\le d_2<m\\ f_{1,\ldots,k:n^*:n}(\boldsymbol{t}_k), & d_2<k\end{cases}$$

(c) Unified Type-III progressive hybrid censoring

$$f^{X_{j:n^*:n},1\le j\le d^*,D_1,D_2,D_3}(\boldsymbol{t}_{d^*},d_1,d_2,d_3)$$

$$=\begin{cases}f_{1,\ldots,n^*:n^*:n}(\boldsymbol{t}_{n^*}), & d_1=n^*\\ \frac{1-F(T_1)}{\gamma_{d_1+1}f(T_1)}f_{1,\ldots,d_1+1:n^*:n}(\boldsymbol{t}_{d_1},T_1), & m\le d_1<n^*\\ f_{1,\ldots,m:n^*:n}(\boldsymbol{t}_m), & d_1<m=d_2\\ \frac{1-F(T_2)}{\gamma_{d_2+1}f(T_2)}f_{1,\ldots,d_2+1:k:n}(\boldsymbol{t}_{d_2},T_2), & k\le d_2<m\\ f_{1,\ldots,k:n^*:n}(\boldsymbol{t}_k), & d_2<k=d_3\\ \frac{1-F(T_3)}{\gamma_{d_3+1}f(T_3)}f_{1,\ldots,d_3+1:k:n}(\boldsymbol{t}_{d_3},T_3), & d_3<k\end{cases}$$

(d) Unified Type-IV progressive hybrid censoring

$$f^{X_{j:n^*:n},1\le j\le d^*,D_1,D_2}(\boldsymbol{t}_{d^*},d_1,d_2)=\begin{cases}f_{1,\ldots,r:n^*:n}(\boldsymbol{t}_r), & d_1=r\\ \frac{1-F(T_1)}{\gamma_{d_1+1}f(T_1)}f_{1,\ldots,d_1+1:n^*:n}(\boldsymbol{t}_{d_1},T_1), & m\le d_1<r\\ f_{1,\ldots,m:n^*:n}(\boldsymbol{t}_m), & d_1<m=d_2\\ \frac{1-F(T_2)}{\gamma_{d_2+1}f(T_2)}f_{1,\ldots,d_2+1:k:n}(\boldsymbol{t}_{d_2},T_2), & k\le d_2<m\\ f_{1,\ldots,k:n^*:n}(\boldsymbol{t}_k), & d_2<k\end{cases} \tag{7.14}$$

7.2. Exponential case: MLEs and its distribution

Progressive hybrid censoring schemes for exponential lifetimes have been first discussed by Childs et al. (2008) and Kundu and Joarder (2006b) as Type-I and Type-II progressive hybrid censoring. Taking into account the comments in Section 7.1.2, it is clear that the MLEs are available in explicit form (provided they exist). We will see that the MLEs of the parameter ϑ (for $Exp(\vartheta)$-distributed lifetimes) can always be written in the form

$$\widehat{\vartheta} = \frac{1}{D_{\text{HCS}}}\text{TTT}_{\text{HCS}},$$

where

$$\text{TTT}_{\text{HCS}} = \sum_{j=1}^{D_{\text{HCS}}}(R_j+1)X_{j:n^*:n} + \gamma_{D_{\text{HCS}}+1}\rho_{\text{HCS}}(X_{D_{\text{HCS}}:n^*:n}, T)$$

and D_{HCS} denote the total time on test statistic and the observed sample size under the implemented hybrid censoring scheme, respectively. ρ_{HCS} is a bivariate function depending on the applied (hybrid) censoring scheme. Thus, the distribution of the MLE mainly depends on both the distribution of TTT_{HCS} given D_{HCS} and the probability mass function of D_{HCS}. We will illustrate that it can be composed of special simple components using the modularization approach of Górny and Cramer (2019b). In the following, we focus on the one-parameter exponential case. Some results regarding the two-parameter model will also be provided. Notice that the two-parameter case can be traced back to the results established in the one-parameter model using the identity in (7.29). One may proceed by analogy with the derivations under Type-I right censoring of progressively Type-II censored data (see p. 229 ff.).

7.2.1 Type-II right censoring of progressively Type-II censored data

In order to get results comparable to the likelihood given in (7.9), we reconsider first a progressively Type-II right censored sample $X_{1:n^*:n}, \ldots, X_{m:n^*:n}$ with $1 \leq m \leq n^*$ observations. Note that these results have already been reported in Section 3.4 but will partially be given here again in order to illustrate the similarities to and the relationships between the various progressive hybrid censoring schemes.

The likelihood function for $Exp(\mu, \vartheta)$-distributed lifetimes under progressive Type-II right censoring is given by

$$\mathscr{L}(\mu, \vartheta; \boldsymbol{x}_m, m) = \prod_{j=1}^{m}\left[\frac{\gamma_j}{\vartheta}e^{-(R_j+1)\frac{x_j-\mu}{\vartheta}}\right] \cdot e^{-\gamma_{m+1}\frac{x_m-\mu}{\vartheta}} \prod_{j=2}^{m} \mathbb{1}_{[x_{j-1},\infty)}(x_j)\mathbb{1}_{[\mu,\infty)}(x_1).$$

Assuming $x_1 < \cdots < x_m$, we get, for μ known, the following expression of the likelihood

$$\mathscr{L}(\vartheta; \boldsymbol{x}_m, m) = \Big[\prod_{j=1}^{m} \gamma_j\Big] \vartheta^{-m} \exp\Big\{-\frac{m\widehat{\vartheta}}{\vartheta}\Big\} \mathbb{1}_{[\mu,\infty)}(x_1), \tag{7.15}$$

with $\widehat{\vartheta} = \frac{1}{m}\Big(\sum_{j=1}^{m}(R_j+1)(x_j-\mu)+\gamma_{m+1}(x_m-\mu)\Big)$. Hence, for $\mu = 0$,

$$\widehat{\vartheta} = \frac{1}{m}\Big(\sum_{j=1}^{m}(R_j+1)X_{j:n^*:n}+\gamma_{m+1}X_{m:n^*:n}\Big) = \frac{1}{m}\cdot \mathsf{TTT}_m, \quad \text{say},$$

denotes the MLE of ϑ. As pointed out in Balakrishnan and Cramer (2014), we get

$$\frac{2m\widehat{\vartheta}}{\vartheta} \sim \chi^2_{2m}, \tag{7.16}$$

or alternatively

$$\mathsf{TTT}_m = m\widehat{\vartheta} \sim \Gamma(\vartheta, m). \tag{7.17}$$

For μ unknown and $m \geq 2$, the likelihood function reads

$$\mathscr{L}(\mu, \vartheta; \boldsymbol{x}_m, m) = \Big[\prod_{j=1}^{m} \gamma_j\Big] \vartheta^{-m} \exp\Big\{-\frac{m\widehat{\vartheta}}{\vartheta}\Big\} \cdot \exp\Big\{-n\frac{x_1-\mu}{\vartheta}\Big\} \mathbb{1}_{[\mu,\infty)}(x_1),$$

where $\widehat{\vartheta} = \frac{1}{m}\Big(\sum_{j=2}^{m}(R_j+1)(x_j-x_1)+\gamma_{m+1}(x_m-x_1)\Big)$. Notice that $D = m \geq 1$ implies that we observe at least the minimum of the sample. Therefore, this yields $\mu \leq X_{1:n^*:n^*}$ (with probability one).

Proceeding as in Balakrishnan and Cramer (2014), the MLEs are given by

$$\widehat{\mu} = X_{1:m:n}, \quad \widehat{\vartheta} = \frac{1}{m}\Big(\sum_{j=2}^{m}(R_j+1)(X_{j:m:n}-X_{1:m:n})+\gamma_{m+1}(X_{m:m:n}-X_{1:m:n})\Big).$$

Then, as pointed out in Viveros and Balakrishnan (1994) (see also Cramer and Kamps, 2001a; Balakrishnan et al., 2001; Balakrishnan and Cramer, 2014), $\widehat{\mu}$ and $\widehat{\vartheta}$ are independent with

$$\widehat{\mu} \sim Exp(\mu, \vartheta/n), \quad \frac{2m\widehat{\vartheta}}{\vartheta} \sim \chi^2_{2m-2}.$$

Notice that this result corresponds to case ⓘⓘ in (7.8).

7.2.2 Type-I right censoring of progressively Type-II censored data

In this section, suppose the progressively Type-II censored sample $X_{1:n^*:n}, \ldots, X_{n^*:n^*:n}$ is subject to Type-I censoring with threshold T. Hence, we observe the sample

$$X_{1:n^*:n}, \ldots, X_{D:n^*:n}$$

where D is defined in (7.3). Before studying the MLEs and the corresponding distributions, we present some general distribution results for exponential lifetimes. Details can be found in Cramer and Balakrishnan (2013).

Given $D = d \in \{1, \ldots, n^* - 1\}$ observations, the conditional density function of the progressively Type-II censored order statistics $X_{1:n^*:n}, \ldots, X_{d:n^*:n}$ is given by

$$f^{X_{j:n^*:n}, 1 \le j \le d | D=d}(\boldsymbol{x}_d) = \frac{\prod_{j=1}^{d+1} \gamma_j}{\vartheta^{d+1} f_{d+1:n^*:n}(T)} \exp\Big\{ -\frac{1}{\vartheta}\Big[\sum_{j=1}^{d} \gamma_j(x_j - x_{j-1}) + \gamma_{d+1}(T - x_d)\Big]\Big\},$$
$$\mu = x_0 \le x_1 \le \cdots \le x_d \le T.$$

Defining the normalized spacings $S_{j,n^*} = \gamma_j(X_{j:n^*:n} - X_{j-1:n^*:n}), j = 1, \ldots, d$, with $X_{0:n^*:n} = \mu$, this leads to the conditional density function of the spacings

$$f^{S_{j,n^*}, 1 \le j \le d | D=d}(\boldsymbol{w}_d) = \frac{\gamma_{d+1} e^{-\gamma_{d+1}(T-\mu)/\vartheta}}{\vartheta f_{d+1:m:n}(T)} \prod_{j=1}^{d} \frac{1}{\vartheta} \exp\Big\{ -\Big(1 - \frac{\gamma_{d+1}}{\gamma_j}\Big)\frac{w_j}{\vartheta}\Big\}, \quad \boldsymbol{w}_d \in \mathcal{W}_d(T), \tag{7.18}$$

with support

$$\mathcal{W}_d(T) = \mathcal{S}_d(1/[\gamma_1(T-\mu)], \ldots, 1/[\gamma_d(T-\mu)])$$
$$= \Big\{ \boldsymbol{w}_d \mid w_j \ge 0, 1 \le j \le d, \sum_{j=1}^{d} \frac{w_j}{\gamma_j} \le T - \mu \Big\}.$$

For $D = n^*$, the conditional density function is given by

$$f^{X_{j:n^*:n}, 1 \le j \le n^* | D=n^*}(\boldsymbol{x}_{n^*}) = \frac{f_{1,\ldots,n^*:n^*:n}(\boldsymbol{x}_{n^*})}{F_{n^*:n^*:n}(T)}, \quad x_n \le T.$$

In this case, the distribution of the spacings has the density function

$$f^{S_{j,n^*}, 1 \le j \le n^* | D=n^*}(\boldsymbol{w}_{n^*}) = \frac{1}{F_{n^*:n^*:n}(T)} \prod_{j=1}^{n^*} \Big(\frac{1}{\vartheta} e^{-w_j/\vartheta}\Big), \quad \boldsymbol{w}_{n^*} \in \mathcal{W}_{n^*}(T). \tag{7.19}$$

Notice that the spacings $S_{1,n^*}, \ldots, S_{d,n^*}$ with $d \in \{2, \ldots, n^*\}$ are not (conditionally) independent even though the joint density function has a factorizable form. This is due to the condition $\sum_{j=1}^{d} \frac{w_j}{\gamma_j} \leq T - \mu$ which restricts the support to a simplex and, thus, links the variables. Following Cramer and Balakrishnan (2013), the marginal density functions of the normalized spacings are given in Theorem 7.3. For Type-I (hybrid) censored sequential order statistics, an analogous result has been established in Burkschat et al. (2016).

Theorem 7.3. *For $1 \leq k \leq d < n^*$, let $f_{d:n^*:n}^{(k)}$ and $F_{d:n^*:n}^{(k)}$ denote the marginal density function and cumulative distribution function of the d-th progressively Type-II censored order statistic from a two-parameter exponential distribution with censoring plan $\mathscr{R}_k = (R_1, \ldots, R_{k-2}, R_{k-1} + R_k + 1, R_{k+1}, \ldots, R_{n^*})$ resulting in the parameters $\gamma_1, \ldots, \gamma_{k-1}, \gamma_{k+1}, \ldots, \gamma_{n^*}$.*

Then, the spacing S_{k,n^*} has the conditional density function given by

$$f^{S_{k,n^*}|D=d}(t) = \frac{f_{d:n^*:n}^{(k)}(T - t/\gamma_k)e^{-t/\vartheta}}{\vartheta f_{d+1:n^*:n}(T)}, \qquad 0 \leq t \leq \gamma_k(T - \mu),$$

and the corresponding cumulative distribution function is given by

$$F^{S_{k,n^*}|D=d}(t) = 1 - \frac{f_{d+1:n^*:n}(T - t/\gamma_k)e^{-t/\vartheta}}{f_{d+1:n^*:n}(T)}, \qquad 0 \leq t \leq \gamma_k(T - \mu).$$

For $D = n^*$ and $1 \leq k \leq n^*$, the spacing S_{k,n^*} has the conditional density function

$$f^{S_{k,n^*}|D=n^*}(t) = \frac{F_{n^*-1:n^*-1:n}^{(k)}(T - t/\gamma_k)e^{-t/\vartheta}}{\vartheta F_{n^*:n^*:n}(T)}, \qquad 0 \leq w \leq \gamma_k(T - \mu), \tag{7.20}$$

and the corresponding cumulative distribution function is given by

$$F^{S_{k,n^*}|D=n^*}(t) = 1 - \frac{F_{n^*:n^*:n}(T - t/\gamma_k)e^{-t/\vartheta}}{F_{n^*:n^*:n}(T)}, \qquad 0 \leq t \leq \gamma_k(T - \mu).$$

According to Theorem 2.53, the conditional cumulative distribution function of the spacings given $D = d \in \{1, \ldots, n^* - 1\}$ can be directly obtained from the density function of the $(d+1)$-th progressively Type-II censored order statistic where an explicit representation is given in (2.50). Therefore, we get for $0 \leq t \leq \gamma_k(T - \mu)$,

$$F^{S_{k,n^*}|D=d}(t) = 1 - \frac{1}{\vartheta f_{d+1:n^*:n}(T)} e^{-t/\vartheta} \left(\prod_{j=1}^{d+1} \gamma_j\right) \sum_{i=1}^{d+1} a_{i,d+1} e^{-\gamma_i(T - t/\gamma_k)/\vartheta}.$$

Example 7.4. For illustration, we consider the spacings of Type-I censored exponential data so that $\mu = 0$ and $n^* = n$. Let $1 \leq d \leq n$ be fixed. Then, for $1 \leq k \leq d \leq n - 1$ and $0 \leq t \leq (n - k + 1)T$, we obtain from (2.9)

$$F^{S_{k,n}|D=d}(t) = 1 - \frac{f_{d+1:n}(T - t/\gamma_k)e^{-t/\vartheta}}{\vartheta f_{d+1:n}(T)}$$

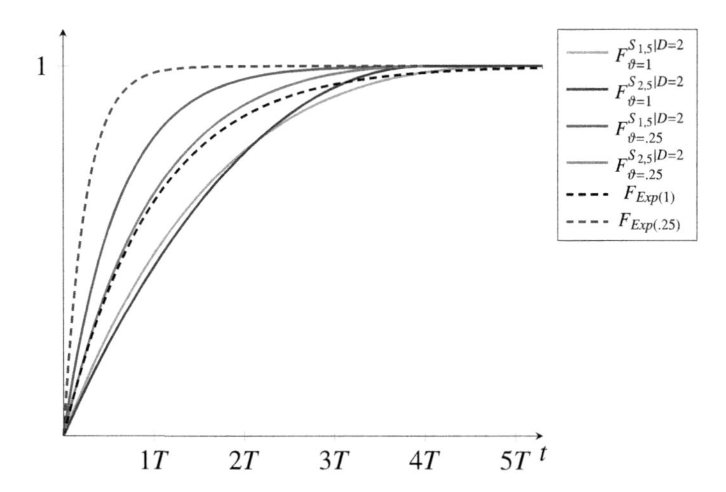

Figure 7.6 Plots of cumulative distribution functions of spacings as given in Example 7.4 for $T = 1$, $d = 2$, $k \in \{1, 2\}$, $n = 5$, and $\vartheta \in \{.25, 1\}$.

$$= 1 - \left[\frac{1 - e^{-(T-t/\gamma_k)/\vartheta}}{1 - e^{-T/\vartheta}}\right]^d \exp\left(-\frac{d-k+1}{n-k+1}\frac{t}{\vartheta}\right). \qquad (7.21)$$

Plots are provided in Figs. 7.6 and 7.7.

Remark 7.5. The expressions of the cumulative distribution functions in Theorem 7.3 enable us to provide conditions ensuring that the normalized spacings under Type-I censoring are stochastically shorter than the normalized spacings of exponential data which are iid exponentially distributed (see Theorem 2.38 and, for progressively Type-II censored data, Theorem 2.67). In particular, for $t \in \mathbb{R}$, we have

$$F^{S_{k,n^*}|D=d}(t) \geq F_{Exp(\vartheta)}(t) \iff f_{d+1:n^*:n}(T - t/\gamma_k) \leq f_{d+1:n^*:n}(T).$$

Recalling that the density functions of (progressively Type-II censored) order statistics from an exponential distribution are unimodal (see, e.g., Cramer, 2004), we can provide conditions ensuring the above inequality. Thus, if T exceeds the mode of $F^{X_{d:n^*:n}}$, then the stochastic ordering of the spacings holds, that is,

$$S_{k,n^*} \mid D = d \leq_{\text{st}} S_{k,n^*}.$$

For order statistics (i.e., $n = n^*$ and $R_i = 0$, $1 \leq i \leq n$), the condition can be explicitly provided. Since the mode of the $(d+1)$-th order statistic from an $Exp(\vartheta)$-distribution is given by $\mathrm{m}^* = -\vartheta \log(1 - d/n)$ this yields the condition

$$-\vartheta \log(1 - d/n) \leq T \iff d \geq n(1 - e^{-T/\vartheta}).$$

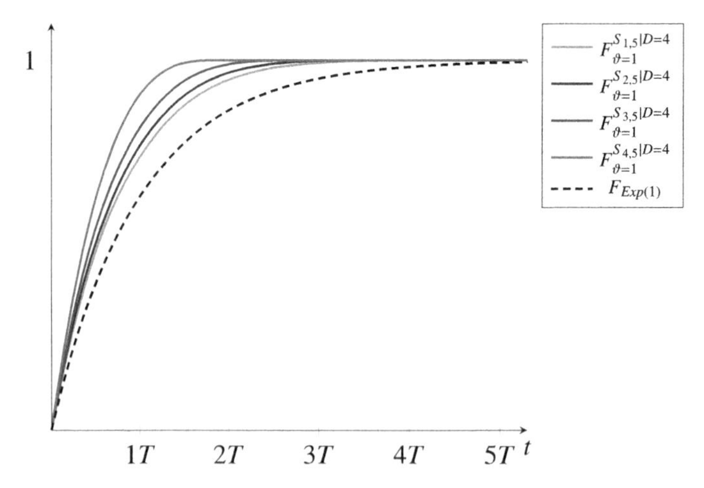

Figure 7.7 Plots of cumulative distribution functions of spacings as given in Example 7.4 for $T=1$, $d=4$, $k \in \{1,2,3,4\}$, $n=5$, and $\vartheta=1$.

Notice that, for $T=\vartheta=1$ and $d=2$, $n=5$, this condition is not satisfied and the spacings are not stochastically ordered (see also Fig. 7.6 where the respective curves of the (conditional) cumulative distribution functions intersect). Choosing $d=4$ and keeping the other values, the condition reads $4 \geq 5(1-e^{-1})$ which obviously is true. Hence, the spacings are stochastically ordered as illustrated in Fig. 7.7, that is, the curves are ordered and do not intersect.

Notice that the reversed stochastic ordering is not possible since the spacings under Type-I censoring have a finite support.

Example 7.6. For $T \to \infty$, it holds that $\Pr(D=n^*) \to 1$ whereas $\Pr(D<n^*) \to 0$. Therefore, for large n^*, the 'relevant' case is given by the condition $D=n^*$. Then, for Type-I censored exponential data,[1] we get from (7.20) the cumulative distribution function

$$F^{S_{k,n}|D=n}(t) = 1 - \frac{F_{n^*:n}(T-t/\gamma_k)e^{-t/\vartheta}}{F_{n:n}(T)} = 1 - \left[\frac{1-e^{-(T-t/\gamma_k)/\vartheta}}{1-e^{-T/\vartheta}}\right]^n e^{-t/\vartheta}, \quad 0 \leq t \leq \gamma_k T,$$

which equals (7.21) with d replaced by n.

7.2.2.1 One-parameter exponential distribution

First, notice that the likelihood function is of type ⓘ in (7.9). Assuming $x_1 < \cdots < x_d$ and $D=d \geq 1$, we get, for μ known with $\mu < T$, as in (7.15)

[1] That is, $n=n^*$ and $R_i=0$, $1 \leq i \leq n$.

$$\mathscr{L}(\vartheta; \boldsymbol{x}_d, d) = \Big[\prod_{j=1}^{d} \gamma_j\Big]\vartheta^{-d} \exp\Big\{-\frac{d\widehat{\vartheta}}{\vartheta}\Big\}\mathbb{1}_{[\mu,\infty)}(x_1),$$

with $\widehat{\vartheta} = \frac{1}{d}\Big(\sum_{j=1}^{d}(R_j+1)(x_j-\mu)+\gamma_{d+1}(T-\mu)\Big)$. Without loss of generality, we assume $\mu=0$ in the following. For general μ, one has to replace $X_{j:n^*:n}$ and T by $X_{j:n^*:n}-\mu$ and $T-\mu$, respectively. Now,

$$\widehat{\vartheta} = \frac{1}{D}\Big(\sum_{j=1}^{D}(R_j+1)X_{j:n^*:n}+\gamma_{D+1}T\Big) \tag{7.22}$$

denotes the MLE of ϑ. It is worth mentioning that we get $\widehat{\vartheta} = \frac{1}{n^*}\sum_{j=1}^{n^*}(R_j+1)X_{j:n^*:n}$ for $D=n^*$.

As mentioned above, the MLE exists only when $D\geq 1$. Thus, the distribution of $\widehat{\vartheta}$ can only be obtained conditionally on the event $D\geq 1$. Although this model seems not have been discussed explicitly in the literature so far, the distribution of the estimator can directly be obtained from results obtained under progressive hybrid censoring by adapting values of parameters. In order to illustrate the derivation of the MLE's distribution, we provide some additional distributional information following the derivations in Cramer and Balakrishnan (2013).

Given $D=d\in\{1,\ldots,n^*\}$, we define the total time on test statistic

$$\mathrm{TTT}_d = \sum_{j=1}^{d}(R_j+1)X_{j:n^*:n}+\gamma_{d+1}T = \sum_{j=1}^{d}\Big(1-\frac{\gamma_{d+1}}{\gamma_j}\Big)S_{j,n^*}+\gamma_{d+1}T, \tag{7.23}$$

so that $\widehat{\vartheta} = \frac{1}{D}\mathrm{TTT}_D$. Notice that TTT_d has a bounded support since $0\leq \mathrm{TTT}_d\leq \gamma_1 T$. The derivation of the (conditional) distribution of TTT_D can be found in Cramer and Balakrishnan (2013). In particular, we get the following theorem where B-spline functions (cf. (C.4)) are involved in the representations of the density functions.

Theorem 7.7. *Let $1\leq d\leq n^*-1$ and denote by $B_{d-1}(\cdot\mid\gamma_{d+1},\ldots,\gamma_1)$ the B-spline with knots $\gamma_{d+1}<\cdots<\gamma_1$ as defined in Appendix C. The density function $f_\vartheta^{\mathrm{TTT}_d\mid D=d}$ is given by*

$$f_\vartheta^{\mathrm{TTT}_d\mid D=d}(t) = \frac{T^d\prod_{j=1}^{d+1}\gamma_j}{d!\vartheta^{d+1}f_{d+1:n^*:n}(T)}B_{d-1}(t\mid\gamma_{d+1}T,\ldots,\gamma_1 T)\,e^{-t/\vartheta},\quad 0\leq t\leq nT.$$

For $d=n^*$, we have

$$f_\vartheta^{\mathrm{TTT}_{n^*}\mid D=n^*}(t) = \frac{T^{n^*}\prod_{j=1}^{n^*}\gamma_j}{n^*!\vartheta^{n^*}F_{n^*:n^*:n}(T)}B_{n^*-1}(t\mid 0,\gamma_{n^*}T,\ldots,\gamma_1 T)\,e^{-t/\vartheta},\quad 0\leq t\leq nT.$$

In particular, the expressions in Theorem 7.7 yield the joint density function of TTT_D and D that can be written as

$$f_\vartheta^{\mathsf{TTT}_D,D}(t,d) = f_\vartheta^{\mathsf{TTT}_d|D=d}(t)\,\mathrm{Pr}_\vartheta(D=d)$$
$$= \frac{T^d\prod_{j=1}^d \gamma_j}{d!\vartheta^d} B_{d-1}(t\mid \gamma_{d+1}T,\ldots,\gamma_1 T)\,e^{-t/\vartheta}, \quad 0\le t, \tag{7.24}$$

with $d\in\{1,\ldots,n^*\}$.

For completion, we present in Theorem 7.8 expressions for the conditional density functions $f_\vartheta^{\mathsf{TTT}_m|D\ge m}$ and $f_\vartheta^{\mathsf{TTT}_m|D<m}$, respectively, with $1\le m\le n^*$. These expressions will be useful in the derivation of the density function of the total time on test statistics under hybrid censoring schemes. First, introducing the counter variable

$$D_| = \sum_{j=1}^m \mathbb{1}_{(-\infty,T]}(X_{j:n^*:n}) = \sum_{j=1}^m \mathbb{1}_{(-\infty,T]}(X_{j:m:n}),$$

we have $\{D_|=m\}=\{D\ge m\}$ so that

$$\mathrm{Pr}_\vartheta(\mathsf{TTT}_m\le t, D\ge m) = \mathrm{Pr}_\vartheta(\mathsf{TTT}_m\le t, D_|=m) = \mathrm{Pr}_\vartheta(\mathsf{TTT}_{D_|}\le t, D_|=m).$$

The probability on the right hand side is of the same type as the probability $\mathrm{Pr}_\vartheta(\mathsf{TTT}_{D_|}\le t, D_|=n^*)=\mathrm{Pr}_\vartheta(\mathsf{TTT}_D\le t, D=n^*)$ where n^* has to be replaced by m. Notice that we use the sample $X_{1:m:n},\ldots,X_{m:m:n}$ (with censoring plan $(R_1,\ldots,R_{m-1},\gamma_m-1)$) which equals the (marginal) sample $X_{1:n^*:n},\ldots,X_{m:n^*:n}$ (with censoring plan $(R_1,\ldots,R_{n^*})$). Therefore, we conclude directly from (7.24) that

$$f_\vartheta^{\mathsf{TTT}_m,D\ge m}(t) = \frac{T^m\prod_{j=1}^m\gamma_j}{m!\vartheta^m} B_{m-1}(t\mid 0,\gamma_m T,\ldots,\gamma_1 T)e^{-t/\vartheta}, \quad 0\le t\le nT$$

(see Cramer and Balakrishnan, 2013). From the identity

$$\mathrm{Pr}_\vartheta(\mathsf{TTT}_m\le t, D<m) = \mathrm{Pr}_\vartheta(\mathsf{TTT}_m\le t) - \mathrm{Pr}_\vartheta(\mathsf{TTT}_m\le t, D\ge m),$$

we get by using (7.17)

$$f_\vartheta^{\mathsf{TTT}_m,D<m}(t) = f_\vartheta^{\mathsf{TTT}_m}(t) - f_\vartheta^{\mathsf{TTT}_m,D\ge m}(t)$$
$$= f_{\Gamma(\vartheta,m)}(t) - \frac{T^m\prod_{j=1}^m\gamma_j}{m!\vartheta^m} B_{m-1}(t\mid 0,\gamma_m T,\ldots,\gamma_1 T)e^{-t/\vartheta},$$

for $0\le t\le nT$, which can be found in Cramer et al. (2016). Thus, we find the following expressions for the conditional density functions. In combination with the expressions provided in Theorem 7.7, these representations of the density functions allow convenient derivation of the density functions of the MLEs under hybrid censoring by utilizing the modularization approach due to Górny and Cramer (2019b).

Theorem 7.8. *Let $1 \le m \le n^*$. Then, for $t \ge 0$,*

$$f_\vartheta^{\mathrm{TTT}_m|D\ge m}(t) = \frac{T^m \prod_{j=1}^m \gamma_j}{m!\vartheta^m F_{m:m:n}(T)} B_{m-1}(t \mid 0, \gamma_m T, \ldots, \gamma_1 T) e^{-t/\vartheta}, \tag{7.25}$$

$$f_\vartheta^{\mathrm{TTT}_m|D<m}(t) = \frac{1}{1 - F_{m:m:n}(T)} \Big(f_{\Gamma(\vartheta,m)}(t) - \frac{T^m \prod_{j=1}^m \gamma_j}{m!\vartheta^m F_{m:m:n}(T)} B_{m-1}(t \mid 0, \gamma_m T, \ldots, \gamma_1 T) e^{-t/\vartheta} \Big).$$

Summing up, the preceding results lead to a representation for the conditional density function of the MLE $\widehat{\vartheta}$ given $\{D_1 \ge 1\} = \{D \ge 1\}$, that is, for $t \ge 0$,

$$\begin{aligned}
f_\vartheta^{\widehat{\vartheta}|D_1\ge 1}(t) &= \frac{1}{\Pr_\vartheta(D \ge 1)} \sum_{d=1}^{n^*} f_\vartheta^{\mathrm{TTT}_d/d|D=d}(t) \Pr_\vartheta(D=d) \\
&= \frac{1}{\Pr_\vartheta(D \ge 1)} \sum_{d=1}^{n^*} d f_\vartheta^{\mathrm{TTT}_d|D=d}(dt) \Pr_\vartheta(D=d) \\
&= \frac{1}{1 - e^{-nT/\vartheta}} \sum_{d=1}^{n^*} \frac{T^d \prod_{j=1}^d \gamma_j}{(d-1)!\vartheta^d} B_{d-1}(dt \mid \gamma_{d+1} T, \ldots, \gamma_1 T) e^{-dt/\vartheta}.
\end{aligned} \tag{7.26}$$

7.2.2.2 Two-parameter exponential distribution

For μ unknown and $D = d \ge 1$, the likelihood function reads

$$\mathscr{L}(\mu, \vartheta; \boldsymbol{x}_d, d) = \Big[\prod_{j=1}^d \gamma_j\Big] \vartheta^{-d} \exp\Big\{-\frac{d\widehat{\vartheta}}{\vartheta}\Big\} \cdot \exp\Big\{-n\frac{x_1 - \mu}{\vartheta}\Big\} \mathbb{1}_{[\mu,\infty)}(x_1),$$

where $\widehat{\vartheta} = \frac{1}{d}\Big(\sum_{j=2}^d (R_j+1)(x_j - x_1) + \gamma_{d+1}(T - x_1)\Big)$. Proceeding as in Balakrishnan and Cramer (2014), the MLEs are given by

$$\widehat{\mu} = X_{1:n^*:n}, \quad \widehat{\vartheta} = \frac{1}{D}\Big(\sum_{j=2}^D (R_j+1)(X_{j:n^*:n} - X_{1:n^*:n}) + \gamma_{D+1}(T - X_{1:n^*:n})\Big).$$

Notice that the MLEs of μ and ϑ also exist if only one observation is available. In the case $D = 1$, we get $\widehat{\mu} = X_{1:n^*:n}$ and $\widehat{\vartheta} = \gamma_2(T - X_{1:n^*:n})$.

By similarity with the one-parameter case, we are now interested in the joint distribution of $S_{1,n^*} = nX_{1:n^*:n}$ and TTT_d^* (conditionally on $D = d \ge 1$) where the modified total time on test statistic[2] is given by

[2] The total time on test is given by $\mathrm{TTT}_d = \mathrm{TTT}_d^* + S_{1,n^*}$

$$\begin{aligned}\mathsf{TTT}_d^* &= \sum_{j=2}^{d}(R_j+1)(X_{j:n^*:n}-X_{1:n^*:n})+\gamma_{d+1}(T-X_{1:n^*:n})\\ &= \sum_{j=2}^{d}\Big(1-\frac{\gamma_{d+1}}{\gamma_j}\Big)S_{j,n^*}+\gamma_{d+1}\Big(T-\mu-\frac{S_{1,n^*}}{n}\Big), \quad d\geq 1. \end{aligned} \tag{7.27}$$

First, notice that TTT_d^* has a bounded support, that is,

$$0\leq \mathsf{TTT}_d^* = \sum_{j=2}^{d}(R_j+1)(X_{j:n^*:n}-X_{1:n^*:n})+\gamma_{d+1}(T-X_{1:n^*:n})\leq \gamma_2(T-\mu).$$

Then, for $D=1$, the distribution of $\widehat{\mu}=X_{1:n^*:n}$ and $\widehat{\vartheta}=\mathsf{TTT}_1^*=\gamma_2(T-X_{1:n^*:n})$ is degenerated. In particular, for $\gamma_2(T-s\wedge T)\leq t$, the joint cumulative distribution function of the MLEs is given by

$$\begin{aligned}&\Pr(X_{1:n^*:n}\leq s, \mathsf{TTT}_D^*\leq t, D=1)\\ &\quad=\Pr(X_{1:n^*:n}\leq s, \gamma_2(T-X_{1:n^*:n})\leq t, X_{1:n^*:n}\leq T< X_{2:n^*:n})\\ &\quad=\Pr(T-t/\gamma_2\leq X_{1:n^*:n}\leq s\wedge T, T<X_{2:n^*:n})\\ &\quad=\Pr(T-t/\gamma_2\leq X_{1:n^*:n}\leq s\wedge T)-\Pr(T-t/\gamma_2\leq X_{1:n^*:n}\leq s\wedge T, X_{2:n^*:n}\leq T)\\ &\quad=F_{1:n}(s\wedge T)-F_{1:n}(T-t/\gamma_2)+F_{1,2:n^*:n}(T-t/\gamma_2, T)-F_{1,2:n^*:n}(s\wedge T, T).\end{aligned}$$

Using (2.52), this simplifies to

$$=\frac{\gamma_1}{\gamma_1-\gamma_2}\overline{F}^{\gamma_2}(T)\Big(\overline{F}^{\gamma_1-\gamma_2}(T-t/\gamma_2)-\overline{F}^{\gamma_1-\gamma_2}(s\wedge T)\Big).$$

If $D\geq 2$, then a (conditional) density function $f_{\mu,\vartheta}^{S_{1,n^*},\mathsf{TTT}_D^*|D=d}$ of S_{1,n^*} and TTT_D^* given $D=d\geq 2$ exists. Respective expressions have been established in Cramer and Balakrishnan (2013) (see also Górny, 2017).

Here, we present an alternative approach that takes the derivation back to the case of a one-parameter exponential distribution. We start with the representation of the total time on test statistic given in terms of spacings (see (7.27)). Furthermore, we know from (2.60) that, for $d\geq 2$,

$$X_{d:n^*:n}=\sum_{j=1}^{d}\frac{S_{j,n^*}}{\gamma_j}=\sum_{j=2}^{d}\frac{S_{j,n^*}}{\gamma_j}+\frac{S_{1,n^*}}{\gamma_1}. \tag{7.28}$$

Consider now the joint cumulative distribution function of S_{1,n^*}, TTT_D^*, and the event $D=d$ with $d\geq 1$. Then, we get from (7.27) and (7.28) as well as the independence of spacings

$$
\begin{aligned}
&\Pr_{\mu,\vartheta}(S_{1,n^*} \le s, \mathsf{TTT}_D^* \le t, D = d)\\
&= \Pr_{\mu,\vartheta}\Bigg(S_{1,n^*} \le s, \sum_{j=2}^{d}\Big(1 - \frac{\gamma_{d+1}}{\gamma_j}\Big)S_{j,n^*} + \gamma_{d+1}\Big(T - \mu - \frac{S_{1,n^*}}{n}\Big) \le t,\\
&\qquad\qquad \sum_{j=1}^{d}\frac{S_{j,n^*}}{\gamma_j} \le T < \sum_{j=1}^{d+1}\frac{S_{j,n^*}}{\gamma_j}\Bigg)\\
&= \int_0^s \Pr_{\mu,\vartheta}\Bigg(\sum_{j=2}^{d}\Big(1 - \frac{\gamma_{d+1}}{\gamma_j}\Big)S_{j,n^*} + \gamma_{d+1}\Big(T - \frac{w}{n} - \mu\Big) \le t,\\
&\qquad\qquad \sum_{j=2}^{d}\frac{S_{j,n^*}}{\gamma_j} \le T - \frac{w}{n} < \sum_{j=2}^{d+1}\frac{S_{j,n^*}}{\gamma_j}\Bigg) f_{\mu,\vartheta}^{S_{1,n^*}}(w)\,dw\\
&= \int_0^s \Pr_{\mu,\vartheta}\Big(\mathsf{TTT}_{d-1}^*(w) \le t, X_{d-1:n^*-1:\gamma_2} \le T_w < X_{d:n^*-1:\gamma_2}\Big) f_{\mu,\vartheta}^{S_{1,n^*}}(w)\,dw\\
&= \int_0^s \Pr_{\mu,\vartheta}\Big(\mathsf{TTT}_{D_w}^*(w) \le t, D_w = d-1\Big) f_{\mu,\vartheta}^{S_{1,n^*}}(w)\,dw,
\end{aligned}
$$

where $X_{1:n^*-1:\gamma_2}, \ldots, X_{d:n^*-1:\gamma_2}$ are (Type-I right censored) progressively Type-II censored order statistics from an $Exp(\mu, \vartheta)$-distribution with threshold $T_w = T - w/n$ and $\mathsf{TTT}_{D_w}^*(w)$ denotes the respective total time on test statistic. As a result the corresponding density functions read

$$
f_{\mu,\vartheta}^{S_{1,n^*},\mathsf{TTT}_D^*,D}(w, v, d) = f_{\mu,\vartheta}^{\mathsf{TTT}_{D_w}^*(w),D_w}(v, d-1) f_{\mu,\vartheta}^{S_{1,n^*}}(w), \tag{7.29}
$$

where the latter density function is based on the parameters $\gamma_2, \ldots, \gamma_d$ and threshold $T_w = T - w/n$. Using (7.24), we get with $S_{1,n^*} \sim Exp(\vartheta)$

$$
\begin{aligned}
f_{\mu,\vartheta}^{S_{1,n^*},\mathsf{TTT}_D^*,D}(w, v, d) = \frac{(T_w - \mu)^{d-1}\prod_{j=2}^{d}\gamma_j}{(d-1)!\vartheta^{d-1}}\\
\times B_{d-2}(v \mid \gamma_{d+1}(T_w - \mu), \ldots, \gamma_2(T_w - \mu))e^{-(v+w)/\vartheta},\\
0 \le w, T_w > \mu, 0 \le v,\ d \in \{2, \ldots, n^*\}.
\end{aligned}
$$

Summing up, this yields the expression in (7.30).

Theorem 7.9. *For $0 \le w \le n(T - \mu)$, $0 \le v \le \gamma_2(T - \mu)$, and $d \in \{2, \ldots, n^*\}$, the joint density function $f_{\mu,\vartheta}^{S_{1,n^*},\mathsf{TTT}_D^*,D}$ of S_{1,n^*}, TTT_D^*, and D is given by*

$$
\begin{aligned}
f_{\mu,\vartheta}^{S_{1,n^*},\mathsf{TTT}_D^*,D}(w, v, d) = \frac{\big(T - \mu - \frac{w}{n}\big)^{d-1}\prod_{j=2}^{d}\gamma_j}{(d-1)!\vartheta^{d}}\\
\times B_{d-2}\Big(v \mid \gamma_{d+1}\Big(T - \mu - \frac{w}{n}\Big), \ldots, \gamma_2\Big(T - \mu - \frac{w}{n}\Big)\Big)e^{-(v+w)/\vartheta}.
\end{aligned} \tag{7.30}
$$

From Theorem 7.9, we deduce immediately the following conditional density functions. For $D = d \in \{2, \ldots, n^* - 1\}$, Cramer and Balakrishnan (2013) have shown that

$$f_{\mu,\vartheta}^{S_{1,n^*},\mathrm{TTT}_D^*|D=d}(w,v) = \frac{\left(T-\mu-\frac{w}{n}\right)^{d-1}\prod_{j=2}^{d+1}\gamma_j}{(d-1)!\vartheta^{d+1}f_{d+1:n^*:n}(T)} \times B_{d-2}\Big(v \mid \gamma_{d+1}\Big(T-\mu-\frac{w}{n}\Big),\ldots,\gamma_2\Big(T-\mu-\frac{w}{n}\Big)\Big)e^{-(v+w)/\vartheta},$$
$$0 \le w \le \gamma_1(T-\mu), 0 \le v \le \gamma_2(T-\mu).$$

For $D = n^*$, we have from (7.27) $\mathrm{TTT}_{n^*}^* = \sum_{j=2}^{n^*} S_{j,n^*}$. Then, (7.30) leads to the expression

$$f_{\mu,\vartheta}^{S_{1,n^*},\mathrm{TTT}_D^*|D=n^*}(w,v) = \frac{\left(T-\mu-\frac{w}{n}\right)^{n^*-1}\prod_{j=2}^{n^*}\gamma_j}{(n^*-1)!\vartheta^{n^*}F_{n^*:n^*:n}(T)} \times B_{n^*-2}\Big(v \mid 0, \gamma_{n^*}\Big(T-\mu-\frac{w}{n}\Big),\ldots,\gamma_2\Big(T-\mu-\frac{w}{n}\Big)\Big)e^{-(v+w)/\vartheta},$$
$$0 \le w \le n(T-\mu), 0 \le v \le \gamma_2(T-\mu).$$

From Theorem 7.9, we get directly the conditional density function of the MLEs given $D \ge 2$, that is,

$$f_{\mu,\vartheta}^{\widehat{\mu},\widehat{\vartheta}|D\ge 2}(w,v) = \frac{ne^{-n(w-\mu)/\vartheta}}{\Pr_{\mu,\vartheta}(D\ge 2)}\sum_{d=2}^{n^*}\frac{d\big(T-w\big)^{d-1}\prod_{j=2}^{d}\gamma_j}{(d-1)!\vartheta^d} \times B_{d-2}\Big(dv \mid \gamma_{d+1}(T-w),\ldots,\gamma_2(T-w)\Big)e^{-dv/\vartheta},$$
$$\mu \le w \le T, 0 \le v \le \gamma_2(T-\mu),$$

where $\Pr_{\mu,\vartheta}(D \ge 2) = \Pr_{\mu,\vartheta}(X_{2:n^*:n} \le T) = F_{\mu,\vartheta;2:n^*:n}(T)$. An expression for the conditional cumulative distribution function can be established when $D \ge 1$ (and $n^* \ge 2$). Then,

$$F_{\mu,\vartheta}^{\widehat{\mu},\widehat{\vartheta}|D\ge 1}(s,t) = \frac{1}{1-e^{-n(T-\mu)/\vartheta}}\Bigg(\Big(F_{1:n}(s\wedge T) - F_{1:n}(T-t/\gamma_2) + F_{1,2:n^*:n}(T-t/\gamma_2, T) - F_{1,2:n^*:n}(s\wedge T, T)\Big)\mathbb{1}_{[\gamma_2(T-s\wedge T),\infty)}(t) + n\sum_{d=2}^{n^*}\int_\mu^s\int_0^{dt}\frac{\big(T-w\big)^{d-1}\prod_{j=2}^{d}\gamma_j}{(d-1)!\vartheta^d} \times B_{d-2}\Big(v \mid \gamma_{d+1}(T-w),\ldots,\gamma_2(T-w)\Big)e^{-n(w-\mu)/\vartheta-v/\vartheta}\,dvdw\Bigg). \tag{7.31}$$

The marginal distributions of the MLEs can also be established. $\widehat{\mu}$ has a two-parameter exponential distribution truncated on the right. The result is exactly the same as that given in Theorem 5.18 under Type-I hybrid censoring, that is,

$$\Pr_{\mu,\vartheta}\left(\widehat{\mu} > t \,\middle|\, D \geq 1\right) = \frac{e^{-n(t-\mu)/\vartheta} - e^{-n(T-\mu)/\vartheta}}{1 - e^{-n(T-\mu)/\vartheta}}, \quad \mu \leq t \leq T.$$

The cumulative distribution function of $\widehat{\vartheta}$ can be obtained from (7.31) by taking $s \to \infty$ (or $s \to T$). Then, we get with the substitutions $z = T - w$ and $y = v/z$

$$\begin{aligned} F_{\mu,\vartheta}^{\widehat{\vartheta}|D\geq 1}(t) = \frac{1}{1 - e^{-n(T-\mu)/\vartheta}} \Bigg(& \frac{\gamma_1}{\gamma_1 - \gamma_2} \left(\overline{F}_{\mu,\vartheta}^{\gamma_1 - \gamma_2}(T - t/\gamma_2) \overline{F}_{\mu,\vartheta}^{\gamma_2}(T) - \overline{F}_{\mu,\vartheta}^{\gamma_1}(T) \right) \mathbb{1}_{[0,\infty)}(t) \\ & + n e^{-n(T-\mu)/\vartheta} \sum_{d=2}^{n^*} \int_0^{T-\mu} e^{nz/\vartheta} \int_0^{dt/z} \frac{z^{d-2} \prod_{j=2}^{d} \gamma_j}{(d-1)!\vartheta^d} \\ & \times B_{d-2}(y \mid \gamma_{d+1}, \ldots, \gamma_2)\, e^{-zy/\vartheta}\, dy dz \Bigg). \end{aligned}$$

7.2.3 Type-I progressive hybrid censoring

From (7.8), we obtain the likelihood function as in (7.9) where w is taken from Table 7.3. In particular, we find the likelihood function

$$\mathscr{L}(\vartheta; \boldsymbol{x}_d, d) = \Big(\prod_{j=1}^{d} \gamma_j\Big) \vartheta^{-d} \exp\Big(-\frac{1}{\vartheta}\Big(\sum_{j=1}^{d}(R_j + 1)x_j + \gamma_{d+1} w\Big)\Big) \tag{7.32}$$

which yields the MLE $\widehat{\vartheta} = \mathrm{TTT}_{D_\mathrm{I}}/D_\mathrm{I}$ (provided $D_\mathrm{I} \geq 1$) with total time on test statistic as in (7.22), that is,

$$\mathrm{TTT}_{D_\mathrm{I}} = \sum_{j=1}^{D_\mathrm{I}} (R_j + 1) X_{j:n^*:n} + \gamma_{D_\mathrm{I}+1} W \tag{7.33}$$

and $D_\mathrm{I} = D \wedge m$, $W = X_{m:n^*:n} \wedge T$. Notice that $\{D_\mathrm{I} \geq 1\} = \{D \geq 1\}$.

By analogy with (7.26) and using the expressions given in Theorems 7.7 and 7.8, this yields the conditional density function of $\widehat{\vartheta}$ given $D \geq 1$, that is, for $t \geq 0$,

$$\begin{aligned} f_{\vartheta}^{\widehat{\vartheta}|D\geq 1}(t) &= \frac{1}{\Pr_\vartheta(D \geq 1)} \sum_{d=1}^{n^*} f_\vartheta^{\mathrm{TTT}_{d\wedge m}/(d\wedge m)|D=d}(t) \Pr_\vartheta(D = d) \\ &= \frac{1}{\Pr_\vartheta(D \geq 1)} \Big(\sum_{d=1}^{m-1} d f_\vartheta^{\mathrm{TTT}_d|D=d}(dt) \Pr_\vartheta(D = d) + m f_\vartheta^{\mathrm{TTT}_m|D\geq m}(mt) \Pr_\vartheta(D \geq m) \Big) \end{aligned}$$

$$= \frac{1}{1-e^{-nT/\vartheta}} \left(\sum_{d=1}^{m-1} \frac{T^d \prod_{j=1}^{d} \gamma_j}{(d-1)!\vartheta^d} B_{d-1}(dt \mid \gamma_{d+1}T, \dots, \gamma_1 T) e^{-dt/\vartheta} \right.$$
$$\left. + \frac{T^m \prod_{j=1}^{m} \gamma_j}{(m-1)!\vartheta^m} B_{m-1}(mt \mid 0, \gamma_m T, \dots, \gamma_1 T) e^{-mt/\vartheta} \right).$$

An expression in terms of shifted gamma density functions can be found in Childs et al. (2008) and Kundu and Joarder (2006b). It has been obtained using the moment generating function approach. The connection between the two alternative representations is shown in Górny and Cramer (2019b).

Remark 7.10. Results for the two-parameter exponential distribution have been established in Cramer and Balakrishnan (2013) and Chan et al. (2015) where the exact (conditional) distribution has been studied. Moreover, Chan et al. (2015) presented bias corrected estimates for both location and scale.

7.2.4 Type-II progressive hybrid censoring

We can proceed by similarity with the derivations under Type-I progressive hybrid censoring. As above, the likelihood function can be written as (7.32) where d and w are taken from Table 7.3. Then, the MLE $\widehat{\vartheta} = \mathrm{TTT}_{D_{\mathrm{II}}}/D_{\mathrm{II}}$ results[3] with total time on test statistic as in (7.22), that is,

$$\mathrm{TTT}_{D_{\mathrm{II}}} = \sum_{j=1}^{D_{\mathrm{II}}} (R_j+1) X_{j:n^*:n} + \gamma_{D_{\mathrm{II}}+1} W \tag{7.34}$$

and $D_{\mathrm{II}} = D \vee m$, $W = X_{m:n^*:n} \vee (T \wedge X_{n^*:n^*:n})$ (see also Table 7.2).

Therefore, we get for $t \geq 0$,

$$f_{\vartheta}^{\widehat{\vartheta}}(t) = \sum_{d=1}^{n^*} f_{\vartheta}^{\mathrm{TTT}_{d\vee m}/(d\vee m)|D=d}(t)\,\Pr_{\vartheta}(D=d)$$
$$= m f^{\mathrm{TTT}_m|D<m}(mt)\,\Pr_{\vartheta}(D<m) + \sum_{d=m}^{n^*} d f_{\vartheta}^{\mathrm{TTT}_d|D=d}(dt)\,\Pr_{\vartheta}(D=d)$$
$$= m f_{\Gamma(\vartheta,m)}(mt) - \frac{T^m \prod_{j=1}^{m} \gamma_j}{(m-1)!\vartheta^m} B_{m-1}(mt \mid 0, \gamma_m T, \dots, \gamma_1 T) e^{-mt/\vartheta}$$
$$+ \sum_{d=m}^{n^*} \frac{T^d \prod_{j=1}^{d} \gamma_j}{(d-1)!\vartheta^d} B_{d-1}(dt \mid \gamma_{d+1}T, \dots, \gamma_1 T) e^{-dt/\vartheta}$$

(see Cramer et al., 2016, Theorem 4.1).

[3] Where $D_{\mathrm{II}} \geq 1$ is ensured by the construction of the Type-II progressive hybrid censoring scheme.

7.2.5 Generalized Type-I progressive hybrid censoring

Generalized Type-I progressive hybrid censoring for exponential lifetimes has been discussed in Cho et al. (2015b) and Górny and Cramer (2016). First, with values of d and w taken from Table 7.3, the likelihood function has the same form as (7.32) so that the MLE is given by $\widehat{\vartheta} = \mathsf{TTT}_{D_{\mathsf{gI}}}/D_{\mathsf{gI}}$ with total time on test statistic as in (7.22), that is,

$$\mathsf{TTT}_{D_{\mathsf{gI}}} = \sum_{j=1}^{D_{\mathsf{gI}}} (R_j + 1) X_{j:n^*:n} + \gamma_{D_{\mathsf{gI}}+1} W$$

and $D_{\mathsf{gI}} = (m \vee D) \wedge k$, $W = (X_{m:n^*:n} \vee T) \wedge X_{k:n^*:n}$. Notice that this hybrid censoring scheme ensures at least k observations.

By analogy with (7.26) and using the expressions given in Theorems 7.7 and 7.8, this yields the conditional density function of $\widehat{\vartheta}$ given $D \geq 1$, that is, for $t \geq 0$,

$$\begin{aligned}
f_\vartheta^{\widehat{\vartheta}}(t) &= \sum_{d=1}^{n^*} f_\vartheta^{\mathsf{TTT}_{(m\vee d)\wedge k}/((m\vee d)\wedge k)|D=d}(t)\,\Pr_\vartheta(D=d) \\
&= k f_\vartheta^{\mathsf{TTT}_m|D<m}(kt)\,\Pr_\vartheta(D<k) + \sum_{d=k}^{m-1} d f_\vartheta^{\mathsf{TTT}_d|D=d}(dt)\,\Pr_\vartheta(D=d) \\
&\quad + m f_\vartheta^{\mathsf{TTT}_m|D\geq m}(mt)\,\Pr_\vartheta(D\geq m) \\
&= k f_{\Gamma(\vartheta,k)}(kt) - \frac{T^k \prod_{j=1}^k \gamma_j}{(k-1)!\vartheta^k} B_{k-1}(kt \mid 0, \gamma_k T, \ldots, \gamma_1 T) e^{-kt/\vartheta} \\
&\quad + \sum_{d=k}^{m-1} \frac{T^d \prod_{j=1}^d \gamma_j}{(d-1)!\vartheta^d} B_{d-1}(dt \mid \gamma_{d+1} T, \ldots, \gamma_1 T) e^{-dt/\vartheta} \\
&\quad + \frac{T^m \prod_{j=1}^m \gamma_j}{(m-1)!\vartheta^m} B_{m-1}(mt \mid 0, \gamma_m T, \ldots, \gamma_1 T) e^{-mt/\vartheta}.
\end{aligned} \tag{7.35}$$

Notice that the last term in the sum representation (7.35) has been included in the sum in Górny and Cramer (2016) since, in that paper, the convention $\gamma_{m+1} = 0$ has been used.

7.2.6 Generalized Type-II progressive hybrid censoring

Generalized Type-II progressive hybrid censoring for exponential lifetimes has been discussed in Górny and Cramer (2016) and Lee et al. (2016a). First, with values of d and w taken from Table 7.3, the likelihood function has the same form as (7.32) so that the MLE is given by $\widehat{\vartheta} = \mathsf{TTT}_{D_{\mathsf{gII}}}/D_{\mathsf{gII}}$ with total time on test statistic as in (7.22), that is,

$$\mathsf{TTT}_{D_{\mathsf{gII}}} = \sum_{j=1}^{D_{\mathsf{gII}}} (R_j + 1) X_{j:n^*:n} + \gamma_{D_{\mathsf{gII}}+1} W$$

and $D_{\mathrm{gII}} = (D_1 \vee m) \wedge D_2$, $W = [X_{m:n^*:n} \vee (T_1 \wedge X_{n^*:n^*:n})] \wedge T_2$. Notice that $\{D_{\mathrm{gII}} \geq 1\} = \{D_2 \geq 1\}$.

Now, since $D_1 \leq D_2$ as $T_1 < T_2$, we find

$$D_{\mathrm{gII}} = (D_1 \vee m) \wedge D_2 = D_1 \vee (D_2 \wedge m) = \begin{cases} D_1, & D_1 \geq m \\ m, & D_1 < m \leq D_2 \\ D_2, & D_2 < m \end{cases}.$$

Similarly, we have

$$[X_{m:n^*:n} \vee (T_1 \wedge X_{n^*:n^*:n})] \wedge T_2 = \begin{cases} X_{n^*:n^*:n}, & D_1 = n^* \\ T_1, & m \leq D_1 < n \\ X_{m:n^*:n}, & D_1 < m \leq D_2 \\ T_2, & D_2 < m \end{cases}.$$

In order to establish a conditional density function of the MLE under generalized Type-II hybrid censoring, we need a representation of $f^{\mathrm{TTT}_m, D_1 < m, D_2 \geq m}$. From the identity, for $t \geq 0$,

$$\mathbf{Pr}_\vartheta(\mathrm{TTT}_m \leq t, D_1 < m, D_2 \geq m) = \mathbf{Pr}_\vartheta(\mathrm{TTT}_m \leq t, D_2 \geq m) - \mathbf{Pr}_\vartheta(\mathrm{TTT}_m \leq t, D_1 \geq m)$$

we get directly

$$f_\vartheta^{\mathrm{TTT}_m, D_1 < m, D_2 \geq m} = f_\vartheta^{\mathrm{TTT}_m, D_2 \geq m} - f_\vartheta^{\mathrm{TTT}_m, D_1 \geq m}$$

where these expressions can be taken from (7.25) with T replaced by T_1 and T_2, respectively.

By analogy with (7.26) and using the expressions given in Theorems 7.7 and 7.8, this yields the conditional density function of $\widehat{\vartheta}$ given $D_2 \geq 1$, that is, for $t \geq 0$,

$$\begin{aligned} f_\vartheta^{\widehat{\vartheta}|D_2 \geq 1}(t) &= \frac{1}{\mathbf{Pr}_\vartheta(D_2 \geq 1)} \sum_{d_1=1}^{n^*} \sum_{d_2=1}^{n^*} f_\vartheta^{\mathrm{TTT}_{(d_1 \vee m) \wedge d_2} / ((d_2 \vee m) \wedge d_2) | D_1 = d_1, D_2 = d_2}(t) \\ &\qquad \times \mathbf{Pr}_\vartheta(D_1 = d_1, D_2 = d_2) \\ &= \frac{1}{\mathbf{Pr}_\vartheta(D_2 \geq 1)} \Bigg(\sum_{d_1=m}^{n^*} \sum_{d_2=1}^{n^*} f_\vartheta^{\mathrm{TTT}_{(d_1 \vee m) \wedge d_2} / ((d_1 \vee m) \wedge d_2) | D_1 = d_1, D_2 = d_2}(t) \\ &\qquad \times \mathbf{Pr}_\vartheta(D_1 = d_1, D_2 = d_2) \\ &\quad + \sum_{d_1=1}^{m-1} \sum_{d_2=m}^{n^*} f_\vartheta^{\mathrm{TTT}_m / m | D_1 = d_1, D_2 = d_2}(t)\, \mathbf{Pr}_\vartheta(D_1 = d_1, D_2 = d_2) \\ &\quad + \sum_{d_1=1}^{m-1} \sum_{d_2=1}^{m-1} f_\vartheta^{\mathrm{TTT}_{(d_1 \vee m) \wedge d_2} / ((d_1 \vee m) \wedge d_2) | D_1 = d_1, D_2 = d_2}(t)\, \mathbf{Pr}_\vartheta(D_1 = d_1, D_2 = d_2) \Bigg) \end{aligned}$$

$$= \frac{1}{\Pr_\vartheta(D_2 \geq 1)} \Bigg(\sum_{d_1=m}^{n^*} \sum_{d_2=1}^{n^*} f_\vartheta^{\mathrm{TTT}_{d_1}/d_1|D_1=d_1,D_2=d_2}(t) \Pr_\vartheta(D_1 = d_1, D_2 = d_2)$$

$$+ \sum_{d_1=1}^{m-1} \sum_{d_2=m}^{n^*} f_\vartheta^{\mathrm{TTT}_m/m|D_1=d_1,D_2=d_2}(t) \Pr_\vartheta(D_1 = d_1, D_2 = d_2)$$

$$+ \sum_{d_1=1}^{m-1} \sum_{d_2=1}^{m-1} f_\vartheta^{\mathrm{TTT}_{d_2}/d_2|D_1=d_1,D_2=d_2}(t) \Pr_\vartheta(D_1 = d_1, D_2 = d_2)\Bigg)$$

$$= \frac{1}{\Pr_\vartheta(D_2 \geq 1)} \Bigg(\sum_{d_1=m}^{n^*} f_\vartheta^{\mathrm{TTT}_{d_1}/d_1|D_1=d_1}(t) \Pr_\vartheta(D_1 = d_1)$$

$$+ f_\vartheta^{\mathrm{TTT}_m/m|D_1<m_1,D_2\geq m}(t) \Pr_\vartheta(D_1 < m, D_2 \geq m)$$

$$+ \sum_{d_2=1}^{m-1} f_\vartheta^{\mathrm{TTT}_{d_2}/d_2|D_2=d_2}(t) \Pr_\vartheta(D_2 = d_2)\Bigg)$$

$$= \frac{1}{1-e^{-nT_2/\vartheta}} \Bigg(\sum_{d_1=m}^{n^*} \frac{T_1^{d_1} \prod_{j=1}^{d_1} \gamma_j}{(d_1-1)!\vartheta^{d_1}} B_{d_1-1}(d_1 t \mid \gamma_{d_1+1} T_1, \ldots, \gamma_1 T_1) e^{-d_1 t/\vartheta}$$

$$+ \frac{T_2^m \prod_{j=1}^{m} \gamma_j}{(m-1)!\vartheta^m} B_{m-1}(mt \mid 0, \gamma_m T_2, \ldots, \gamma_1 T_2) e^{-mt/\vartheta}$$

$$- \frac{T_1^m \prod_{j=1}^{m} \gamma_j}{(m-1)!\vartheta^m} B_{m-1}(mt \mid 0, \gamma_m T_1, \ldots, \gamma_1 T_1) e^{-mt/\vartheta}$$

$$+ \sum_{d_2=1}^{m-1} \frac{T_2^{d_2} \prod_{j=1}^{d_2} \gamma_j}{(d_2-1)!\vartheta^{d_2}} B_{d_2-1}(d_2 t \mid \gamma_{d_2+1} T_2, \ldots, \gamma_1 T_2) e^{-d_2 t/\vartheta}\Bigg).$$

7.2.7 Unified progressive hybrid censoring

The MLE of ϑ under unified progressive hybrid censoring schemes (see Górny and Cramer, 2018b) can be established in the same manner as under unified hybrid censoring (see Section 5.5.2). In particular, the MLE is connected to the total time in test by $\widehat{\vartheta} = \mathrm{TTT}_{D_{\mathrm{HCS}}}/D_{\mathrm{HCS}}$ provided $D_{\mathrm{HCS}} \geq 1$. $\mathrm{TTT}_{D_{\mathrm{HCS}}}$ is constructed as in (5.28) for hybrid censoring schemes taking additionally into account the progressively withdrawn items, that is,

total time on test = sum of observed failure times
+ sum of censoring times of progressively censored items
+ no. of right censored items × stopping time.

The distribution of the MLE can be derived along the lines of the previously discussed progressive hybrid censoring schemes. In particular, the representations of the four types of density functions

$$f_{\vartheta}^{\mathrm{TTT}_d/d,D=d}, \quad f_{\vartheta}^{\mathrm{TTT}_m|D<m}, \quad f_{\vartheta}^{\mathrm{TTT}_m|D\geq m}, \quad f_{\vartheta}^{\mathrm{TTT}_m,D_i<m,D_{i+1}\geq m},$$

are used and then combined appropriately applying the modularization approach (i.e., the law of total probability) proposed in Górny and Cramer (2018b). At this point it should be noted that the observed sample size D_{HCS} under progressive hybrid censoring has the same representation as that one under hybrid censoring. In particular, the order statistics $X_{j:n}$ have to be replaced by $X_{j:n^*:n}$ in the definition of D_{HCS}. This means that the functional form is identical. For convenience, the expressions have been summarized in Table 5.1.

7.2.8 Progressive hybrid censoring for multiple samples

Multi-sample situations under progressive hybrid censoring have been discussed in Górny and Cramer (2020b) (Type-I) and Jansen et al. (2022) (Type-II). They considered data composed of k independent Type-I and Type-II progressive hybrid censored experiments, respectively. They allowed different thresholds $T_1, \ldots, T_k$ as well as different censoring parameters $m_1, \ldots, m_k$ in each sub-experiment. In should be noted that Górny and Cramer (2020b) discussed the model of sequential order statistics which covers progressively Type-II censored data as a special case. In the following, we will consider k progressively Type-II censored samples as given in Model 7.11.

Model 7.11 (Multi-sample progressive censoring). *Let $k \in \mathbb{N}$, $n_1, \ldots, n_k \in \mathbb{N}$, $n_1^*, \ldots, n_k^* \in \mathbb{N}$ with $n_i^* \leq n_i$, $1 \leq i \leq k$.*

Then, k samples of progressively Type-II censored order statistics with censoring plans $\mathscr{R}_i = (R_{i,1}, \ldots, R_{i,n_i^*})$ and parameters $\gamma_{i,j}$, $1 \leq j \leq n_i^*$, $1 \leq i \leq k$, as well as its realization are denoted by

$$X_{i,1:n_i^*:n_i}, \ldots, X_{i,n_i^*:n_i^*:n_i} \quad \text{and} \quad \boldsymbol{x}_i = (x_{i,1}, \ldots, x_{i,n_i^*}), \qquad 1 \leq i \leq k,$$

respectively.

This model has been introduced in Balakrishnan et al. (2001) (for the more general model of sequential order statistics, see also Cramer and Kamps, 2001a,b).

Multi-sample Type-I progressive hybrid censoring

Given Model 7.11, the *multi-sample Type-I progressive hybrid censoring scheme* is generated from $k \geq 2$ independently conducted Type-I progressive hybrid censored experiments with parameters $m_i \leq n_i^*$ and thresholds T_i, $1 \leq i \leq k$. The termination of each censoring experiment is accomplished independently of the other $k-1$ experiments. Given prefixed threshold times $T_1, \ldots, T_k > 0$ with $F^{\leftarrow}(0) < T_i$, $1 \le i \le k$, the i-th sample of *Type-I progressive hybrid censored order statistics* is given by

$$X_{i,1:n_i^*:n_i}, \ldots, X_{i,D_{\mathrm{I},i}:n_i^*:n_i}, \tag{7.36}$$

where the counter variable

$$D_i = \sum_{j=1}^{n_i^*} \mathbb{1}_{(-\infty,T_i]}(X_{i,j:n_i^*:n_i})$$

defines the number of observed failures $D_{\mathrm{I},i} = D_i \wedge m_i$ until threshold T_i, $1 \le i \le k$. The stopping time T_i^* of the i-th sub-experiment is given by

$$T_i^* = X_{i,m_i:n_i^*:n_i} \wedge T_i, \quad 1 \le i \le k,$$

so that the stopping time of the experiment is given by $\max\{T_i^* \mid 1 \le i \le k\}$. The total number of observations is denoted by $D_{\mathrm{I},\bullet k} = \sum_{i=1}^k D_{\mathrm{I},i}$.

As pointed out in Górny and Cramer (2020b), inferential and distributional results obtained in the single sample case can be utilized in the multi-sample setting. For $Exp(\vartheta)$-distributed lifetimes, the MLE of ϑ is given by

$$\widehat{\vartheta} = \frac{1}{D_{\mathrm{I},\bullet k}} \sum_{i=1}^k \mathsf{TTT}_{i,D_{\mathrm{I},i}} \tag{7.37}$$

provided that $D_{\mathrm{I},\bullet k} \ge 1$. $\mathsf{TTT}_{i,D_{\mathrm{I},i}}$ denotes the total time on test statistic in the i-th sub-experiment given by (see (7.33))

$$\mathsf{TTT}_{i,D_{\mathrm{I},i}} = \sum_{j=1}^{D_{\mathrm{I},i}} (R_{i,j}+1) X_{i,j:n_i^*:n_i} + \gamma_{D_{\mathrm{I},i}+1} X_{i,D_{\mathrm{I},i}:n_i^*:n_i} \wedge T_i, \ 1 \le i \le k.$$

If the i-th subsample is empty, we have $D_{\mathrm{I},i} = 0$ and, thus, $\mathsf{TTT}_{i,0} = n_i T_i$. From (7.37), we get that the MLE is a weighted sum of the MLEs in the subsamples, that is,

$$\widehat{\vartheta} = \sum_{i=1}^k \omega_i \widehat{\vartheta}_i \quad \text{with } \omega_i = \frac{D_{\mathrm{I},i}}{D_{\mathrm{I},\bullet k}}, \ \widehat{\vartheta}_i = \frac{1}{D_{\mathrm{I},i}} \mathsf{TTT}_{i,D_{\mathrm{I},i}}, \ 1 \le i \le k.$$

Notice that, in order to exist, we need only one observation in the sample. In particular, for all $\vartheta > 0$,

$$\Pr_\vartheta(D_{\mathrm{I},\bullet k} \ge 1) = 1 - \Pr_\vartheta(D_{\mathrm{I},i} = 0, 1 \le i \le k) = 1 - \exp\Big\{ - \sum_{i=1}^k n_i T_i/\vartheta \Big\} \xrightarrow{k\to\infty} 1,$$

so that the MLE exists asymptotically with probability 1. From Lemma 7.2, the expected sample size is given by

$$E(D_{\mathrm{I},\bullet k}) = \sum_{i=1}^{k}\sum_{j=1}^{m_i} F_{i,j:m_i:n_i}(T_i).$$

Notice that $\lim_{T_i\to\infty} F_{i,j:m_i:n_i}(T_i) = 1$, $1 \le j \le m_i$, $1 \le i \le k$, so that $E(D_{\mathrm{I},\bullet k})$ tends to $m_{\bullet k}$ when T_i tends to infinity, $1 \le i \le k$.

As illustrated in Górny and Cramer (2020b), the exact (conditional) distribution of the MLE $\widehat{\vartheta}$ can be established using results from the single sample case. However, the resulting expressions are quite messy so that we do not present details here.

Remark 7.12. Jia and Guo (2017) considered exact inference for exponentially distributed lifetimes under multiply Type-I censoring with possibly different thresholds $T_1, \ldots, T_k$. This model can be seen as a particular case of multi-sample Type-I progressive hybrid censoring. Notice that, in this model, k independent Type-I censored single observations are available. The model has been discussed for Weibull distributions in Wang (2014), Jia and Guo (2018), and Jia et al. (2018).

Multi-sample Type-II progressive hybrid censoring

The extension of Type-II progressive hybrid censoring to multiple samples has been presented in Jansen et al. (2022). Assuming Model 7.11, each subsample is subject to Type-II progressive hybrid censoring so the resulting data reads

$$X_{i,1:n_i^*:n_i}, \ldots, X_{i,D_{\mathrm{II},i}:n_i^*:n_i}, \quad 1 \le i \le k, \tag{7.38}$$

where $D_{\mathrm{II},i} = D_i \vee m_i$, $1 \le i \le k$. Therefore, the total sample size $D_{\mathrm{II},\bullet k} = \sum_{i=1}^{k} D_{\mathrm{II},i}$ is at least $m_{\bullet k} = \sum_{i=1}^{k} m_i$. As for multi-sample Type-I progressive hybrid censoring, the MLE of ϑ can be obtained for exponentially distributed lifetimes. It is given by

$$\widehat{\vartheta} = \frac{1}{D_{\mathrm{II},\bullet k}} \sum_{i=1}^{k} \mathsf{TTT}_{i,D_{\mathrm{II},i}}$$

with total time on test statistics (see (7.34))

$$\mathsf{TTT}_{i,D_{\mathrm{II},i}} = \sum_{j=1}^{D_{\mathrm{II},i}} (R_{i,j}+1) X_{i,j:n_i^*:n_i} + \gamma_{D_{\mathrm{I},i}+1} X_{i,D_{\mathrm{II},i}:n_i^*:n_i} \vee (T_i \wedge X_{i,n_i^*:n_i^*:n_i}), \ 1 \le i \le k.$$

Furthermore, exact distributions can be established. For details, we refer to Jansen et al. (2022).

7.2.9 Type-I censoring of sequential order statistics

Burkschat et al. (2016) have applied Type-I censoring to sequential order statistics $X_*^{(1)}, \ldots, X_*^{(n)}$ with parameters $\gamma_1, \ldots, \gamma_n > 0$ and cumulative distribution function F (see Section 2.5; for the multi-sample case, see Górny and Cramer, 2020b). The presentation includes the derivation of the fundamental distributions as well as inference for exponential distributions. As above, the counter variable

$$D = \sum_{j=1}^{n} \mathbb{1}_{(-\infty, T]}(X_*^{(j)})$$

plays a crucial role in the analysis. Moreover, the following result for conditional distributions of sequential order statistics (see also Burkschat and Navarro, 2013) is important in the derivations.

Lemma 7.13. *Let $d \in \{1, \ldots, n\}$ and $0 \le x_j \le T$, $1 \le j \le d$. If $d < n$, then*

$$\Pr(X_*^{(j)} > x_j, 1 \le j \le d \mid D = d) = \Pr(X_*^{(j)} > x_j, 1 \le j \le d \mid X_*^{(d+1)} = T).$$

If $d = n$, then

$$\Pr(X_*^{(j)} > x_j, 1 \le j \le n \mid D = n) = \Pr(X_*^{(j)} > x_j, 1 \le j \le n \mid X_*^{(n)} \le T).$$

As for Type-I (progressive) hybrid censoring, the distribution of the spacings of the data Type-I hybrid censored data

$$X_*^{(1)} \wedge T, \ldots, X_*^{(n)} \wedge T$$

can be utilized to establish expressions for the (conditional) joint density function of the spacings given D. The formulas have the same form as those obtained for Type-I progressive hybrid censoring (see (7.18) and (7.19)). However, the parameters $\gamma_1, \ldots, \gamma_n$ may be arbitrary positive real values in this case and the marginal density functions and cumulative distribution function have to be replaced by those of the corresponding sequential order statistics. As pointed out in Cramer and Kamps (2003) and Cramer (2006), these functions are special Meijer's G-functions (see Mathai, 1993) and have a complicated structure.

For exponential distributions, the distribution of the total time on test statistic as in (7.23) can be established. This has been done in Burkschat et al. (2016) using the following theorem which can be found in, e.g., Adell and Sangüesa (2005) (see also Karlin et al., 1986; Ignatov and Kaishev, 1989).

Theorem 7.14. *Let $d \in \mathbb{N}$ and $x_0, x_1, \ldots, x_m \in \mathbb{R}$ with $x_0 \le x_1 \le \cdots \le x_m$ and $x_0 < x_m$. Let $\delta_i = x_i - x_{i-1}$, $i = 1, \ldots, m$. Moreover, let $U_{1:m} \le \cdots \le U_{m:m}$ be order statistics based on an* *iid sample of random variables $U_1, \ldots, U_m$ from a standard uniform distribution and let*

$$V^m_{(\delta_1,\ldots,\delta_m)} = \delta_1 U_{m:m} + \cdots + \delta_m U_{1:m}.$$

Then, the density function $f^{V^m_{(\delta_1,\ldots,\delta_m)}+x_0}$ of $V^m_{(\delta_1,\ldots,\delta_m)} + x_0$ is given by the m-th order B-spline with knots $x_0, x_1, \ldots, x_m$, that is,

$$f^{V^m_{(\delta_1,\ldots,\delta_m)}+x_0}(t) = B_{m-1}(t \mid x_0, \ldots, x_m), \quad t \in \mathbb{R}.$$

This enables the derivation of the (conditional) cumulative distribution function of the MLE of ϑ as given in Theorem 7.15. Note that the MLE exists for $D \geq 1$. Furthermore, due to equal γ's, a density function of the MLE

$$\widehat{\vartheta} = \frac{1}{D}\left(\sum_{j=1}^{D}(\gamma_j - \gamma_{j+1})X_*^{(j)} + \gamma_{D+1}T\right) = \frac{1}{D}\cdot \mathrm{TTT}_D,$$

given $D \geq 1$ may not exist as shown in Theorem 7.15 (see also Remark 7.16).

Theorem 7.15. *Let $\gamma_1, \ldots, \gamma_n > 0$, $\gamma_{n+1} = 0$, and $p \in \{0, \ldots, n-1\}$ with $\gamma_1 = \cdots = \gamma_{p+1} \neq \gamma_{p+2}$. Then, the cumulative distribution function of $\widehat{\vartheta}$, conditionally on the event $D \geq 1$, is given by*

$$\Pr_\vartheta(\widehat{\vartheta} \leq t \mid D \geq 1) = \frac{e^{-\gamma_1 T/\vartheta}}{1 - e^{-\gamma_1 T/\vartheta}} \sum_{d=1}^{p} \mathbb{1}_{(0,dt]}(\gamma_1 T)\frac{\gamma_1^d}{d!}\left(\frac{T}{\vartheta}\right)^d$$
$$+ \frac{\gamma_1^p}{1 - e^{-\gamma_1 T/\vartheta}} \sum_{d=p+1}^{n} \frac{\prod_{j=p+1}^{d}\gamma_j}{d!}\left(\frac{T}{\vartheta}\right)^d \int_0^{dt} B_d(s \mid \gamma_{d+1}T, \ldots, \gamma_1 T)e^{-s/\vartheta}\, ds, \; v \geq 0.$$

Remark 7.16. If $\gamma_1 \neq \gamma_2$, i.e., $p = 0$ in the preceding theorem, then the distribution of $\widehat{\vartheta}$, given $D \geq 1$, has the following density function

$$f_\vartheta^{\widehat{\vartheta}|D\geq 1}(t) = \frac{1}{1 - e^{-\gamma_1 T/\vartheta}} \sum_{d=1}^{n} \frac{\prod_{j=1}^{d}\gamma_j}{(d-1)!}\left(\frac{T}{\vartheta}\right)^d B_{d-1}(dt \mid \gamma_{d+1}T, \ldots, \gamma_1 T)e^{-dt/\vartheta}, \quad t \geq 0,$$

which is exactly the expression given in (7.26) (with n^* replaced by n). Notice that $\gamma_1 > \gamma_2$ holds under progressive censoring.

If all γ's are equal, that is, $\gamma_1 = \cdots = \gamma_n$, we have $p = n - 1$ and the cumulative distribution function can be written as

$$\Pr_\vartheta(\widehat{\vartheta} \leq t \mid D \geq 1)$$
$$= \frac{1}{1 - e^{-\gamma_1 T/\vartheta}}\left(\vartheta \sum_{d=1}^{n-1} \mathbb{1}_{(0,dt]}(\gamma_1 T) f_{\Gamma(\vartheta,d+1)}(\gamma_1 T) + F_{\Gamma(\vartheta,n)}\Big(\min\{nt, \gamma_1 T\}\Big)\right), \quad t \geq 0.$$

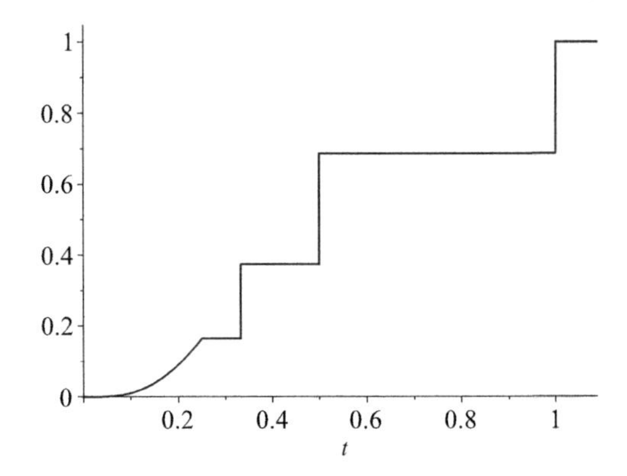

Figure 7.8 Plot of $\Pr_\vartheta(\widehat{\vartheta} \leq t \mid D \geq 1)$ with $\vartheta = 0.5, T = 1, n = 4, \gamma_1 = 1$.

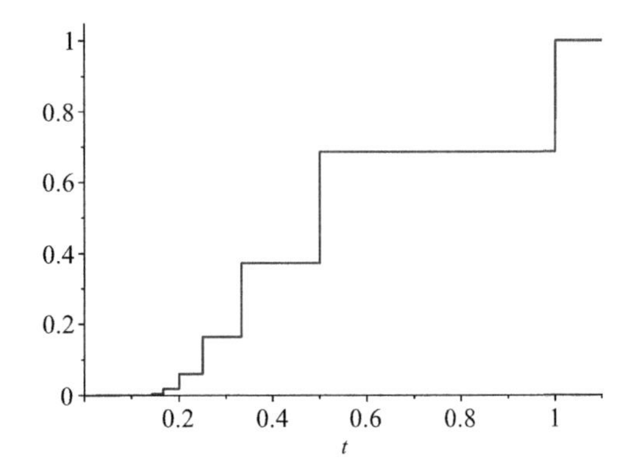

Figure 7.9 Plot of $\Pr_\vartheta(\widehat{\vartheta} \leq t \mid D \geq 1)$ with $\vartheta = 0.5, T = 1, n = 8, \gamma_1 = 1$.

Plots of the cumulative distribution function on the interval $[0, 1]$ for selected values of n and γ_1 are provided in Figs. 7.8–7.11. In particular, they illustrate the jumps of the cumulative distribution function at $\gamma_1 T/d$, $d = 1, \ldots, n$.

Remark 7.17. From the preceding results, it follows that similar results can be established for Type-I hybrid censored sequential order statistics.

7.2.10 Bayesian inference

Bayesian inference under progressive hybrid censoring schemes with exponentially distributed life times can be conducted by analogy with the presentation for hybrid censoring schemes (see Sections 5.3.3 and 5.4.2). Initiated by Kundu and Joarder (2006b) for Type-I progressive hybrid censoring, Bayesian inference has been widely discussed

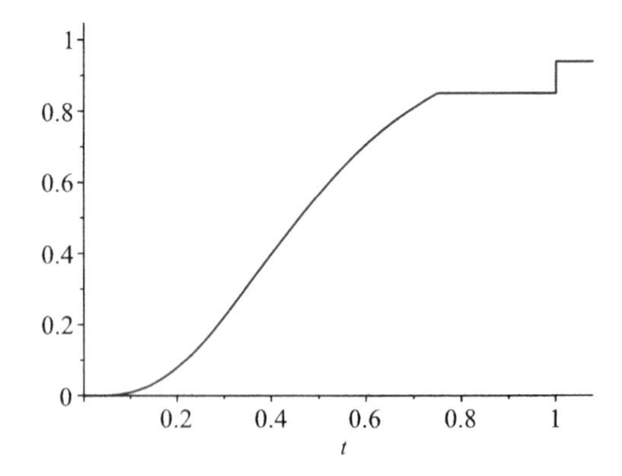

Figure 7.10 Plot of $\Pr_\vartheta(\widehat{\vartheta} \le t \mid D \ge 1)$ with $\vartheta = 0.5, T = 1, n = 4, \gamma_1 = 3$.

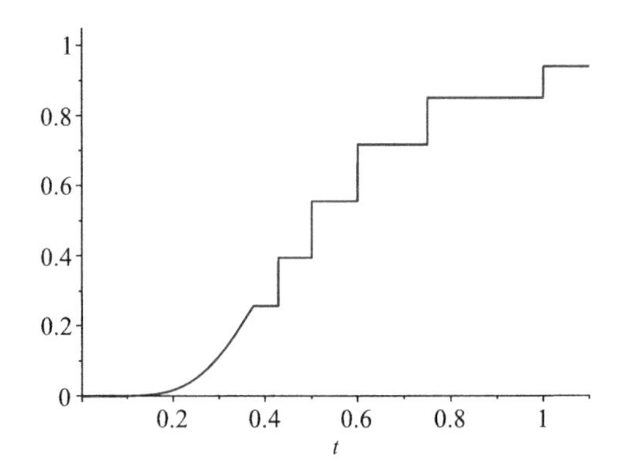

Figure 7.11 Plot of $\Pr_\vartheta(\widehat{\vartheta} \le t \mid D \ge 1)$ with $\vartheta = 0.5, T = 1, n = 8, \gamma_1 = 3$.

for exponentially distributed life times using the prior as in (5.18), that is,

$$\pi(\vartheta) = \frac{\lambda^\beta}{\Gamma(\beta)} \vartheta^{-(\beta+1)} e^{-\lambda/\vartheta}, \quad \vartheta > 0,$$

with hyper-parameters $\beta > 0$ and $\lambda > 0$. For the choice $\beta = \lambda = 0$, the prior in (5.18) becomes a non-informative prior. Then, from the likelihood function in (7.32), the posterior density function $p(\cdot \mid \text{data})$ of ϑ is obtained as

$$p(\vartheta \mid \text{data}) = \frac{(\mathrm{TTT}_{d^*} + \lambda)^{d^*+\beta}}{\Gamma(d^* + \beta)} \vartheta^{-(d^*+\beta+1)} e^{-(\mathrm{TTT}_{d^*}+\lambda)/\vartheta}, \quad \vartheta > 0,$$

where the total time on test TTT_{d^*} is defined in (7.33) and d^* is the realization of D_{HCS}. As under Type-I hybrid censoring, the Bayesian estimator of ϑ under the squared-error loss function is simply obtained as

$$\widehat{\vartheta}_{B} = \frac{\mathsf{TTT}_{D_{HCS}} + \lambda}{D_{HCS} + \beta - 1}$$

provided $D_{HCS} + \beta > 1$. Comparing these results with those presented under Type-I hybrid censoring in Section 5.3.3, the similarity is striking. Clearly, the representation is almost the same by choosing the appropriate version of the total time on test statistic $\mathsf{TTT}_{D_{HCS}}$. Therefore, we abstain from presenting further details and refer to the respective references where similar results can be found. For Type-I progressive hybrid censoring, we refer to Kundu and Joarder (2006b) and Lin et al. (2011, 2013) whereas Lin et al. (2011, 2013) and considered Type-II progressive hybrid censoring. Further progressive hybrid censoring schemes have been studied by Mohie El-Din et al. (2017).

Remark 7.18. Bayesian inference for the two-parameter exponential distribution under various hybrid censoring has been discussed in Kundu et al. (2013). The presentation includes Type-I progressive hybrid censoring. Similar results have been obtained under generalized Type-I/Type-II progressive hybrid censoring in Seo and Kim (2017) and Seo and Kim (2018).

7.2.11 Confidence intervals

Exact confidence intervals

Exact (conditional) confidence intervals can be established using the method of pivoting the cumulative distribution function as illustrated in Section 5.3.2.1. As for hybrid censoring schemes, the stochastic monotonicity of the cumulative distribution function $F_{\vartheta}^{\widehat{\vartheta}}$ in ϑ has to hold as well as the limit conditions in Condition 5.13 need to be checked. This has been done for various progressive hybrid censoring schemes. For the sake of completeness, we summarize the results and provide the respective references for further (technical) details. Note that, under progressive Type-II censoring, exact confidence intervals can be directly established using the χ^2-distribution (see (7.16) and Balakrishnan and Cramer, 2014, p. 380).

In the following, $\widehat{\vartheta}$ denotes the MLE under the considered progressive hybrid censoring scheme. A general account to the stochastic monotonicity requirement is given by van Bentum and Cramer (2019).

(a) Type-I progressive hybrid censoring: The stochastic monotonicity has been established in Burkschat et al. (2016) in terms of Type-I hybrid censored sequential order statistics (see also Remark 1 in van Bentum and Cramer, 2019). Furthermore, for $t > 0$, the necessary limits are given by

$$\lim_{\vartheta\to 0} F_\vartheta^{\widehat{\vartheta}|D_I\geq 1}(t) = 1, \quad \lim_{\vartheta\to\infty} F_\vartheta^{\widehat{\vartheta}|D_I\geq 1}(t) = \begin{cases} 0, & 0 < t \leq \gamma_2 T \\ \frac{t-\gamma_2 T}{\gamma_1 T - \gamma_2 T}, & \gamma_2 T < t \leq \gamma_1 T \\ 1, & \gamma_1 T < t \end{cases}. \tag{7.39}$$

This shows that we face the same problem as under Type-I hybrid censoring (see Fig. 5.7 and Lemma 5.17) so that similar comments apply to the present situation. The problem has also been studied in Childs et al. (2008). The case of multiple Type-I progressive hybrid censoring has been resolved in Górny and Cramer (2020b).

(b) Type-II progressive hybrid censoring: For Type-II progressive hybrid censoring, the stochastic monotonicity can be taken from van Bentum and Cramer (2019). Furthermore, by following the same arguments as in Górny and Cramer (2016), we have

$$\lim_{\vartheta\to 0} F_\vartheta^{\widehat{\vartheta}}(t) = 1, \quad \lim_{\vartheta\to\infty} \overline{F}_\vartheta^{\widehat{\vartheta}}(t) = 0, \tag{7.40}$$

so that the pivoting method can be directly applied. This construction has also been proposed in Childs et al. (2008) and in Jansen et al. (2022) for multiple samples.

(c) Generalized Type-I progressive hybrid censoring: The stochastic monotonicity of the MLE has been discussed in Example 3 in van Bentum and Cramer (2019). Górny and Cramer (2016) showed that the limits are as in (7.40). Further discussions have been provided in Cho et al. (2015b).

(d) Generalized Type-II progressive hybrid censoring: The proof of the stochastic monotonicity of the MLE has been presented in Example 2 in van Bentum and Cramer (2019). The limits of the (conditional) cumulative distribution function have been obtained in Górny and Cramer (2016) as

$$\lim_{\vartheta\to 0} F_\vartheta^{\widehat{\vartheta}|D_{gII}\geq 1}(t) = 1, \quad \lim_{\vartheta\to\infty} F_\vartheta^{\widehat{\vartheta}|D_{gII}\geq 1}(t) = \begin{cases} 0, & 0 < t \leq \gamma_2 T \\ \frac{t\wedge(\gamma_1 T_2)-\gamma_2 T_2}{\gamma_1 T_2 - \gamma_2 T_2}, & \gamma_2 T_2 < t \leq \gamma_1 T_2 \end{cases}$$

which is the same as (7.39) with T replaced by the threshold T_2. The approach has also been discussed in Lee et al. (2016a).

(e) Unified progressive hybrid censoring: Clearly, the pivoting approach can also be applied under unified progressive hybrid censoring. However, it seems that the assumptions (i.e., stochastic monotonicity, limits of the cumulative distribution function) have not been checked so far.

Approximative confidence intervals

For larger sample sizes, approximative confidence intervals may be more appropriate since the computational effort to determine the exact confidence interval is quite large. Such proposals are often discussed in the literature by applying a normal approximation (see, e.g., Gupta and Kundu, 1998; Sundberg, 2001; Kundu and Joarder, 2006a; Balakrishnan and Kundu, 2013; Arabi Belaghi and Noori Asl, 2019). Confidence intervals based on normal approximations of the MLE under progressive hybrid censoring schemes have been proposed in Lee et al. (2016a).

In the multiple sample case, Górny and Cramer (2020b) and Jansen et al. (2022) have proposed normal and χ^2-approximations. First, consider the Type-I progressive hybrid censoring scheme in (7.36). Denoting by z_p the p-quantile of a standard normal distribution, an approximated $(1-\alpha)$-confidence interval for ϑ is given by

$$\left[\widehat{\vartheta} - z_{1-\alpha/2}\widehat{\vartheta}/\sqrt{D_{\bullet k}}, \widehat{\vartheta} + z_{1-\alpha/2}\widehat{\vartheta}/\sqrt{D_{\bullet k}}\right].$$

This confidence interval has the asymptotic level $1-\alpha$ (for $k \to \infty$).

An alternative approximated confidence interval may be obtained by assuming limits $T_i \to \infty$, $1 \le i \le k$, that is, the Type-I censoring process is neglected (see Cohen, 1995, Section 4.2.3). Then, $2m_{\bullet k}\widehat{\vartheta}/\vartheta$ has approximately a χ^2-distribution with $2m_{\bullet k}$ degrees of freedom (see Cramer and Kamps, 2001a). Therefore, the interval

$$[2m_{\bullet k}\widehat{\vartheta}/\chi^2_{2m_{\bullet k};1-\alpha/2}, 2m_{\bullet k}\widehat{\vartheta}/\chi^2_{2m_{\bullet k},\alpha/2}]$$

may also serve as an approximative confidence interval where $\chi^2_{r,p}$ denotes the p-quantile of a χ^2 distribution with r degrees of freedom. Notice that this may lead to reasonable estimates only when the true value of ϑ is significantly less than the thresholds employed in the experiments.

For the Type-II progressive hybrid model in (7.38), Jansen et al. (2022) pointed out that, for small thresholds T_i, the MLE corresponds almost to the MLE in the case of a progressively Type-II censored sample with sample sizes $m_1, \ldots, m_k$. Therefore, the MLE is approximately χ^2-distributed with $2m^* = 2m_{\bullet k}$ degrees of freedom (see Balakrishnan et al., 2014, Theorem 12.1.1). If the thresholds are considered to be large, then the MLE will approximately be χ^2-distributed with $2m_* = 2n^*_{\bullet k}$ degrees of freedom so that the approximative confidence interval

$$CI^\chi = [2m_*\widehat{\vartheta}/\chi^2_{2m_*,1-\alpha/2}, 2m_*\widehat{\vartheta}/\chi^2_{2m_*,\alpha/2}]$$

results. Of course, the situation may be diverse for each subsamble so that in one sample the threshold may be considered as small (then use m_i) whereas it may be interpreted as large in another case (thus, use n^*_j). In such a setting the degrees of freedom can be defined by

$$2m_* = 2 \sum_{T_i \text{ small}} m_i + 2 \sum_{T_i \text{ large}} n_i^*.$$

In all the mentioned models, Bayesian or bootstrap intervals may be constructed and can be found in many publications. For relevant references, we refer to Balakrishnan and Kundu (2013) and references cited therein. Moreover, many of the above cited references discuss such procedures.

7.3. Progressive hybrid censored data: other cases

Progressive hybrid censoring has also been considered for various lifetime distributions. But, as pointed out in Section 7.1 and illustrated for the exponential distribution in Section 7.2, the modularization approach guides the path of how the MLEs can be computed under arbitrary progressive hybrid censoring schemes, too. In fact, for a lifetime cumulative distribution function $F_{\boldsymbol{\theta}}$ with density function $f_{\boldsymbol{\theta}}$ and parameter $\boldsymbol{\theta} \in \Theta \subseteq \mathbb{R}^s$, the likelihood function is given in (7.9), that is,

$$\mathscr{L}(\boldsymbol{\theta}; \boldsymbol{x}_d, d) = \prod_{j=1}^{d} \gamma_j \Big(\prod_{j=1}^{d} f_{\boldsymbol{\theta}}(x_j)(1 - F_{\boldsymbol{\theta}}(x_j))^{R_j} \Big) (1 - F_{\boldsymbol{\theta}}(w))^{\gamma_{d+1}}$$

with appropriate choices of d and w as given in Table 7.3.

7.3.1 Weibull distribution

In order to illustrate the similarity to hybrid censoring as shown in Section 6.2.1, we present some details for Type-I progressive hybrid censoring which has been studied first in Mokhtari et al. (2011). Then, the log-likelihood function for *Weibull*(ϑ, β)-distributions is given by

$$\mathscr{L}^*(\vartheta, \beta; \boldsymbol{x}_d, d) = c_d^* + d \log(\beta/\vartheta) + (\beta - 1) \sum_{j=1}^{d} \log x_j - \frac{1}{\vartheta} \sum_{j=1}^{d} (R_j + 1) x_j^{\beta} - \frac{\gamma_{d+1}}{\vartheta} w^{\beta}. \quad (7.41)$$

From Balakrishnan and Kateri (2008), it follows that the log-likelihood function has a unique maximum which is attained by $\widehat{\vartheta}, \widehat{\beta}$ where

$$\widehat{\vartheta} = \widehat{\vartheta}(\widehat{\beta}) = \frac{1}{d} \Big(\sum_{j=1}^{d} (R_j + 1) x_j^{\widehat{\beta}} + \gamma_{d+1} w^{\widehat{\beta}} \Big)$$

and $\widehat{\beta}$ is the unique solution of the equation

$$0 = \frac{d}{\beta} - \frac{\sum_{j=1}^{d}(R_j+1)x_j^{\beta}\log x_j + \gamma_{d+1}w^{\beta}\log w}{\widehat{\vartheta}(\beta)} + \sum_{j=1}^{d}\log x_j.$$

By similarity with the hybrid censoring case, iterative procedures like the Newton-Raphson method or fixpoint procedures may be used to determine the estimates. In order to find appropriate starting values, Lin et al. (2009) and Mokhtari et al. (2011) proposed approximate MLEs by proceeding as in Kundu (2007). According to the procedure outlined in Section 5.2, the concept of approximate MLEs has to be applied to the likelihood function in (7.41). Notice that this idea had previously been applied to progressively Type-II censored data by, e.g., Balasooriya et al. (2000), Balasooriya and Balakrishnan (2000), Balakrishnan et al. (2004a) (see also Balakrishnan and Cramer, 2014, Section 12.9.2, for further references). For Weibull distributed data, a transformation of the data to a location-scale family of extreme value distributions and an appropriate linearization solves the problem (see also Balasooriya et al., 2000). The respective linearized equations are given by

$$\begin{aligned}
&-\sum_{j=1}^{d}(R_j+1)(a_j - b_j(\beta\log x_j - \log\vartheta)) \\
&\qquad\qquad -\gamma_{d+1}(a_d - b_d(\beta\log w - \log\vartheta)) + n - d = 0 \\
&-d - \sum_{j=1}^{d}(R_j+1)(a_j - b_j(\beta\log x_j - \log\vartheta))(\beta\log x_j - \log\vartheta) \\
&\qquad + \sum_{j=1}^{d}R_j(\beta\log x_j - \log\vartheta) + \gamma_{d+1}(\beta\log w - \log\vartheta) \\
&\qquad\qquad -\gamma_{d+1}(a_d - b_d(\beta\log w - \log\vartheta))(\beta\log w - \log\vartheta) = 0
\end{aligned} \tag{7.42}$$

where

$$a_j = 1 + \log(q_j)\big[-\log(-\log q_j)\big], \quad b_j = -\log q_j, \qquad 1 \le j \le d$$

Notice that Eqs. (7.42) are a simplified version of the equations given in Mokhtari et al. (2011) which can be interpreted as a weighted version of (6.2). The result has also been presented in Lin et al. (2009) in a sightly different form. Mokhtari et al. (2011) proposed to choose the values $q_1, \ldots, q_d$ as

$$q_j = \prod_{i=1}^{j}\frac{\gamma_i}{\gamma_i+1} \qquad 1 \le j \le d, \tag{7.43}$$

Table 7.4 Quantities in the derivation of approximate MLEs for the Weibull parameters under Type-I progressive hybrid censoring.

quantity	Type-I progressive hybrid
w	$x_m \wedge T$
d	$D \wedge m$
ζ	$\dfrac{\sum_{j=1}^{d}(R_j+1)b_j \log x_j + \gamma_{d+1} b_d \log w}{\sum_{j=1}^{d}(R_j+1)b_j + \gamma_{d+1} b_d}$
τ	$\dfrac{\sum_{j=1}^{d}(R_j+1)a_j + \gamma_{d+1} a_d - (n-d)}{\sum_{j=1}^{d}(R_j+1)b_j + \gamma_{d+1} b_d}$
δ	$\sum_{j=1}^{d}(R_j+1)a_j(\log x_j - \zeta) + \gamma_{d+1} a_d(\log w - \zeta)$ $- \sum_{j=1}^{d} R_j(\log x_j - \zeta) - \gamma_{d+1}(\log w - \zeta)$ $-2\tau \sum_{j=1}^{d} b_j(\log x_j - \zeta) - 2\gamma_{d+1}\tau b_d(\log w - \zeta)$
η	$\sum_{j=1}^{d}(R_j+1)b_j(\log x_j - \zeta)^2 + \gamma_{d+1} b_d(\log w - \zeta)^2$

which result from means of progressively Type-II censored order statistics from a standard uniform distribution (see Balakrishnan and Cramer (2014), Theorem 7.2.3, as well as Corollary 2.65). Note that Lin et al. (2009) proposed a different choice for q_d when $d < m$, that is, time censoring is applied at T. In that case, they used

$$q_d^* = \frac{q_d + q_d + 1}{2}$$

with q_{d-1}, q_d as in (7.43). This definition of the proportions is in accordance with the definition under Type-I and Type-I hybrid censoring (see (6.3)).

Eqs. (7.42) can be explicitly solved for the parameters ϑ and β yielding the approximate MLEs. Using the notation in Table 7.4, the equations in (6.2) can be written as

$$\log \vartheta = \zeta\beta - \tau, \qquad \beta^2 - \frac{\delta}{\eta}\beta - \frac{d}{\eta} = 0, \tag{7.44}$$

where $\eta > 0$. Notice that the equations have the same form as under hybrid censoring (see (6.4)) but with different expressions for ζ, τ, δ, and η as given in Table 7.4. Solving the quadratic equation (7.44) for the parameters yields the estimators

$$\widehat{\vartheta}^* = e^{\zeta\widehat{\beta}^* - \tau}, \qquad \widehat{\beta}^* = \frac{\delta + \sqrt{\delta^2 + 4d\eta}}{2\eta} = \frac{2d}{-\delta + \sqrt{\delta^2 + 4d\eta}}.$$

Furthermore, the similarities for Bayesian inference are also striking. By analogy with Kundu (2007), assuming an inverse gamma prior $\mathsf{I}\Gamma(b, a)$ and density function π_1

for ϑ as in (6.5) and an arbitrary prior π_2 for β with a support in $(0,\infty)$ (independent of ϑ), the joint density function is given by (see also Mokhtari et al., 2011 and Lin et al., 2012 for a different parametrization)

$$
\begin{aligned}
f(\text{data},\beta,\vartheta) &= f(\boldsymbol{x}_d,d,\beta,\vartheta)\\
&= e^{c_d^*}\beta^d\vartheta^{-(a+d+1)}\Big(\prod_{j=1}^{d}x_j\Big)^{\beta-1}\exp\Big(-\frac{1}{\vartheta}\Big(\sum_{j=1}^{d}(R_j+1)x_j^\beta+\gamma_{d+1}w^\beta+b\Big)\Big)\pi_2(\beta).
\end{aligned}
$$

Notice that Kundu (2007) as well as Mokhtari et al. (2011) and Lin et al. (2012) used a gamma prior due to a different parametrization of the Weibull distribution. Then, the (conditional) distribution of ϑ given β and the data is given by an inverse gamma distribution

$$
\mathsf{I}\Gamma\Big(\sum_{j=1}^{d}(R_j+1)x_j^\beta+\gamma_{d+1}w^\beta+b,\,a+d\Big).
$$

For $d=0$, we get $\mathsf{I}\Gamma(nT^\beta+b,a)$.

Now, the same machinery as proposed by Kundu (2007) for Type-I hybrid censoring can be applied (see Mokhtari et al., 2011; Lin et al., 2012 as well as Section 6.2.1). In particular, assuming that the prior π_2 of β is supposed log-concave, Gibbs sampling procedures can be utilized to compute Bayes estimates of β and ϑ. Clearly, the conditional density function of β given ϑ and the data is also log-concave (cf. Mokhtari et al., 2011). Additionally, Lin et al. (2012) discussed Lindley's as well as Tierney and Kadane's approximation.

Finally, it should be noted that the same approaches sketched above can be applied to other progressive hybrid censoring schemes using the ideas presented in Section 7.1.2. Therefore, we do not provide further details here.

Remark 7.19. Similar methods and approaches have been discussed for other progressive hybrid censoring schemes. Likelihood inference for Type-I generalized progressively hybrid censored data has been considered by Cho et al. (2015b) assuming two-parameter Weibull distributions as well as two-parameter Rayleigh distributions (see also Cho et al., 2015a; Shoaee and Khorram, 2016). As an alternative to maximum likelihood estimates, Zhu (2020) discussed least squares, weighted least squares, and maximum product of spacings estimates

7.3.2 Further distributions

Progressive hybrid censoring schemes have been applied to other distributions as well. Most of the results have been established for Type-I progressive hybrid censoring. The derivations are in the spirit of those presented in Section 6.3 for hybrid censoring schemes by taking into account the comments provided in Section 7.1.2. For details in this direction and further reading, we refer the reader to the following rather incomplete list of references.

exponential-type distributions: Tian et al. (2015), Tian et al. (2016), Panahi (2017), Goyal et al. (2020)
generalized exponential distribution: Hemmati and Khorram (2013), Sen et al. (2020)
Weibull-type distributions: Salem and Abo-Kasem (2011), Nassar and Abo-Kasem (2016b), Maswadah (2022)
extreme value distribution: Alma and Arabi Belaghi (2016)
gamma-type distributions: Elshahhat and Elemary (2021)
logistic type distributions: Azizpour and Asgharzadeh (2018)
Lindley-type distributions: Basu et al. (2018)
Burr-type distribution: Cho et al. (2015b), Kayal et al. (2017), Noori Asl et al. (2017), Singh et al. (2019), Kohansal (2020), Nagy et al. (2021)
normal distribution: Alma and Arabi Belaghi (2016)
lognormal distribution: Sen et al. (2018c)
Chen distribution: Kayal et al. (2022)
Kumaraswamy distributions: Kohansal and Nadarajah (2019), Sultana et al. (2020), Tu and Gui (2020)
Gompertz distribution: Mohie El-Din et al. (2019)
Maxwell distributions: Tomer and Panwar (2015)
bathtub-shaped distribution: Zhu (2021b), Liu and Gui (2021)

CHAPTER 8

Information measures

Contents

8.1. Introduction

In this chapter, we provide results for some information measures. The main focus is on Fisher information about parameters contained in a (progressive) hybrid censored sample (see Section 8.2). For completeness, we present some few results on entropy, Kullback-Leibler divergence, and Pitman closeness in the subsequent sections.

8.2. Fisher information

Before discussing Fisher information, we would like to point out that, in order to establish the subsequent representations, usually technical assumptions have to be imposed on the density functions and cumulative distribution functions ensuring that mathematical operations (like exchanging integration and differentiation) are allowed. Such regularity conditions can be found in many references discussing Fisher information. For the sake of brevity, we will not refer to specific prerequisites and conditions in the following, but only provide the corresponding results. For technical details, proofs, as well as elaborated collections of regularity conditions, we refer the reader to, e.g., Efron and Johnstone (1990), Lehmann and Casella (1998), Escobar and Meeker (2001), Lehmann and Romano (2005), Wang and He (2005), Balakrishnan et al. (2008c), Burkschat and Cramer (2012), Balakrishnan and Cramer (2014, Chapter 9) as well as to the original references provided.

Hybrid Censoring Know-How
https://doi.org/10.1016/B978-0-12-398387-9.00016-7

The (expected) Fisher information about a (single) parameter $\theta \in \Theta \subseteq \mathbb{R}$ contained in a random variable $X \sim F_\theta$ with density function f_θ is defined by

$$\mathcal{I}(X;\theta) = E_\theta\left[\left(\frac{\partial}{\partial\theta}\log f_\theta(X)\right)^2\right]. \tag{8.1}$$

This quantity measures the amount of information that a random variable X contains about an unknown parameter θ of a distribution serving as a model for X. For details about interpretation and statistical use, we refer to, e.g., Rao (1973) or Ly et al. (2017).

Under mild conditions (see, e.g., Lehmann and Casella, 1998, p. 116), an important alternative formula involving second partial derivatives of the log-likelihood is given by

$$\mathcal{I}(X;\theta) = -E_\theta\left[\frac{\partial^2}{\partial\theta^2}\log f_\theta(X)\right]. \tag{8.2}$$

From (8.1), the Fisher information in a sample of independent random variables $X_1, \ldots, X_n$ is given by

$$\mathcal{I}(\mathbf{X};\theta) = \mathcal{I}(X_1,\ldots,X_n;\theta) = \sum_{i=1}^{n}\mathcal{I}(X_i;\theta) \tag{8.3}$$

which equals $n\mathcal{I}(X_1;\theta)$ in case of identical distributions.

A useful representation of the Fisher information in terms of the hazard rate $h_\theta = f_\theta/(1-F_\theta)$ of the baseline distribution has been established by Efron and Johnstone (1990), that is, under certain regularity conditions one gets (see also Lehmann and Casella, 1998, p. 145)

$$\mathcal{I}(X;\theta) = E_\theta\left[\left(\frac{\partial}{\partial\theta}\log h_\theta(X)\right)^2\right].$$

In a location or scale family, the Fisher information about the parameter θ simplifies considerably in the sense that it could be directly computed from the Fisher information of the standard member F.

Remark 8.1. Given some regularity conditions, the following results hold:

(a) If F_θ, $\theta \in \mathbb{R}$, forms a location family, then $\mathcal{I}(X;\theta) = \mathcal{I}(X;0)$ showing that the Fisher information in θ does not depend on the location parameter;

(b) If F_θ, $\theta > 0$, forms a scale family, then $\mathcal{I}(X;\theta) = \theta^{-2}\mathcal{I}(X;1)$ so that the Fisher information in θ is proportional to θ^{-2}.

The result in Remark 8.1 is quite useful in deriving the Fisher information for particular distributions. For location or scale families with standard member F and corresponding hazard rate h, expression (8.18) also has a simpler form, that is, for a location family,

$$\mathscr{I}(X;0) = \int_{-\infty}^{\infty} \left\{ \frac{h'(x)}{h(x)} \right\}^2 f(x)\, dx. \tag{8.4}$$

For a scale family, the corresponding formula reads

$$\mathscr{I}(X;1) = \int_{-\infty}^{\infty} \left\{ 1 + x\frac{h'(x)}{h(x)} \right\}^2 f(x)\, dx. \tag{8.5}$$

For a family of cumulative distribution functions $F_{\boldsymbol{\theta}}$, $\boldsymbol{\theta} \in \Theta \subseteq \mathbb{R}^p$, having a multiple parameter $\boldsymbol{\theta} = (\theta_1, \ldots, \theta_p)$, the respective quantity is the so-called Fisher information matrix (for regularity conditions, see, e.g., Lehmann and Casella, 1998, p. 124). Let $[\![\cdot]\!]$ be defined by the matrix $[\![a]\!] = a \cdot a^T$, $a \in \mathbb{R}^p$, where a^T denotes the transposed vector of a. Moreover, let $\frac{\partial}{\partial \boldsymbol{\theta}} \log h_{\boldsymbol{\theta}} = (\frac{\partial}{\partial \theta_i} \log h_{\boldsymbol{\theta}})_{1 \le i \le p}$ and $\boldsymbol{\theta} = (\theta_1, \ldots, \theta_p)$, where $h_{\boldsymbol{\theta}} = f_{\boldsymbol{\theta}}/(1 - F_{\boldsymbol{\theta}})$ denotes the hazard rate of $F_{\boldsymbol{\theta}}$.

Given some regularity conditions, Efron and Johnstone (1990) pointed out that the Fisher information matrix about $F_{\boldsymbol{\theta}}$ in a single observation X can be expressed in terms of the hazard rate $h_{\boldsymbol{\theta}}$ of $F_{\boldsymbol{\theta}}$, i.e.,

$$\mathscr{I}(X;\boldsymbol{\theta}) = E_{\boldsymbol{\theta}}\left[\left[\!\left[\frac{\partial}{\partial \boldsymbol{\theta}} \log h_{\boldsymbol{\theta}}(X) \right]\!\right] \right],$$

i.e., for $1 \le i, j \le p$ and $\boldsymbol{\theta} \in \Theta$, the components are given by

$$\Big(\mathscr{I}(X;\boldsymbol{\theta})\Big)_{ij} = \mathscr{I}_{ij}(X;\boldsymbol{\theta}) = E_{\boldsymbol{\theta}}\left[\left\{ \frac{\partial}{\partial \theta_i} \log h_{\boldsymbol{\theta}}(X) \right\} \left\{ \frac{\partial}{\partial \theta_j} \log h_{\boldsymbol{\theta}}(X) \right\} \right].$$

Notice that the Fisher information matrix is symmetric so that, for a two-dimensional parameter, it is sufficient to provide expressions for $\mathscr{I}_{11}(X;\boldsymbol{\theta})$, $\mathscr{I}_{22}(X;\boldsymbol{\theta})$, and $\mathscr{I}_{12}(X;\boldsymbol{\theta})$.

As above, these expressions simplify in case of a location-scale family with location parameter μ and scale parameter ϑ. Given some regularity conditions, the Fisher information matrix in a location-scale family is given by

$$\mathscr{I}_{11}(X;\mu,\vartheta) = \vartheta^2 \cdot \mathscr{I}_{11}(X;0,1) = \vartheta^2 E\left[\left\{ \frac{h'(X)}{h(X)} \right\}^2 \right],$$

$$\mathscr{I}_{22}(X;\mu,\vartheta) = \frac{1}{\vartheta^2} \cdot \mathscr{I}_{22}(X;0,1) = \frac{1}{\vartheta^2} E\left[\left\{ 1 + X\frac{h'(X)}{h(X)} \right\}^2 \right],$$

$$\mathscr{I}_{12}(X;\mu,\vartheta) = \mathscr{I}_{12}(X;0,1) = -E\left[\frac{h'(X)}{h(X)} \left\{ 1 + X\frac{h'(X)}{h(X)} \right\} \right],$$

where $h = f/(1 - F)$ denotes the hazard rate of F and the expectation is calculated w.r.t. $f = f_{0,1}$.

In order to derive expressions for Fisher information in particular distributional models, the following observation is quite useful. Suppose F_θ is a cumulative distribution function from a one-parameter exponential family given by

$$F_\theta(x) = 1 - e^{-\eta(\theta)d(x)}, \quad x \in (\alpha, \omega), \tag{8.6}$$

with $-\infty \leq \alpha < \omega \leq \infty$, $d(\alpha+) = \lim_{x\to\alpha+} d(x) = 0$, $d(\omega-) = \lim_{x\to\omega-} d(x) = \infty$, where d is non-decreasing and differentiable, and η is positive and twice differentiable. Then, as pointed out Lehmann and Casella (1998, p. 115),

$$\mathscr{I}(X;\theta) = \mathscr{I}(X;\eta(\theta))[\eta'(\theta)]^2 \tag{8.7}$$

so that

$$\mathscr{I}(X;\theta) = \frac{[\eta'(\theta)]^2}{\eta^2(\theta)}.$$

In particular, we get for an *Exp*(ϑ)-distribution

$$\mathscr{I}(X;\vartheta) = \frac{1}{\vartheta^2}. \tag{8.8}$$

For a *Weibull*(ϑ^β, β)-distribution (with density function as in (B.1)), the Fisher information matrix in a single random variable is given by (cf. (8.21))

$$\mathscr{I}_{11}(X;\vartheta,\beta) = \frac{\beta^2}{\vartheta^2}, \quad \mathscr{I}_{22}(X;\vartheta,\beta) = \frac{\pi^2}{6\beta^2} + \frac{(\gamma-1)^2}{\beta^2}, \quad \mathscr{I}_{12}(X;\vartheta,\beta) = \frac{\gamma-1}{\vartheta}$$

where $\gamma = 0.57721\ldots$ denotes Euler-Mascheroni's constant (see Gupta and Kundu, 2006; Gradshteyn and Ryzhik, 2007, Section 8.36).

The above concepts will now be applied to ordered and censored data. Fisher information in order statistics and, more generally, in ordered and censored data has been considered by many authors (see, e.g., Mehrotra et al., 1979; Gertsbakh, 1995; Park, 1994, 1996, 2003; Zheng and Gastwirth, 2000; Nagaraja and Abo-Eleneen, 2003; David and Nagaraja, 2003; Park and Zheng, 2004; Wang and He, 2005; Zheng and Park, 2005; Arnold et al., 2008; Burkschat and Cramer, 2012). A review has been provided by Zheng et al. (2009).

It should be noted that many results have been established for Fisher information in this kind of data. For brevity, we will restrict ourselves to Fisher information in censored ordered data $X_{(1)} \leq \cdots \leq X_{(D_{\mathrm{HCS}})}$ where the observed sample size D_{HCS} is (non-)random and the random variables are the first D_{HCS} (progressively Type-II censored) order statistics. We start the discussion with Type-II (Section 8.2.1) and Type-I (Section 8.2.2) censored samples of $X_1, \ldots, X_n$ and the situation of a progressively Type-II censored sample (Section 8.2.3). Afterwards, Fisher information in presence of (progressive) hybrid censoring is discussed.

8.2.1 Fisher information in Type-II (right) censored samples

The Fisher information $\mathscr{I}_{1,\ldots,m:n}(\theta) = \mathscr{I}(X_{1:n}, \ldots, X_{m:n}; \theta)$ in a Type-II censored sample $X_{1:n}, \ldots, X_{m:n}$ can be obtained directly from (8.2) by considering the corresponding joint density function of the first m order statistics (see (2.3)), that is,

$$\mathscr{I}_{1,\ldots,m:n}(\theta) = -\sum_{i=1}^{m} E\left[\frac{\partial^2}{\partial\theta^2} \log f_\theta(X_{i:n})\right] - (n-m)E\left[\frac{\partial^2}{\partial\theta^2} \log(1 - F_\theta(X_{i:n}))\right].$$

According to Park (2003), an expression in terms of the baseline hazard rate holds (see also Zheng, 2001; Escobar and Meeker, 2001), that is,

$$\begin{aligned}\mathscr{I}_{1,\ldots,m:n}(\theta) &= \sum_{j=1}^{m} E_\theta\left[\left(\frac{\partial}{\partial\theta} \log h_\theta(X_{j:n})\right)^2\right] \\ &= \int_{-\infty}^{\infty} \left\{\frac{\partial}{\partial\theta} \log h_\theta(x)\right\}^2 \sum_{j=1}^{m} f_{j:n;\theta}(x)\, dx,\end{aligned} \tag{8.9}$$

where $f_{j:n;\theta}$, $1 \le j \le n$, denote the marginal density functions of order statistics based on f_θ.

The basic idea in the derivation of identity (8.9) is the Markov property and the corresponding product representation of the density function. For illustration, we sketch the proof here. Note that the same trick has been utilized in other models, too (e.g., for progressive Type-II censoring, Type-I (progressive) hybrid censoring). From Theorem 2.25, the joint density function of order statistics $X_{1:n}, \ldots, X_{m:n}$ can be written as

$$f_{1,\ldots,m:n}(\boldsymbol{t}_m) = f_{1:n}(t_1) \prod_{j=2}^{m} f_{j|j-1:n}(t_j \mid t_{j-1}). \tag{8.10}$$

The Fisher information in $X_{1:n}$ is given by (see Park, 1996)

$$\mathscr{I}_{1:n}(\theta) = E_\theta\left[\left(\frac{\partial}{\partial\theta} \log h_\theta(X_{1:n})\right)^2\right].$$

Furthermore, the average Fisher information in $X_{j:n}$ given $X_{j-1:n}$ is obtained as

$$\mathscr{I}_{j|j-1:n}(\theta) = E_\theta\Big(\mathscr{I}(X_{j:n}|X_{j-1:n}; \theta)\Big)$$

Since $X_{j:n}$ given $X_{j-1:n} = t_{j-1}$ has a left truncated cumulative distribution function $F(\cdot \mid t_{j-1}) = (F(\cdot) - F(t_{j-1}))/(1 - F(t_{j-1}))$ with hazard rate h on (t_{j-1}, ∞) (see Theorem 2.12), we find with Efron and Johnstone (1990) that

$$\mathscr{I}(X_{j:n}|X_{j-1:n}=t_{j-1};\theta)=E_\theta\left[\left(\frac{\partial}{\partial\theta}\log h_\theta(X_{j:n})\right)^2\Big|X_{j-1:n}=t_{j-1}\right].$$

Thus, we finally arrive at

$$\mathscr{I}_{j|j-1:n}(\theta)=E_\theta\left[\left(\frac{\partial}{\partial\theta}\log h_\theta(X_{j:n})\right)^2\right].$$

Furthermore, Park (2003) showed the identities

$$\mathscr{I}_{1,\ldots,m:n}(\theta)=n\int_{-\infty}^{\infty}\left\{\frac{\partial}{\partial\theta}\log h_\theta(x)\right\}^2(1-F_{m:n-1;\theta}(x))f_\theta(x)\,dx \tag{8.11}$$

and

$$\mathscr{I}_{1,\ldots,m:n}(\theta)=E\left[\int_{-\infty}^{X_{m:n-1}}\left\{\frac{\partial}{\partial\theta}\log h_\theta(x)\right\}^2 f_\theta(x)\,dx\right].$$

Expressions for Fisher information given particular distributions or families of distributions (like location-scale families) can be directly taken from the results on progressively Type-II censored order statistics choosing $R_i=0$ and $\gamma_i=n-i+1$, $1\le i\le m$. Therefore, for brevity, we do not present details here and refer to Section 8.2.3. We provide only the Fisher information for exponential and Weibull distributions, respectively.

For an exponential distribution $Exp(\vartheta)$, the Fisher information in the first m order statistics is given by

$$\mathscr{I}_{1,\ldots,m:n}(\theta)=\frac{m}{\vartheta^2}. \tag{8.12}$$

Expression (8.20) is useful to establish a representation for the Fisher information included in the first m order statistics $X_{1:n},\ldots,X_{m:n}$ from an *Weibull*$(1,\beta)$-distribution. After some rearrangements, this yields the following expression ($1\le m\le n-1$)

$$\mathscr{I}_{1,\ldots,m:n}(\beta)=\frac{m\pi^2}{6\beta^2}+\frac{1}{\beta^2}\sum_{i=1}^{m}(-1)^{m+i}\binom{n}{i-1}\binom{n-i-1}{n-m-1}(\gamma-1+\log(n-i+1))^2. \tag{8.13}$$

For a two-parameter *Weibull*(ϑ^β, β)-distribution with density function as in (B.1), one gets the Fisher information matrix

$$\begin{aligned}
\mathscr{I}_{11;1,\ldots,m:n}(\vartheta,\beta) &= m\frac{\beta^2}{\vartheta^2},\\
\mathscr{I}_{22;1,\ldots,m:n}(\vartheta,\beta) &= \frac{1}{\beta^2}\mathscr{I}_{1,\ldots,m:m:n}(1) \quad \text{(see (8.13))},\\
\mathscr{I}_{12;1,\ldots,m:n}(\vartheta,\beta) &= \frac{(\gamma-1)m}{\vartheta} + \frac{1}{\vartheta}\sum_{i=1}^{m}(-1)^{m+i}\binom{n}{i-1}\binom{n-i-1}{n-m-1}\log(n-i+1).
\end{aligned} \tag{8.14}$$

8.2.2 Fisher information in Type-I censored random variables

Park (2003) has established a representation of the Fisher information in single Type-I right censored observation X at the threshold T. With the corresponding indicator $D = \mathbb{1}_{(-\infty,T]}(X)$, one gets

$$\mathscr{I}(X \wedge T, D; \theta) = E_\theta\left[\left(\frac{\partial}{\partial\theta}\log h_\theta(X)\right)^2 \mathbb{1}_{(-\infty,T]}(X)\right] = \int_{-\infty}^{T}\left\{\frac{\partial}{\partial\theta}\log h_\theta(x)\right\}^2 f_\theta(x)\,dx$$

(see also Gertsbakh, 1995; Gupta et al., 2004; Wang and He, 2005). For a Type-I censored sample $X_1, \ldots, X_n$ with time censoring at T and $D = \sum_{j=1}^{n}\mathbb{1}_{(-\infty,T]}(X_j)$, the Fisher information[1]

$$\mathscr{I}_{D\wedge n}(\theta) = \mathscr{I}(X_{1:n}, \ldots, X_{D:n}, D; \theta)$$

can be obtained directly from (8.3) leading to

$$\mathscr{I}_{D\wedge n}(\theta) = n\int_{-\infty}^{T}\left\{\frac{\partial}{\partial\theta}\log h_\theta(x)\right\}^2 f_\theta(x)\,dx$$

and

$$\begin{aligned}
\mathscr{I}_{D\wedge n}(\theta) &= nE\left[\left\{\frac{\partial}{\partial\theta}\log h_\theta(X_1)\right\}^2 \mathbb{1}_{(-\infty,T]}(X_1)\right]\\
&= \sum_{j=1}^{n} E\left[\left\{\frac{\partial}{\partial\theta}\log h_\theta(X_{j:n})\right\}^2 \mathbb{1}_{(-\infty,T]}(X_{j:n})\right].
\end{aligned} \tag{8.15}$$

[1] The observed sample size in a Type-I (or hybrid) censored sample is a function of the random counter D and the parameters of the hybrid censoring scheme. In the following, we use this particular function as a subscript to denote the corresponding Fisher information. For instance, we write the subscript $D \wedge n$ for Type-I censoring, $D_{\text{I}} = D \wedge m$ for Type-I hybrid censoring, and $D_{\text{II}} = D \vee m$ for Type-II hybrid censoring. This allows both a compact and intuitive notation for the Fisher information in such samples.

In case of an exponential distribution $Exp(\vartheta)$, we have $h_\vartheta(x) = 1/\vartheta$ so that the corresponding Fisher information is given by

$$\mathscr{I}_{D\wedge n}(\theta) = \frac{n}{\vartheta^2}F\Big(\frac{T}{\vartheta}\Big) = \frac{1}{\vartheta^2}ED, \tag{8.16}$$

where F is the cumulative distribution function of a standard exponential distribution. For the Weibull distribution, one gets (see Gupta and Kundu, 2006; Park et al., 2008)

$$\begin{aligned}
\mathscr{I}_{11;D\wedge n}(\vartheta,\beta) &= \frac{n\beta^2}{\vartheta^2}\Big(1-e^{-(T/\vartheta)^\beta}\Big),\\
\mathscr{I}_{22;D\wedge n}(\vartheta,\beta) &= \frac{n}{\beta^2}\int_0^{(T/\vartheta)^\beta}(1+\log y)^2e^{-y}\,dy,\\
\mathscr{I}_{12;D\wedge n}(\vartheta,\beta) &= -\frac{n}{\vartheta}\int_0^{(T/\vartheta)^\beta}(1+\log y)e^{-y}\,dy.
\end{aligned} \tag{8.17}$$

8.2.3 Fisher information under progressive Type-II censoring

Results on Fisher information $\mathscr{I}^{\mathscr{R}}_{1,\ldots,m:n}(\theta)$ in progressively Type-II censored samples have been summarized in Balakrishnan and Cramer (2014, Chapter 9). In the following section, we will present a selection of results involved in the discussion of progressive hybrid censoring. From the definition of Fisher information and the joint density function of a progressively Type-II censored sample $\mathbf{X}^{\mathscr{R}} \sim F_\theta^{\mathbf{X}^{\mathscr{R}}}$, the Fisher information about the parameter θ can be written as

$$\mathscr{I}^{\mathscr{R}}_{1,\ldots,m:m:n}(\theta) = \mathscr{I}(\mathbf{X}^{\mathscr{R}};\theta) = \sum_{i=1}^{m} E\Big[-\frac{\partial^2}{\partial\theta^2}\log f_\theta(X_{i:m:n}) - R_i\frac{\partial^2}{\partial\theta^2}\log(1-F_\theta(X_{i:m:n}))\Big]$$

provided that all derivatives as well as the expected values exist. Under some regularity conditions, the Fisher information about a single parameter θ in a progressively Type-II censored sample $\mathbf{X}^{\mathscr{R}}$ is given by

$$\mathscr{I}^{\mathscr{R}}_{1,\ldots,m:m:n}(\theta) = E\Big[-\frac{\partial^2}{\partial\theta^2}\log f_\theta^{\mathscr{R}}(\mathbf{X}^{\mathscr{R}})\Big],$$

where $f_\theta^{\mathscr{R}}$ denotes the density function of $\mathbf{X}^{\mathscr{R}}$ (see, for example, Lehmann and Casella, 1998). The expected Fisher information in a progressively Type-II censored data has been considered in Zheng and Park (2004) who obtained an expression of the Fisher information in terms of the hazard rate function which is similar to a representation for non-censored data established by Efron and Johnstone (1990). More information as well as details on this representation and regularity conditions are provided by Burkschat and Cramer (2012). Throughout this section, let $f_{j:m:n;\theta}$, $1 \le j \le m$, denote the marginal density functions of progressively Type-II censored order statistics based on f_θ.

A proof of the following theorem can be found in Zheng and Park (2004) and Balakrishnan and Cramer (2014, Chapter 9).

Theorem 8.2. *Under some regularity conditions, the Fisher information about a single parameter θ in $\mathbf{X}^{\mathscr{R}}$ is given by*

$$\mathscr{I}^{\mathscr{R}}_{1,\ldots,m:m:n}(\theta) = \int_{-\infty}^{\infty} \left\{ \frac{\partial}{\partial\theta} \log h_\theta(x) \right\}^2 \sum_{j=1}^{m} f_{j:m:n;\theta}(x)\, dx, \tag{8.18}$$

where $h_\theta = f_\theta/(1-F_\theta)$ denotes the hazard rate of the cumulative distribution function F_θ.

The Fisher information in a sample of progressively Type-II right censored order statistics can be obtained in the same way by replacing m by $r \le m$, that is,

$$\mathscr{I}^{\mathscr{R}}_{1\ldots r:m:n}(\theta) = \int_{-\infty}^{\infty} \left\{ \frac{\partial}{\partial\theta} \log h_\theta(x) \right\}^2 \sum_{j=1}^{r} f_{j:m:n;\theta}(x)\, dx.$$

In a location or scale family, the Fisher information about the parameter θ simplifies considerably in the sense that it could be directly computed from the Fisher information of the standard member F.

Remark 8.3. By similarity with Remark 8.1, the following results hold:

(a) If F_θ, $\theta \in \mathbb{R}$, forms a location family, then $\mathscr{I}^{\mathscr{R}}_{1,\ldots,m:m:n}(\theta) = \mathscr{I}^{\mathscr{R}}_{1,\ldots,m:m:n}(0)$ showing that the Fisher information in θ does not depend on the location parameter;

(b) If F_θ, $\theta > 0$, forms a scale family, then $\mathscr{I}^{\mathscr{R}}_{1,\ldots,m:m:n}(\theta) = \theta^{-2}\mathscr{I}^{\mathscr{R}}_{1,\ldots,m:m:n}(1)$ so that the Fisher information in θ is proportional to θ^{-2}.

By similarity with the derivations after Remark 8.1, the result in Remark 8.3 yields the following expressions for location or scale families with standard member F and corresponding hazard rate h (cf. (8.4) and (8.5)): for a location family, one has

$$\mathscr{I}^{\mathscr{R}}_{1,\ldots,m:m:n}(0) = \int_{-\infty}^{\infty} \left\{ \frac{h'(x)}{h(x)} \right\}^2 \sum_{j=1}^{m} f_{j:m:n}(x)\, dx.$$

For a scale family, the corresponding formula reads

$$\mathscr{I}^{\mathscr{R}}_{1,\ldots,m:m:n}(1) = \int_{-\infty}^{\infty} \left\{ 1 + x\frac{h'(x)}{h(x)} \right\}^2 \sum_{j=1}^{m} f_{j:m:n}(x)\, dx.$$

For a family of cumulative distribution functions $F_{\boldsymbol{\theta}}$, $\boldsymbol{\theta} \in \Theta \subseteq \mathbb{R}^p$, having a multiple parameter $\boldsymbol{\theta} = (\theta_1, \ldots, \theta_p)$, Fisher information in progressively Type-II censored order statistics has been addressed first by Ng et al. (2004) for Weibull, log-normal, extreme value, and normal distributions, respectively. Beside computing it directly from the definition of the Fisher information matrix, they employed the missing information principle to calculate the Fisher information in a progressively Type-II censored sample (see Section 8.2.3.1). Dahmen et al. (2012) established a multi-parameter version of the hazard rate representation of the Fisher information matrix. For details on regularity conditions, we refer to Dahmen et al. (2012), Burkschat and Cramer (2012), and Balakrishnan and Cramer (2014, p. 204).

Theorem 8.4. *Let $[\![\cdot]\!]$ be defined by the matrix $[\![a]\!] = a \cdot a^T$, $a \in \mathbb{R}^p$, where a^T denotes the transposed vector of a. Moreover, let $\frac{\partial}{\partial \boldsymbol{\theta}} \log h_{\boldsymbol{\theta}} = (\frac{\partial}{\partial \theta_i} \log h_{\boldsymbol{\theta}})_{1 \le i \le p}$ and $\boldsymbol{\theta} = (\theta_1, \ldots, \theta_p)$, where $h_{\boldsymbol{\theta}} = f_{\boldsymbol{\theta}}/(1 - F_{\boldsymbol{\theta}})$ denotes the hazard rate of $F_{\boldsymbol{\theta}}$.*

Given some regularity conditions, the Fisher information matrix about $F_{\boldsymbol{\theta}}$ in the sample $\mathbf{X}^{\mathscr{R}}$ can be expressed in terms of the hazard rate $h_{\boldsymbol{\theta}}$ of $F_{\boldsymbol{\theta}}$, i.e.,

$$\mathscr{I}^{\mathscr{R}}_{1,\ldots,m:m:n}(\boldsymbol{\theta}) = \int_{-\infty}^{\infty} \left[\!\!\left[\frac{\partial}{\partial \boldsymbol{\theta}} \log h_{\boldsymbol{\theta}}(x) \right]\!\!\right] \sum_{s=1}^{m} f_{s:m:n;\boldsymbol{\theta}}(x)\, dx,$$

i.e., for $1 \le i, j \le p$ and $\boldsymbol{\theta} \in \Theta$, the components are given by

$$\left(\mathscr{I}^{\mathscr{R}}_{1,\ldots,m:m:n}(\boldsymbol{\theta})\right)_{ij} = \mathscr{I}_{ij}(\mathbf{X}^{\mathscr{R}}; \boldsymbol{\theta}) = \int_{-\infty}^{\infty} \left\{ \frac{\partial}{\partial \theta_i} \log h_{\boldsymbol{\theta}}(x) \right\} \left\{ \frac{\partial}{\partial \theta_j} \log h_{\boldsymbol{\theta}}(x) \right\} \sum_{s=1}^{m} f_{s:m:n;\boldsymbol{\theta}}(x)\, dx.$$

Remark 8.5. Using the hazard representation of Fisher information in progressively Type-II censored order statistics, Abo-Eleneen (2007) established a representation of Fisher information in terms of Fisher information of minima. In particular, using the density function of the cumulative distribution function given in (2.51), this yields

$$\mathscr{I}^{\mathscr{R}}_{1,\ldots,m:m:n}(\theta) = \sum_{r=1}^{m} c_{r-1} \sum_{j=1}^{r} \frac{1}{\gamma_j} a_{j,r} \mathscr{I}_{1:\gamma_j}(\theta).$$

This representation is quite useful for computational purposes, particularly, when numerical integration is necessary like in the case of the normal distribution. In order to compute the Fisher information in a sample $\mathbf{X}^{\mathscr{R}}$, we have to compute only the Fisher information $\mathscr{I}_{1:\gamma_j}(\theta)$ in the minima $X_{1:\gamma_j}$, $1 \le j \le m$. An alternative representation has been established in Balakrishnan and Cramer (2014):

$$\mathscr{I}^{\mathscr{R}}_{1,\ldots,m:m:n}(\boldsymbol{\theta}) = \sum_{j=1}^{m} \mathscr{I}_{1:\gamma_j}(\boldsymbol{\theta}) \left(\sum_{r=j}^{m} \prod_{i=1, i \ne j}^{r} \frac{\gamma_i}{\gamma_i - \gamma_j} \right).$$

Notice that the weights in this representation are related to minimal signatures of progressively Type-II censored order statistics (see Cramer and Navarro, 2015, 2016).

In the multi-parameter case, similar representations hold also in terms of the Fisher information matrices $\mathscr{I}_{1:\gamma_j}(\boldsymbol{\theta})$ of minima.

Dahmen et al. (2012) established an expression for the Fisher information matrix in location-scale families. Let $f_{s:m:n} = f_{s:m:n;0,1}$ be the density function of a progressively Type-II censored order statistic $X_{s:m:n}$ from the standard member F. Then, the following representation holds. It follows directly from the identity $f_{s:m:n;\mu,\vartheta}(x) = \frac{1}{\vartheta} \cdot f_{s:m:n}\big((x - \mu)/\vartheta\big)$, $x \in \mathbb{R}$, $1 \le s \le m$, and Theorem 8.4 (see also (8.4) and (8.5)).

Theorem 8.6. *Given some regularity conditions, the Fisher information matrix in a location-scale family is given by*

$$\mathscr{I}_{11}(\mathbf{X}^{\mathscr{R}}; \mu, \vartheta) = \vartheta^2 \cdot \mathscr{I}_{11}(\mathbf{X}^{\mathscr{R}}; 0, 1) = \vartheta^2 \int_{-\infty}^{\infty} \left\{ \frac{h'(x)}{h(x)} \right\}^2 g(x)\, dx,$$

$$\mathscr{I}_{22}(\mathbf{X}^{\mathscr{R}}; \mu, \vartheta) = \frac{1}{\vartheta^2} \cdot \mathscr{I}_{22}(\mathbf{X}^{\mathscr{R}}; 0, 1) = \frac{1}{\vartheta^2} \int_{-\infty}^{\infty} \left\{ 1 + x \frac{h'(x)}{h(x)} \right\}^2 g(x)\, dx,$$

$$\mathscr{I}_{12}(\mathbf{X}^{\mathscr{R}}; \mu, \vartheta) = \mathscr{I}_{12}(\mathbf{X}^{\mathscr{R}}; 0, 1) = -\int_{-\infty}^{\infty} \frac{h'(x)}{h(x)} \left\{ 1 + x \frac{h'(x)}{h(x)} \right\} g(x)\, dx,$$

where $g = \sum_{s=1}^{m} f_{s:m:n}$ *and* $h = f/(1 - F)$ *denotes the hazard rate of* F.

8.2.3.1 Fisher information via missing information principle

As an alternative to the above approaches, Ng et al. (2002) proposed the missing information principle (see Louis, 1982; Tanner, 1993) to calculate expected Fisher information. They interpreted the progressively Type-II censored order statistics $\mathbf{X}^{\mathscr{R}} = (X_{1:m:n}, \ldots, X_{m:m:n})$ as observed information and the random vector $\boldsymbol{W} = (\boldsymbol{W}_1, \ldots, \boldsymbol{W}_m)$ of progressively censored random variables with $\boldsymbol{W}_j = (W_{j1}, \ldots, W_{jR_j})$ denoting the random variables corresponding to units withdrawn in the j-th step of the progressive censoring procedure as missing data. Thus, merging $\mathbf{X}^{\mathscr{R}}$ and $\boldsymbol{W}$ yields the complete data $\mathbf{X} = (\mathbf{X}^{\mathscr{R}}, \boldsymbol{W})$. Ng et al. (2002) established the conditional density function of $\boldsymbol{W}_j$, given $\mathbf{X}^{\mathscr{R}}$. For a proof, see also Balakrishnan and Cramer (2014, p. 207).

Theorem 8.7. *Given* $\mathbf{X}_j^{\mathscr{R}} = \boldsymbol{x}_j = (x_1, \ldots, x_j)$, *the conditional density function of* W_{jk}, $k \in \{1, \ldots, R_j\}$, *is given by*

$$f^{W_{jk}|\mathbf{X}_j^{\mathscr{R}}}(w \mid \boldsymbol{x}_j) = f^{W_{jk}|X_{j:m:n}}(w|x_j) = \frac{f(w)}{1 - F(x_j)}, \qquad w > x_j,$$

and W_{jk} *and* $W_{j\ell}$, $k \neq \ell$, *are conditionally independent given* $X_{j:m:n} = x_j$.

Representing the observed, complete, and missing information by $\mathscr{I}(\mathbf{X}^{\mathscr{R}}; \theta)$, $\mathscr{I}((\mathbf{X}^{\mathscr{R}}, \boldsymbol{W}); \theta)$, and $\mathscr{I}(\boldsymbol{W}|\mathbf{X}^{\mathscr{R}}; \theta)$, respectively, these quantities can be related by the missing information principle. The complete information is given by

$$\mathscr{I}((\mathbf{X}^{\mathscr{R}}, \boldsymbol{W}); \theta) = \mathscr{I}(\mathbf{X}; \theta) = n\mathscr{I}(X_1; \theta).$$

Defining the Fisher information about θ in a random variable W progressively censored at the time of the j-th failure, given $X_{j:m:n}$, by

$$\mathcal{I}^{(j)}(W|X_{j:m:n};\theta) = -E\Big[\frac{\partial^2}{\partial\theta^2}\log f_\theta^{W|X_{j:m:n}}(W \mid X_{j:m:n})\Big],$$

we get the expected Fisher information of $\boldsymbol{W}$, given $\mathbf{X}^{\mathscr{R}}$, as

$$\mathcal{I}(\boldsymbol{W}|\mathbf{X}^{\mathscr{R}};\theta) = \sum_{j=1}^{m} R_j \mathcal{I}^{(j)}(\boldsymbol{W}_j|X_{j:m:n};\theta).$$

Hence, the expected Fisher information can be written as

$$\mathcal{I}^{\mathscr{R}}_{1,\dots,m:m:n}(\theta) = \mathcal{I}(\mathbf{X};\theta) - \sum_{j=1}^{m} R_j \mathcal{I}^{(j)}(\boldsymbol{W}_j|X_{j:m:n};\theta).$$

This relation is utilized in the computation of maximum likelihood estimates via the EM-algorithm.

8.2.3.2 Fisher information for particular distributions

(a) For a one-parameter exponential family as in (8.6), we get by using (8.7)

$$\mathcal{I}^{\mathscr{R}}_{1,\dots,m:m:n}(\theta) = \frac{m[\eta'(\theta)]^2}{\eta^2(\theta)}$$

(see Balakrishnan and Cramer, 2014, Chapter 9) so that, for all cumulative distribution functions F_θ of the form (8.6), the Fisher information in the progressively censored sample $\mathbf{X}^{\mathscr{R}}$ does not depend on the censoring plan $\mathscr{R}$. In particular, we get for an $Exp(\vartheta)$-distribution (cf. (8.8))

$$\mathcal{I}^{\mathscr{R}}_{1,\dots,m:m:n}(\vartheta) = \frac{m}{\vartheta^2}. \tag{8.19}$$

(b) Using (8.7), the Fisher information in the shape parameter of a Weibull distribution and that in the scale parameter of an extreme value distribution (Type-I) are related. Suppose $\mathbf{X}^{\mathscr{R}}$ forms a progressively Type-II censored sample from a *Weibull*$(1,\beta)$-distribution. Then, $Y_{j:m:n} = \log X_{j:m:n}$, $1 \le j \le m$, is a progressively Type-II censored sample from an extreme value distribution with scale parameter β and cumulative distribution function $F_\beta(t) = 1 - e^{-e^{\beta t}}$, $t \in \mathbb{R}$ (see Balakrishnan et al., 2008c; Dahmen et al., 2012). Therefore, it is sufficient to calculate the Fisher information for the extreme value distribution. Using the results in Theorem 8.6, this yields for a *Weibull*$(1, \beta)$-distribution (with density function as in (B.1))

$$\mathscr{I}^{\mathscr{R}}_{1,\ldots,m:m:n}(\beta) = \beta^{-2}\mathscr{I}^{\mathscr{R}}_{1,\ldots,m:m:n}(1) = \frac{m\pi^2}{6\beta^2} + \frac{1}{\beta^2}\sum_{s=1}^{m} c_{s-1}\sum_{i=1}^{s}\frac{a_{i,s}}{\gamma_i}(\gamma - 1 + \log\gamma_i)^2 \tag{8.20}$$

(see also Cramer and Ensenbach, 2011). Different expressions are given in Balakrishnan et al. (2008c) and Ng et al. (2004) (see also Burkschat and Cramer, 2012, for expressions in terms of generalized order statistics). For a two-parameter *Weibull*(ϑ^β, β)-distribution (with density function as in (B.1)) one gets the Fisher information matrix

$$\begin{aligned}
\mathscr{I}_{11}(\mathbf{X}^{\mathscr{R}}; \vartheta, \beta) &= \beta^2\mathscr{I}_{11}(\mathbf{X}^{\mathscr{R}}; \vartheta, 1) = m\frac{\beta^2}{\vartheta^2},\\
\mathscr{I}_{22}(\mathbf{X}^{\mathscr{R}}; \vartheta, \beta) &= \mathscr{I}^{\mathscr{R}}_{1,\ldots,m:m:n}(\beta) = \frac{1}{\beta^2}\mathscr{I}^{\mathscr{R}}_{1,\ldots,m:m:n}(1) \quad \text{(see (8.20))},\\
\mathscr{I}_{12}(\mathbf{X}^{\mathscr{R}}; \vartheta, \beta) &= \frac{1}{\vartheta}\mathscr{I}_{12}(\mathbf{X}^{\mathscr{R}}; 0, 1)\\
&= \frac{(\gamma - 1)m}{\vartheta} + \frac{1}{\vartheta}\sum_{s=1}^{m} c_{s-1}\sum_{i=1}^{s}\frac{a_{i,s}}{\gamma_i}\log\gamma_i.
\end{aligned} \tag{8.21}$$

Escobar and Meeker (1986) suggested an algorithm to compute the Fisher information in the Weibull case which can also be applied in the progressive censoring setting. The extreme value distribution of Type-II (see Definition B.14) is considered in Cramer and Davies (2018).

(c) Since the scale Laplace distribution forms an exponential family as in (8.6), the Fisher information is constant as in (8.19). For the location parameter, Burkschat and Cramer (2012) established the following representation of the Fisher information

$$\mathscr{I}^{\mathscr{R}}_{1,\ldots,m:m:n}(0) = \sum_{r=1}^{m} c_{r-1}\sum_{j=1}^{r} a_{j,r}\,\kappa(\gamma_j - 2),$$

where

$$\kappa(d) = \begin{cases}\frac{1}{d}\left(1 - \left(\frac{1}{2}\right)^d\right), & d \neq 0,\\ \log(2), & d = 0.\end{cases}$$

(d) For a location family of logistic distributions with location parameter $\mu \in \mathbb{R}$ given by

$$f_\mu(x) = \frac{e^{-(x-\mu)}}{(1 + e^{-(x-\mu)})^2}, \quad x \in \mathbb{R},$$

Burkschat and Cramer (2012) established the following explicit representation:

$$\mathscr{I}^{\mathscr{R}}_{1,\ldots,m:m:n}(0) = \sum_{r=1}^{n}\prod_{j=1}^{r}\frac{\gamma_j}{\gamma_j+2}. \tag{8.22}$$

(e) For normal distributions, explicit representations are not available. Computational results for selected censoring plans are provided by Balakrishnan et al. (2008c) and Abo-Eleneen (2007, 2008).

(f) Dahmen et al. (2012) established explicit expressions for the Fisher information in a two-parameter family of Lomax distributions with cumulative distribution function

$$F_{q,\vartheta}(x) = 1-(1+\vartheta x)^{-q}, \quad x>0,$$

and parameters $q, \vartheta > 0$. They found that

$$\mathscr{I}_{11}(\mathbf{X}^{\mathscr{R}}; q,\vartheta) = \frac{m}{q^2},$$

$$\mathscr{I}_{22}(\mathbf{X}^{\mathscr{R}}; q,\vartheta) = \frac{1}{\vartheta^2}\sum_{s=1}^{m}\prod_{j=1}^{s}\left(1-\frac{2}{q\gamma_j+2}\right),$$

$$\mathscr{I}_{12}(\mathbf{X}^{\mathscr{R}}; q,\vartheta) = \frac{1}{q\vartheta}\sum_{s=1}^{m}\prod_{j=1}^{s}\left(1-\frac{1}{q\gamma_j+1}\right).$$

Notice that, for $q=1$, $\mathscr{I}_{22}(\mathbf{X}^{\mathscr{R}}; 1,\vartheta)$ equals the expression of the Fisher information for the logistic distribution as given in (8.22).

8.2.4 Fisher information under hybrid censoring schemes

8.2.4.1 Type-I hybrid censoring

For a Type-I hybrid censored sample $X_{1:n},\ldots,X_{D_{\text{I}}:n}$ with time censoring at T and $D=\sum_{j=1}^{n}\mathbb{1}_{(-\infty,T]}(X_{j:n})$ as in (4.2), Wang and He (2005) and Park et al. (2008) established representations of the Fisher information

$$\mathscr{I}_{D\wedge m}(\theta) = \mathscr{I}(X_{1:n},\ldots,X_{D_{\text{I}}:n}, D_{\text{I}};\theta)$$

similar to those in (8.9) and (8.11)obtained under Type-II censoring (for Type-I censoring, see (8.15)), that is,

$$\mathscr{I}_{D\wedge m}(\theta) = n\int_{-\infty}^{T}\left\{\frac{\partial}{\partial\theta}\log h_\theta(x)\right\}^2(1-F_{m:n-1}(x))f_\theta(x)\,dx$$

and

$$\begin{aligned}\mathscr{I}_{D\wedge m}(\theta) &= \int_{-\infty}^{T}\left\{\frac{\partial}{\partial\theta}\log h_\theta(x)\right\}^2\sum_{j=1}^{m} f_{j:n;\theta}(x)\,dx\\ &= \sum_{j=1}^{m} E\left[\left\{\frac{\partial}{\partial\theta}\log h_\theta(X_{j:n})\right\}^2 \mathbb{1}_{(-\infty,T]}(X_{j:n})\right],\end{aligned} \tag{8.23}$$

which reduces to (8.15) when $m = n$ holds. Thus, in comparison to the Fisher information in order statistics $X_{1:n}, \dots, X_{m:n}$, the integration area has been bounded by the threshold T. These expressions have been utilized to establish results for particular distributions.

For an $Exp(\vartheta)$-distribution, the hazard rate $h_\vartheta(x) = 1/\vartheta$ and (8.23) yield directly the following expression for the Fisher information

$$\mathscr{I}_{D\wedge m}(\vartheta) = \frac{1}{\vartheta^2}\sum_{i=1}^{m} F_{i:n}\Big(\frac{T}{\vartheta}\Big) \tag{8.24}$$

where $F_{i:n}$ is the cumulative distribution function of the i-th order statistic from a standard exponential distribution. Notice that, according to (4.11),

$$\mathscr{I}_{D\wedge m}(\vartheta) = \frac{1}{\vartheta^2}E(D\wedge m) = \frac{1}{\vartheta^2}ED_{\mathsf{I}}$$

that is, the Fisher information can be written as the product of the Fisher information in a single exponential random variable, that is, $1/\vartheta^2$ (see (8.19)), and the expected value of the random sample size D_{I}.

In case of the Weibull distribution, Park et al. (2008) presented the following results:

$$\begin{aligned}\mathscr{I}_{11;D\wedge m}(\vartheta,\beta) &= \frac{\beta^2}{\vartheta^2}\sum_{i=1}^{m} F_{i:n}\Big((T/\vartheta)^\beta\Big),\\ \mathscr{I}_{22;D\wedge m}(\vartheta,\beta) &= \frac{1}{\beta^2}\sum_{i=1}^{m}\int_0^{(T/\vartheta)^\beta}(1+\log y)^2 f_{i:n}(y)\,dy,\\ \mathscr{I}_{12;D\wedge m}(\vartheta,\beta) &= -\frac{1}{\vartheta}\sum_{i=1}^{m}\int_0^{(T/\vartheta)^\beta}(1+\log y) f_{i:n}(y)\,dy.\end{aligned} \tag{8.25}$$

$f_{i:n}$ and $F_{i:n}$ denote the density function and cumulative distribution function of the i-th order statistic from an $Exp(1)$-distribution.

Remark 8.8. In the following, it is useful to write the Fisher information in a (progressive) hybrid censored sample in terms of the values of the respective counter D_{HCS}.

Using the representation of the density function of $X_{1:n}, \ldots, X_{D_{\mathrm{I}}:n}$ and D (cf. (4.18)), that is

$$\phi_{d;\theta}(t_1, \ldots, t_n) = f_\theta^{X_{j:n}, 1\le j\le D_{\mathrm{I}}, D}(\boldsymbol{t}_d, d) = \begin{cases} \frac{1-F(T)}{(n-d)f(T)} f_{1,\ldots,d+1:n;\theta}(\boldsymbol{t}_d, T), & d \in \{1, \ldots, m-1\} \\ f_{1,\ldots,m:n;\theta}(\boldsymbol{t}_m), & d \in \{m, \ldots, n\} \end{cases},$$

we find

$$\begin{aligned} \mathscr{I}_{D\wedge m}(\theta) = -E_\theta\left[\frac{\partial^2}{\partial\theta^2}\log f_{1,\ldots,m:n;\theta}(X_{1:n}, \ldots, X_{m:n})\mathbb{1}_{\{m,\ldots,n\}}(D)\right] \\ -\sum_{d=0}^{m-1} E_\theta\left[\frac{\partial^2}{\partial\theta^2}\log\phi_{d;\theta}(X_{1:n}, \ldots, X_{n:n})\mathbb{1}_{\{d\}}(D)\right] \end{aligned} \quad (8.26)$$

For $m = n$, the case of a Type-I censored sample is covered in this expression.

8.2.4.2 *Type-II hybrid censoring and a general account to Fisher information in hybrid censoring schemes*

As pointed out in Chapter 4, the likelihood functions in hybrid censoring models are composed of only two types ⓘ and ⓘⓘ which correspond to a Type-I and a Type-II setting, respectively. A similar idea can be used to establish a general account to decompose the Fisher information into basic modules (as it has been done before). We illustrate this idea which apparently has first been utilized in Park and Balakrishnan (2009) for Type-II progressive hybrid censoring and subsequently been applied in many settings (see, e.g., Sen et al., 2018a). Notice that we use a different approach which can be easily extended to progressive hybrid censoring schemes. We illustrate this concept for Type-II hybrid censoring first and apply it to other hybrid censoring schemes in the next section.

In case of Type-II hybrid censoring, the joint density function of $X_{1:n}, \ldots, X_{D_{\mathrm{II}}:n}$ and D (cf. (4.30)) can be written as

$$\begin{aligned} \phi_{d;\theta}(t_1, \ldots, t_n) &= f^{X_{j:n}, 1\le j\le D_{\mathrm{II}}, D}(\boldsymbol{t}_{d\vee m}, d) \\ &= \begin{cases} f_{1,\ldots,m:n;\theta}(\boldsymbol{t}_m), & d < m \\ \frac{1-F_\theta(T)}{(n-d)f_\theta(T)} f_{1,\ldots,d+1:n;\theta}(\boldsymbol{t}_d, T), & d \in \{m, \ldots, n-1\} \\ f_{1,\ldots,n:n;\theta}(\boldsymbol{t}_n), & d = n \end{cases}. \end{aligned} \quad (8.27)$$

Thus, using (8.2), we can write the Fisher information

$$\mathscr{I}_{D\vee m}(\theta) = \mathscr{I}(X_{1:n}, \ldots, X_{D_{\mathrm{II}}:n}, D_{\mathrm{II}}; \theta)$$

in the Type-II hybrid censored sample as

$$
\begin{aligned}
\mathscr{I}_{D\vee m}(\theta) &= -E_\theta\left[\frac{\partial^2}{\partial\theta^2}\log\phi_{D;\theta}(X_{1:n},\ldots,X_{n:n})\right]\\
&= -E_\theta\left[\frac{\partial^2}{\partial\theta^2}\log\phi_{d;\theta}(X_{1:n},\ldots,X_{n:n})\mathbb{1}_{\{0,\ldots,m-1\}}(D)\right]\\
&\quad -\sum_{d=m}^{n}E_\theta\left[\frac{\partial^2}{\partial\theta^2}\log\phi_{d;\theta}(X_{1:n},\ldots,X_{n:n})\mathbb{1}_{\{d\}}(D)\right]\\
&= -E_\theta\left[\frac{\partial^2}{\partial\theta^2}\log f_{1,\ldots,m:n;\theta}(X_{1:n},\ldots,X_{m:n})\mathbb{1}_{\{0,\ldots,m-1\}}(D)\right]\\
&\quad -\sum_{d=m}^{n}E_\theta\left[\frac{\partial^2}{\partial\theta^2}\log\phi_{d;\theta}(X_{1:n},\ldots,X_{n:n})\mathbb{1}_{\{d\}}(D)\right]\\
&= -E_\theta\left[\frac{\partial^2}{\partial\theta^2}\log f_{1,\ldots,m:n;\theta}(X_{1:n},\ldots,X_{m:n})\right]\\
&\quad +E_\theta\left[\frac{\partial^2}{\partial\theta^2}\log f_{1,\ldots,m:n;\theta}(X_{1:n},\ldots,X_{m:n})\mathbb{1}_{\{m,\ldots,n\}}(D)\right]\\
&\quad -\sum_{d=0}^{n}E_\theta\left[\frac{\partial^2}{\partial\theta^2}\log\phi_{d;\theta}(X_{1:n},\ldots,X_{n:n})\mathbb{1}_{\{d\}}(D)\right]\\
&\quad +\sum_{d=0}^{m-1}E_\theta\left[\frac{\partial^2}{\partial\theta^2}\log\phi_{d;\theta}(X_{1:n},\ldots,X_{n:n})\mathbb{1}_{\{d\}}(D)\right]\\
&= \mathscr{I}_{1,\ldots,m:n}(\theta)+\mathscr{I}_{D\wedge n}(\theta)-\mathscr{I}_{D\wedge m}(\theta),
\end{aligned}
$$

where we have used (8.26) in the last step. Thus, we have proved the following theorem. Notice that the approach is independent of the particular representation of the Fisher information. The result holds provided that the Fisher information in a Type-I and Type-II censored sample based on the cumulative distribution function F_θ exists. Furthermore, the above derivation shows that this connection also holds in case of multiple parameters so that the identity is also satisfied for the Fisher information matrix for a parameter vector $\boldsymbol{\theta} \in \Theta \subseteq \mathbb{R}^p$.

Theorem 8.9. *The Fisher information in a Type-II hybrid censored sample can be obtained as*

$$\mathscr{I}_{D\vee m}(\theta) = \mathscr{I}_{1,\ldots,m:n}(\theta)+\mathscr{I}_{D\wedge n}(\theta)-\mathscr{I}_{D\wedge m}(\theta). \tag{8.28}$$

For an *Exp*(ϑ)-distribution, (8.12), (8.16), (8.24), and (8.28) yield directly the following expression for the Fisher information in a Type-II hybrid censored sample (see

Park et al., 2008)

$$\mathscr{I}_{D\vee m}(\vartheta) = \frac{m}{\vartheta^2} + \frac{n}{\vartheta^2}F\Big(\frac{T}{\vartheta}\Big) - \frac{1}{\vartheta^2}\sum_{i=1}^{m} F_{i:n}\Big(\frac{T}{\vartheta}\Big)$$
$$= \frac{m}{\vartheta^2} + \frac{1}{\vartheta^2}\sum_{i=m+1}^{n} F_{i:n}\Big(\frac{T}{\vartheta}\Big), \tag{8.29}$$

where $F_{i:n}$ is the cumulative distribution function of the i-th order statistic from a standard exponential distribution. From (4.27), we find $\mathscr{I}_{D\vee m}(\vartheta) = \frac{1}{\vartheta^2}ED_{\mathsf{II}}$ which has the same structure as the representation in (8.24) obtained under Type-I hybrid censoring.

Clearly, the above approach can also be utilized for other hybrid censoring schemes. The respective formulas are reported in the following section. In particular, we illustrate how the Fisher information can be decomposed.

Remark 8.10. As pointed out in Park and Balakrishnan (2009), the Fisher information in 'minima' and 'maxima' satisfies the rule

$$\mathscr{I}_{S\vee T}(\theta) = \mathscr{I}_S(\theta) + \mathscr{I}_T(\theta) - \mathscr{I}_{S\wedge T}(\theta)$$

which may remind someone of the inclusion-exclusion formula for two sets.

8.2.4.3 Further hybrid censoring schemes

In the following, we present expressions for the corresponding Fisher information in hybrid censoring schemes composed of simpler (hybrid) censoring schemes. For illustration, we provide the corresponding expressions in case of an *Exp*(ϑ)-distribution only (in these expressions, $F_{i:n}$ denotes the cumulative distribution function of the i-th order statistic from a standard exponential distribution). For other distributions like Weibull distributions, the expressions can be directly obtained from the formulas given in (8.14), (8.17), and (8.25). We start with generalized hybrid censoring discussed in Park and Balakrishnan (2009).

(a) Generalized Type-I hybrid censoring with $D_{\mathsf{gI}} = (k \vee D) \wedge m = k \vee (D \wedge m)$ and $k < m$:

$$\mathscr{I}_{(k\vee D)\wedge m}(\theta) = \mathscr{I}_{1\ldots k:n}(\theta) + \mathscr{I}_{D\wedge m}(\theta) - \mathscr{I}_{D\wedge k}(\theta)$$

which is (8.28) with obvious changes. For an *Exp*(ϑ)-distribution, this yields directly the following expression for the Fisher information (see Park and Balakrishnan, 2009)

$$\mathscr{I}_{(k\vee D)\wedge m}(\vartheta) = \frac{k}{\vartheta^2} + \frac{1}{\vartheta^2}\sum_{i=k+1}^{m} F_{i:n}\Big(\frac{T}{\vartheta}\Big).$$

(b) Generalized Type-II hybrid censoring with $D_{\text{gII}} = (D_1 \vee m) \wedge D_2$ and $T_1 < T_2$:

$$\mathcal{I}_{(D_1 \vee m)\wedge D_2}(\theta) = \mathcal{I}_{D_2 \wedge m}(\theta) + \mathcal{I}_{D_1 \wedge n}(\theta) - \mathcal{I}_{D_1 \wedge m}(\theta).$$

Using (8.28), this can alternatively be written as

$$\mathcal{I}_{(D_1 \vee m)\wedge D_2}(\theta) = \mathcal{I}_{D_1 \vee m}(\theta) + \mathcal{I}_{D_2 \wedge n}(\theta) - \mathcal{I}_{D_2 \vee m}(\theta).$$

For an $Exp(\vartheta)$-distribution, this yields directly the following expression for the Fisher information (cf. Park and Balakrishnan, 2009)

$$\mathcal{I}_{(D_1 \vee m)\wedge D_2}(\vartheta) = \frac{1}{\vartheta^2}\sum_{i=1}^{m} F_{i:n}\left(\frac{T_2}{\vartheta}\right) + \frac{1}{\vartheta^2}\sum_{i=m+1}^{n} F_{i:n}\left(\frac{T_1}{\vartheta}\right).$$

Fisher information in Type-I and Type-II unified hybrid censoring schemes has been discussed in Park and Balakrishnan (2012).

(a) Type-I unified hybrid censoring with $D_{\text{uI}} = (D_1 \wedge m) \vee (D_2 \wedge k)$ and $k < m$, $T_1 < T_2$:

$$\mathcal{I}_{(D_1 \wedge m)\vee(D_2 \wedge k)}(\theta) = \mathcal{I}_{D_1 \wedge m}(\theta) + \mathcal{I}_{D_2 \wedge k}(\theta) - \mathcal{I}_{D_1 \wedge k}(\theta).$$

For an $Exp(\vartheta)$-distribution, this yields directly the following expression for the Fisher information (cf. Park and Balakrishnan, 2012)

$$\mathcal{I}_{(D_1 \wedge m)\vee(D_2 \wedge k)}(\vartheta) = \frac{1}{\vartheta^2}\sum_{i=1}^{k} F_{i:n}\left(\frac{T_2}{\vartheta}\right) + \frac{1}{\vartheta^2}\sum_{i=k+1}^{m} F_{i:n}\left(\frac{T_1}{\vartheta}\right).$$

(b) Type-II unified hybrid censoring with $D_{\text{uII}} = (D_1 \vee m) \wedge (D_2 \vee k)$ and $k < m$, $T_1 < T_2$:

$$\begin{aligned}\mathcal{I}_{(D_1 \vee m)\wedge(D_2 \vee k)}(\theta) &= \mathcal{I}_{D_1 \vee m}(\theta) + \mathcal{I}_{D_2 \vee k}(\theta) - \mathcal{I}_{D_2 \vee m}(\theta)\\ &= \mathcal{I}_{1\ldots k:n}(\theta) + \mathcal{I}_{D_1 \wedge n}(\theta) - \mathcal{I}_{D_1 \wedge m}(\theta) - \mathcal{I}_{D_2 \wedge k}(\theta) + \mathcal{I}_{D_2 \wedge m}(\theta).\end{aligned}$$

The last identity follows directly from (8.28). For an $Exp(\vartheta)$-distribution, this yields directly the following expression for the Fisher information (cf. Park and Balakrishnan, 2012)

$$\mathcal{I}_{(D_1 \vee m)\wedge(D_2 \vee k)}(\vartheta) = \frac{k}{\vartheta^2} + \frac{1}{\vartheta^2}\sum_{i=k+1}^{m} F_{i:n}\left(\frac{T_2}{\vartheta}\right) + \frac{1}{\vartheta^2}\sum_{i=m+1}^{n} F_{i:n}\left(\frac{T_1}{\vartheta}\right).$$

(c) Type-III unified hybrid censoring with $D_{\text{uIII}} = (k \wedge D_3) \vee ((m \vee D_1) \wedge D_2)$ and $k < m$, $T_1 < T_2 < T_3$:

$$\begin{aligned}\mathcal{I}_{(k\wedge D_3)\vee((m\vee D_1)\wedge D_2)}(\theta) &= \mathcal{I}_{D_3 \wedge k}(\theta) + \mathcal{I}_{(D_1 \vee m)\wedge D_2}(\theta) - \mathcal{I}_{D_2 \wedge k}(\theta)\\ &= \mathcal{I}_{D_1 \wedge n}(\theta) + \mathcal{I}_{D_2 \wedge m}(\theta) - \mathcal{I}_{D_1 \wedge m}(\theta) + \mathcal{I}_{D_3 \wedge k}(\theta) - \mathcal{I}_{D_2 \wedge k}(\theta).\end{aligned}$$

For an $Exp(\vartheta)$-distribution, this yields directly the following expression for the Fisher information (cf. Park and Balakrishnan, 2012)

$$\mathcal{I}_{(k\wedge D_3)\vee((m\vee D_1)\wedge D_2)}(\vartheta) = \frac{1}{\vartheta^2}\sum_{i=1}^{k} F_{i:n}\Big(\frac{T_3}{\vartheta}\Big) + \frac{1}{\vartheta^2}\sum_{i=k+1}^{m} F_{i:n}\Big(\frac{T_2}{\vartheta}\Big) + \frac{1}{\vartheta^2}\sum_{i=m+1}^{n} F_{i:n}\Big(\frac{T_1}{\vartheta}\Big).$$

(d) Type-IV unified hybrid censoring with $D_{\mathsf{uIV}} = (r \wedge D_1) \vee ((k \vee D_2) \wedge m)$ and $k < m < r$, $T_1 < T_2$:

$$\begin{aligned}\mathcal{I}_{(r\wedge D_1)\vee((k\vee D_2)\wedge m)}(\theta) &= \mathcal{I}_{r\wedge D_1}(\theta) + \mathcal{I}_{(k\vee D_2)\wedge m)}(\theta) - \mathcal{I}_{D_1\wedge m}(\theta)\\ &= \mathcal{I}_{1\dots k:n}(\theta) + \mathcal{I}_{D_1\wedge r}(\theta) - \mathcal{I}_{D_1\wedge m}(\theta) + \mathcal{I}_{D_2\wedge m}(\theta) - \mathcal{I}_{D_2\wedge k}(\theta).\end{aligned}$$

For an $Exp(\vartheta)$-distribution, this yields directly the following expression for the Fisher information (cf. Park and Balakrishnan, 2012)

$$\mathcal{I}_{(r\wedge D_1)\vee((k\vee D_2)\wedge m)}(\vartheta) = \frac{k}{\vartheta^2} + \frac{1}{\vartheta^2}\sum_{i=k+1}^{m} F_{i:n}\Big(\frac{T_2}{\vartheta}\Big) + \frac{1}{\vartheta^2}\sum_{i=m+1}^{r} F_{i:n}\Big(\frac{T_1}{\vartheta}\Big).$$

8.2.5 Fisher information under progressive hybrid censoring

Results on the Fisher information $\mathcal{I}^{\mathscr{R}}_{D\wedge m}(\theta)$ about a single parameter θ in a Type-I progressively hybrid censored sample have been established in Park et al. (2011) when the absolutely continuous population cumulative distribution function F_θ, $\theta \in \Theta \subseteq \mathbb{R}$, has a continuous density function with support contained in the positive real line. For Type-I progressively hybrid censored data, they noticed that

$$\mathcal{I}^{\mathscr{R}}_{D\wedge m}(\theta) = \sum_{i=1}^{m} \mathcal{I}(X_{1:\gamma_i} \wedge T; \theta)$$

using the structure of the joint density function. This leads to the following result (cf. (8.18)) which extends expressions of Wang and He (2005) and Park et al. (2008) for Type-I hybrid censored data.

Theorem 8.11. *The Fisher information about θ in a Type-I progressively hybrid censored sample is given by*

$$\mathcal{I}^{\mathscr{R}}_{D\wedge m}(\theta) = \int_0^T \Big\{\frac{\partial}{\partial\theta}\log h_\theta(x)\Big\}^2 \sum_{i=1}^{m} f_{i:m:n;\theta}(x)\,dx.$$

Notice that this expression is very similar to the Fisher information in a progressively Type-II censored sample as given in (8.18). In fact, the integration area is restricted to the interval $[0, T]$. For an $Exp(\vartheta)$-distribution, this expression simplifies to

$$\mathscr{I}^{\mathscr{R}}_{D\wedge m}(\vartheta) = \frac{1}{\vartheta^2}\sum_{i=1}^{m} F_{i:m:n}\Big(\frac{T}{\vartheta}\Big), \tag{8.30}$$

where $F_{i:m:n} = F_{i:n^*:n}$ denotes the cumulative distribution function of the i-th progressively Type-II censored order statistic from a standard exponential distribution. The similarity to the expression in case of Type-I hybrid censoring in (8.24) is striking. Moreover, we get from Lemma 7.2, $\mathscr{I}^{\mathscr{R}}_{D\wedge m}(\vartheta) = \frac{1}{\vartheta^2}ED_{\text{I}}$ as under Type-I hybrid censoring.

For Type-II progressively hybrid censored data, Park et al. (2011) established a result for the Fisher information $\mathscr{I}^{\mathscr{R}}_{D\vee m}(\theta)$ in the Type-II progressively hybrid censored data based on the (complete) progressively Type-II censored sample as in (7.2), that is,

$$X_{1:m+R_m:n} \le \cdots \le X_{m:m+R_m:n} \le X_{m+1:m+R_m:n} \le \cdots \le X_{m+R_m:m+R_m:n}.$$

In the following, we consider the more general setting as given in (7.2), i.e.,

$$X_{1:n^*:n}, \ldots, X_{n^*:n^*:n}$$

with censoring plan $\mathscr{R} = (R_1, \ldots, R_{n^*})$ and baseline sample size $n = \sum_{j=1}^{n^*}(R_j + 1)$.

Now, we can proceed by analogy with hybrid censoring schemes so that we get the following expressions. Notice that the respective counters are defined w.r.t. the data in (7.2), that is, for instance $D = \sum_{j=1}^{n^*} \mathbb{1}_{(-\infty, T]}(X_{j:m:n})$.

Thus, we obtain for Type-II progressive hybrid censoring by analogy with (8.28) the identity

$$\mathscr{I}^{\mathscr{R}}_{D\vee m}(\theta) = \mathscr{I}^{\mathscr{R}}_{1,\ldots,m:n^*:n}(\theta) + \mathscr{I}^{\mathscr{R}}_{D\wedge n^*}(\theta) - \mathscr{I}^{\mathscr{R}}_{D\wedge m}(\theta), \tag{8.31}$$

where $\mathscr{I}^{\mathscr{R}}_{D\wedge n^*}(\theta)$ denotes the Fisher information in the Type-I censored original sample $X_{1:n^*:n}, \ldots, X_{n^*:n^*:n}$.

In case of an $Exp(\vartheta)$-distribution, we get from (8.19) and (8.30)

$$\begin{aligned}\mathscr{I}^{\mathscr{R}}_{D\vee m}(\vartheta) &= \frac{m}{\vartheta^2} + \frac{1}{\vartheta^2}\sum_{i=1}^{n^*} F_{i:n^*:n}\Big(\frac{T}{\vartheta}\Big) - \frac{1}{\vartheta^2}\sum_{i=1}^{m} F_{i:n^*:n}\Big(\frac{T}{\vartheta}\Big)\\ &= \frac{m}{\vartheta^2} + \frac{1}{\vartheta^2}\sum_{i=m+1}^{n^*} F_{i:n^*:n}\Big(\frac{T}{\vartheta}\Big),\end{aligned} \tag{8.32}$$

where $F_{i:n^*:n}$ is the cumulative distribution function of the progressively Type-II censored order statistic $X_{i:n^*:n}$ from a standard exponential distribution. In fact, expression (8.32) shows that the gain of information using the Type-II hybrid censoring procedure is given by $\frac{1}{\vartheta^2}\sum_{i=m+1}^{n^*} F_{i:n^*:n}\left(\frac{T}{\vartheta}\right)$ which is increasing in T and bounded by the maximum gain $(n^*-m)/\vartheta^2$. Finally, we have

$$\mathcal{I}^{\mathcal{R}}_{D\vee m}(\vartheta) \xrightarrow{T\to\infty} \frac{n^*}{\vartheta^2} = \mathcal{I}^{\mathcal{R}}_{1,\ldots,n^*:n^*:n}(\vartheta).$$

Remark 8.12. In case of the particular sampling situation in (7.2), $\mathcal{I}^{\mathcal{R}}_{D\wedge n^*}(\theta)$ can be interpreted as the Fisher information in the Type-I censored 'complete' progressively Type-II censored data with maximum observed sample size $n^* = m + R_m$. Thus, Eq. (8.31) reads

$$\mathcal{I}^{\mathcal{R}}_{D\vee m}(\theta) = \mathcal{I}^{\mathcal{R}}_{1,\ldots,m:m:n}(\theta) + \mathcal{I}_{D\wedge m+R_m}(\theta) - \mathcal{I}_{D\wedge m}(\theta).$$

In Park et al. (2011), the term $\mathcal{I}_{D\wedge m+R_m}(\theta) - \mathcal{I}_{D\wedge m}(\theta)$ was incorrectly interpreted as $\mathcal{I}(T;\theta) - \mathcal{I}_{D_1\wedge n-R_m}(\theta)$, that is, the difference between the Fisher information in Type-I censored data and that in Type-I hybrid censored data (with sample size $n-R_m$). Thus their result on Fisher information in a Type-II progressive hybrid censored sample is in error. Although the formula is similar to (8.31), they identified the random variables in the extended sample (see (7.2)) with the order statistics $X_{n-R_m:n},\ldots,X_{n:n}$ of the original sample which is not true since progressive censoring has been employed to the data before. The mistake seems to be caused by a notation clash in Childs et al. (2008) who introduced the extended sample and used an order statistic notation. Unfortunately, the same mistake has been copied in other papers (e.g., Sen et al., 2018b for generalized Type-II progressive hybrid censoring and Almohaimeed, 2017 who discussed entropy of Type-II progressive hybrid censored data).

As has been done for hybrid censoring schemes in Section 8.2.4.3, expressions for the Fisher information in various progressive hybrid censoring schemes can be established. For brevity, we present only the results for generalized hybrid censoring schemes (see also Sen et al., 2018b where the flaw sketched in Remark 8.12 is also present). The necessary adjustments are obvious from the results in hybrid censoring so that we omit the details. The results for an $Exp(\vartheta)$-distribution are easily obtained from those in Section 8.2.4.3 by replacing the cumulative distribution function $F_{i:n}$ by that of the respective cumulative distribution function $F_{i:n^*:n}$ of the i-th progressively Type-II censored order statistic.

(a) Generalized Type-I progressive hybrid censoring with $D_{\mathsf{gI}} = (k\vee D)\wedge m = k\vee(D\wedge m)$ and $k<m$:

$$\mathcal{I}^{\mathcal{R}}_{(k\vee D)\wedge m}(\theta) = \mathcal{I}^{\mathcal{R}}_{1,\ldots,k:n^*:n}(\theta) + \mathcal{I}^{\mathcal{R}}_{D\wedge m}(\theta) - \mathcal{I}^{\mathcal{R}}_{D\wedge k}(\theta).$$

(b) Generalized progressive Type-II hybrid censoring with $D_{\mathsf{gII}} = (D_1\vee m)\wedge D_2$ and $T_1<T_2$:

$$\mathcal{I}^{\mathcal{R}}_{(D_1\vee m)\wedge D_2}(\theta) = \mathcal{I}^{\mathcal{R}}_{D_2\wedge m}(\theta) + \mathcal{I}^{\mathcal{R}}_{D_1\wedge n}(\theta) - \mathcal{I}^{\mathcal{R}}_{D_1\wedge m}(\theta).$$

8.3. Entropy

The joint differential entropy (Shannon entropy) of a random vector $\boldsymbol{X}$ with density function $f^{\mathbf{X}}$ is defined via the expectation

$$\mathscr{H}(\mathbf{X}) = -E\log f^{\mathbf{X}}(\mathbf{X}) \tag{8.33}$$

(see Cover and Thomas, 2006). The joint differential entropy (8.33) can be seen as a measure of uncertainty in the sample $\mathbf{X}$ which measures uniformity of the density $f^{\mathbf{X}}$. On the other hand, the negative entropy can be interpreted as a measure of concentration and, thus, measures the information in $\boldsymbol{X}$ (see Soofi, 2000). A connection to Fisher information is the so-called isoperimetric inequality for entropies (see Dembo and Cover, 1991): The Fisher information $\mathscr{I}(\boldsymbol{X};\theta)$ can be bounded from below by

$$\mathscr{I}(\mathbf{X};\theta) \geq 2\pi e\, m \exp\left\{-\frac{2}{m}\mathscr{H}(\mathbf{X})\right\}.$$

This bound may be useful if the Fisher information is hard to compute. Further information on entropy and information measures can be found in, e.g., Kullback (1959) and Cover and Thomas (2006). A representation of the entropy in terms of the hazard rate $h = f/(1-F)$ has been established by Teitler et al. (1986), that is,

$$\mathscr{H}(X) = 1 - E\log h(X). \tag{8.34}$$

The entropy in order statistics is discussed in Wong and Chen (1990) who found that the entropy in an ordered sample of iid random variables $X_1, \ldots, X_n$ is given by

$$\mathscr{H}_{1\ldots n:n} = n\mathscr{H}(X) - \log n!\,. \tag{8.35}$$

Here, $n\mathscr{H}(X)$ denotes the entropy in the iid sample. This illustrates that the ordering process influences the value of the entropy considerably. The entropy in a Type-II censored sample $X_{1:n}, \ldots, X_{m:n}$ has been presented in Park (2005) (see also Ebrahimi et al., 2004; Park, 1995). It is given by

$$\mathscr{H}_{1,\ldots,m:n} = m - \log\frac{n!}{(n-m)!} - \sum_{j=1}^{m} E\log h(X_{j:n}), \tag{8.36}$$

where $h = f/(1-F)$ denotes the population hazard rate function. The derivation is based on the factorization of the density function given in (8.10). Notice that, for $m = n$, (8.36) reduces to (8.35) by taking into account (8.34). For an *Exp*(ϑ)-distribution, the entropy is given by

$$\mathscr{H}_{1,\ldots,m:n} = m - \log\frac{n!}{(n-m)!} + m\log\vartheta.$$

Balakrishnan et al. (2007a) established the following representation of the entropy of a progressively censored sample (see also Abo-Eleneen, 2011; Cramer and Bagh, 2011). Notice that the Fisher information can also be expressed in terms of the hazard rate function (see (8.18)). The entropy in a progressively Type-II censored sample $\mathbf{X}^{\mathscr{R}}$ is given by

$$\mathscr{H}^{\mathscr{R}}_{1,\ldots,m:n^*:n} = m - \log c(\mathscr{R}) - \sum_{j=1}^{m} E \log h(X_{j:m:n}),$$

where $h = f/(1-F)$ and $c(\mathscr{R}) = \prod_{j=1}^{m} \gamma_j$ denote the population hazard rate function and the normalizing constant of the density function, respectively. For an $Exp(\vartheta)$-distribution, the entropy is given by

$$\mathscr{H}^{\mathscr{R}}_{1,\ldots,m:n^*:n} = m - \log c(\mathscr{R}) + m \log \vartheta.$$

The entropy in a Type-I censored sample $X_1, \ldots, X_n$ is given by

$$\mathscr{H}(X_1 \wedge T, \ldots, X_n \wedge T) = nF(T) - nE\Big(\log h(X_1) \mathbb{1}_{(-\infty,T]}(X_1)\Big) \tag{8.37}$$

using that the entropy in an iid sample is the sum of the entropies of the random variable. Here, we use that the entropy of $X_1 \wedge T$ is given by

$$\mathscr{H}(X_1 \wedge T) = F(T) - E\Big(\log h(X_1) \mathbb{1}_{(-\infty,T]}(X_1)\Big).$$

For an $Exp(\vartheta)$-distribution, the entropy in a Type-I censored sample with sample size n is given by

$$\mathscr{H}(X_1 \wedge T, \ldots, X_n \wedge T) = n(1 + \log \vartheta) F\Big(\frac{T}{\vartheta}\Big),$$

where F denotes the cumulative distribution function of a standard exponential distribution.

The entropy in (progressive) hybrid censored samples has been studied in Morabbi and Razmkhah (2009) and Almohaimeed (2017). For a Type-I hybrid censored sample, they found that

$$\mathscr{H}_{D\wedge m} = \sum_{j=1}^{m} \Bigg[\Big(1 - \log(n-j+1)\Big)\Big(1 - \binom{n}{j-1} \overline{F}^{n-j+1}(T)\Big) - E\Big(\log h(X_{j:n}) \mathbb{1}_{(-\infty,T]}(X_{j:n})\Big)\Bigg]. \tag{8.38}$$

Notice that $\mathscr{H}_{D\wedge n}$ measures the entropy in an <u>ordered</u> Type-I censored sample which is different from the entropy in the iid sample given in (8.37). This reflects that ordering of the sample affects the entropy of the sample. For an $Exp(\vartheta)$-distribution, the entropy in a Type-I hybrid censored sample with sample size n is given by

$$\mathscr{H}_{D\wedge m} = \sum_{j=1}^{m} \left[\Big(1 - \log(n-j+1)\Big)\Big(1 - \binom{n}{j-1}\overline{F}^{n-j+1}\Big(\frac{T}{\vartheta}\Big)\Big) + (\log\vartheta) F_{j:n}\Big(\frac{T}{\vartheta}\Big)\right]$$

where $F_{j:n}$ and $\overline{F}$ denote the cumulative distribution function of the j-th order statistic and the survival function of a standard exponential distribution, respectively.

Clearly, the entropy in a Type-II hybrid censored sample can be obtained along the lines of the Fisher information as presented on page 267. For illustration, we find with the representation of the density function in (8.27) by performing the same steps in the derivation

$$\begin{aligned}
\mathscr{H}_{D\vee m} &= -E\big[\log h_{D;\theta}(X_{1:n}, \ldots, X_{n:n})\big] \\
&= -E\big[\log f_{1,\ldots,m:n;\theta}(X_{1:n}, \ldots, X_{m:n})\big] \\
&\quad + E\big[\log f_{1,\ldots,m:n;\theta}(X_{1:n}, \ldots, X_{m:n}) \mathbb{1}_{\{m,\ldots,n\}}(D)\big] \\
&\quad - \sum_{d=0}^{n} E\big[\log \phi_{d;\theta}(X_{1:n}, \ldots, X_{n:n}) \mathbb{1}_{\{d\}}(D)\big] \\
&\quad + \sum_{d=0}^{m-1} E\big[\log \phi_{d;\theta}(X_{1:n}, \ldots, X_{n:n}) \mathbb{1}_{\{d\}}(D)\big] \\
&= \mathscr{H}_{1,\ldots,m:n} + \mathscr{H}_{D\wedge n} - \mathscr{H}_{D\wedge m}.
\end{aligned}$$

Using the same approach, expressions for the entropy of other hybrid censoring schemes may be obtained by analogy with those established for the Fisher information. Finally, it should be noted that the same technique can be applied to progressive hybrid censored data. In case of a Type-I progressive hybrid censored sample, an expression is obtained by analogy with Morabbi and Razmkhah (2009) taking into account density functions of progressively Type-II censored order statistics:

$$\mathscr{H}^{\mathscr{R}}_{D\wedge m} = \sum_{j=1}^{m} \left[\Big(1 - \log\gamma_j\Big)\Big(1 - \kappa_j \overline{F}^{\gamma_j}(T)\Big) - E\Big(\log h(X_{j:m:n}) \mathbb{1}_{(-\infty,T]}(X_{j:m:n})\Big)\right],$$

where $\kappa_j = \prod_{i=1}^{j-1} (\gamma_i/(\gamma_i - \gamma_j))$, $1 \le j \le m$, replace the binomial coefficients.[2] Expressions for the entropy under other progressive hybrid censoring schemes can be established along the line for the entropy in hybrid censoring schemes and the Fisher information, respectively so that we omit details here.

[2] Notice that the expressions for the coefficients κ_j given in Almohaimeed (2017) are in error.

Remark 8.13. It should be mentioned that estimation of the entropy has been considered based on data subject to various (progressive) hybrid censoring schemes and several population distributions. So far, the papers use estimates of the distribution parameters and estimate the population entropy by replacing the parameters in the entropy by the corresponding estimators (e.g., maximum likelihood estimators, Bayes estimators, etc.). A selection of references is given by the following list: Cho et al. (2014), Cho et al. (2015a), Lee (2017, 2020).

8.4. Kullback-Leibler information

The Kullback-Leibler information serves as a measure to compare two distributions. For two random samples $\boldsymbol{X}$ and $\boldsymbol{Y}$ with population density functions f and g, it is defined as (see Kullback, 1959)

$$\mathscr{I}(f \parallel g) = \mathscr{I}(f^{\boldsymbol{X}} \parallel g^{\boldsymbol{Y}}) = \int f^{\boldsymbol{X}}(\boldsymbol{x}) \log \frac{f^{\boldsymbol{X}}(\boldsymbol{x})}{g^{\boldsymbol{Y}}(\boldsymbol{x})} d\boldsymbol{x}.$$

Kullback-Leibler information w.r.t. censored data has been discussed in many papers which particularly covers applications to goodness of fit tests. For brevity, we present only some selected results and expressions.

Park (2005) has discussed the Kullback-Leibler information in Type-II censored data with two population density functions f and f_0, respectively, and established an approximation in terms of the entropy. A hazard rate representation has been presented in Park (2014), that is,

$$\mathscr{I}(f_{1,\dots,m:n} \parallel g_{1,\dots,m:n}) = \int \left(\frac{h_g(x)}{h_f(x)} - \log \frac{h_g(x)}{h_f(x)} - 1 \right) \sum_{j=1}^{m} f_{j:n}(x)\, dx \tag{8.39}$$

with hazard rates $h_g = g/(1-G)$ and $h_f = f/(1-F)$. This result is based on representation established in Park and Shin (2014), i.e.,

$$\mathscr{I}(f \parallel g) = \int \left(\frac{h_g(x)}{h_f(x)} - \log \frac{h_g(x)}{h_f(x)} - 1 \right) f(x) dx.$$

The Kullback-Leibler information in two Type-I censored random variables X and Y has been considered in Park and Shin (2014) leading to the expression

$$\mathscr{I}(f^{X\wedge T} \parallel f^{Y\wedge T}) = \int \mathbb{1}_{(-\infty,T]}(x) f^X(x) \log \frac{f^X(x)}{f^Y(x)} dx + (1 - F^X(T)) \log \frac{1 - F^X(T)}{1 - F^Y(T)}.$$

A similar measure is considered in Lim and Park (2007). Alternatively, it can be written in the form

$$\mathscr{I}(f^{X\wedge T}\,\|f^{Y\wedge T}) = \int \mathbb{1}_{(-\infty,T]}(x)\left(\frac{h_g(x)}{h_f(x)} - \log\frac{h_g(x)}{h_f(x)} - 1\right)f(x)dx.$$

Balakrishnan et al. (2007a) and Habibi Rad et al. (2011) have used Kullback-Leibler information based on progressively Type-II censored samples.

$$\mathscr{I}_{\mathscr{R}}(f\,\|\,g) = -\mathscr{H}^{\mathscr{R}}_{1,\ldots,m:m:n} - \int_S f^{\mathbf{X}^{\mathscr{R}}}(\boldsymbol{x})\log g^{\mathbf{Y}^{\mathscr{R}}}(\boldsymbol{x})d\boldsymbol{x}$$

to construct goodness-of-fit tests where $\mathscr{H}^{\mathscr{R}}_{1,\ldots,m:m:n}$ is the entropy of $P^{\mathbf{X}^{\mathscr{R}}}$ as in (8.33). The Kullback-Leibler information in favor of an iid sample $Y_1, \ldots, Y_m$ from a population with population density function f against a progressively Type-II censored sample $X_{1:m:n}, \ldots, X_{m:m:n}$ from the same population has been discussed in Cramer and Bagh (2011). In Balakrishnan and Cramer (2014), the distance is considered for two samples of progressively Type-II censored order statistics with the same sample size m, but possibly different censoring plans $\mathscr{R}$ and $\mathscr{S}$ and original sample sizes $\gamma_1(\mathscr{R})$ and $\gamma_1(\mathscr{S})$. Then, the Kullback-Leibler information reads

$$\mathscr{I}(f^{\mathbf{X}^{\mathscr{R}}}\,\|f^{\mathbf{X}^{\mathscr{S}}}) = \log c(\mathscr{R}) - \log c(\mathscr{S}) - m + \sum_{j=1}^{m}\frac{\gamma_j(\mathscr{S})}{\gamma_j(\mathscr{R})}.$$

Kullback-Leibler information in hybrid censored data has been studied in Park (2016) leading to the following integral expression for Type-I hybrid censoring which is (8.39) restricted to the interval $(-\infty, T]$:

$$\mathscr{I}(f_{D\wedge m}\,\|\,g_{D\wedge m}) = \int \mathbb{1}_{(-\infty,T]}(x)\left(\frac{h_g(x)}{h_f(x)} - \log\frac{h_g(x)}{h_f(x)} - 1\right)\sum_{j=1}^{m} f_{j:n}(x)\,dx.$$

Using the same approach as has been applied to Fisher information and entropy, expressions for other hybrid censoring schemes can be established. For instance, for Type-II hybrid censoring one gets

$$\begin{aligned}\mathscr{I}(f_{D\vee m}\,\|\,g_{D\vee m}) &= \mathscr{I}(f_{1,\ldots,m:n}\,\|\,g_{1,\ldots,m:n}) + \mathscr{I}(f_{D\wedge n}\,\|\,g_{D\wedge n}) - \mathscr{I}(f_{D\wedge m}\,\|\,g_{D\wedge m})\\ &= \mathscr{I}(f_{1,\ldots,m:n}\,\|\,g_{1,\ldots,m:n}) + n\mathscr{I}(f^{X\wedge T}\,\|\,g^{X\wedge T}) - \mathscr{I}(f_{D\wedge m}\,\|\,g_{D\wedge m})\end{aligned}$$

(see Park, 2016). Alizadeh Noughabi and Chahkandi (2018) extended the approach of Park (2005) to Type-I hybrid censored data and proposed goodness of fit-tests based on the Kullback-Leibler information measure (see Chapter 11).

8.5. Pitman closeness

Pitman closeness is a measure to compare the closeness of two estimators $\widehat{\theta}_1$, $\widehat{\theta}_2$ to the estimated parameter θ by means of the probability

$$\pi_\theta(\widehat{\theta}_1, \widehat{\theta}_2) = \Pr_\theta\left(|\widehat{\theta}_1 - \theta| < |\widehat{\theta}_2 - \theta|\right). \tag{8.40}$$

It was introduced in 1937 by Pitman (1937) who presented some applications with regard to the population median as well as estimates, e.g., of the variance in a normal population and of location and scale parameters of an exponential distribution. A comprehensive discussion including many examples has been provided in the monograph by Keating et al. (1993).

The notion of Pitman closeness has been introduced by Balakrishnan et al. (2009a) to determine the closest order statistic in a sample of size n to a given population quantile (for the median, see Balakrishnan et al., 2009b). Given a quantile ξ_p of the baseline distribution, they were interested in identifying an order statistic $X_{\ell:n}$ such that

$$\pi_{p;\ell,i} = \Pr(|X_{\ell:n} - \xi_p| < |X_{i:n} - \xi_p|) \geq \frac{1}{2} \quad \text{for all } i \in \{1, \dots, n\} \setminus \{\ell\}.$$

Volterman et al. (2013b) adapted this idea to two independent progressively Type-II censored samples $X_{1:m:n}^{\mathscr{R}}, \dots, X_{m:m:n}^{\mathscr{R}}$ and $Y_{1:s:k}^{\mathscr{S}}, \dots, Y_{s:s:k}^{\mathscr{S}}$ from the same cumulative distribution function F but possibly different censoring plans $\mathscr{R}$ and $\mathscr{S}$. For a quantile ξ_p, they introduced the Pitman closeness probabilities

$$\pi_{p;i,j}^{\mathscr{R},\mathscr{S}}(\xi_p) = \Pr(|X_{i:m:n}^{\mathscr{R}} - \xi_p| < |Y_{j:s:k}^{\mathscr{S}} - \xi_p|), \quad 1 \leq i \leq m, 1 \leq j \leq s.$$

The above idea has been further discussed by Balakrishnan et al. (2010). They introduced the notion of simultaneous-closeness probability (SCP) following a proposal of Blyth (1972). Replacing the order statistics by progressively Type-II censored order statistics, the same idea has been introduced into the framework of progressively Type-II censored order statistics by Volterman et al. (2013a). The simultaneous-closeness probability of $X_{i:m:n}$, $i \in \{1, \dots, m\}$, among the order statistics $X_{1:m:n}, \dots, X_{m:m:n}$ in the estimation of a population parameter θ is defined as

$$\pi_{\theta;i:m:n}(\theta) = \Pr\left(|X_{i:m:n} - \theta| < \min_{j \neq i} |X_{j:m:n} - \theta|\right).$$

An explicit expression for the SCP in terms of the cumulative distribution function has been established in Volterman et al. (2013a). Furthermore, some particular baseline distributions like exponential, uniform, and normal distributions have been discussed as well as extensive computations for selected censoring plan have been provided. Simultaneous Pitman closeness has been applied by Volterman et al. (2013b) to measure the distance of predictors of a future sample of progressively Type-II censored order statistics. Volovskiy (2018) applied Pitman closeness to compare maximum observed likelihood predictors and maximum likelihood predictors.

So far, the paper by Davies (2021) seems to be the only one that has studied Pitman closeness for (Type-I) hybrid censoring schemes. Based on the probability in (8.40), MLEs of an $Exp(\vartheta)$-distribution are compared when either the threshold T or the number m of failures in the Type-I hybrid censoring scheme are modified. An analogous discussion comparing MLEs for ϑ under Type-I and Type-II censoring has been provided by Balakrishnan and Davies (2013).

Denoting by $D_{\mathsf{I}}(m, T)$ the number of observed failures and by

$$\widehat{\vartheta}_{m,T} = \frac{1}{D_{\mathsf{I}}(m, T)}\Bigg(\sum_{j=1}^{D_{\mathsf{I}}(m,T)} X_{j:n} + (n - D_{\mathsf{I}}(m, T))W\Bigg)$$

the MLE under the Type-I hybrid censoring scheme with m failures to observe and threshold T, the following comparisons are discussed. Notice that, since $D_{\mathsf{I}}(m, T) = 0$ is possible, conditional probabilities are considered. Moreover, $D_{\mathsf{I}}(m, T) \geq 1$ implies both $D_{\mathsf{I}}(m, T^*) \geq 1$ for $T < T^*$ and $D_{\mathsf{I}}(k, T) \geq 1$ for $m < k$ so that all the MLEs exists under the condition $D_{\mathsf{I}}(m, T) \geq 1$.

(1) $\widehat{\vartheta}_{m,T}$ and $\widehat{\vartheta}_{m,T^*}$ with $T < T^*$ leading to

$$\pi_{\vartheta;T,T^*} = \Pr_\theta\Big(|\widehat{\vartheta}_{m,T} - \vartheta| < |\widehat{\vartheta}_{m,T^*} - \vartheta| \mid D_{\mathsf{I}}(m, T) \geq 1\Big).$$

Assuming $\vartheta = 1$, Davies (2021) reported that the MLE with the larger threshold T^* was up to one exception always Pitman closer than the estimator based on T (given that the estimators are different).

(2) $\widehat{\vartheta}_{m,T}$ and $\widehat{\vartheta}_{k,T^*}$ with $m < k$ leading to

$$\pi_{\vartheta;m,k} = \Pr_\theta\Big(|\widehat{\vartheta}_{m,T} - \vartheta| < |\widehat{\vartheta}_{k,T} - \vartheta| \mid D_{\mathsf{I}}(m, T) \geq 1\Big).$$

For $\vartheta = 1$, the numerical results in Davies (2021) indicate that $\widehat{\vartheta}_{k,T}$, that is, the MLE based on the larger number of failures is frequently closer than the other one, given the estimators are different.

The results are reasonable in the sense that both a larger threshold and a larger number of failures to observe should result in a gain of information (e.g., more observations, a longer test duration) which should provide a 'better' estimate of the mean.

CHAPTER 9

Step-stress testing

Contents

9.1. Introduction

Products that are tested in industrial experiments are often extremely reliable leading to large mean times to failure under normal operating conditions. Therefore, adequate information about the lifetime distributions and the associated parameters may be quite difficult to obtain using conventional life-testing experiments. For this reason, the units under test are subjected to higher operational demands than under normal operating conditions. Such methods are widely used and are known as accelerated life-testing (ALT). Several approaches are adapted in accelerated life-testing like, e.g., constant high stress level, progressive stress, or step-wise increasing stress levels. Such a procedure will enable the reliability experimenter to assess the effects of stress factors such as load, pressure, temperature, and voltage on the lifetimes of experimental units. This kind of accelerated life-test usually will reduce the time to failure of specimens, thus resulting in more failures than under normal operating conditions. Data collected from an accelerated life-test then needs to be extrapolated to estimate the parameters of the lifetime distribution under normal operating conditions. In the literature, several models have been proposed to connect the stress levels in the ALT-environment to the parameters of the original distribution. A popular model that will be assumed in the following is the cumulative exposure model introduced by Sedyakin (1966) and discussed further by Bagdonavičius (1978) and Nelson (1980, 1990). As pointed out in Cramer (2021), the observed data can be interpreted as order statistics, but under different model assumptions. Consider a simple step-stress test with stress levels s_0 and s_1 and a single stress-change time τ (as depicted in Fig. 9.1). The lifetime cumulative distribution function on levels s_j is denoted by $F_j, j=0,1$. Under the cumulative exposure model, changing the stress from s_0 to s_1 entails the transition of the lifetime distribution

Hybrid Censoring Know-How
https://doi.org/10.1016/B978-0-12-398387-9.00017-9

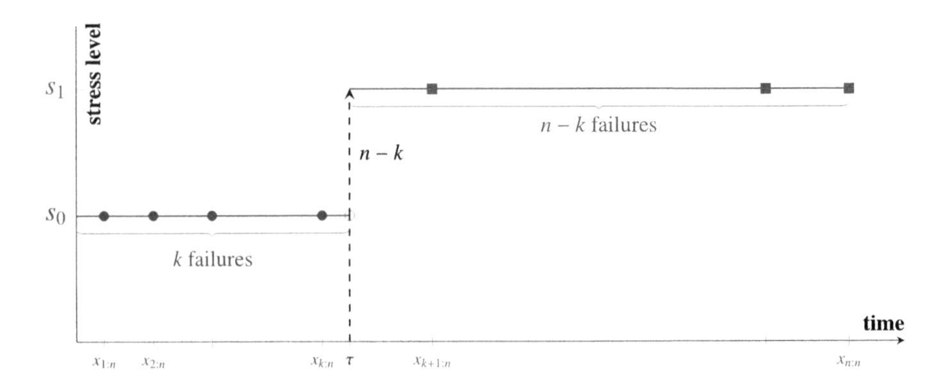

Figure 9.1 Simple step-stress model with change time τ.

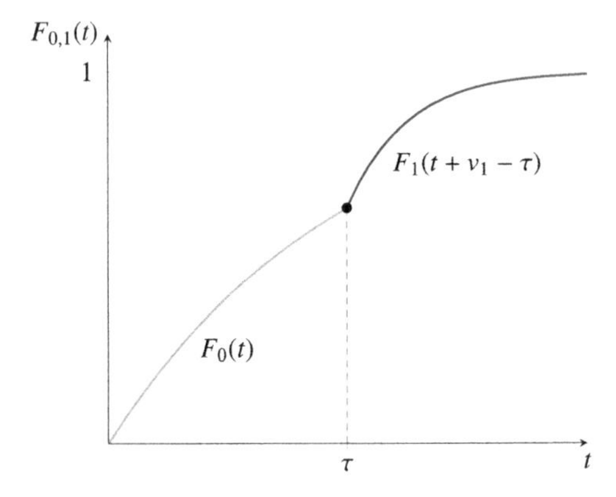

Figure 9.2 Construction of cumulative distribution function $F_{0,1}$ in a cumulative exposure model with stress change-time τ satisfying Eq. (9.1).

on stress level s_1 from $F_0(t)$ to $F_1(t+\nu_1-\tau)$ where ν_1 is the solution of the equation

$$F_0(\tau)=F_1(\nu_1) \tag{9.1}$$

which ensures that the cumulative distribution function of an item on test is continued continuously with F_1 after τ. In particular, the resulting cumulative distribution function is given by (see also Fig. 9.2)

$$F_{0,1}(t)=\begin{cases} F_0(t), & t\leq\tau \\ F_1(t+\nu_1-\tau), & \tau<t<\infty \end{cases}. \tag{9.2}$$

Detailed accounts to accelerated life-testing are provided by Nelson and Meeker (1978), Meeker and Hahn (1985), Nelson (1990), Meeker and Escobar (1998), and

Bagdonavičius and Nikulin (2002). Results for the exponential distribution are summarized in Basu (1995).

A special type of accelerated life-testing is step-stress testing. In such a testing environment, the experimenter can choose different conditions at various intermediate stages of the life-test as follows. n identical units are placed on a life-test at an initial stress level of s_1. Then, at pre-fixed times $\tau_1 < \cdots < \tau_k$, the stress levels are changed to $s_2, \ldots, s_{k+1}$, respectively. The special case of two stress levels s_1 and s_2, wherein the stress change occurs at a pre-fixed time τ, is called simple step-stress experiment. This situation is depicted in Fig. 9.1 for observed failure times $x_{1:n} \leq \cdots \leq x_{n:n}$. Further details can be found in Nelson (1990, Chapter 10) and Kundu and Ganguly (2017) (see also Balakrishnan and Cramer, 2014, 2023).

The simple step-stress model has been discussed extensively in the literature. Comprehensive reviews of work on step-stress models have been provided by Gouno and Balakrishnan (2001), Tang (2003), Balakrishnan (2009), Balakrishnan and Cramer (2014), and Kundu and Ganguly (2017). In the following, we point out results on step-stress models related to hybrid censoring. Moreover, we assume throughout that the stress changes are modeled by a cumulative exposure model.

From the construction of a step-stress model it is clear that inference based on the observations up to time τ, is the same as inference based on a (progressive) Type-I censored sample with time threshold $T = \tau$. Therefore, all the results available for inference on Type-I hybrid censoring can be utilized for inference on stress level s_0 (see also the comments in Balakrishnan and Cramer, 2014, p. 487).

9.2. Step-stress models under censoring

In this section, we sketch the derivation of the joint density function in a step-stress model with a single stress change time τ. In the following, we consider step stress data

$$X_{1:n}^{(\tau)}, \ldots, X_{n:n}^{(\tau)} \tag{9.3}$$

which then is subject to censoring. From a distributional point of view, this means that the joint distribution of the data is now given by that of a step stress data instead of pure order statistics' data. Assuming absolutely continuous cumulative distribution functions F_0 and F_1 with density functions f_0 and f_1 as well as the cumulative exposure model, the joint density function is given by the following construction. For some fixed $k \in \{0, \ldots, n\}$ and cumulative distribution function F_0, the joint density function of order statistics and the observed number of failures on stress level s_0,

$$D_\tau = \sum_{j=1}^{n} \mathbb{1}_{(-\infty,\tau]}(X_{j:n}),$$

can be written as

$$
\begin{aligned}
f_{0,0}^{X_{1:n},\ldots,X_{n:n},D_\tau}(\boldsymbol{t}_n, k) &= \Pr_0(D_\tau = k) f_0^{X_{1:n},\ldots,X_{k:n}|D_\tau=k}(\boldsymbol{t}_k) f_0^{X_{k+1:n},\ldots,X_{n:n}|D_\tau=k}(t_{k+1},\ldots,t_n) \\
&= f_0^{X_{1:n},\ldots,X_{k:n},D_\tau}(\boldsymbol{t}_k, k) f_0^{X_{k+1:n},\ldots,X_{n:n}|D_\tau=k}(t_{k+1},\ldots,t_n)
\end{aligned} \tag{9.4}
$$

by using the block independence property of order statistics (see Theorem 2.28). Applying the cumulative exposure model means that the conditional density function $f_0^{X_{k+1:n},\ldots,X_{n:n}|D_\tau=k}$ (based on F_0) in (9.4) is replaced by the same expression based on the cumulative distribution function F_1 appropriately adapted to the stress changing time τ as given in (9.2).

In particular, we have for $t_1 \leq \cdots \leq t_k \leq \tau < t_{k+1} \leq \cdots \leq t_n$ (see Remark 2.29)

$$
f_0^{X_{1:n},\ldots,X_{k:n},D_\tau}(\boldsymbol{t}_k, k) = \frac{n!}{(n-k)!}(1 - F_0(\tau))^{n-k} \prod_{j=1}^{k} f_0(t_j),
$$

$$
f_0^{X_{k+1:n},\ldots,X_{n:n}|D_\tau=k}(t_{k+1},\ldots,t_n) = (n-k)! \prod_{j=k+1}^{n} \frac{f_0(t_j)}{1 - F_0(\tau)},
$$

$$
f_1^{X_{k+1:n},\ldots,X_{n:n}|D_\tau=k}(t_{k+1},\ldots,t_n) = (n-k)! \prod_{j=k+1}^{n} \frac{f_1(t_j + v_1 - \tau)}{1 - F_0(\tau)},
$$

with v_1 as the solution of Eq. (9.1), that is, $F_0(\tau) = F_1(v_1)$. Furthermore, the first expression is defined as 1 for $k = 0$ whereas the other two are defined as 1 for $k = n$. With these definitions, the formulas can be used for any $k \in \{0,\ldots,n\}$.

As a consequence of the application of the cumulative exposure model, we get the joint density function of the respective ordered random variables as, for $t_1 \leq \cdots \leq t_n$,

$$
f_{0,1}^{X_{1:n},\ldots,X_{n:n},D_\tau}(\boldsymbol{t}_n, k) = n! \prod_{j=1}^{k} f_0(t_j) \prod_{j=k+1}^{n} f_1(t_j + v_1 - \tau). \tag{9.5}
$$

Notice that, for $k \in \{0,\ldots,n\}$, $t_1 \leq \cdots \leq t_k \leq \tau < t_{k+1} \leq \cdots \leq t_n$. Furthermore, the joint density function in (9.5) yields the likelihood function in case of a complete step-stress data set.

Introducing a second threshold $T > \tau$ (which is involved in the hybrid censoring schemes) with indicator D as in (2.12), we get the joint density function of $X_{1:n},\ldots,X_{n:n}$ and D_τ, D as, for $t_1 \leq \cdots \leq t_n$,

$$
f_{0,1}^{X_{1:n},\ldots,X_{n:n},D_\tau,D}(\boldsymbol{t}_n, k, d) = n! \prod_{j=1}^{k} f_0(t_j) \prod_{j=k+1}^{n} f_1(t_j + v_1 - \tau) \tag{9.6}
$$

where k and d satisfy $t_k \leq \tau < t_{k+1}$ and $t_d \leq T < t_{d+1}$. The representation (9.6) is crucial to derive the subsequent density functions. At a first glance, the right hand side of (9.6) seems to be independent of d, but notice that the condition is imposed on its support, that is, for given d, the condition $t_d \leq T < t_{d+1}$ must be satisfied.

Remark 9.1. In the following, we consider an (hybrid) censored random sample

$$X_{1:n}^{(\tau)}, \ldots, X_{D_{\mathsf{HCS}}:n}^{(\tau)}$$

of the step-stress data given in (9.3), where the random sample size D_{HCS} is determined by the (hybrid) censoring procedure (see, e.g., Table 5.1 for expressions resulting from particular hybrid censoring schemes). Since the sample is generated in the same manner as for a sample of order statistics $X_{1:n}, \ldots, X_{n:n}$, the construction of the joint density function or the likelihood function is along the same lines. Moreover, the value of D_τ can be either determined from the sample $X_{1:n}^{(\tau)}, \ldots, X_{D_{\mathsf{HCS}}:n}^{(\tau)}$ or is not observed in the experiment. In the latter case, there is no information on the distributions involved after the stress changing time τ. This happens when $D_{\mathsf{HCS}} \leq D_\tau$ or, equivalently, $X_{D_{\mathsf{HCS}}:n}^{(\tau)} \leq \tau$. In order to simplify the notation, we suppress the value of D_τ (in the following usually denoted by k) in the representation of the likelihood.

Furthermore, due to the construction of the model, it is clear that the model equals a standard (hybrid) censored model when $D_\tau \geq D_{\mathsf{HCS}}$. In this case, the stress change does not influence the observed values in the sense that we do not have measurements after the stress change. As a result, the estimates can be obtained directly from the results obtained for the iid model under the particular censoring scheme. If $D_\tau < D_{\mathsf{HCS}}$ then we have to take into account the observed value of D_τ. For convenience, we use the notation ⓒ (censored) and ⓞ (observed) for the cases $D_\tau \geq D_{\mathsf{HCS}}$ and $D_\tau < D_{\mathsf{HCS}}$, respectively.

9.2.1 Type-II censoring

Step-stress data under Type-II censoring has been considered in Xiong (1998) and Balakrishnan et al. (2007b) assuming (exponential) cumulative distribution functions $F_{\boldsymbol{\theta}_0}$, $F_{\boldsymbol{\theta}_1}$ on each stress-level (see Fig. 9.3). As a result, one observes the failure data $X_{1:n}^{(\tau)}, \ldots, X_{m:n}^{(\tau)}$ for some fixed $m \in \{1, \ldots, n\}$. Then, from (9.5), the likelihood function is given by

$$\mathscr{L}(\boldsymbol{\theta}_0, \boldsymbol{\theta}_1 \mid \boldsymbol{x}_m, m) = \frac{n!}{(n-m)!} \times \begin{cases} \Big(1 - F_{\boldsymbol{\theta}_1}(x_m + v_1 - \tau)\Big)^{n-m} \prod_{j=1}^{k} f_{\boldsymbol{\theta}_0}(t_j) \prod_{j=k+1}^{m} f_{\boldsymbol{\theta}_1}(x_j + v_1 - \tau), & \text{ⓞ } D_\tau = k \in \{0, \ldots, m-1\} \\ (1 - F_{\boldsymbol{\theta}_0}(x_m))^{n-m} \prod_{j=1}^{m} f_{\boldsymbol{\theta}_0}(x_j), & \text{ⓒ } D_\tau = k \in \{m, \ldots, n\} \end{cases}$$

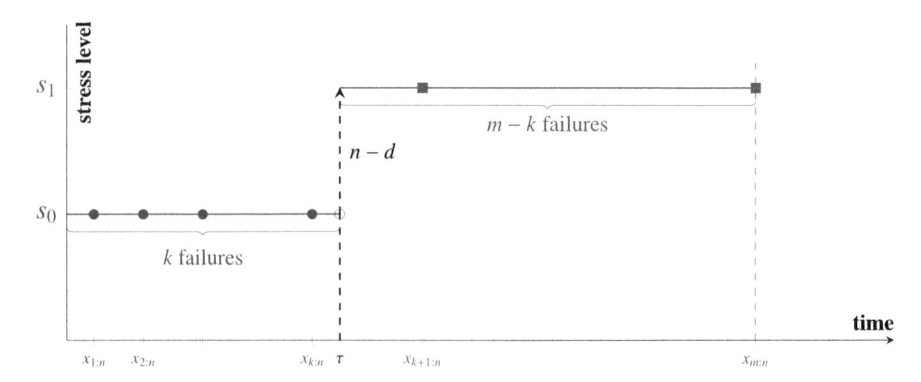

Figure 9.3 Simple step-stress model with change time τ under Type-II censoring (with $X_{m:n} > \tau$).

For $Exp(\vartheta_i)$-distributions, $i = 1, 2$, this leads to the expression

$$\begin{aligned}
&\mathscr{L}(\vartheta_0, \vartheta_1 \mid \boldsymbol{x}_m, m)\\
&= \frac{n!}{(n-m)!}\begin{cases} \vartheta_0^{-k}\vartheta_1^{-(m-k)}e^{-\mathrm{TTT}_{0,k}/\vartheta_0 - \mathrm{TTT}_{1,k}/\vartheta_1}, & \circledcirc\, D_\tau = k \in \{0,\dots,m-1\}\\ \vartheta_0^{-m}e^{-\mathrm{TTT}_{0,m}/\vartheta_0}, & \copyright\, D_\tau = k \in \{m,\dots,n\}\end{cases}\\
&= \frac{n!}{(n-m)!}\vartheta_0^{-k}\vartheta_1^{-(m-k)}e^{-\mathrm{TTT}_{0,k}/\vartheta_0 - \mathrm{TTT}_{1,k}/\vartheta_1}
\end{aligned}$$

where[1]

$$\mathrm{TTT}_{1,k} = \sum_{j=k+1}^{m}(x_j - \tau) + (n-m)(x_m - \tau), \qquad \mathrm{TTT}_{0,k} = \sum_{j=1}^{k\wedge m} x_j + (n-k)w_0$$

and

$$w_0 = \begin{cases} \tau, & k \in \{0,\dots,m-1\}\\ x_m, & k \in \{m,\dots,n\}\end{cases}.$$

Remark 9.2. Notice that the observations $x_{k+1},\dots,x_m$ used in $\mathrm{TTT}_{1,k}$ are realizations of the random variables $X_{k+1:n}^{(\tau)},\dots,X_{m:n}^{(\tau)}$. This is important in the notation of the respective estimators and, of course, in the derivation of the respective distributions. Therefore, $\widehat{\vartheta}_0$ is a function of $X_{1:n},\dots,X_{k:n}$, whereas $\widehat{\vartheta}_1$ is a function of $X_{k+1:n},\dots,X_{n:n}$ (provided the MLE exists).

Clearly, the MLE for ϑ_0 and ϑ_1 exist only iff $D_\tau > 0$ and $D_\tau < m$, respectively. Thus, we get by standard arguments the MLEs

[1] For convenience, we use in the following the convention that $\sum_{j=\ell}^{\nu}\cdots = 0$ when $\ell > \nu$.

(a) for $D_\tau = 0$ ⓞ: $\widehat{\vartheta}_1 = \frac{1}{m}\mathrm{TTT}_{1,0} = \frac{1}{m}\Big(\sum_{j=1}^{m}(X_{j:n}^{(\tau)} - \tau) + (n-m)(X_{m:n}^{(\tau)} - \tau)\Big)$,

(b) for $0 < D_\tau < m$ ⓞ: $\widehat{\vartheta}_0 = \frac{1}{D_\tau}\mathrm{TTT}_{0,D_\tau}$, $\widehat{\vartheta}_1 = \frac{1}{m-D_\tau}\mathrm{TTT}_{1,D_\tau}$,

(c) for $D_\tau \geq m$ ©: $\widehat{\vartheta}_0 = \frac{1}{m}\mathrm{TTT}_{0,m} = \frac{1}{m}\Big(\sum_{j=1}^{k} X_{j:n} + (n-m)X_{m:n}\Big)$

The choice $m = n$ yields the MLEs for the complete sample case. Notice that $\widehat{\vartheta}_0$ equals the estimator under Type-I hybrid censoring with stopping time $T^* = X_{m:n} \wedge \tau$. Thus, its (conditional) distribution can be taken from the results presented in Chapter 5. The distribution of $\widehat{\vartheta}_1$ can be obtained by conditioning on D_τ. Here, we get that the distribution of

$$\widehat{\vartheta}_1 \mid D_\tau < m$$

is a mixture of scaled χ^2-distributions.

Theorem 9.3. *Let $D_\tau < m$ and $\widehat{\vartheta}_1$ be the MLE of ϑ_1. Then, the conditional density function of $\widehat{\vartheta}_1$, given $0 \leq D_\tau < m$, has the form*

$$f^{\widehat{\vartheta}_1 | D_\tau < m}(t) = \frac{2}{\vartheta_1 \overline{F}_{m:n}(\tau)} \sum_{d=0}^{m-1} (m-d)\,\Pr(D_\tau = d) f_{\chi^2_{2m-2d}}\Big(\frac{2(m-d)t}{\vartheta_1}\Big), \quad t \in \mathbb{R}.$$

The probabilities $\Pr(D_\tau = d)$ *are given in* (4.4) *with* T *replaced by* τ.

Remark 9.4. **(a)** As for censored data from an iid sample, the above results can be used to construct exact confidence intervals etc. The stochastic monotonicity of the MLE has been established in Balakrishnan and Iliopoulos (2010).

(b) A summary of results on Type-II censored step-stress is provided by Balakrishnan (2009, Section 3). Balakrishnan and Han (2008) and Ganguly and Kundu (2016) discussed step-stress testing for exponentially distributed lifetimes in presence of competing risks and Type-II censoring.

9.2.2 Type-I censoring

The situation of a step-stress model under Type-I censoring with threshold $T > \tau$ is depicted in Fig. 9.4. This model has been discussed by Balakrishnan et al. (2009c) for exponentially distributed lifetimes. From (9.5), we can now derive the joint density function under Type-I censoring. Notice that, by construction, the step-stress data satisfies the block independence property but, of course, with different distributions. Since by assumption $\tau < T$, we have $D_\tau \leq D$ with D as in (2.12). D_τ is only observable when $D_\tau < D$. Therefore, we get for $t_1 \leq \cdots \leq t_n$ and $0 \leq d \leq n$,

$$f_{0,1}^{X_{1:n}^{(\tau)},\dots,X_{D:n}^{(\tau)},D_\tau,D}(\boldsymbol{t}_d, k, d) = \frac{n!}{(n-d)!} \tag{9.7}$$

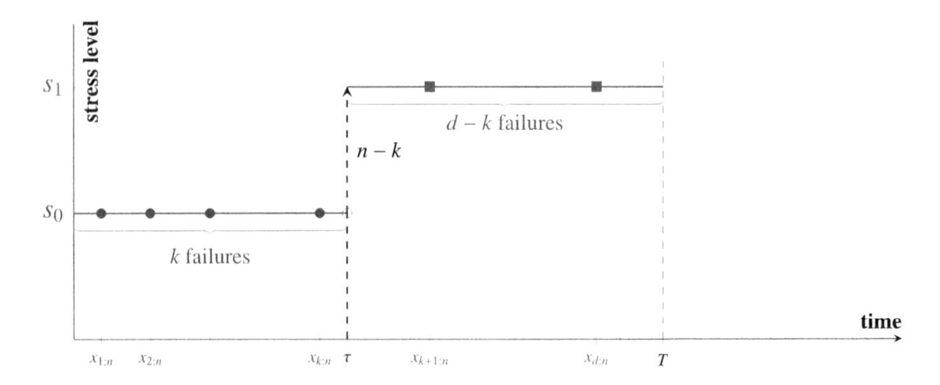

Figure 9.4 Simple step-stress model with change time τ and time threshold $T > \tau$.

$$\times \begin{cases} \prod_{j=1}^{n} f_0(t_j), & ⓒ\, k = d = n \\ \Big(\prod_{j=1}^{d} f_0(t_j)\Big)\Big(1 - F_1(T + \nu_1 - \tau)\Big)^{n-d}, & ⓒ\, k \geq d < n \\ \Big(\prod_{j=1}^{k} f_0(t_j)\Big)\Big(\prod_{j=k+1}^{n} f_1\big(t_j + \nu_1 - \tau\big)\Big), & ⓞ\, k < d = n \\ \Big(\prod_{j=1}^{k} f_0(t_j)\Big)\Big(\prod_{j=k+1}^{d} f_1\big(t_j + \nu_1 - \tau\big)\Big)\Big(1 - F_1(T + \nu_1 - \tau)\Big)^{n-d}, & ⓞ\, k < d < n \end{cases}.$$

This yields the likelihood function

$$\begin{aligned} &\mathscr{L}(\boldsymbol{\theta}_0, \boldsymbol{\theta}_1 \mid \boldsymbol{x}_d, d) \\ &\quad = \frac{n!}{(n-d)!}\Big(\prod_{j=1}^{k} f_0(x_j) \prod_{j=k+1}^{d} f_1\big(x_j + \nu_1 - \tau\big)\Big)\Big(1 - F_1(T + \nu_1 - \tau)\Big)^{n-d} \end{aligned}$$

where $x_1 \leq \cdots \leq x_d \leq T$ and k as in (9.7).

For $Exp(\vartheta_i)$-distributions with $\vartheta_i > 0$, $i = 1, 2$, this yields the likelihood function

$$\mathscr{L}(\vartheta_0, \vartheta_1 \mid \boldsymbol{x}_d, d) = \frac{n!}{(n-d)!} \vartheta_0^{-k} \vartheta_1^{-(d-k)} e^{-\mathsf{TTT}_{0,k}/\vartheta_0 - \mathsf{TTT}_{1,k}/\vartheta_1},$$

where

$$\mathsf{TTT}_{1,k,d} = \sum_{j=k+1}^{d} (x_j - \tau) + (n-d)(T - \tau), \qquad \mathsf{TTT}_{0,k} = \sum_{j=1}^{k} x_j + (n-k)\tau$$

denote total time on test statistics on stress levels s_1 and s_0, respectively. Notice that $\mathsf{TTT}_{0,n} = \sum_{j=1}^{n} x_j$.

As under Type-II censoring, the MLEs of ϑ_0 and ϑ_1 exist only iff $D_\tau > 0$ and $D_\tau < D$, respectively. Thus, we get by standard arguments the MLEs

(a) for $D_\tau = 0 < D$ ◎: $\widehat{\vartheta}_1 = \frac{1}{D}\mathsf{TTT}_{1,0,D} = \frac{1}{D}\Big(\sum_{j=1}^{D}(X_{j:n}^{(\tau)} - \tau) + (n-D)(T-\tau)\Big)$,
(b) for $0 < D_\tau < D$ ◎: $\widehat{\vartheta}_0 = \frac{1}{D_\tau}\mathsf{TTT}_{0,D_\tau}$, $\widehat{\vartheta}_1 = \frac{1}{D-D_\tau}\mathsf{TTT}_{1,D_\tau,D}$,
(c) for $D_\tau \geq D$ ©: $\widehat{\vartheta}_0 = \frac{1}{D}\mathsf{TTT}_{0,D} = \frac{1}{D}\Big(\sum_{j=1}^{D} X_{j:n} + (n-D)X_{D:n}\Big)$.

Notice that $\widehat{\vartheta}_0$ equals the estimator under Type-I censoring with stopping time $T^* = \tau$.

Clearly, the distribution of the estimators can only be obtained conditionally on the assumption that they exist. Balakrishnan et al. (2009c) utilized a (conditional) moment generating function-approach to establish explicit representations. If both MLEs are supposed to exist then one has to condition on the event

$$A = \{0 < D_\tau < D\}$$

ensuring that both D_τ and $D - D_\tau$ are positive. If one is only interested in one of the MLEs, then one can restrict the discussion to the events $D_\tau > 0$ (for ϑ_0) and $D > D_\tau$ (for ϑ_1), respectively. The distribution of $\widehat{\vartheta}_0$ can be directly taken from the discussion of Type-I censoring with T replaced by τ. For $\widehat{\vartheta}_1$, we can write

$$\begin{aligned}
&\Pr(X_{j:n}^{(\tau)} \leq t_j, 1 \leq j \leq D \mid D > D_\tau) \\
&= \frac{1}{\Pr(D > D_\tau)} \sum_{k=0}^{n-1} \sum_{d=k+1}^{n} \Pr(X_{j:n}^{(\tau)} \leq t_j, 1 \leq j \leq d, D_\tau = k, D = d) \\
&= \frac{1}{\Pr(D > D_\tau)} \sum_{k=0}^{n-1} \sum_{d=k+1}^{n} \int_{-\infty}^{t_1} \int_{x_1}^{t_2} \cdots \int_{x_{d-1}}^{t_d} f^{X_{j:n}^{(\tau)}, 1 \leq j \leq d, D_\tau, D}(\boldsymbol{x}_d, k, d)\, d\boldsymbol{x}_d
\end{aligned}$$

where the density function can be taken from (9.6). Furthermore, the joint probability mass function of D_τ and D is needed, that is,

$$\begin{aligned}
\Pr(D_\tau = k, D = d) &= \binom{n}{k}\binom{n-k}{d-k} \\
&\quad \times F_{\boldsymbol{\theta}_0}^{k}(\tau)\big(F_{\boldsymbol{\theta}_1}(T + \nu_1 - \tau) - F_{\boldsymbol{\theta}_0}(\tau)\big)^{d-k}\big(1 - F_{\boldsymbol{\theta}_1}(T + \nu_1 - \tau)\big)^{n-d},
\end{aligned}$$

for $(k, d) \in \{(i, j) \mid 1 \leq i \leq j \leq n - 1\}$ for general $F_{\boldsymbol{\theta}_0}, F_{\boldsymbol{\theta}_1}$. Finally, the probability of the event A is given by (cf. Balakrishnan et al., 2009c)

$$\Pr(A) = 1 - \big(1 - F_{\boldsymbol{\theta}_0}(\tau)\big)^n - \big(1 - F_{\boldsymbol{\theta}_1}(T + \nu_1 - \tau) + F_{\boldsymbol{\theta}_0}(\tau)\big)^n + \big(1 - F_{\boldsymbol{\theta}_1}(T + \nu_1 - \tau)\big)^n. \tag{9.8}$$

From (9.8), we get for exponentially distributed lifetimes with mean ϑ_i, $i = 1, 2$,

$$\Pr(A) = 1 + e^{-n\tau/\vartheta_0}(e^{-n(T-\tau)/\vartheta_1} - 1) - \big(1 + e^{-\tau/\vartheta_0}(e^{-(T-\tau)/\vartheta_1} - 1)\big)^n. \tag{9.9}$$

Using these results, the (conditional) distribution of the MLEs can be derived. A derivation of the (conditional) joint density function $f^{\widehat{\vartheta}_0,\widehat{\vartheta}_1|A}$ is given in Górny and Cramer (2020a) which results in a representation in terms of B-spline functions. For $\boldsymbol{t}_2 \in [0,\infty)^2$, the joint density function of the MLEs $\widehat{\vartheta}_0$ and $\widehat{\vartheta}_1$ conditionally on the event A is given by

$$f^{\widehat{\vartheta}_0,\widehat{\vartheta}_1|A}(\boldsymbol{t}_2) = \frac{1}{\Pr(A)} \sum_{k=1}^{n-1} \sum_{d=k+1}^{n} k(d-k)\binom{n}{k}\binom{n-k}{d-k}\left(\frac{\tau}{\vartheta_0}\right)^k \left(\frac{T-\tau}{\vartheta_1}\right)^{d-k}$$
$$\times B_{k-1}(kt_1 \mid (n-k)\tau, \dots, n\tau)\, e^{-kt_1/\vartheta_0}$$
$$\times B_{d-k-1}((d-k)t_2 \mid (n-d)(T-\tau), \dots, (n-k)(T-\tau))\, e^{-(d-k)t_2/\vartheta_1}$$

with $\Pr(A)$ as in (9.9). The (conditional) marginal density functions are given by:

(a) The conditional density function of the MLE $\widehat{\vartheta}_0$ in terms of B-spline functions is given by (for $t \geq 0$)

$$f^{\widehat{\vartheta}_0|A}(t) = \frac{1}{\Pr(A)} \sum_{k=1}^{n-1} k \binom{n}{k}\left(1 - e^{-(n-k)(T-\tau)/\vartheta_0}\right)\left(\frac{\tau}{\vartheta_0}\right)^k$$
$$\times B_{k-1}(kt \mid (n-k)\tau_1, \dots, n\tau)\, e^{-kt/\vartheta_0}.$$

(b) The density function $f^{\widehat{\vartheta}_1|A}$ in terms of B-spline functions is given by (for $t \geq 0$)

$$f^{\widehat{\vartheta}_1|A}(s) = \frac{1}{\Pr(A)} \sum_{k=1}^{n-1} \binom{n}{k}\left(1 - e^{-\tau/\vartheta_0}\right)^k e^{-(n-k)\tau/\vartheta_0} \sum_{d=k+1}^{n} (d-k)\left(\frac{T-\tau}{\vartheta_1}\right)^{d-k}\binom{n-k}{d-k}$$
$$\times B_{d-k-1}\big((d-k)t \mid (n-d)(T-\tau), \dots, (n-k)(T-\tau)\big)e^{-(d-k)t/\vartheta_1}.$$

Remark 9.5. **(a)** As for censored data from an iid sample, the above results can be used to construct exact confidence intervals etc. The stochastic monotonicity of the MLEs has been established in Balakrishnan and Iliopoulos (2010) (see also Balakrishnan, 2009, Section 4). A multi-sample model has been proposed in Kateri et al. (2010).

(b) Weibull distributions have been studied in Kateri and Balakrishnan (2008). Log-normally distributed lifetimes have been discussed in Balakrishnan et al. (2009d) whereas results for uniform distributions can be found in Górny and Cramer (2020a).

(c) Type-I censored step stress data in presence of competing risks has been analyzed by Han and Balakrishnan (2010).

(d) Multiple step-stress models under time constraint have been considered in Balakrishnan and Alam (2019) (Birnbaum-Saunders distribution) and Chen and Gui (2021) (two-parameter Rayleigh distribution).

9.3. Step stress models under hybrid censoring

Using the same tools and applying the same methodology as under Type-I and Type-II censoring, exact (conditional) distributional results can be established using the modularization approach (or the moment generating function method). Therefore, we do not present further details here and restrict ourselves to some basic results. Note that similar comments apply to progressive hybrid censoring schemes (see also the comments in Section 9.3.3).

9.3.1 Type-I hybrid censoring

Balakrishnan and Xie (2007a) proposed a step-stress model for exponential lifetimes in presence of Type-I hybrid censoring with some threshold T exceeding the stress changing time τ (see also Balakrishnan, 2009, Section 5). Their model is based on the following assumptions:

(1) m $(\leq n)$ and $0 < \tau < T$ are fixed in advance where τ denotes the fixed time at which the stress level is changed from s_0 to s_1;

(2) T denotes a fixed maximum time allowed for the experiment;

(3) $X_{1:n}^{(\tau)} \leq \cdots \leq X_{n:n}^{(\tau)}$ denote the ordered failure times of the n units under test (assuming a cumulative exposure model);

(4) $T^* = X_{m:n} \wedge T$ denotes the random time when the life-testing experiment is terminated. $D_{\mathsf{I}} = D \wedge m$ denotes the number of observed failures.

For step-stress model under Type-I hybrid censoring with threshold $T > \tau$ we get the following results. As mentioned above, D_τ can only be observed when $D_\tau < D_{\mathsf{I}} = D \wedge m$. Notice that $X_{m:n} < \tau$ implies that $D_\tau \geq m$. Thus, for $t_1 \leq \cdots \leq t_n$ and $0 \leq d \leq n$,

$$f_{0,1}^{X_{1:n}^{(\tau)},\ldots,X_{D_{\mathsf{I}}:n}^{(\tau)},D_\tau,D_{\mathsf{I}}}(\boldsymbol{t}_d, k, d) = \frac{n!}{(n-d)!}$$

$$\times \begin{cases} \prod_{j=1}^{m} f_0(t_j)(1-F_0(t_m))^{n-m}, & \text{ⓒ}\, k \geq d = m \\ \left(\prod_{j=1}^{d} f_0(t_j)\right)\left(1-F_1(T+v_1-\tau)\right)^{n-d}, & \text{ⓒ}\, k \geq d < m \\ \left(\prod_{j=1}^{k} f_0(t_j)\right)\left(\prod_{j=k+1}^{d} f_1\left(t_j+v_1-\tau\right)\right)\left(1-F_1(t_m+v_1-\tau)\right)^{n-m}, & \text{ⓞ}\, k < d = m \\ \left(\prod_{j=1}^{k} f_0(t_j)\right)\left(\prod_{j=k+1}^{d} f_1\left(t_j+v_1-\tau\right)\right)\left(1-F_1(T+v_1-\tau)\right)^{n-d}, & \text{ⓞ}\, k < d < m \end{cases}.$$

This yields the likelihood function

$$\mathscr{L}(\boldsymbol{\theta}_0, \boldsymbol{\theta}_1 \mid \boldsymbol{x}_d, d) = \frac{n!}{(n-d)!}$$

$$\times \begin{cases} \prod_{j=1}^{m} f_{\boldsymbol{\theta}_0}(x_j)(1-F_{\boldsymbol{\theta}_0}(x_m))^{n-m} & \text{ⓒ}\, d = m \\ \left(\prod_{j=1}^{d} f_{\boldsymbol{\theta}_0}(x_j)\right)\left(1-F_{\boldsymbol{\theta}_1}(T+v_1-\tau)\right)^{n-d}, & \text{ⓒ}\, d < m \\ \left(\prod_{j=1}^{k} f_{\boldsymbol{\theta}_0}(x_j)\right)\left(\prod_{j=k+1}^{d} f_{\boldsymbol{\theta}_1}\left(x_j+v_1-\tau\right)\right)\left(1-F_{\boldsymbol{\theta}_1}(w_1+v_1-\tau)\right)^{n-d}, & \text{ⓞ}\, k < d \end{cases}$$

where $x_1 \leq \cdots \leq x_d \leq T$ and

$$w_1 = \begin{cases} x_m, & d = m \\ T, & d \in \{0, \ldots, m-1\} \end{cases}.$$

For $Exp(\vartheta_i)$-distributions, $i = 1, 2$, this leads to the expression

$$\mathcal{L}(\vartheta_0, \vartheta_1 \mid \boldsymbol{x}_d, d) = \frac{n!}{(n-d)!} \begin{cases} \vartheta_0^{-m} e^{-\mathrm{TTT}_{0,d}/\vartheta_0}, & \copyright\ d = m \\ \vartheta_0^{-d} e^{-\mathrm{TTT}_{0,d}/\vartheta_0 - \mathrm{TTT}_{1,d,d}/\vartheta_1}, & \copyright\ d < m \\ \vartheta_0^{-k} \vartheta_1^{-(d-k)} e^{-\mathrm{TTT}_{0,k}/\vartheta_0 - \mathrm{TTT}_{1,k,d}/\vartheta_1}, & \circledcirc\ k < d \end{cases}$$

where

$$\mathrm{TTT}_{1,k,d} = \sum_{j=k+1}^{d} (x_j - \tau) + (n-d)(w_1 - \tau), \quad \mathrm{TTT}_{0,k} = \sum_{j=1}^{k} x_j + (n-k) w_0$$

and

$$w_0 = \begin{cases} x_m, & d = m \\ \tau, & d \in \{0, \ldots, m-1\} \end{cases}.$$

The corresponding MLEs are given by

(a) for $D_\tau = 0 < D_{\mathrm{I}}$ $\circledcirc$: $\widehat{\vartheta}_1 = \frac{1}{D} \mathrm{TTT}_{1,0,D} = \frac{1}{D} \Big(\sum_{j=1}^{D} (X_{j:n}^{(\tau)} - \tau) + (n-D)(W_1 - \tau) \Big)$,

(b) for $0 < D_\tau < D_{\mathrm{I}}$ $\circledcirc$: $\widehat{\vartheta}_0 = \frac{1}{D_\tau} \mathrm{TTT}_{0,D_\tau}$, $\widehat{\vartheta}_1 = \frac{1}{D - D_\tau} \mathrm{TTT}_{1,D_\tau,D}$,

(c) for $D_\tau \geq D_{\mathrm{I}}$ $\copyright$: $\widehat{\vartheta}_0 = \frac{1}{D} \mathrm{TTT}_{0,D} = \frac{1}{D} \Big(\sum_{j=1}^{D} X_{j:n} + (n-D) W_0 \Big)$.

9.3.2 Type-II hybrid censoring

Type-II hybrid censoring for step-stress data has been addressed in Balakrishnan and Xie (2007b) (see also Balakrishnan, 2009, Section 5). Using similar arguments as for Type-I hybrid censoring, the joint density function is obtained as follows. As before, D_τ can only be observed when $D_\tau < D_{\mathrm{II}} = D \vee m$. Thus, for $t_1 \leq \cdots \leq t_n$ and $m \leq d \leq n$, we get the following expressions:

(a) $d = m$: Here, the life test may be terminated at T or at the m-th failure time so that the density function reads

$$f_{0,1}^{X_{1:n}^{(\tau)}, \ldots, X_{D_{\mathrm{I}}:n}^{(\tau)}, D_\tau, D_{\mathrm{II}}}(\boldsymbol{t}_d, k, d) = \frac{n!}{(n-d)!} \tag{9.10}$$

$$\times \begin{cases} \left(\prod_{j=1}^{m} f_0(t_j)\right)\left(1 - F_1(T + v_1 - \tau)\right)^{n-m}, & ©\, k \geq d \\ & t_m \leq T \\ \left(\prod_{j=1}^{m} f_0(t_j)\right)\left(1 - F_1(t_m + v_1 - \tau)\right)^{n-m}, & ©\, k \geq d \\ & t_m > T \\ \left(\prod_{j=1}^{k} f_0(t_j)\right)\left(\prod_{j=k+1}^{m} f_1(t_j + v_1 - \tau)\right)\left(1 - F_1(t_m + v_1 - \tau)\right)^{n-m}, & ©\, k < d \\ & t_m \leq T \\ \left(\prod_{j=1}^{k} f_0(t_j)\right)\left(\prod_{j=k+1}^{m} f_1(t_j + v_1 - \tau)\right)\left(1 - F_1(T + v_1 - \tau)\right)^{n-m}, & ©\, k < d \\ & t_m > T \end{cases}.$$

(b) $m < d \leq n$: Now, the experiment is terminated either at T (when $d < n$) or at the last failure time (that is $d = n$; this case does not need to be noted separately since the following formulas yield the right value for $d = n$). Therefore, we have

$$f_{0,1}^{X_{1:n}^{(\tau)},\ldots,X_{D_{\text{II}}:n}^{(\tau)},D_\tau,D_{\text{II}}}(\boldsymbol{t}_d, k, d) = \frac{n!}{(n-d)!} \tag{9.11}$$

$$\times \begin{cases} \left(\prod_{j=1}^{d} f_0(t_j)\right)\left(1 - F_1(T + v_1 - \tau)\right)^{n-d}, & ©\, d \leq k \\ \left(\prod_{j=1}^{k} f_0(t_j)\right)\left(\prod_{j=k+1}^{d} f_1(t_j + v_1 - \tau)\right)\left(1 - F_1(T + v_1 - \tau)\right)^{n-d}, & ©\, k < d \end{cases}.$$

The different cases given in (9.10) and (9.11) can be combined by introducing the notation

$$w_1 = \begin{cases} x_m, & d = m, x_m > T \\ T, & d = m, x_m \leq T \\ T, & d \in \{m+1, \ldots, n-1\} \\ x_n, & d = n \end{cases}.$$

Therefore, the likelihood function is obtained as

$$\mathcal{L}(\boldsymbol{\theta}_0, \boldsymbol{\theta}_1 \mid \boldsymbol{x}_d, d)$$
$$= \frac{n!}{(n-d)!}\left(\prod_{j=1}^{k} f_{\boldsymbol{\theta}_0}(t_j)\right)\left(\prod_{j=k+1}^{d} f_{\boldsymbol{\theta}_1}(t_j + v_1 - \tau)\right)\left(1 - F_{\boldsymbol{\theta}_1}(w_1 + v_1 - \tau)\right)^{n-d}$$

with $D_\tau = k$ determined from $x_1 \leq \cdots \leq x_d$.

For $Exp(\vartheta_i)$-distributions, $i = 1, 2$, this leads to the expression

$$\mathcal{L}(\vartheta_0, \vartheta_1 \mid \boldsymbol{x}_d, d) = \frac{n!}{(n-d)!}\vartheta_0^{-k}\vartheta_1^{-(d-k)}e^{-\text{TTT}_{0,k}/\vartheta_0 - \text{TTT}_{1,k,d}/\vartheta_1},$$

Table 9.1 Types of likelihood functions for step-stress testing under (hybrid) censoring schemes with stress change time τ. d denotes the realization of the observed sample size D^* whereas $D_\tau = k$ denotes the number of failures observed under stress level s_0. Values of D_τ marked by † are not observable since the experiment is terminated before τ.

censoring scheme	parameters			
	D_τ=k	D_{HCS}=d	w_0	w_1
Type-I	$0, \dots, n-1$	$k, \dots, n-1$		
	n	n		
Type-II	$0, \dots, m-1$	m	τ	
	$m^\dagger, \dots, n^\dagger$	m	x_m	
Type-I hybrid	$0, \dots, d-1$	$1, \dots, m-1$	τ	T
	$0, \dots, d-1$	m	τ	x_m
	$d^\dagger$	$1, \dots, m-1$	τ	T
	$m^\dagger$	m	x_m	
Type-II hybrid	$0, \dots, m-1$	$m\,(x_m > T)$		x_m
	$0, \dots, d^\dagger$	$m, \dots, n-1\,(x_m \le T)$		T
	$0, \dots, n$	n		x_n

where

$$\mathsf{TTT}_{1,k,d} = \sum_{j=k+1}^{d} (x_j - \tau) + (n-d)(w_1 - \tau), \qquad \mathsf{TTT}_{0,k} = \sum_{j=1}^{k} x_j + (n-k)\tau.$$

The corresponding MLEs are given by

(a) for $D_\tau = 0 < D_{\text{II}}$ ⓞ: $\widehat{\vartheta}_1 = \frac{1}{D_{\text{II}}}\mathsf{TTT}_{1,0,D_{\text{II}}} = \frac{1}{D}\Big(\sum_{j=1}^{D_{\text{II}}}(X_{j:n}^{(\tau)} - \tau) + (n - D_{\text{II}})(W_1 - \tau)\Big)$,

(b) for $0 < D_\tau < D_{\text{II}}$ ⓞ: $\widehat{\vartheta}_0 = \frac{1}{D_\tau}\mathsf{TTT}_{0,D_\tau}$, $\widehat{\vartheta}_1 = \frac{1}{D_{\text{II}} - D_\tau}\mathsf{TTT}_{1,D_\tau,D_{\text{II}}}$,

(c) for $D_\tau \ge D_{\text{II}}$ ©: $\widehat{\vartheta}_0 = \frac{1}{D_{\text{II}}}\mathsf{TTT}_{0,D_{\text{II}}} = \frac{1}{D}\Big(\sum_{j=1}^{D_{\text{II}}} X_{j:n} + (n - D_{\text{II}})W_0\Big)$.

Table 9.1 summarizes the choices for w_0 and w_1 used in the particular censoring schemes.

9.3.3 Further hybrid censoring schemes

As for hybrid censored iid data, more complex hybrid censoring schemes can be studied for step-stress data along the same lines as the above sketched Type-I/Type-II (hybrid) censoring schemes. This includes likelihood inference, Bayesian inference, confidence intervals, prediction, competing risks, etc. Of course, the analysis becomes more complicated due to the increasing number of different cases to be taken into account. However, in principle, the analysis follows the same path and uses the same techniques and algorithms. Therefore, we do not present further details here and refer to the respective literature. Subsequently, we present a selection of papers dealing with such extensions, knowing that the published literature is growing very rapidly and a complete coverage is hardly possible.

Exact inference for a simple step-stress model with generalized Type-I hybrid censored data from the exponential distribution has been considered in Shafay (2016b). Lin et al. (2019) established results for a log-location-scale family of lifetime distributions under Type-I hybrid censoring. Progressive Type-I hybrid censored exponential step-stress data has been studied by Ling et al. (2009). For Weibull distributions, we refer to Ismail (2014). Wang and Gui (2021) investigated generalized progressive hybrid censored step-stress accelerated dependent competing risks model for Marshall-Olkin bivariate Weibull distribution. Competing risks models have been studied in Mao et al. (2021).

CHAPTER 10

Applications in reliability

Contents

10.1. Introduction

In this chapter, we present some applications of (progressive) hybrid censoring schemes to particular data settings discussed in reliability. We start with a discussion of competing risks models which is followed by some comments on inference in stress-strength models under hybrid censored data. Then, we sketch some problems on optimal designs which addresses the problem to choose optimal design parameters (like threshold T etc.). Finally, reliability acceptance sampling plans are briefly considered.

10.2. Competing risks analysis

In competing risk modeling, it is assumed that a unit may fail due to several causes of failure. For two competing risks, the lifetime of the i-th unit is given by

$$X_i = \min\{X_{1i}, X_{2i}\}, \quad i = 1, \ldots, n,$$

where X_{ji} denotes the latent failure time of the i-th unit under the j-th cause of failure, $j = 1, 2$. We assume that the latent failure times are independent, where $X_{ji} \sim F_j$ with density function f_j, $j = 1, 2$, $i = 1, \ldots, m$. This ensures that the distributions of the latent failure times entirely determine the competing risks model (see, e.g., Tsiatis, 1975; Crowder, 2001, and Pintilie, 2006). Furthermore, it is assumed that the cause of each failure is known so that the available data are given by

$$(X_1, C_1), (X_2, C_2), \ldots, (X_n, C_n), \tag{10.1}$$

Hybrid Censoring Know-How
https://doi.org/10.1016/B978-0-12-398387-9.00018-0

where $C_i = 1$ if the i-th failure is due to the first cause and $C_i = 2$ otherwise. The observed data is denoted by $(x_1, c_1), (x_2, c_2), \ldots, (x_n, c_n)$. Further, we define the indicators

$$\delta_{i,j} = \mathbb{1}_{\{j\}}(C_i) = \begin{cases} 1, & C_i = j \\ 0, & \text{else} \end{cases}.$$

Thus, the random variables

$$N_1 = \sum_{i=1}^{n} \mathbb{1}_{\{1\}}(C_i) = \sum_{i=1}^{n} \mathbb{1}_{(-\infty, X_{2i}]}(X_{1i}) \quad \text{and} \quad N_2 = n - N_1 = \sum_{i=1}^{n} \mathbb{1}_{\{2\}}(C_i)$$

describe the number of failures due to the first and the second cause of failure, respectively. Observe that both N_1 and N_2 are binomials with sample size n and probability of success $\Pr(X_{11} \leq X_{21})$ and $1 - \Pr(X_{11} \leq X_{21})$, respectively.

Then,

$$\begin{aligned}\Pr(X_i \leq x_i, C_i = 1) &= \Pr(X_i \leq x_i, X_{1i} \leq X_{2i}) = \int \mathbb{1}_{(-\infty, x_i]}(x) P(x \leq X_{2i}) f_1(x)\, dx \\ &= \int \mathbb{1}_{(-\infty, x_i]}(x)(1 - F_2(x)) f_1(x)\, dx.\end{aligned}$$

Similarly, $\Pr(X_i \leq x_i, C_i = 2) = \int \mathbb{1}_{(-\infty, x_i]}(x)(1 - F_1(x)) f_2(x)\, dx$ so that the joint density function is given by

$$f^{X_i, C_i}(x_i, c_i) = \prod_{j=1}^{2} h_j^{\delta_{i,j}}(x_i) \cdot \prod_{j=1}^{2} (1 - F_j(x_i)),$$

where $h_j = f_j/(1 - F_j)$ denotes the hazard rate of F_j, $j = 1, 2$. The joint density function (w.r.t. the measure $\lambda^n \otimes \#^n$) of an iid sample is given by

$$f^{\mathbf{X}, \mathbf{C}}(\mathbf{x}_n, \mathbf{c}_n) = \prod_{i=1}^{n} \left[\prod_{j=1}^{2} h_j^{\delta_{i,j}}(x_i) \cdot \prod_{j=1}^{2} (1 - F_j(x_i)) \right]. \tag{10.2}$$

In general, the density function under Type-II censoring can not be established in a closed form expression. However, assuming proportional hazard rates for the underlying cumulative distribution functions, that is,

$$F_1 = F, \quad 1 - F_2 = (1 - F)^a \text{ for some } a > 0 \tag{10.3}$$

and for some cumulative distribution function F, the situation simplifies considerably (see also comments in Cox, 1959). In this case, we have $h_1 = h$ and $h_2 = ah$.

Assumption 10.1. *The subsequent derivations in this section are subject to the assumption of a proportional hazard rate model with cumulative distribution functions as in* (10.3) *for some* $a > 0$.

Under Assumption 10.1, (10.2) reads

$$
\begin{aligned}
f^{\mathbf{X},\mathbf{C}}(\boldsymbol{x}_n, \boldsymbol{c}_n) &= \left[\prod_{i=1}^{n} a^{\delta_{i,2}}\right] \prod_{i=1}^{n} \left(f(x_i) \cdot (1 - F(x_i))^a\right) \\
&= \frac{a^{n_2}}{(a+1)^n} \cdot (a+1)^n \prod_{i=1}^{n} \left(f(x_i) \cdot (1 - F(x_i))^a\right) \\
&= \frac{a^{n_2}}{(a+1)^n} \prod_{i=1}^{n} \psi_a(x_i),
\end{aligned}
$$

where ψ_a is a proper density function with cumulative distribution function $\Psi_a = 1 - (1-F)^{a+1}$. Notice that $n_2 = \sum_{i=1}^{n} \delta_{i,2}$, $\Pr(X_{11} \leq X_{21}) = 1/(a+1)$, and that

$$
P(\boldsymbol{C}_n = \boldsymbol{c}_n) = \frac{a^{n_2}}{(a+1)^n} = \frac{a^{\sum_{i=1}^{n} \delta_{i,2}}}{(a+1)^n}, \qquad \boldsymbol{c}_n \in \{1, 2\}^n,
$$

denotes the probability mass function of $\boldsymbol{C}_n$.

Under Type-II censoring (w.r.t. the first component), the sample in (10.1) under competing risks can be seen as the order statistics combined with the respective concomitants,[1] that is,

$$
\left(X_{1:n}, C_{[1:n]}\right), \left(X_{2:n}, C_{[2:n]}\right), \ldots, \left(X_{m:n}, C_{[m:n]}\right).
$$

Under the proportional hazards rate model (10.3), the corresponding density function is given by

$$
\begin{aligned}
f_{1\ldots m:n;C}(\boldsymbol{x}_m, \boldsymbol{c}_m) &= \frac{a^{m_2}}{(a+1)^m} \frac{n!}{(n-m)!} \left(1 - \Psi_a(x_m)\right)^{n-m} \prod_{i=1}^{m} \psi_a(x_i) \\
&= \frac{a^{m_2}}{(a+1)^m} \times \psi_{1,\ldots,m:n;a}(\boldsymbol{x}_m)
\end{aligned} \tag{10.4}
$$

where $m_2 = \sum_{i=1}^{m} \delta_{i,2}$ and $\psi_{1,\ldots,m:n;a}$ denotes the density function of the first m order statistics from the cumulative distribution function Ψ_a.[2] Furthermore, we get for the joint cumulative distribution function

$$
F_{1,\ldots,m:n;C}(\boldsymbol{x}_m, \boldsymbol{c}_m) = \frac{a^{m_2}}{(a+1)^m} \times \Psi_{1,\ldots,m:n;a}(\boldsymbol{x}_m). \tag{10.5}
$$

[1] For the notion of concomitants, see David (1973) and David and Nagaraja (1998).

[2] For brevity, we denote by $\boldsymbol{C}_d$ (and $\boldsymbol{c}_d$) the first d causes of failure observed (no matter which censoring scheme has been applied).

Remark 10.2. For progressively Type-II censored data under competing risks

$$\left(X_{1:m:n}, C_{[1:m:n]}\right), \left(X_{2:m:n}, C_{[2:m:n]}\right), \dots, \left(X_{m:m:n}, C_{[m:m:n]}\right),$$

one gets under Assumption 10.1 the joint density function

$$\begin{aligned} f_{1,\dots,m:m:n;C}(\boldsymbol{x}_m, \boldsymbol{c}_m) &= \frac{a^{m_2}}{(a+1)^m}\left(\prod_{i=1}^m \gamma_i\right)\prod_{i=1}^m \left(\psi_a(x_i)\left(1-\Psi_a(x_i)\right)^{R_i}\right) \\ &= \frac{a^{m_2}}{(a+1)^m} \times \psi_{1,\dots,m:m:n;a}(\boldsymbol{x}_m) \end{aligned} \tag{10.6}$$

for $x_1 < \cdots < x_m$ (see Kundu et al., 2004). $\psi_{1,\dots,m:m:n;a}$ denotes the density function of a progressively Type-II censored sample based on the cumulative distribution function Ψ_a.

Representations (10.4) and (10.6) illustrate that we can replace the density function f in the standard model by the density function ψ_a under competing risks. This observation is quite important since it allows to simplify the discussion of competing risks data under (progressive) hybrid censoring significantly. In particular, the joint density functions (and, thus, the likelihood functions) can be taken directly from Chapters 4 and 7, respectively. Therefore, we illustrate the procedure for Type-I and Type-II hybrid censoring and provide references for the other (progressive) hybrid censoring schemes.

10.2.1 Type-I hybrid censoring scheme

From the preceding comments, it is clear that we can utilize the results presented in Section 4.3 by taking into account the competing risks model. In order to incorporate the information about the competing risks, we need to consider the random counters

$$D_{I,1} = \sum_{i=1}^{D_I} \mathbb{1}_{\{1\}}(C_i) = \sum_{i=1}^{D_I} \mathbb{1}_{(-\infty, X_{2i}]}(X_{1i}) \quad \text{and} \quad D_{I,2} = D_I - D_{I,1} = \sum_{i=1}^{D_I} \mathbb{1}_{\{2\}}(C_i)$$

which obviously satisfy the equation $D_I = D_{I,1} + D_{I,2}$. Notice that $(D_I, \boldsymbol{C}_{D_I})$ determines the random counters $D_{I,1}$ and $D_{I,2}$ uniquely.

Since a formal derivation of the joint cumulative distribution function and density function is missing in the literature, we provide some details here. We proceed as for the derivation of the cumulative distribution function under Type-I hybrid censoring (see p. 128). Let $t_1, \dots, t_m \in \mathbb{R}$. From the law of total probability, we get

$$\begin{aligned} \Pr(X_{j:n} \le t_j, \boldsymbol{C}_d = \boldsymbol{c}_d, 1 \le j \le D_I) &= \Pr(X_{j:n} \le t_j, \boldsymbol{C}_m = \boldsymbol{c}_m, 1 \le j \le m, D \ge m) \\ &\quad + \sum_{d=0}^{m-1} \Pr(X_{j:n} \le t_j, 1 \le j \le d, \boldsymbol{C}_d = \boldsymbol{c}_d, D = d). \end{aligned}$$

Scenario ①: We get by (4.3) and (10.5)

$$\begin{aligned}\Pr(X_{j:n} \le t_j, 1 \le j \le m, \boldsymbol{C}_m = \boldsymbol{c}_m, D \ge m) &= F_{1,\dots,m:n;C}(\boldsymbol{t}_m \wedge T, \boldsymbol{c}_m)\\ &= \frac{a^{m_2}}{(a+1)^m} \times \Psi_{1,\dots,m:n;a}(\boldsymbol{t}_m \wedge T),\end{aligned}$$

where $m_2 = \sum_{i=1}^m \mathbb{1}_{\{2\}}(c_i)$. Notice that this yields for $t_j \to \infty$, $j = 1, \dots, m$,

$$\Pr(\boldsymbol{C}_m = \boldsymbol{c}_m, D \ge m) = \frac{a^{m_2}}{(a+1)^m} \times \Psi_{1,\dots,m:n;a}(T^{*m}). \tag{10.7}$$

Furthermore, we get

$$\begin{aligned}&\Pr(X_{j:n} \le t_j, 1 \le j \le m, D_{\text{I},2} = m_2, D \ge m)\\ &\quad = \binom{m}{m_2} \frac{a^{m_2}}{(a+1)^m} \times \Psi_{1,\dots,m:n;a}(\boldsymbol{t}_m \wedge T), \quad m_2 \in \{0, \dots, m\}.\end{aligned} \tag{10.8}$$

This representation shows that $X_{j:n}$, $1 \le j \le m$, and $D_{\text{I},2}$ given $D \ge m$ are independent.

Scenario ②: First, we get $\big(1 - F(T)\big)^{(a+1)n} \mathbb{1}_{[T,\infty)}(\min_{1 \le j \le m} t_j)$ for $d = 0$. Supposing $d \in \{1, \dots, m-1\}$, we find by analogy with the derivations on p. 128 and $d_2 = \sum_{i=1}^d \mathbb{1}_{\{2\}}(c_i)$ that

$$\begin{aligned}&\Pr(X_{j:n} \le t_j, 1 \le j \le d, \boldsymbol{C}_d = \boldsymbol{c}_d, D = d)\\ &\quad = \mathbb{1}_{[T,\infty)}(\min_{d+1 \le j \le m} t_j) \Big\{ F_{1,\dots,d:n;C}(\boldsymbol{t}_d \wedge T, \boldsymbol{c}_d)\\ &\qquad - \sum_{c_{d+1}=1}^{2} F_{1,\dots,d+1:n;C}(\boldsymbol{t}_d \wedge T, T, \boldsymbol{c}_{d+1}) \Big\}\end{aligned}$$

From (10.5), we get with $d_2 = \sum_{i=1}^d \delta_{i,2}$

$$\begin{aligned}&= \mathbb{1}_{[T,\infty)}(\min_{d+1 \le j \le m} t_j) \Big\{ \frac{a^{d_2}}{(a+1)^d} \times \Psi_{1,\dots,d:n;a}(\boldsymbol{t}_d \wedge T)\\ &\qquad - \sum_{c_{d+1}=1}^{2} \frac{a^{d_2 + \delta_{d+1,2}}}{(a+1)^{d+1}} \times \Psi_{1,\dots,d+1:n;a}(\boldsymbol{t}_d \wedge T, T) \Big\}\\ &= \mathbb{1}_{[T,\infty)}(\min_{d+1 \le j \le m} t_j) \frac{a^{d_2}}{(a+1)^d} \Big\{ \Psi_{1,\dots,d+1:n;a}(\boldsymbol{t}_d \wedge T) - \Psi_{1,\dots,d:n;a}(\boldsymbol{t}_d \wedge T, T) \Big\}\\ &= \mathbb{1}_{[T,\infty)}(\min_{d+1 \le j \le m} t_j) \frac{a^{d_2}}{(a+1)^d} \binom{n}{d} (1 - \Psi_a(T))^{n-d} \Psi_{1,\dots,d:d;a}(\boldsymbol{t}_d \wedge T).\end{aligned}$$

Notice that this expression also holds for $d=0$ with $\Psi_{1,\dots,0:0;a} \equiv 1$. Furthermore, as in (10.7), we find for $t_j \to \infty$, $j=1,\dots,d$,

$$\Pr(\mathbf{C}_d = \mathbf{c}_d, D=d) = \frac{a^{d_2}}{(a+1)^d}\binom{n}{d}(1-\Psi_a(T))^{n-d}\Psi_{1,\dots,d:d;a}(T^{*d}). \tag{10.9}$$

Moreover, we get

$$\begin{aligned}
&\Pr(X_{j:n} \le t_j, 1 \le j \le m, D_{I,2} = d_2, D=d)\\
&= \binom{d}{d_2}\mathbb{1}_{[T,\infty)}(\min_{d+1\le j\le m} t_j)\frac{a^{d_2}}{(a+1)^d}\binom{n}{d}(1-\Psi_a(T))^{n-d}\Psi_{1,\dots,d:d;a}(\mathbf{t}_d \wedge T),\\
&\qquad d_2 \in \{0,\dots,d\}.
\end{aligned} \tag{10.10}$$

As for $D \ge m$, this representation shows that $X_{j:n}$, $1 \le j \le d$, and $D_{I,2}$ given $D=d$ are independent.

The joint density function in (4.18) yields now the following result for the joint density function under competing risks

$$f^{X_{j:n},1\le j\le D_I,\mathbf{C}_{D_I},D}(\mathbf{t}_d,\mathbf{c}_d,d) = \frac{a^{d_2}}{(a+1)^d}\begin{cases}\frac{1-\Psi_a(T)}{(n-d)\psi_a(T)}\psi_{1,\dots,d+1:n;a}(\mathbf{t}_d,T), & d\in\{1,\dots,m-1\}\\ \psi_{1,\dots,m:n;a}(\mathbf{t}_m), & d=m\end{cases}.$$

Furthermore, the above considerations can be applied to find the probability mass function of $D_{I,1}$ and $D_{I,2}$, respectively. First, notice that

$$\Psi_{1,\dots,m:n;a}(T^{*m}) = \Pr(X_{a,j:n} \le T, 1\le j\le m) = \Pr(X_{a,m:n} \le T) = \Psi_{m:n;a}(T),$$

where $\Psi_{m:n;a}$ denotes the cumulative distribution function of the m-th order statistic $X_{a,j:n}$ with population cumulative distribution function Ψ_a. Then, from (10.7) and (10.9), we get

$$\Pr(D_{I,1} = i_1, D \ge m) = \binom{m}{i_1}\frac{a^{m-i_1}}{(a+1)^m}\Psi_{m:n;a}(T), \tag{10.11}$$

and for $0 \le i_1 \le d \in \{0,\dots,m-1\}$,

$$\Pr(D_{I,1} = i_1, D=d) = \binom{n}{d}\binom{d}{i_1}\frac{a^{d-i_1}}{(a+1)^d}(1-\Psi_a(T))^{n-d}\Psi_{d:d;a}(T). \tag{10.12}$$

Using the law of total probability, this yields directly an expression for the probabilities $\Pr(D_{I,1} = i_1)$, $i_1 \ge 0$, and, thus, the probability mass function of $D_{I,1}$. For instance, this shows that

$$\Pr(D_{I,1} = 0) = \sum_{d=0}^{m-1}\binom{n}{d}\frac{a^d}{(a+1)^d}(1-\Psi_a(T))^{n-d}\Psi_a^d(T) + \frac{a^m}{(a+1)^m}\Psi_{m:n;a}(T).$$

This expression can be found in Kundu and Gupta (2007) in case of exponential distributions. Furthermore, we get with $D \sim bin(n, \Psi_a(T))$ the conditional probability mass functions

$$\Pr(D_{I,1} = i_1 \mid D \geq m) = \binom{m}{i_1}\Big(\frac{1}{a+1}\Big)^{i_1}\Big(1 - \frac{1}{a+1}\Big)^{m-i_1}, \qquad i_1 \in \{0, \dots, m\},$$
$$\Pr(D_{I,1} = i_1 \mid D = d) = \binom{d}{i_1}\Big(\frac{1}{a+1}\Big)^{i_1}\Big(1 - \frac{1}{a+1}\Big)^{d-i_1}, \qquad i_1 \in \{0, \dots, d\},$$

showing that $D_{I,1} \mid D \geq m \sim bin(m, 1/(a+1))$ and $D_{I,1} \mid D = d \sim bin(d, 1/(a+1))$, $d = 1, \dots, m-1$.

10.2.1.1 Exponentially distributed lifetimes

In case of $Exp(\vartheta_1)$- and $Exp(\vartheta_2)$-distributed lifetimes, one has $a = \vartheta_1/\vartheta_2$ so that

$$\Psi_a(t) = 1 - \exp\Big(-\Big(\frac{1}{\vartheta_1} + \frac{1}{\vartheta_2}\Big)t\Big), \quad t \geq 0, \tag{10.13}$$

is the cumulative distribution function of an $Exp(\frac{1}{\vartheta_1} + \frac{1}{\vartheta_2})$-distribution. Therefore, from (5.3), the likelihood function reads

$$\begin{aligned}\mathscr{L}(\vartheta_1, \vartheta_2; \boldsymbol{x}_d, d, d_2) &= \frac{\vartheta_1^{d_2}\vartheta_2^{d-d_2}}{(\vartheta_1 + \vartheta_2)^d} \times \frac{n!}{(n-d)!}\Big(\frac{1}{\vartheta_1} + \frac{1}{\vartheta_2}\Big)^d \\ &\quad \times \exp\Big\{-\Big(\frac{1}{\vartheta_1} + \frac{1}{\vartheta_2}\Big)\Big[\sum_{j=1}^{d} x_j + (n-d)x_m \wedge T\Big]\Big\} \\ &= \frac{n!}{(n-d)!\vartheta_1^{d-d_2}\vartheta_2^{d_2}} e^{-\mathsf{TTT}_d\left(\frac{1}{\vartheta_1} + \frac{1}{\vartheta_2}\right)},\end{aligned} \tag{10.14}$$

for $0 \leq x_1 \leq \cdots \leq x_d \leq T$ (see Kundu and Gupta, 2007), where the total time on test TTT_d is defined as in (5.5), that is, for $x_1 \leq \cdots \leq x_m$,

$$\mathsf{TTT}_d = \sum_{j=1}^{d} x_j + (n-d)x_m \wedge T, \quad d = \sum_{j=1}^{m} \mathbb{1}_{(-\infty, T]}(x_j).$$

Obviously, the likelihood function in (10.14) factorizes into two functions of the same type as in (5.4) so that the MLEs of ϑ_1 and ϑ_2 are given by

$$\widehat{\vartheta}_1 = \frac{1}{D_{I,1}}\mathsf{TTT}_{D_I}, \quad \widehat{\vartheta}_2 = \frac{1}{D_{I,2}}\mathsf{TTT}_{D_I} \tag{10.15}$$

which has first been shown in Kundu and Gupta (2007). Notice that the MLE $\widehat{\vartheta}_j$ exists only iff $D_{I,j} > 0$, $j = 1, 2$.

In order to derive the (conditional) distributions of the MLEs, we use the probability mass functions of the random counters $D_{I,j}$, $j=1,2$, given in (10.11) and (10.12) as well as the representations given in (10.8) and (10.10). Thus, we find for $t \geq 0$

$$\begin{aligned}
&P_{\vartheta_1,\vartheta_2}(\widehat{\vartheta}_1 \leq t, D_{I,1} > 0) \\
&\quad = \sum_{i=1}^{m} \Pr_{\vartheta_1,\vartheta_2}(\widehat{\vartheta}_1 \leq t, D_{I,1} = i, D \geq m) + \sum_{d=1}^{m-1}\sum_{i=1}^{d} \Pr_{\vartheta_1,\vartheta_2}(\widehat{\vartheta}_1 \leq t, D_{I,1} = i, D = d) \\
&\quad = \sum_{i=1}^{m} \Pr_{\vartheta_1,\vartheta_2}(\mathrm{TTT}_m \leq it, D_{I,1} = i, D \geq m) \\
&\qquad + \sum_{d=1}^{m-1}\sum_{i=1}^{d} \Pr_{\vartheta_1,\vartheta_2}(\mathrm{TTT}_d \leq it, D_{I,1} = i, D = d)
\end{aligned}$$

Using that $X_{j:n}$, $1 \leq j \leq d$, and $D_{I,2}$ given $D=d$ are independent (the same for $d=m$ and $D \geq m$; see (10.8) and (10.10)), this leads to

$$\begin{aligned}
&= \sum_{i=1}^{m} \Pr_{\vartheta_1,\vartheta_2}(D_{I,1} = i, D \geq m) \Pr_{\vartheta_1,\vartheta_2}(\mathrm{TTT}_m \leq it \mid D \geq m) \\
&\quad + \sum_{d=1}^{m-1}\sum_{i=1}^{d} \Pr_{\vartheta_1,\vartheta_2}(D_{I,1} = i, D = d) \Pr_{\vartheta_1,\vartheta_2}(\mathrm{TTT}_d \leq it \mid D = d) \\
&= \sum_{i=1}^{m} \Pr_{\vartheta_1,\vartheta_2}(D_{I,1} = i, D \geq m) F_{\vartheta_1,\vartheta_2}^{\mathrm{TTT}_m \mid D \geq m}(it) \\
&\quad + \sum_{d=1}^{m-1}\sum_{i=1}^{d} \Pr_{\vartheta_1,\vartheta_2}(D_{I,1} = i, D = d) F_{\vartheta_1,\vartheta_2}^{\mathrm{TTT}_d \mid D = d}(it).
\end{aligned}$$

As a result, we arrive at the joint density function

$$\begin{aligned}
f_{\vartheta_1,\vartheta_2}^{\widehat{\vartheta}_1 \mid D_{I,1}>0}(t) = \frac{1}{\Pr_{\vartheta_1,\vartheta_2}(D_{I,1} > 0)} \Bigg(&\sum_{i=1}^{m} i \Pr_{\vartheta_1,\vartheta_2}(D_{I,1} = i, D \geq m) f_{\vartheta_1,\vartheta_2}^{\mathrm{TTT}_m \mid D \geq m}(it) \\
&+ \sum_{d=1}^{m-1}\sum_{i=1}^{d} i \Pr_{\vartheta_1,\vartheta_2}(D_{I,1} = i, D = d) f_{\vartheta_1,\vartheta_2}^{\mathrm{TTT}_d \mid D = d}(it) \Bigg), \quad t \geq 0,
\end{aligned}$$

where $f_{\vartheta_1,\vartheta_2}^{\mathrm{TTT}_m \mid D \geq m}$ and $f_{\vartheta_1,\vartheta_2}^{\mathrm{TTT}_d \mid D = d}$ can be taken from Theorem 5.7 with exponential distributions with cumulative distribution function as in (10.13), that is, $f_{d+1:n}$ and $F_{m:n}$ have to be replaced by $\psi_{d+1:n;a}$ and $\Psi_{m:n;a}$, respectively. Moreover, ϑ has to be replaced by $\left(\frac{1}{\vartheta_1} + \frac{1}{\vartheta_2}\right)^{-1}$.

A representation in terms of shifted gamma distributions can be found in Kundu and Gupta (2007). Using Górny (2017), the B-spline representation can be rewritten in terms of shifted gamma density functions. Notice that the preceding approach can also be used to find representations for $P_{\vartheta_1,\vartheta_2}(\widehat{\vartheta}_2 \leq t, D_{\text{I},2} > 0)$ and $P_{\vartheta_1,\vartheta_2}(\widehat{\vartheta}_1 \leq t_1, \widehat{\vartheta}_2 \leq t_2, D_{\text{I},1} > 0, D_{\text{I},2} > 0)$ using the same techniques.

Clearly, the result can also be used to find Bayesian estimates of the parameters as has been done Kundu and Gupta (2007) assuming independent inverse gamma priors as in (5.18) for the parameters ϑ_1 and ϑ_2. This yields the Bayes estimates under squared error loss (cf. (5.20))

$$\widehat{\vartheta}_{1,\text{B}} = \frac{\text{TTT}_{D_\text{I}} + h_1}{D_{\text{I},1} + \beta_1 - 1}, \qquad \widehat{\vartheta}_{2,\text{B}} = \frac{\text{TTT}_{D_\text{I}} + h_2}{D_{\text{I},2} + \beta_2 - 1},$$

where β_1, h_1 and β_2, h_2 are the hyper-parameters of the priors. Furthermore, statistical intervals can be constructed. Kundu and Gupta (2007) have proposed the application of asymptotical and bootstrap intervals as well as credibility intervals. Following the ideas of Kundu and Basu (2000), the pivoting method sketched in Section 5.3.2.1 has been utilized to construct approximate confidence intervals. In order to construct them, the second (unknown) parameter ϑ_i is replaced by the corresponding maximum likelihood estimate. For instance, one considers the equations (cf. (5.15))

$$F_{\vartheta_1,\widehat{\vartheta}_{2,\text{obs}}}^{\widehat{\vartheta}_1|D_{\text{I},1}>0}(\widehat{\vartheta}_{1,\text{obs}}) = 1 - \frac{\alpha}{2}, \qquad F_{\vartheta_1,\widehat{\vartheta}_{2,\text{obs}}}^{\widehat{\vartheta}_1|D_{\text{I},1}>0}(\widehat{\vartheta}_{1,\text{obs}}) = \frac{\alpha}{2}$$

to construct an approximate (conditional) confidence interval for ϑ_1 ($\widehat{\vartheta}_{j,\text{obs}}$ denotes the maximum likelihood estimate of ϑ_j, $j = 1, 2$, obtained for the observed data). In order to ensure that this method works, one has to check the monotonicity in ϑ_1 (for any given ϑ_2) and the respective limits for $\vartheta_1 \to 0$ and $\vartheta_1 \to \infty$ (cf. Balakrishnan et al., 2014; van Bentum and Cramer, 2019). Iliopoulos (2015) has pointed out for general Type-I hybrid censored competing risks data that it can not be ensured that the above equations have a solution (although the stochastic monotonicity can be established). Similar arguments can be utilized to show the respective results for Type-I hybrid censored competing risks data.

10.2.1.2 Weibull distributed lifetimes

Weibull distributed lifetimes under competing risks have been discussed in Bhattacharya et al. (2014b) assuming identical shape parameters and different scale parameters. Thus, these distributions form a proportional hazards family as in (10.3) so that Assumption 10.1 is satisfied with $a = \vartheta_1/\vartheta_2$ and

$$\Psi_a(t) = 1 - \exp\Big(-\Big(\frac{1}{\vartheta_1} + \frac{1}{\vartheta_2}\Big)t^\beta\Big), \qquad t \geq 0.$$

Notice that Bhattacharya et al. (2014b) considered the parametrization $h_j = 1/\vartheta_j$, $j = 1, 2$. Therefore, we can apply the results obtained in Section 10.2.1 under the proportional hazard rate assumption. First, let $\eta(\beta) = \sum_{i=1}^{d} x_j^{\beta} + (n-d)w^{\beta}$. Proceeding as for the exponential distribution, we find the likelihood function (cf. (5.1))

$$\mathscr{L}(\vartheta_1, \vartheta_2, \beta; \boldsymbol{x}_d, d, d_2) = \frac{\vartheta_1^{d_2}\vartheta_2^{d-d_2}}{(\vartheta_1+\vartheta_2)^d} \times \frac{n!}{(n-d)!}\Big(\prod_{j=1}^{d} \psi_a(x_j)\Big)(1-\Psi_a(w))^{n-d})$$

$$= \frac{n!}{(n-d)!\vartheta_1^{d-d_2}\vartheta_2^{d_2}}\Big(\prod_{i=1}^{d} x_i\Big)^{\beta-1} \exp\Big\{-\Big(\frac{1}{\vartheta_1}+\frac{1}{\vartheta_2}\Big)\eta(\beta)\Big\},$$

which can be written in the form

$$= \frac{n!}{(n-d)!}\Big(\prod_{i=1}^{d} x_i\Big)^{\beta} \nu_{d-d_2}(\vartheta_1, \beta)\nu_{d_2}(\vartheta_2, \beta)$$

with $\nu_s(\vartheta, \beta) = \vartheta^{-s}\exp\big(-\eta(\beta)/\vartheta\big)$. Then, given $\beta > 0$, $\nu_s(\cdot, \beta)$ is maximized by (cf. (6.1)) $\eta(\beta)/s$ so that the likelihood function is bounded by

$$\mathscr{L}(\vartheta_1, \vartheta_2, \beta; \boldsymbol{x}_d, d, d_2) \le \frac{n!}{(n-d)!}(d-d_2)^{d-d_2} d_2^{d_2}\Big(\prod_{i=1}^{d} x_i\Big)^{\beta-1} \eta(\beta)^{-d}e^{-d}.$$

Equality holds iff $\vartheta_1 = \eta(\beta)/(d-d_2)$, $\vartheta_2 = \eta(\beta)/d_2$. Taking logarithms, we get for the log-likelihood function

$$\mathscr{L}^*(\vartheta_1, \vartheta_2, \beta; \boldsymbol{x}_d, d, d_2) \le \text{const} + \beta\sum_{i=1}^{d} \log x_i - d\log\eta(\beta).$$

The part depending on β, that is, $\beta\sum_{i=1}^{d} \log x_i - d\log\eta(\beta)$ is exactly the same as under Type-I hybrid censoring. Therefore, the comments provided in Section 6.2.1 directly apply to the present setting, too (e.g., existence and uniqueness of the MLEs).

Similar comments apply to the methods of approximate maximum likelihood estimation and Bayesian inference as elaborated in detail in Bhattacharya et al. (2014b).

Remark 10.3. The same approach can be applied to Type-I progressive hybrid censored data as sketched in Remark 10.2. Type-I progressive hybrid censoring of exponentially distributed lifetimes has been addressed in Kundu and Joarder (2006a). Respective results for Weibull data can be found in Wu et al. (2017).

10.2.2 Type-II hybrid censoring scheme

Given the proportional hazards model in Assumption 10.1, the basic results can be obtained by similarity to those established under Type-I hybrid censoring presented in

Section 10.2.1 by using the counter D_{II} instead of D_{I}. Therefore, we do not present details here. The joint density function in (4.30) yields the following expression for the joint density function under competing risks

$$f^{X_{j:n},1\leq j\leq D_{\text{II}},\mathbf{C}_{D_{\text{II}}},D}(\mathbf{t}_d,\mathbf{c}_d,d)=\frac{a^{d_2}}{(a+1)^d}\begin{cases}\psi_{1,\dots,m:n;a}(\mathbf{t}_m), & d<m\\ \frac{1-\Psi_a(T)}{(n-d)\psi_a(T)}\psi_{1,\dots,d+1:n;a}(\mathbf{t}_d,T), & d\in\{m,\dots,n-1\}\\ \psi_{1,\dots,n:n;a}(\mathbf{t}_m), & d=n\end{cases},$$

where $d_2=\sum_{i=1}^{d}\mathbb{1}_{\{2\}}(c_i)$.

Analysis of Type-II hybrid censored competing risks data from exponential distributions has been discussed in Koley et al. (2017) (see also Koley, 2018). Taking into account the preliminary comments in Section 5.2 as well as the derivations in Section 10.2.1, we can directly utilize the respective results. The MLEs are given by

$$\widehat{\vartheta}_1=\frac{1}{D_{\text{II},1}}\text{TTT}_{D_{\text{II}}},\quad \widehat{\vartheta}_2=\frac{1}{D_{\text{II},2}}\text{TTT}_{D_{\text{II}}}$$

which is obviously the same as (10.15) with D_{I}, $D_{\text{I},j}$ replaced by D_{II}, $D_{\text{II},j}$, respectively. Consequently, the distribution of the MLEs as well as other results like confidence intervals or Bayesian inference can be carried out in the same manner. For instance, the cumulative distribution function of $\widehat{\vartheta}_1$ can be derived as under Type-I hybrid censoring using the law of total probability, that is,

$$\begin{aligned}P_{\vartheta_1,\vartheta_2}&(\widehat{\vartheta}_1\leq t,D_{\text{II},1}>0)\\ &=\sum_{i=1}^{m}\Pr_{\vartheta_1,\vartheta_2}(\widehat{\vartheta}_1\leq t,D_{\text{II},1}=i,D<m)+\sum_{d=m}^{n-1}\sum_{i=1}^{d}\Pr_{\vartheta_1,\vartheta_2}(\widehat{\vartheta}_1\leq t,D_{\text{II},1}=i,D=d)\\ &\quad+\sum_{i=1}^{n}\Pr_{\vartheta_1,\vartheta_2}(\widehat{\vartheta}_1\leq t,D_{\text{II},1}=i,D=n)\end{aligned}$$

Performing similar manipulations as for Type-I hybrid censoring, this yields the joint density function

$$\begin{aligned}f_{\vartheta_1,\vartheta_2}^{\widehat{\vartheta}_1|D_{\text{II},1}>0}(t)=&\frac{1}{\Pr_{\vartheta_1,\vartheta_2}(D_{\text{II},1}>0)}\bigg(\sum_{i=1}^{m}i\Pr_{\vartheta_1,\vartheta_2}(D_{\text{II},1}=i,D<m)f_{\vartheta_1,\vartheta_2}^{\text{TTT}_m|D<m}(it)\\ &+\sum_{d=m}^{n-1}\sum_{i=1}^{d}i\Pr_{\vartheta_1,\vartheta_2}(D_{\text{II},1}=i,D=d)f_{\vartheta_1,\vartheta_2}^{\text{TTT}_d|D=d}(it)\\ &+\sum_{i=1}^{n}i\Pr_{\vartheta_1,\vartheta_2}(D_{\text{II},1}=i,D=d)f_{\vartheta_1,\vartheta_2}^{\text{TTT}_n|D=n}(it)\bigg),\quad t\geq 0,\end{aligned}$$

where $f_{\vartheta_1,\vartheta_2}^{\mathrm{TTT}_d|D=d}$, $d=m,\ldots,n$, and $f_{\vartheta_1,\vartheta_2}^{\mathrm{TTT}_m|D<m}$ can be taken from Theorem 5.7 and Theorem 5.21 with exponential distributions with cumulative distribution function as in (10.13), that is, $f_{d+1:n}$ and $F_{m:n}$ have to be replaced by $\psi_{d+1:n;a}$ and $\Psi_{m:n;a}$, respectively. In particular, ϑ has to be replaced by $\left(\frac{1}{\vartheta_1}+\frac{1}{\vartheta_2}\right)^{-1}$.

The probabilities $\Pr_{\vartheta_1,\vartheta_2}(D_{\mathrm{II},1}=i, D=d)$ can be taken from (10.12). Furthermore,

$$\Pr_{\vartheta_1,\vartheta_2}(D_{\mathrm{II},1}=i, D<m)=\sum_{d=0}^{m}\binom{n}{d}\binom{d}{i}\frac{a^{d-i}}{(a+1)^d}(1-\Psi_a(T))^{n-d}\Psi_{d:d;a}(T).$$

Remark 10.4. Similarly, Type-II progressive hybrid censored data from exponential distributions can be considered. Obviously, the cumulative distribution functions or density functions of the order statistics $X_{1:n}, X_{2:n}, \ldots$ have to be replaced by the corresponding expressions for the progressively Type-II censored order statistics $X_{1:n^*:n}, X_{2:n^*:n}, \ldots$.

10.2.3 Generalized (progressive) hybrid censoring schemes

Generalized Type-I hybrid censored exponential data has been considered in Mao et al. (2014), Iliopoulos (2015), and Lee et al. (2016b). Wang and Li (2022) extended the situation in the sense that for some observations the cause of failure is unknown. The case of generalized Type-I progressive hybrid censoring has been addressed in Koley and Kundu (2017). Wang (2018) studied Weibull competing risks data under generalized Type-I progressive hybrid censoring. In Singh et al. (2021), two-parameter Rayleigh competing risks data under generalized progressive hybrid censoring have been studied. Dependent competing risks in the presence of progressive hybrid censoring using Marshall–Olkin bivariate Weibull distribution are investigated in Feizjavadian and Hashemi (2015) (see also Samanta and Kundu, 2021 as well as Shi and Wu, 2016 for Gompertz distributions). A Burr XII model in presence of Type-I progressive hybrid censoring has been analyzed in Hashemi and Azar (2011). Since the shape parameter of the Burr XII-distribution is supposed known, the results are immediately obtained from the exponential distribution by a data transformation. A discussion of Gompertz distributions can be found in Wu et al. (2016).

The model with partially observed causes of failures introduced in Wang and Li (2022) has been studied for Weibull distributed lifetimes in Wang et al. (2020), for a general family of inverted exponentiated distributions in Lodhi et al. (2021), and for Kumaraswamy distribution in Mahto et al. (2022a).

Likelihood inference under generalized Type-II hybrid censoring has been investigated by Mao and Shi (2020). Abushal et al. (2022) considered partially observed caused of failure with Lomax distributed lifetimes under generalized progressive Type-II hybrid censoring.

So far, other hybrid censoring schemes and lifetime distributions, respectively, do not seem to have been discussed in the literature.

10.3. Stress-strength models

In a reliability context, let a random variable Y describe the strength of a unit subjected to certain stress represented by the random variable X. The unit fails when the stress X exceeds the strength Y. Thus, the probability

$$R = \Pr(X < Y)$$

may serve as a measure for the reliability of the unit (see, e.g., Tong, 1974, 1975a,b; Beg, 1980; Constantine et al., 1986; Cramer and Kamps, 1997; Cramer, 2001, and Kotz et al., 2003). Stress-strength models have been widely investigated in the literature. For a detailed account on models, inferential results, and applications, we refer to Kotz et al. (2003). In particular, Chapter 7 of this monograph provides an extensive survey on applications and examples of stress-strength models.

Inference for R has been based on various assumptions regarding the underlying data. For instance, for two independent samples of $X_1, \ldots, X_{n_1}$ and $Y_1, \ldots, Y_{n_2}$, Birnbaum (1956) showed that the Mann-Whitney statistic

$$U = \sum_{i=1}^{n_1} \sum_{j=1}^{n_2} \mathbb{1}_{(X_i,\infty)}(Y_j)$$

yields an unbiased nonparametric estimator of R, i.e.,

$$\widehat{R} = \frac{1}{n_1 n_2} U.$$

Since this pioneering work, many results have been obtained for various models and assumptions. For an extensive review, we refer to Kotz et al. (2003).

In the following, we are interested in the estimation of R based on two independent (progressive) hybrid censored samples. Thus, using the respective MLEs one gets the MLE of the stress-strength reliability. For two exponential samples with means ϑ_1 and ϑ_2, one has to estimate the quantity

$$R = \frac{\vartheta_2}{\vartheta_1 + \vartheta_2} \tag{10.16}$$

which first seemed to be discussed in Asgharzadeh and Kazemi (2014) for Type-I hybrid censored data. Clearly, the explicit expressions of the MLEs can be plugged into the probability in (10.16) by considering any kind of (progressive) hybrid censored data using the estimates previously obtained from various concepts. Related material can be found in Mirjalili et al. (2016).

Weibull distributions with identical shape but different scale parameters lead to a stress-strength reliability as in (10.16). As above, plug-in estimators can be used to estimate R which includes MLEs, approximate MLEs, as well as Bayesian estimates (for complete data, see Kundu and Gupta, 2006). For Type-I hybrid censored Weibull data, respective results can be found in Asgharzadeh et al. (2015a). Systems with non-identical component strengths have been studied by Çetinkaya (2021). A multicomponent stress-strength model with dependent stress and strength components has been discussed in Bai et al. (2018). Weibull distributed lifetimes under Type-II progressively hybrid censoring have been discussed in Shoaee and Khorram (2016).

Two independent samples from two-parameter Chen distributions under Type-I hybrid censoring have been discussed in Pundir and Gupta (2018). Notice that the stress-strength reliability has the same form as in (10.16). Lomax distributions under Type-II hybrid censoring have been considered in Yadav et al. (2019b). For Kumaraswamy distributions, we refer to Kohansal and Nadarajah (2019). Burr Type XII distributions have been addressed in Kohansal (2020). Garg and Kumar (2021) focused on generalized inverted exponential distribution.

10.4. Optimal designs

Since hybrid censoring schemes impose requirements on sample sizes as well as on test duration, it is natural to ask for optimal choices of these quantities w.r.t. some given criteria of interest. Additionally, the censoring plan $\mathscr{R}$ can be considered under progressive hybrid censoring. In the literature, several concepts and ideas have been applied to find optimal hybrid censoring schemes.

Ebrahimi (1988) seems to be the first paper discussing optimal Type-I hybrid censoring. He considers exponential lifetimes and a cost function approach with a cost function

$$\mathscr{C}(n) = c_1 X_{r:n} \wedge T + c_2 n + c_3$$

involving constants c_1 (cost per unit of time during the test is being conducted), c_2 (the cost per item), and c_3 (fixed cost of the life test). Clearly, the expected costs are given by (see Lemma 4.5)

$$\begin{aligned} E\mathscr{C}(n) &= c_1 E(X_{r:n} \wedge T) + c_2 n + c_3 \\ &= c_1 T(1 - F_{m:n}(T)) + c_1 \int_{-\infty}^{T} t f^{X_{m:n}}(t)\, dt + c_2 n + c_3. \end{aligned} \tag{10.17}$$

Ebrahimi (1988) studied properties of the expected cost function in terms of the baseline sample size n. Thus, he was interested in determining an optimal sample size n while

r and T are considered as prespecified. Furthermore, some results for other (known) lifetime distributions are provided.

For a fixed sample size n, one might also consider different choices of r and T in Type-I hybrid censoring and ask which combination of r and T subject to $1 \le r \le n$ and $0 < T < \infty$ would be preferable. For brevity, such a hybrid censoring scheme will be denoted by (r, T). In order to compare two different schemes (r_1, T_1) and (r_2, T_2), it is reasonable that (r_1, T_1) is better than (r_2, T_2) if, for example, (r_1, T_1) provides more information than (r_2, T_2) about the model parameters. Thus, it is important to define an information measure for a given censoring scheme.

Remark 10.5. Given a sample size n, the expected cost function in (10.17) is obviously increasing in r and T, respectively. Therefore, this particular cost function is minimized choosing r and T as small as possible. Consequently, this kind of cost function is not an appropriate measure to design a reasonable experiment.

If only a single parameter is present, one can choose a measure which is proportional to the asymptotic variance of the unknown parameter (see Zhang and Meeker, 2005). Choosing the Fisher information as an information criterion, we get from (8.24) that

$$\mathscr{I}_{D\wedge m}(\theta) = \frac{1}{\vartheta^2} \sum_{i=1}^{m} F_{i:n}\left(\frac{T}{\vartheta}\right)$$

is increasing in both m and T (independently of ϑ). A similar observation holds for Type-II hybrid censoring using (8.29). Therefore, the criteria make only sense by introducing additional constraints like in (10.20) or by penalizing long test durations as has been done in (10.19). A general account to such problems has been developed in Bhattacharya et al. (2020) discussing optimal experimental designs for Type-I and Type-II hybrid censoring in case of exponentially distributed lifetimes.

Kundu (2007) and Banerjee and Kundu (2008) have provided such a discussion for the case of Weibull distribution for Type-I and Type-II hybrid censored data, respectively. In practice, it is quite common to consider choosing the 'optimal censoring scheme' from a class of possible censoring schemes. But, if the lifetime distribution has more than one parameter, then there is no unique and natural way to define the information. Some of the common choices that have been discussed in the literature are the trace and determinant of the Fisher information matrix. But, unfortunately, they are not scale-invariant which leads to undesirable behavior (see, e.g., Gupta and Kundu, 2006). One way to define the information measure for a particular sampling scheme is as the inverse of the asymptotic variance of the estimator of the p-th-quantile obtained from that particular censoring scheme (see Zhang and Meeker, 2005). For the Weibull distribution, the p-th-quantile is given by

$$\xi_p = \big(-\vartheta \log(1-p)\big)^{1/\beta} = q(\vartheta, \beta), \quad \text{say.}$$

Denoting by $[\mathscr{I}_{**;D\wedge m}(\vartheta, \beta)]$ the Fisher information matrix under Type-I hybrid censoring (see (8.25)), the asymptotic variance is given by

$$\mathrm{Var}_{m,T}(p) = \left(\tfrac{\partial}{\partial \vartheta} q(\vartheta, \beta), \tfrac{\partial}{\partial \beta} q(\vartheta, \beta)\right)[\mathscr{I}_{**;D\wedge m}(\vartheta, \beta)]^{-1} \begin{pmatrix} \frac{\partial}{\partial \vartheta} q(\vartheta, \beta) \\ \frac{\partial}{\partial \beta} q(\vartheta, \beta) \end{pmatrix}.$$

The respective information measure is given by the inverse integrated variance, that is,

$$\mathscr{I}_{\vartheta,\beta}(n, m, T) = \left[\int_0^1 \mathrm{Var}_{m,T}(p)dp\right]^{-1}. \tag{10.18}$$

As noted by Kundu (2007), a solution with maximum information is given by $m = n$ and $T = \infty$ which clearly is a reasonable choice. Therefore, the measure is not of much use without imposing additional constraints on the experiment. Therefore, the test duration has been taken into account to construct a cost measure as

$$\mathscr{I}_{\vartheta,\beta}(n, m, T) - c_1 E(X_{m:n} \wedge T), \tag{10.19}$$

where c_1 denotes the cost per unit time of the experiment. Notice that such an approach has been discussed in a Bayesian framework for Type-I censored exponential distribution by Yeh (1994) and Lin et al. (2002). Alternatively one may introduce the bound

$$E(X_{m:n} \wedge T) \leq T_* \tag{10.20}$$

for some given maximum test duration T_*. This account has been discussed in Kundu (2007) who illustrated it by an example. For Type-II hybrid censoring, Banerjee and Kundu (2008) utilized these ideas. Additionally, a weighted information measure of (10.18)

$$\mathscr{I}_{\vartheta,\beta;\omega}(n, m, T) = \left[\int_0^1 \mathrm{Var}_{m,T}(p)\omega(p)\, dp\right]^{-1}$$

with a weight function ω defined on $(0, 1)$ has been proposed.

In order to find optimal values for n, m, T, several cost function approaches have been proposed in Bhattacharya et al. (2014a) incorporating the Fisher information matrix. For instance, they considered the cost functions

$$\begin{aligned} \mathscr{C}_1(n, m, T) &= c_1 n + c_2 ED_1 + c_3 E(X_{m:n} \wedge T) + c_4\, \mathrm{trace}([\mathscr{I}_{**;D\wedge m}(\vartheta, \beta)]), \\ \mathscr{C}_2(n, m, T) &= c_1 n + c_2 ED_1 + c_3 E(X_{m:n} \wedge T) + c_4 \mathscr{I}_{\vartheta,\beta}(n, m, T). \end{aligned} \tag{10.21}$$

Alternatively, they addressed a constrained optimization problem

$$\begin{aligned} &\underset{n,m,T}{\text{minimize}}\ [\mathscr{I}_{\vartheta,\beta}(n, m, T)]^{-1} \\ &\text{subject to } c_1 n + c_2 ED_1 + c_3 E(X_{m:n} \wedge T) \leq c_0. \end{aligned}$$

Remark 10.6. Optimal censoring designs for lognormal distributions have been addressed in Dube et al. (2011). Generalized exponential distributions under Type-II hybrid censoring have been studied in Sen et al. (2018d). Mahto et al. (2022b) discussed the problem for Gumbel Type-II distributions.

In progressive hybrid censoring, additionally the censoring plan $\mathscr{R}$ has been included in the optimization process. Such an approach has been proposed in Bhattacharya and Pradhan (2017) for Type-I progressive hybrid censoring who investigated both a cost function minimization and a constrained minimization approach. As above, a cost function of the form in (10.21) has been considered, that is,

$$\Psi(\mathscr{R}, T) = c_2 ED_1 + c_3 E(X^{\mathscr{R}}_{m:m:n} \wedge T) + c_4 \int_0^1 \mathrm{Var}_{\mathscr{R},T}(p)\, dp, \tag{10.22}$$

where $T \in (0, \infty)$ and

$$\mathscr{C}_{n,m} = \left\{\mathscr{R} = (R_1, \dots, R_m) \,\middle|\, R_i \in \mathbb{N}_0, 1 \le i \le m, \sum_{j=1}^{m} R_j = n - m\right\}$$

denotes the set of admissible censoring plans (see Balakrishnan and Cramer, 2014, Eq. (1.1)). The optimization problem has been attacked by the variable neighborhood search algorithm proposed by Hansen and Mladenović (1999). For an application of this strategy to find optimal censoring plans in progressive censoring, we refer to Bhattacharya et al. (2016) (see also Bhattacharya, 2020 for a general discussion on multicriteria optimization under progressive censoring).

Bhattacharya and Pradhan (2017) also studied the minimization problem

$$\begin{aligned} &\underset{(\mathscr{R},T)\in\mathscr{C}_{n,m}\times(0,\infty)}{\text{minimize}} \quad \Psi(\mathscr{R}, T) \\ &\text{subject to } c_2 ED_1 + c_3 E(X_{m:n} \wedge T) \le c_0, \end{aligned}$$

where $\Psi(\mathscr{R}, T) = \int_0^1 \mathrm{Var}_{\mathscr{R},T}(p)\, dp$ is the objective function depending on the censoring plan $\mathscr{R}$ (with fixed m, n) and the threshold T. Note that this is a particular case of (10.22) with $c_2 = c_3 = 0$ and $c_4 = 1$. Finally, maximization of the trace of the Fisher information (so-called T-optimality) has been addressed w.r.t. the set $\mathscr{C}_{n,m}$ and a threshold T. They showed that, under certain conditions imposed on the population distribution, optimal plans are so-called one-step censoring plans (for related results in progressive censoring, see, e.g., Balakrishnan et al., 2008c; Burkschat, 2008; Cramer and Ensenbach, 2011; Balakrishnan and Cramer, 2014). As an example for right censoring as an optimal censoring plan, they presented the Lomax distribution. Notice that the results are along the lines of that one under progressive Type-II censoring given in Dahmen et al. (2012). Further, they claimed that first step censoring provides an optimal censoring plan for Weibull distributions. However, this result seems to be in error since the required monotonicity properties of the criterion are not satisfied for two-parameter Weibull distributions. Furthermore, minimization of the trace and the determinant of the Fisher information matrix have been studied in Bhattacharya and Pradhan (2017) and computational results have been provided.

Similar investigations have been established for generalized exponential distributions in Sen et al. (2020). Lognormal distributions have been studied in Sen et al. (2018c).

10.5. Reliability acceptance sampling plans

Acceptance sampling plans play a key role in reliability analysis and quality control. An acceptance sampling plan uses life-test procedures to suggest rules for either accepting or rejecting a lot of units based on the observed lifetime data from the sample. The topic has been extensively discussed in statistical quality control where various assumptions are made w.r.t. distributions and data. For lifetime data, related references are Hosono and Kase (1981), Kocherlakota and Balakrishnan (1986), Balasooriya (1995), Fertig and Mann (1980), Schneider (1989), Balasooriya et al. (2000), Balakrishnan and Aggarwala (2000, Chapter 11), Fernández (2005), and Balakrishnan and Cramer (2014, Chapter 22). A detailed introduction may be found in, e.g., Montgomery (2013). Acceptance sampling plans are also called reliability sampling plans when the acceptance sampling procedure is based on lifetime data. Since we are discussing only lifetime data, these terms will be used synonymously throughout this section.

In any acceptance sampling plan, two kinds of risks exist, one is the producer's risk (denoted by α) and the other is the consumer's risk (denoted by β). For data $\boldsymbol{X} = (X_1, \ldots, X_m)$, a lot is accepted if

(a) $\phi(\mathbf{X}) \geq b_l$ or $\phi(\mathbf{X}) \leq b_u$ (one-sided plans), or

(b) $b_l \leq \phi(\mathbf{X}) \leq b_u$ (two-sided plans),

where ϕ and the bounds b_l and b_u are determined such that the probability requirements

$$\begin{aligned} &\Pr(\text{ Accept lot }) = \Pr(\phi(\mathbf{X}) \geq b_l) \geq 1-\alpha \quad \text{and} \\ &\Pr(\text{ Reject lot }) = \Pr(\phi^*(\mathbf{X}) \leq b_u) \leq \beta \end{aligned} \tag{10.23}$$

are satisfied for functions ϕ, ϕ^* and bounds b_l, b_u.

10.5.1 Sampling plans for the exponential distribution

In case of exponential distributions, the problem can be formulated as follows. Based on the data from the life-test, we want to test the following hypotheses on the mean lifetime ϑ:

$$H_0 : \vartheta = \vartheta_0 \quad \text{vs.} \quad H_1 : \vartheta = \vartheta_1 (< \vartheta_0).$$

A decision rule should be chosen such that the following two conditions are satisfied:

$$\Pr_{\vartheta_0}(\text{ Accept } H_0) \geq 1-\alpha \quad \text{and} \quad \Pr_{\vartheta_1}(\text{ Accept } H_0) \leq \beta. \tag{10.24}$$

Inspired by results of Epstein (1954), Jeong and Yum (1995) (for considering only producers' risk under Type-I censoring, see Spurrier and Wei, 1980) proposed a decision rule based on the MLE $\widehat{\vartheta}$ given Type-I censored data, that is, H_0 is accepted if

$$\widehat{\vartheta} > c.$$

Then, according to (10.24), an acceptance sampling plan (r, c) for given T, α, and β is specified by the risk requirements

$$\begin{aligned}
\Pr_{\vartheta_0}(\text{ Accept } H_0) &= \Pr_{\vartheta_0}(\widehat{\vartheta} > c) \geq 1-\alpha, \\
\Pr_{\vartheta_1}(\text{ Accept } H_0) &= \Pr_{\vartheta_1}(\widehat{\vartheta} > c) \leq \beta.
\end{aligned}$$

These probabilities can be obtained using the results in Chapter 5 (see also Bartholomew, 1963).

Jeong et al. (1996) developed hybrid sampling plans (r, T) for the exponential distribution which meet the producer's and consumer's risks simultaneously given that n, α, and β are known. Jeong et al. (1996) proposed the following acceptance sampling plan for given r and T,

$$\text{accept } H_0 \text{ if } X_{r:n} > T.$$

Now, in view of (10.24), r and T are obtained from the conditions

$$\begin{aligned}
\Pr_{\vartheta_0}(\text{ Accept } H_0) &= \Pr_{\vartheta_0}(X_{r:n} > T) = 1 - F_{r:n;\vartheta_0}(T) \geq 1-\alpha, \\
\Pr_{\vartheta_1}(\text{ Accept } H_0) &= \Pr_{\vartheta_1}(X_{r:n} > T) = 1 - F_{r:n;\vartheta_1}(T) \leq \beta.
\end{aligned}$$

Jeong et al. (1996) provided an algorithm for determining r and T for a given n, α, and β, and also provided extensive tables of values of r and T for different choices of n, α, and β. Related results for Weibull distribution with known shape parameter have been established by Kim and Yum (2011).

Bhattacharya and Aslam (2019) constructed sampling plans for exponentially distributed lifetimes using an estimate of the lifetime performance index C_L (see Montgomery, 2013; Laumen and Cramer, 2015) which, for an $Exp(\vartheta)$-distribution, is given by

$$C_L = 1 - \frac{L}{\vartheta}$$

with lower specification limit L. Using the exact distribution of the MLE $\widehat{\vartheta}$ under Type-I hybrid censoring (similarly for Type-II hybrid censoring) and $\widehat{C}_L = 1 - \frac{L}{\widehat{\vartheta}}$, the lot is accepted if

$$\widehat{C}_L \geq b_l.$$

The operating characteristic (OC)-curve is given by

$$L(p) = \Pr\left(\widehat{C}_L \geq b_l \mid p\right) = \Pr\left(\widehat{\vartheta} \geq \frac{\vartheta \log(1-p)}{1-b_l}\right),$$

where p is called the lifetime non-conforming rate. It measures whether the lifetime X of the product does not exceed its pre-specified lower quality specification L, that is,

$$p = \Pr(X < L).$$

Note that, for exponential lifetimes, it holds that $C_L = 1 + \log(1-p)$. The optimum (minimum) sample size n has been obtained as solution of a constrained minimization problem with constraints $L(p_\alpha) \geq 1-\alpha$, $L(p_\beta) \leq \beta$ given α, β, p_α, p_β. An extension to so-called generalized multiple dependent state sampling plans has been discussed in Bhattacharya and Aslam (2020).

10.5.2 Sampling plans for the Weibull distribution

For location-scale families as given in (3.2), the Lieberman-Resnikoff procedure is employed to construct the decision rules ϕ, ϕ^* given in (10.23) (see Lieberman and Resnikoff, 1955; Kocherlakota and Balakrishnan, 1986; Schneider, 1989). For some value k, the decision rule ϕ_k is constructed via estimates $\widehat{\mu}$ and $\widehat{\vartheta}$ of the location and scale parameters μ and ϑ. In particular, ϕ_k exhibits the form

$$\phi_k(\mathbf{X}) = \widehat{\mu} - k\widehat{\vartheta},$$

where k denotes the acceptance constant. Therefore, for a lower specification limit b_l, a lot is accepted when $\phi_k(\mathbf{X}) = \widehat{\mu} - k\widehat{\vartheta} \geq b_l$.

Some results in this direction have been obtained for Weibull distributed lifetimes (or the extreme value distribution by a log-transformation of the data). Inspired by Schneider (1989), Bhattacharya et al. (2015) used a normal approximation of the OC-curve

$$\begin{aligned} L(p) \equiv L(p; n, k) &= \Pr(\widehat{\mu} - k\widehat{\vartheta} > L') \\ &= \Pr\left(U > \frac{\vartheta(\xi_p + k)}{v}\right), \quad p \in (0,1), \end{aligned}$$

under Type-I hybrid censoring when the lifetimes are supposed Weibull distributed random variables. Here, ξ_p is the p-th quantile of the standard extreme value distribution,

$$U = \frac{\widehat{\mu} - k\widehat{\vartheta} - (\mu - k\vartheta)}{v},$$

and v^2 denotes the asymptotic variance of $\widehat{\mu} - k\widehat{\vartheta}$. v^2 can be expressed in terms of the expected Fisher information, that is,

$$v^2 = \ell^{11}(\boldsymbol{\theta}) + \ell^{22}(\boldsymbol{\theta}) - 2k\ell^{12}(\boldsymbol{\theta})$$

where $\boldsymbol{\theta} = (\mu, \vartheta)$ and $\ell^{ij}(\boldsymbol{\theta})$ denote the elements of the inverse of the expected Fisher information matrix for the extreme value distribution (see also (8.25) as well as Park et al., 2008 for representations in terms of the Weibull distribution). Bhattacharya et al. (2015) argued that U is asymptotically normal distributed so that

$$L(p; n, k) \approx 1 - \Phi\Big(\frac{\vartheta(\xi_p + k)}{v}\Big)$$

where Φ is the cumulative distribution function of the standard normal distribution. Given $(p_\alpha, 1-\alpha)$ and (p_β, β) as points on the OC-curve, values of n and k are obtained from the equations

$$1 - \Phi\Big(\frac{\vartheta(\xi_{p_\alpha} + k)}{v}\Big) = 1 - \alpha, \quad 1 - \Phi\Big(\frac{\vartheta(\xi_{p_\beta} + k)}{v}\Big) = \beta$$

or alternatively (see Schneider, 1989)

$$\frac{\vartheta(\xi_{p_\alpha} + k)}{v} = z_\alpha, \qquad \frac{\vartheta(\xi_{p_\beta} + k)}{v} = z_{1-\beta}$$

where z_α denotes the α-quantile of the standard normal distribution. Unlike in the case of Type-II censored data discussed in Schneider (1989), these equations will generally depend on the parameters μ and ϑ, so that values for these parameters have to be assumed in order to determine a value of the sample size n (this is implicitly assumed in Section 2.3 in Bhattacharya et al., 2015). In particular, they found that

$$k = \frac{\xi_{p_\alpha} z_{1-\beta} - \xi_{p_\beta} z_\alpha}{z_\alpha - z_{1-\beta}}$$

and the sample size n has to be determined from the equation

$$\frac{1}{\vartheta^2}\left(\frac{z_\alpha - z_{1-\beta}}{\xi_{p_\alpha} - \xi_{p_\beta}}\right)^2 v^2 = 1.$$

An algorithm has been proposed to find n. Additionally, they conducted Monte Carlo simulations and provided tables for small sample sizes.

Using expressions for the asymptotic Fisher information under Type-II hybrid censoring, they provided results for acceptance sampling plans in case of Type-II hybrid censoring. Sen et al. (2018d) extended these results to generalized hybrid censoring adopting the same methodology. In fact, it needs only to incorporate the appropriately chosen estimators and the corresponding expression of the asymptotic variance. This quantity is obtained from the respective expression of the expected Fisher information matrix. Clearly, the approach can also be applied to other hybrid censoring plans using the connections to the expected Fisher information presented in Section 8.2.4. A warranty cost approach has been discussed in Chakrabarty et al. (2021).

10.5.3 Bayesian sampling plans for exponentially distributed lifetimes

Acceptance sampling plans have also been studied from a decision theoretic point of view where most results have been obtained in the area of Bayesian variable sampling plans. Given a sample $\boldsymbol{X}$, the lot is accepted (1) or rejected (0) using a decision function δ given by

$$\delta(\boldsymbol{x}) = \begin{cases} 1, & \widehat{\theta} \geq c \\ 0, & \widehat{\theta} < c \end{cases}$$

with an appropriate statistic $\widehat{\theta}$ and a value c. $\boldsymbol{x}$ denotes the observed sample, that is, the realization of $\boldsymbol{X}$. Then, the Bayes risk of δ given some reasonable loss function is considered to find the optimal Bayesian sampling plan, that is, the optimal Bayesian sampling plan minimizes the Bayes risk of decision function $\delta(\boldsymbol{X})$ under the given loss function for some prior distribution on the parameter space.

For the exponential distribution with mean $1/\lambda$ under Type-I and Type-II hybrid censoring, such an approach has been discussed in Lin et al. (2008, 2010a) (see also Yeh, 1988; for Type-I censoring, see Yeh, 1994; Lin et al., 2010b). In particular, they considered a gamma prior $\Gamma(1/a, b)$ and a polynomial (quadratic) loss function

$$\mathcal{L}(\delta(\mathbf{X}), \lambda, n) = nC_s + \delta(\mathbf{X}) \sum_{j=0}^{p} a_j \lambda^j + (1 - \delta(\mathbf{X}))C_r$$

with $p = 2$, where C_s is the cost of inspecting an item, C_r is the loss of rejecting the lot, and the polynomial $\sum_{j=0}^{p} a_j \lambda^j$ is the loss of accepting the lot. Furthermore, they assumed that $\sum_{j=0}^{p} a_j \lambda^j$ is positive for $\lambda > 0$. They obtained integral representations of the Bayes risk

$$R(n, m, T, \xi) = E\mathcal{L}(\delta(\mathbf{X}), \lambda, n) = E_\lambda\Big[E_{\mathbf{X}|\lambda}\mathcal{L}(\delta(\mathbf{X}), \lambda, n)\Big]$$

under both Type-I and Type-II hybrid censoring.

Remark 10.7. **(a)** An alternative decision rule under Type-I hybrid censoring has been studied in Chen et al. (2004). Liang and Yang (2013) pointed out that these two Bayesian sampling plans are not optimal from a theoretic decision point of view since the proposed decision functions are not the Bayes decision functions.

They considered the decision theoretic approach and derived the optimal Bayesian sampling plan based on sufficient statistics under a general loss function. Other loss functions have been considered in Prajapati et al. (2019a,b). So-called curtailed Bayesian sampling has been proposed in Chen et al. (2021).

(b) The case of generalized Type-II hybrid censoring has been investigated in Prajapati et al. (2020). Yang et al. (2017) studied a particular case of generalized Type-I hybrid censoring (with $m = n$) called modified Type-II hybrid censoring.

(c) Two-parameter exponential distributions under Type-I hybrid censoring have recently been discussed in Prajapat et al. (2021).

(d) Bayesian sampling plans under progressive Type-I hybrid censoring have been studied in Lin and Huang (2012) and Lin et al. (2011, 2013) (see also the comments in Liang, 2014).

CHAPTER 11

Goodness-of-fit tests

Contents

11.1. Introduction

So far, we have presented many inferential results for several forms of censoring by assuming different lifetime distributions for the observed censored data. Even though these inferential results have been shown to be optimal, they remain so only if the assumed distribution is indeed the true distribution underlying these data. So, a natural problem of interest is then to carry out a formal goodness-of-fit test to validate the model assumption made on the observed censored data.

A number of omnibus goodness-of-fit tests have been developed in this regard over the last several decades. For example, Kolmogorov-Smirnov, Cramér-von Mises and Anderson-Darling tests are some of the prominent ones. In the complete sample situation, these omnibus goodness-of-fit tests are, respectively, based on the following statistics:

$$D_n = \sup_{x \in \mathbb{R}} |F_n(x) - F(x)|, \tag{11.1}$$

$$W^2 = \int_{-\infty}^{\infty} (F_n(x) - F(x))^2 \, dF(x) \tag{11.2}$$

and

$$A^2 = n \int_{-\infty}^{\infty} (F_n(x) - F(x))^2 \, w(x) \, dF(x), \tag{11.3}$$

where $w(\cdot)$ is a weight function; in all the above, F is the presumed cumulative distribution function while F_n is the empirical distribution function (EDF) defined as

$$F_n(x) = \frac{\text{No. of sample observations} \leq x}{n} = \frac{1}{n} \sum_{i=1}^{n} \mathbb{1}_{(-\infty,x]}(X_i), \tag{11.4}$$

Hybrid Censoring Know-How
https://doi.org/10.1016/B978-0-12-398387-9.00019-2

where $\mathbb{1}$ denotes the indicator function. Evidently, Cramér-von Mises test statistic in (11.2) is a special case of Anderson-Darling test in (11.3) when the weight function is chosen to be $w \equiv 1$. While the Anderson-Darling test in (11.3) belongs to the general class of quadratic EDF statistics, and that any weight function $w(\cdot)$ may be used, the choice of $w(\cdot) = \frac{1}{F(\cdot)(1-F(\cdot))}$ is made by Anderson and Darling (1954) in order to place more weight on observations in the tails of the distribution. This results in the familiar form of Anderson-Darling statistic as

$$A^2 = n \int_{-\infty}^{\infty} \frac{(F_n(x) - F(x))^2}{F(x)\,(1 - F(x))} dF(x).$$

Some tables of critical values, as well as convenient approximations, of critical values have been presented in the literature.

One important aspect is that when the distribution F that is tested for contains only location and scale parameters, the asymptotic theory of EDF statistics depends only on the form of F, but not on the parameters; see Stephens (1986). This, however, is not the case when the distribution involves a shape parameter which is indeed the situation with most lifetime distributions considered in reliability analysis. To assist in the process, Chen and Balakrishnan (1995) proposed a general omnibus goodness-of-fit procedure, based on the use of EDF statistics for normality following an estimated transformation. This method has been shown to be quite efficient for validating the assumption of many lifetime models.

The methods mentioned above are all applicable when the available sample is complete. Their adaptation for the conventional forms of Type-II and Type-I censored data have been discussed rather extensively in the literature. Interested readers may refer to the books by D'Agostino and Stephens (1986), Huber-Carol et al. (2002), Voinov et al. (2013), and Nikulin and Chimitova (2017) for pertinent details.

But, for the cases of progressive and hybrid censored data, relatively far fewer methods are available. In this chapter, we present a brief review of some recent developments in this direction.

11.2. Progressive censoring

Suppose we observed a progressively Type-II right censored sample $(X_{1:m:n}, \ldots, X_{m:m:n})$ using a specific progressive censoring plan $\mathscr{R} = (R_1, \ldots, R_m)$.

11.2.1 Tests based on spacings

Balakrishnan et al. (2002b) were the first to develop a test of exponentiality based on progressively Type-II right censored data. They used the spacings result in Theorem 2.67

to propose a test statistic of the form

$$\mathcal{T} = \frac{1}{m-1} \sum_{i=1}^{m-1} \frac{\sum_{j=1}^{i} S_{j,m}^{\mathscr{R}}}{S^{\mathscr{R}}}, \tag{11.5}$$

where $S_{j,m}^{\mathscr{R}}$ are the normalized spacings

$$S_{j,m}^{\mathscr{R}} = \gamma_j \left(X_{j:m:n}^{\mathscr{R}} - X_{j-1:m:n}^{\mathscr{R}} \right), \text{ with } X_{0:m:n}^{\mathscr{R}} \equiv 0,$$

and $S^{\mathscr{R}} = \sum_{j=1}^{m} S_{j,m}^{\mathscr{R}}$. They then discussed the exact null distribution of $\mathcal{T}$ in (11.5) and showed that under H_0 (i.e., the observed sample is from $Exp(\vartheta)$-distribution), $E(\mathcal{T}) = \frac{1}{2}$ and $\mathrm{Var}(\mathcal{T}) = \frac{1}{12(m-1)}$, and that the null distribution tends to normality quite rapidly as m increases.

It should be mentioned that the test statistic $\mathcal{T}$ was suggested by Tiku (1980) for complete and Type-II censored samples, while a detailed power study in that case was carried out by Balakrishnan (1983).

Example 11.1. Consider the progressively Type-II right censored data on times to breakdown of an insulating fluid presented earlier in Example 3.53. There, we estimated the mean lifetime of the insulating fluid by assuming an $Exp(\vartheta)$-distribution for the observed data. Now, to validate this exponential assumption, with $n = 19$ and $m = 18$, we find the value of $\mathcal{T}$ in (11.5) from the given data to be

$$\mathcal{T} = \frac{220.070}{508.821} = 0.433,$$

yielding an approximate p-value (using normal approximation) of

$$2\Phi\left(\frac{0.433 - \frac{1}{2}}{\sqrt{\frac{1}{84}}} \right) = 0.536.$$

Therefore, the observed data do not provide enough evidence against the assumed $Exp(\vartheta)$-distribution for the observed progressively Type-II right censored sample. Balakrishnan and Lin (2003) presented an algorithm for determining the exact critical values and p-values of this test. In fact, using their algorithm, these authors determined the exact p-value for the test in this particular example to be 0.544, which is quite close to the normal approximate value 0.536, even though m is as small as 8 in this case!

By evaluating the performance of several EDF-based tests and those based on regression methods (in the complete sample situation), Spinelli and Stephens (1987) concluded that Cramér-von Mises and Shapiro-Wilk tests (Shapiro and Wilk, 1972) performed the best. For this reason, Balakrishnan et al. (2002b) evaluated the relative performance of the test based on $\mathcal{T}$ in (11.5) with Cramér-von Mises test based on

$$A^2 = -\frac{1}{m-1}\sum_{i=1}^{m}(2i-1)\{\ln Z_{i,m} + \ln(1-Z_{i,m})\} - m,$$

where $Z_{i,m} = 1 - \exp\left\{-\frac{mS_{i,m}^{\mathcal{R}}}{S^{\mathcal{R}}}\right\}$, for $i = 1, \ldots, m$, and Shapiro-Wilk test based on

$$W^2 = \frac{\left(\frac{S^{\mathcal{R}}}{m}\right)^2}{\sum_{i=1}^{m} S_{i,m}^{\mathcal{R}2}};$$

in the above, large value of A^2 forms the critical region while that of W^2 is two-sided.

Independently, Marohn (2002) also considered the statistic $\mathcal{T}$ in (11.5) along with Kolmogorov-Smirnov statistic and Fisher's κ-statistic and made a comparison of these tests. Döring and Cramer (2019) discussed an adjustment to the Kolmogorov-Smirnov test, and Wang (2008) studied the statistic of the form

$$\mathcal{T}^* = -2\sum_{i=1}^{m}\log\left\{\frac{\sum_{j=1}^{i} S_{j,m}^{\mathcal{R}}}{S^{\mathcal{R}}}\right\}.$$

While all the above results are for either scaled-exponential or two-parameter exponential distribution, Balakrishnan et al. (2004b) generalized their previous work and presented a goodness-of-fit test for a general location-scale distribution. Their test statistic is of the form

$$\mathcal{T} = \frac{\sum_{i=2}^{m-1}(m-i)\left\{\frac{X_{i:m:n}^{\mathcal{R}} - X_{i-1:m:n}^{\mathcal{R}}}{\mu_{i:m:n}^{\mathcal{R}} - \mu_{i-1:m:n}^{\mathcal{R}}}\right\}}{(m-2)\sum_{i=2}^{m}\left\{\frac{X_{i:m:n}^{\mathcal{R}} - X_{i-1:m:n}^{\mathcal{R}}}{\mu_{i:m:n}^{\mathcal{R}} - \mu_{i-1:m:n}^{\mathcal{R}}}\right\}}, \tag{11.6}$$

where $\mu_{i:m:n}^{\mathcal{R}}$ is the mean of the i-th progressively Type-II censored order statistic for the corresponding standard density function f, which is evidently location and scale invariant. These authors also discussed an approximation of the null distribution as well as the power function, and its mean and variance under H_0 for a specified distribution $F_{\mu,\vartheta}$ for the data; see also Lee and Lee (2019) for a Lorenz curve based test in this general case and Pakyari (2021) for a Gini index-based test.

Example 11.2. Let us consider the progressively Type-II right censored data on log-times to breakdown of an insulating fluid considered earlier in Example 3.56. In this case, with $n = 19$, $m = 8$ and the progressive censoring plan as $\mathcal{R} = (0, 0, 3, 0, 3, 0, 0, 5)$, and assuming extreme value, $EV_1(\mu', \vartheta')$, distribution for these data, Balakrishnan et al.

(2004b) found from (11.6) that $\mathcal{T} = 0.5396$, with a corresponding normal-approximate p-value of 0.7341. Hence, we can conclude that the observed data do not provide enough evidence against the assumed $EV_I(\mu', \vartheta')$-distribution for the observed progressively Type-II right censored data on the log-times to breakdown and so against the assumed Weibull, $Weibull(\vartheta^{\beta}, \beta)$, distribution for the original times to breakdown of the insulating fluid.

11.2.2 Information measures-based tests

Balakrishnan et al. (2007a) were the first authors to propose a goodness-of-fit test for exponentiality based on progressively Type-II censored samples using Kullback-Leibler information measure. Since then, considerable amount of work has been done in this direction. Alizadeh Noughabi (2017) modified the test of Balakrishnan et al. (2007a) to make it scale-invariant, while Habibi Rad et al. (2011) discussed a more general Kullback-Leibler information-based goodness-of-fit tests for log-normal, Pareto and Weibull distributions. Some other known information measures, such as Rényi entropy, cumulative residual entropy and cumulative Kullback-Leibler information, have also been used effectively in developing goodness-of-fit tests; see, for example, Kohansal and Rezakhah (2013), Baratpour and Habibi Rad (2012, 2016), and Park and Pakyari (2015).

11.2.3 Transformation-based tests

By proceeding along the lines of Chen and Balakrishnan (1995) mentioned earlier in the Introduction, some estimated transformation-based goodness-of-fit tests have been discussed, with probability integral transformation being used to approximate the estimated cumulative distribution functions at the observed progressively Type-II censored observations by the corresponding progressively Type-II censored sample from the uniform distribution. Then, EDF statistics as well as different forms of distance statistics have been made use of in carrying out a test for uniformity on the so-transformed data; see, for example, Pakyari and Balakrishnan (2012, 2013) and Torabi et al. (2017).

11.2.4 Tests based on Kaplan-Meier estimates

The Kaplan-Meier estimator of the survival function, based on a progressively Type-II right censored sample, is known to be

$$S_{\mathsf{KM}}(t) = \prod_{i:X_{i:m:n} \le t} \left(1 - \frac{1}{\gamma_i}\right).$$

By comparing this with the parametric estimate of the survival function of $Exp(\vartheta)$-distribution, Döring and Cramer (2019) suggested some goodness-of-fit tests based on the absolute difference between the two survival functions as well as based on the estimate of hazard rate.

11.2.5 Empirical comparison

An elaborate empirical power study of all the above-mentioned tests for exponentiality has been carried out by Döring and Cramer (2019). For this purpose, they considered a wide array of distributions including some symmetric alternatives, two-component mixtures (with multi-modality) and several skewed distributions. In addition, they also considered different choices of progressive censoring plans so as to evaluate the sensitivity of the tests with respect to different forms of censoring in the data. Based on such a detailed empirical study, they made the following key observations:

(1) Many of the tests possess high power against some alternatives even for effective sample sizes as small as $m = 5$;

(2) Some of the tests, like $\mathcal{T}$ and $\mathcal{T}^*$, possess high average power over all progressive censoring plans considered, while some others, like W^2 and A^2, demonstrated weak average performance;

(3) Some tests, like A^2, showed highly varying power performance, with very high power some times and very low power other times, depending on the form of progressive censoring (such as more censoring early on or later on in the experiment);

(4) By examining the minimal power of all the tests considered over all the alternatives, they observed that some tests (like A^2) are quite unstable;

(5) Overall, some statistics (like $\mathcal{T}$ and $\mathcal{T}^*$ and Kaplan-Meier-based tests) possess good stable power values, though not always having highest power values;

(6) The Kaplan-Meier-based test and the modified Kolmogorov-Smirnov-type test considered by Döring and Cramer (2019) turned out to be the most stable statistics among different censoring schemes.

11.2.6 Further issues

It is important to observe that all the test procedures mentioned in Section 11.2.5 are for progressively Type-II right censored samples, and that no formal tests seem to have been developed and studied for progressive Type-I censoring case. Though the tests in this case will need to be conditional on the observed data, it will be of interest to develop such tests and to examine their properties.

Another issue to be noted is that most of the tests have been developed specifically for testing exponentiality (as they are based on some specific property or result for the exponential case). It will, therefore, be of great interest to consider goodness-of-fit tests for other important lifetime distributions such as Weibull. Following that, a detailed empirical power study of all such tests, like to one diligently carried out by Döring and Cramer (2019) for the exponential case, would be highly valuable!

11.3. Hybrid censoring

Even though many goodness-of-fit tests have been discussed in the literature for the case of Type-II censoring and to a lesser extent for Type-I censoring, almost no work exists on goodness-of-fit based on hybrid censored samples. Banerjee and Pradhan (2018) recently discussed Kolmogorov-Smirnov tests based on Type-I and Type-II hybrid censoring scheme.

Their methodology, for an arbitrary distribution F with its parameters completely specified, involves the use of probability integral transformation on the observed hybrid censored data and then testing for uniformity of the so-transformed data based on Kolmogorov-Smirnov test statistics in (11.1). For the case of *Uniform*$(0, 1)$-distribution, as the Kolmogorov-Smirnov statistic in (11.1) takes on the form

$$D_n = \sup_n |F_n(x) - x|,$$

Banerjee and Pradhan (2018) made use of the suggestions of Koziol and Byar (1975) and Dufour and Maag (1978) to consider the following forms of test statistics for Type-I and Type-II censored samples:

$$KS_{\mathsf{I}} = \sup_{x\in[0,1]} \sqrt{n}|F_n(xT_0) - xT_0| \tag{11.7}$$

and

$$KS_{\mathsf{II}} = \sup_{x\in[0,1]} \sqrt{n}|D_{n,\mathsf{II}}(x)|, \tag{11.8}$$

where F_n is the empirical distribution function as defined in (11.4), T_0 is the termination time after the probability integral transformation of the data, and

$$\begin{aligned}\sqrt{n}D_{n,\mathsf{II}}(x) &= \sqrt{n}(F_n(xT_{r:n}) - xT_{r:n})\\ &\stackrel{d}{=} \sqrt{\frac{r-1}{n}}\left\{\sqrt{r-1}\,(F_{r-1}(x) - x)\right\} - x\sqrt{n}\left(T_{r:n} - \frac{r-1}{n}\right)\end{aligned}$$

with $T_{r:n}$ representing the r-th failure time following the transformation of the data and F_{r-1} is the cumulative distribution function of $U_1, \dots, U_{r-1} \stackrel{\text{iid}}{\sim}$ *Uniform*$(0, 1)$ which are supposed independent of $T_{r:n}$.

Based on the test statistics in (11.7) and (11.8), Banerjee and Pradhan (2018) proposed the following test statistics for Type-I and Type-II hybrid censoring scheme situations:

$$KS_{\mathsf{I\text{-}HCS}} = \sup_{x\in[0,1]} \sqrt{n}|F_n(xT_{[m]}) - xT_{[m]}|$$

and

$$KS_{\text{II-HCS}} = \sup_{x \in [0,1]} \sqrt{n} |F_n(xT_{[M]}) - xT_{[M]}|,$$

where $T_{[m]} = \min\{T_0, T_{r:n}\}$ and $T_{[M]} = \max\{T_0, T_{r:n}\}$ correspond to transformed termination times (after the probability integral transformation of the data) under Type-I and Type-II hybrid censoring scheme, respectively. These authors then discussed the asymptotic properties of these test statistics and the determination of their critical values.

Example 11.3. This illustrative example, presented by Banerjee and Pradhan (2018), involved the data on lifetimes of $n = 101$ aluminum coupons given by Birnbaum and Saunders (1958). After fitting the two-parameter gamma distribution on these data and determining the MLEs of the shape and rate parameters to be 11.8 and $\frac{1}{118.76}$, respectively, Banerjee and Pradhan (2018) then carried out the Kolmogorov-Smirnov goodness-of-fit tests in (11.3) and (11.4) based on Type-I and Type-II hybrid censoring scheme data produced with the choices of $T_0 = 0.55$ and $r = 55$. They determined the values of the test statistics to be 0.5522 and 0.5809 and the corresponding p-values to be 0.7653 and 0.7522, thus providing strong evidence in favor of the assumed two-parameter gamma distribution for these hybrid censored lifetime data. Note, however, that the mull hypothesis tested here is a simple hypothesis of gamma distribution with the MLEs being regarded as the true values of the parameters.

It also must be mentioned here that the described goodness-of-fit tests would require large sample sizes as they are based on EDF-type statistics. Hence, there is certainly a need for the development of some other forms of goodness-of-fit tests that could also be useful when the sample sizes are small, which is often the case in practice!

A test for exponentiality based on Type-I hybrid censored data has been proposed by Alizadeh Noughabi and Chahkandi (2018). They (see their Eq. (9)) provided a non-parametric estimate of the entropy under Type-I hybrid censoring, as derived in Morabbi and Razmkhah (2009) (see (8.38)). This estimator is then used to estimate the Kullback-Leibler distance which serves as a goodness-of-fit test statistic; see Park (2005) for an analogue for the case of Type-II censored data. A review on entropy-based goodness-of-fit tests has been provided by Alizadeh Noughabi and Mohtashami Borzadaran (2020).

Remark 11.4. There also exist two early works on goodness-of-fit testing problems for hybrid censored data that should be mentioned here.

(a) Santner and Tenga (1984) discussed tests for the hypothesis that the observed censored data are from some restricted families of distributions, namely, Increasing Failure rate (IFR) and Decreasing Failure Rate (DFR) models, with the alternative simply nullifying the assumption of that restricted family. For this purpose, they then used Kolmogorov-Smirnov type statistics to measure the distance between the empirical cumulative hazard function and the total time on test statistics with the corresponding ones from the greatest convex minorant distribution. After discussing properties of their test statistics for conventional Type-II censoring, the considered the Type-I hybrid censoring scheme and took the scaling factor to be close to 0 so that the behavior of the test statistics is close to those under the usual Type-II censoring. It is important to emphasize here that Santner and Tenga (1984), instead of examining the power performance of their tests under different alternatives, focus on the least favorable configuration for their procedures.

(b) Das and Nag (2002) have discussed a test for exponentiality, within the Weibull family, against the IFR/DFR alternatives through a test of hypothesis for the shape parameter, for the case of Type-I hybrid censoring. More specifically, by considering the Weibull distribution with density function

$$f_{\vartheta,\beta}(x) = \frac{\beta}{\vartheta^{\beta}} x^{\beta-1} e^{-(x/\vartheta)^{\beta}}, \quad x > 0, \ \vartheta, \beta > 0,$$

they considered testing exponentiality through a test of $H_0 : \beta = 1$ against the one-sided alternatives $H_1 : \beta < 1$ (for DFR) and $H_1 : \beta > 1$ (for IFR), respectively, by the use of the test statistic

$$D_{r,\alpha} = D(t_{r,\alpha}) = \sum_{i=1}^{n} \mathbb{1}_{(-\infty, t_{r,\alpha}]}(X_i),$$

where $t_{r,\alpha} = -\log(1 - p_{r,\alpha})$, with $p_{r,\alpha}$ being that value for which the binomial tail probability of at least r failures (out of n units tested) is α. Of course, under the DFR alternative, the critical region would correspond to large values of $D_{r,\alpha}$. Das and Nag (2002) have then presented the optimal values of r and $t_{r,\alpha}$ for different choices of the level of significance α. However, this test procedure has been developed under the strong assumption that the scale parameter ϑ is known (set to be 1, without loss of any generality). Observe, in this case, that the test simply reduces to a test concerning the probability of failure in a binomial setting. Moreover, the method can not be adapted to a plug-in type procedure with an estimate of ϑ as the scale parameter ϑ is required in the design stage itself.

A final mention needs to be made to the work of Zhu (2020, 2021a) in which a parametric bootstrap goodness-of-fit test has been discussed based on progressively hybrid censored data.

CHAPTER 12

Prediction methods

Contents

12.1. Introduction

In many practical problems, one often would like to use data to predict a (future) observation from the same population indicating that prediction problems are extensively discussed in lifetime analysis. Based on an informative sample $X_1, \ldots, X_m$, one wishes to predict the outcome of a random variable Y or random vector $\boldsymbol{Y}$. This random variable may be a (censored) observation from the same experiment (one-sample prediction) or may be part of an independent future sample (two-sample prediction). In the first setup, Y and the X-sample will be correlated.

In the following, we consider two scenarios. Given a (progressive) hybrid censored sample, we are interested in the prediction of the lifetime of a hybrid censored item, that is, we would like to predict the outcome of a future order statistic. Moreover, we address the problem of predicting random variables in a future sample $Y_1, \ldots, Y_r$. Of course, it is important which distributional assumption is put on that sample. Point and interval prediction are both discussed.

12.2. Point prediction

12.2.1 General concepts in point prediction

Before turning to specific prediction problems, some general concepts of point prediction and the resulting predictors are briefly introduced (see also Balakrishnan and Cramer, 2014, Chapter 16). We assume an informative sample $\boldsymbol{Y} = (Y_1, \ldots, Y_m)$ used to predict a single observation Y. A crucial tool in the construction of predictors is the joint distribution of Y and $\boldsymbol{Y}$ or the conditional distribution of Y given $\boldsymbol{Y}$. Furthermore, we assume in this chapter that the baseline cumulative distribution function of the data depends on a parameter $\boldsymbol{\theta} \in \Theta$.

Hybrid Censoring Know-How
https://doi.org/10.1016/B978-0-12-398387-9.00020-9

The best unbiased predictor (BUP) $\widehat{\pi}_{\mathsf{BU}}(\boldsymbol{Y})$ is given by the conditional expectation

$$\widehat{\pi}_{\mathsf{BU}}(\boldsymbol{Y}) = E_{\boldsymbol{\theta}}(Y|\boldsymbol{Y}).$$

If $\boldsymbol{\theta}$ is known, the expectation and thus the BUP can be calculated directly from the conditional distribution of Y given $\boldsymbol{Y}$. Unknown parameters can be replaced by appropriate estimates leading to a plug-in predictor. If $\boldsymbol{\theta}$ is unknown, then a result due to Ishii (1978) and reported in Takada (1981) can be applied: $\widehat{\pi}_{\mathsf{BU}}(\boldsymbol{Y})$ is BUP iff

$$E_{\boldsymbol{\theta}}\big([Y - \widehat{\pi}_{\mathsf{BU}}(\boldsymbol{Y})] \cdot \varepsilon_0(\boldsymbol{Y})\big) = 0 \quad \text{for all } \boldsymbol{\theta}$$

for any squared integrable estimator $\varepsilon_0(\boldsymbol{Y})$ which is an unbiased estimator of zero. This characterization result is similar to that one for best unbiased estimators (see Lehmann and Casella, 1998, Theorem 1.7).

An often used concept in point prediction is linear prediction. A linear predictor is a linear function of the observations $Y_1, \ldots, Y_m$:

$$\widehat{\pi}_{\mathsf{L}}(\boldsymbol{Y}) = \sum_{j=1}^{m} c_j Y_j$$

with given weights $c_1, \ldots, c_m \in \mathbb{R}$. $\widehat{\pi}_{\mathsf{L}}(\boldsymbol{Y})$ is said to be unbiased if its expectation equals $E_{\boldsymbol{\theta}} Y$ for every choice of the parameter. Best linear unbiased prediction is widely used since it can be applied in various contexts. A best linear unbiased predictor (BLUP) is defined as the linear unbiased predictor of Y which minimizes the (standardized) mean squared error. The mathematical treatment is presented in, e.g., Goldberger (1962) or Christensen (2020). In location-scale families (see (3.2)) with unknown parameter $\boldsymbol{\theta} = (\mu, \vartheta)^T$, the BLUP is given by the expression

$$\widehat{\pi}_{\mathsf{LU}}(\boldsymbol{Y}) = \widehat{\mu}_{\mathsf{LU}} + \alpha_Y \widehat{\vartheta}_{\mathsf{LU}} + \boldsymbol{\varpi}^T V^{-1}(\boldsymbol{Y} - \widehat{\mu}_{\mathsf{LU}} \mathbf{1} - \widehat{\vartheta}_{\mathsf{LU}} \boldsymbol{\alpha}),$$

where $\alpha_Y = EY$, $\boldsymbol{\alpha} = E\boldsymbol{Y}$, $V = \mathrm{Cov}(\boldsymbol{Y})$, $\boldsymbol{\varpi} = \mathrm{Cov}(Y, \boldsymbol{Y})$, and $\widehat{\mu}_{\mathsf{LU}}$ and $\widehat{\vartheta}_{\mathsf{LU}}$ are the BLUEs of μ and ϑ, respectively. The corresponding best linear equivariant predictor (BLEP) has a similar representation, in which the BLUEs of the parameters are replaced by the BLEEs $\widehat{\mu}_{\mathsf{LE}}$, $\widehat{\vartheta}_{\mathsf{LE}}$ of the parameters (see Balakrishnan et al., 2008b):

$$\widehat{\pi}_{\mathsf{LE}}(\boldsymbol{Y}) = \widehat{\mu}_{\mathsf{LE}} + \alpha_Y \widehat{\vartheta}_{\mathsf{LE}} + \boldsymbol{\varpi}^T V^{-1}(\boldsymbol{Y} - \widehat{\mu}_{\mathsf{LE}} \mathbf{1} - \widehat{\vartheta}_{\mathsf{LE}} \boldsymbol{\alpha}).$$

An alternative concept proposed by Kaminsky and Rhodin (1985) is maximum likelihood prediction. Given the informative sample $\boldsymbol{y}$, the so-called predictive likelihood function (PLF)

$$\mathscr{L}(\boldsymbol{\theta}, y; \boldsymbol{y}) = f_{\boldsymbol{\theta}}^{Y,\boldsymbol{Y}}(y, \boldsymbol{y})$$

is considered and maximized simultaneously with regard to both the observation y to be predicted and the parameter $\boldsymbol{\theta}$. A solution $(\widehat{\boldsymbol{\theta}}_{\mathsf{p}}, \widehat{\pi}_{\mathsf{ML}}(\boldsymbol{y}))$ of the maximization problem

$$\sup_{(y,\boldsymbol{\theta})} \mathscr{L}(\theta, y; \boldsymbol{y})$$

defines the maximum likelihood predictor (MLP) $\widehat{\pi}_{\mathsf{ML}}(\boldsymbol{Y})$. The solution $\widehat{\boldsymbol{\theta}}_{\mathsf{p}}$ for $\boldsymbol{\theta}$ is called the predictive maximum likelihood estimator of $\boldsymbol{\theta}$. Kaminsky and Rhodin (1985) applied this approach to predict an order statistic $X_{s:n}$ given a Type-II right censored sample $X_{1:n}, \ldots, X_{m:n}$ with $1 \leq m < s \leq n$. Since explicit expressions of the maximum likelihood predictor $\widehat{\pi}_{\mathsf{ML}}(\boldsymbol{Y})$ will only be available in exceptional cases, Basak and Balakrishnan (2009) used a Taylor approximation of order one to linearize the likelihood equations generated by the PLF. The resulting solution $\widehat{\pi}_{\mathsf{AM}}(\boldsymbol{y})$ is called approximate maximum likelihood predictor (AMLP).

The median unbiased predictor (MUP) $\widehat{\pi}_{\mathsf{MU}}(\boldsymbol{Y})$ is defined via the generalized median condition

$$\Pr_{\boldsymbol{\theta}}(\widehat{\pi}_{\mathsf{MU}}(\boldsymbol{Y}) \leq Y) = \Pr_{\boldsymbol{\theta}}(\widehat{\pi}_{\mathsf{MU}}(\boldsymbol{Y}) \geq Y). \tag{12.1}$$

This idea was employed by Takada (1991) for Type-II censored samples. He showed that the MUP leads to a smaller value of Pitman's measure of closeness than the BLUP. The median $\widehat{\pi}_{\mathsf{CM}}(\boldsymbol{Y})$ of the conditional distribution $f^{Y|\boldsymbol{Y}}$ is called conditional median predictor (CMP). Since it obviously satisfies (12.1), it is also a MUP.

Finally, Bayesian methods are also considered in prediction. Suppose the population cumulative distribution function is given by $F_{\boldsymbol{\theta}}$ with density function $f_{\boldsymbol{\theta}}$ and that the prior distribution is defined by the prior density function $\pi_{\boldsymbol{a}}$ with some hyper-parameter $\boldsymbol{a}$. $\pi_{\boldsymbol{a}}^*(\cdot \mid \boldsymbol{y})$ denotes the posterior density function, given $\boldsymbol{Y} = \boldsymbol{y}$. We illustrate the idea for samples of order statistics since this model is close to the models discussed in hybrid censoring.

(a) One-sample set-up

Given a sample $X_{1:n}, \ldots, X_{m:n}$ of order statistics, we are interested in predicting a future outcome $X_{s+m:n}$, $s > 0$, from the same sample. The procedure is based on the predictive distribution of $X_{s+m:n}$, given the observations $\boldsymbol{X}_m = (X_{1:n}, \ldots, X_{m:n})$. Then, the density function of $X_{s+m:n}$, given $\boldsymbol{X}_m = \boldsymbol{x}_m$, is given by

$$f_{\boldsymbol{\theta}}^{X_{s+m:n}|\boldsymbol{X}_m}(t \mid \boldsymbol{x}_m) = f_{\boldsymbol{\theta}}^{X_{s+m:n}|X_{m:n}}(t \mid \boldsymbol{x}_m), \quad x_1, \ldots, x_m, t \in \mathbb{R}.$$

Due to the Markov property of order statistics, this conditional density function depends only on the largest observation x_m. The predictive density function of $X_{s+m:n}$, given $\boldsymbol{X}_m = \boldsymbol{x}_m$, is defined via

$$f_{s+m}(t \mid \boldsymbol{x}_m) = \int f_{\boldsymbol{\theta}}^{X_{s+m:n}|X_{m:n}}(t \mid \boldsymbol{x}_m)\pi_{\boldsymbol{a}}^*(\boldsymbol{\theta} \mid \boldsymbol{x}_m)\, d\boldsymbol{\theta}, \quad \boldsymbol{x}_m \in \mathbb{R}^m, t \in \mathbb{R}, \tag{12.2}$$

where $\pi_{\boldsymbol{a}}^*(\boldsymbol{\theta} \mid \boldsymbol{x}_m)$ denotes the posterior density function. In fact, from Theorem 2.12, the conditional density function $f_{\boldsymbol{\theta}}^{X_{s:n}|X_{m:n}}(\cdot \mid \boldsymbol{x}_m)$ can be written as (see (2.11), $t > x_m$)

$$f_{\boldsymbol{\theta}}^{X_{s+m:n}|X_{m:n}}(t \mid \boldsymbol{x}_m) = s\binom{n-m}{s}\left(\frac{F_{\boldsymbol{\theta}}(t) - F_{\boldsymbol{\theta}}(x_m)}{1 - F_{\boldsymbol{\theta}}(x_m)}\right)^{s-1}(1 - G_{\boldsymbol{\theta},x_m}(t))^{n-s+m} g_{\boldsymbol{\theta},x_m}(t),$$

where $G_{\boldsymbol{\theta},x_m}(\cdot) = 1 - [1 - F_{\boldsymbol{\theta}}(\cdot)]/[1 - F_{\boldsymbol{\theta}}(x_m)]$ is a left-truncated cumulative distribution function, and $g_{\boldsymbol{\theta},x_m}$ denotes the corresponding density function.

(b) Two-sample set-up

Given the informative sample $X_{1:n}, \ldots, X_{m:n}$ of order statistics, the goal of this method is to predict outcomes of an independent future sample $Y_{1:N} \leq \cdots \leq Y_{N:N}$ from the same population. The procedure is based on the predictive distribution of $Y_{k:N}$, given the informative sample $\boldsymbol{X}_m = (X_{1:n}, \ldots, X_{m:n})$. Then, the density function of $Y_{s:N}$ is given by

$$f_{\boldsymbol{\theta}}^{Y_{s:N}}(y) = s\binom{N}{s} f_{\boldsymbol{\theta}}(y)(1 - F_{\boldsymbol{\theta}}(y))^{s-1} F_{\boldsymbol{\theta}}^{N-s}(y), \quad y \in \mathbb{R}.$$

For a posterior density function $\pi_{\boldsymbol{a}}^*(\cdot \mid \boldsymbol{x}_m)$, the predictive density function of $Y_{s:N}$ given $\boldsymbol{X}_m = \boldsymbol{x}_m$ is defined via

$$f_s(y \mid \boldsymbol{x}_m) = \int f_{\boldsymbol{\theta}}^{Y_{s:N}}(y)\pi_{\boldsymbol{a}}^*(\boldsymbol{\theta} \mid \boldsymbol{x}_m)\,d\boldsymbol{\theta}, \quad y \in \mathbb{R}. \tag{12.3}$$

Notice that $\boldsymbol{X}_m$ and $Y_{s:N}$ are supposed independent random variables.

Under squared-error loss, the Bayes predictive estimator of $X_{s:n}$ (or $Y_{s:N}$) is defined as the expected value of the predictive distribution, that is,

$$\widehat{\pi}_{\mathsf{BA}}(\boldsymbol{X}_m) = \int y f_s(y \mid \boldsymbol{X}_m)\,dy. \tag{12.4}$$

Under absolute error loss, the median $\widehat{\pi}_{\mathsf{BAM}}(\boldsymbol{X}^{\mathscr{R}})$ of the predictive distribution is the Bayes predictor of $X_{s+m:n}$ and $Y_{s:N}$, respectively.

12.2.2 Point prediction under Type-I hybrid censoring

As pointed out in the preceding section, the PLF is an important tool in the construction of predictors. In the following, we consider a hybrid censored sample $X_{1:n}, \ldots, X_{D_{\mathsf{HCS}}:n}$ where D_{HCS} denotes the observed sample size.[1]

[1] We do not specify the particular hybrid censoring scheme so far. As has been pointed out in Chapter 4, a hybrid censored sample is almost specified by the random counter D_{HCS} which of course depends on the particular censoring plan.

In the one-sample prediction, we would like to predict a hybrid censored observation $X_{D_{\mathsf{HCS}}+s:n}$ (where $D_{\mathsf{HCS}} + s \le n$) based on the informative sample $X_{1:n}, \ldots, X_{D_{\mathsf{HCS}}:n}$, D_{HCS}. Thus, an expression of the joint density function of $f_{\boldsymbol{\theta}}^{X_{1:n},\ldots,X_{D_{\mathsf{HCS}}:n},X_{D_{\mathsf{HCS}}+s:n},D_{\mathsf{HCS}}}$ is needed. Expressions can be established by analogy with the derivations presented in Chapter 6.

We illustrate the derivation of the density function for Type-I hybrid censoring by proceeding by analogy with the arguments presented on p. 128. Let $t_1, \ldots, t_m, x \in \mathbb{R}$ and consider the Type-I hybrid censored informative sample $X_{1:n}, \ldots, X_{D_\mathsf{I}:n}$ with effectively observed sample size D_I as well as a future observation $X_{D_\mathsf{I}+s:n}$. Then, the joint cumulative distribution function of the informative sample and the observation to be predicted is given by

$$\begin{aligned}\Pr(X_{j:n} \le t_j, 1 \le j \le D_\mathsf{I}, X_{D_\mathsf{I}+s:n} \le x) &= \Pr(X_{j:n} \le t_j, 1 \le j \le m, D \ge m, X_{m+s:n} \le x)\\ &\quad + \sum_{d=0}^{m-1} \Pr(X_{j:n} \le t_j, 1 \le j \le d, D = d, X_{d+s:n} \le x).\end{aligned}$$

As before, we distinguish two scenarios depending on the censoring effectively applied.

Scenario ①: The experiment is terminated by the m-th failure time. Thus, we get by (4.3)

$$\begin{aligned}&\Pr(X_{j:n} \le t_j, 1 \le j \le m, D \ge m, X_{m+s:n} \le x)\\ &\quad = \Pr(X_{j:n} \le t_j, 1 \le j \le m-1, X_{m:n} \le t_m \wedge T, X_{m+s:n} \le x)\\ &\quad = F_{1,\ldots,m,s:n}(\boldsymbol{t}_{m-1}, t_m \wedge T, x),\end{aligned}$$

Scenario ②: The experiment is terminated by the threshold T. First, we get for $d = 0$, $s = 1$, and $T \le x$

$$\Pr(D = 0, X_{s:n} \le x) = \Pr(T < X_{1:n}, X_{1:n} \le x) = F_{1:n}(x) - F_{1:n}(T).$$

For $d = 0$ and $s \ge 2$, we find for $T \le x$

$$\Pr(D = 0, X_{s:n} \le x) = \Pr(T < X_{1:n}, X_{s:n} \le x) = F_{s:n}(x) - F_{1,s:n}(T).$$

Let $d \in \{1, \ldots, m-1\}$ and $s \ge 2$. Then,

$$\begin{aligned}&\Pr(X_{j:n} \le t_j, 1 \le j \le d, D = d, X_{d+s:n} \le x)\\ &\quad = \Pr(X_{j:n} \le t_j, 1 \le j \le m, X_{d:n} \le T < X_{d+1:n}, X_{d+s:n} \le x)\\ &\quad = \Pr(X_{j:n} \le t_j, 1 \le j \le d-1, X_{d:n} \le t_d \wedge T, X_{d+1:n} \ge T,\\ &\qquad\qquad T \le \min_{d+1 \le j \le m} t_j, X_{d+s:n} \le x)\end{aligned}$$

$$
\begin{aligned}
&= \mathbb{1}_{[T,\infty)}(\min_{d+1\le j\le m} t_j)\Big\{\Pr(X_{j:n}\le t_j,\,1\le j\le d-1,\,X_{d:n}\le t_d\wedge T,\,X_{d+s:n}\le x)\\
&\qquad - \Pr(X_{j:n}\le t_j,\,1\le j\le d-1,\,X_{d:n}\le t_d\wedge T,\,X_{d+1:n}\le T,\,X_{d+s:n}\le x)\Big\}\\
&= \mathbb{1}_{[T,\infty)}(\min_{d+1\le j\le m} t_j)\Big\{F_{1,\dots,d,d+s:n}(\boldsymbol{t}_{d-1},\,t_d\wedge T,\,x)\\
&\qquad\qquad - F_{1,\dots,d+1,d+s:n}(\boldsymbol{t}_{d-1},\,t_d\wedge T,\,T,\,x)\Big\}\\
&= \mathbb{1}_{[T,\infty)}(\min_{d+1\le j\le m} t_j)\Big\{F_{1,\dots,d,d+s:n}(\boldsymbol{t}_d\wedge T,x) - F_{1,\dots,d+1,d+s:n}(\boldsymbol{t}_d\wedge T,\,T,\,x)\Big\}.
\end{aligned}
$$

For $d\in\{1,\dots,m-1\}$ and $s=1$, we get by similar derivations

$$
\begin{aligned}
&\Pr(X_{j:n}\le t_j,\,1\le j\le d,\,D=d,\,X_{d+1:n}\le x)\\
&\qquad = \mathbb{1}_{[T,\infty)}(\min_{d+1\le j\le m} t_j)\Big\{F_{1,\dots,d+1:n}(\boldsymbol{t}_d\wedge T,x) - F_{1,\dots,d+1:n}(\boldsymbol{t}_d\wedge T,\,x\wedge T)\Big\}.
\end{aligned}
$$

As a result, we find the following representation of the density function depending on the values of $D_{\rm I}$ and s:

(a) $D_{\rm I}=m,\ s\ge 1$:

$$
f^{X_{1:n},\dots,X_{m:n},X_{m+s:n},D_{\rm I}}(\boldsymbol{t}_m,x,m) = f_{1,\dots,m,m+s:n}(\boldsymbol{t}_m,x),\qquad t_1\le\cdots\le t_m\le x.
$$

(b) $D_{\rm I}\in\{1,\dots,m-1\},\ s=1$:

$$
f^{X_{1:n},\dots,X_{d:n},X_{d+1:n},D_{\rm I}}(\boldsymbol{t}_d,x,d) = f_{1,\dots,d+1:n}(\boldsymbol{t}_d,x),\qquad t_1\le\cdots\le t_d\le T<x. \tag{12.5}
$$

(c) $D_{\rm I}\in\{1,\dots,m-1\},\ s\ge 2$: for $t_1\le\cdots\le t_d\le T<x$, we get from (2.7)

$$
\begin{aligned}
f^{X_{1:n},\dots,X_{d:n},X_{d+s:n},D_{\rm I}}(\boldsymbol{t}_d,x,d) &= \int_T^x f_{1,\dots,d+1,d+s:n}(\boldsymbol{t}_d,z,x)\,dz \qquad (12.6)\\
&= \frac{n!}{(s-1)!(n-d-s)!}\,(F(x)-F(T))^{s-1}\,(1-F(x))^{n-d-s}\left(\prod_{i=1}^d f(t_i)\right)f(x).
\end{aligned}
$$

For $s=1$, the expression in (12.6) coincides with that one given in (12.5), so that we do not need to distinguish these cases later on.

Summing up, for $D_{\rm I}=d>0$, the predictive likelihood is given by

$$
\begin{aligned}
\mathscr{L}(x_{d+s},\theta\mid \boldsymbol{x}_d,d) &= f^{X_{1:n},\dots,X_{d:n},X_{d+s:n},D_{\rm I}}(\boldsymbol{x}_d,x,d) = \frac{n!}{(s-1)!(n-d-s)!} \qquad (12.7)\\
&\quad\times (F_\theta(x_{d+s})-F_\theta(w))^{s-1}\,(1-F_\theta(x_{d+s}))^{n-d-s}\left(\prod_{i=1}^d f_\theta(x_i)\right)f_\theta(x_{d+s}),
\end{aligned}
$$

with d and w taken from Table 4.7.

Remark 12.1. **(a)** Denoting by G_T, H_T (with density functions g_T and h_T) the left/right truncated cumulative distribution functions defined by $G_T(x) = (F(x) - F(T))/(1 - F(x))$, $x \geq T$, and $H_T(x) = F(x)/F(T)$, $x \leq T$, we can write (12.6) as

$$\begin{aligned} &f^{X_{1:n},\ldots,X_{d:n},X_{d+s:n},D_{\text{I}}}(\boldsymbol{t}_d, x, d) \\ &= s\binom{n-d}{s} G_T^{s-1}(x)\,(1 - G_T(x))^{n-d-s} g_T(x) \cdot d!\left(\prod_{i=1}^{d} h_T(t_i)\right) \\ &\qquad \times \binom{n}{d}(1 - F(T))^{n-d} F^d(T) \\ &= \Pr(D = d) \cdot g_{T;s:n-d}(x) \cdot h_{T;1,\ldots,d:d}(\boldsymbol{t}_d), \quad t_1, \ldots, t_d \leq T \leq x. \end{aligned}$$

This factorization result is not surprising since it corresponds to the block independence result established in Iliopoulos and Balakrishnan (2009), that is, $X_{d+s:n}$ and $X_{1:n}, \ldots, X_{d:n}$ given $D = d$ are stochastically independent with the given right and left truncated distributions. In view of this connection, the above result can also be obtained from the block independence result applied to this particular data situation.

(b) Notice that, from (12.6), the conditional density function $f^{X_{d+s:n}|X_{1:n},\ldots,X_{d:n},D_{\text{I}}}$ is given by

$$\begin{aligned} &f^{X_{d+s:n}|X_{1:n},\ldots,X_{d:n},D_{\text{I}}}(x \mid \boldsymbol{x}_d, d) \qquad (12.8) \\ &= s\binom{n-d}{s} (F(x) - F(w))^{s-1} (1 - F(x))^{n-d-s} (1 - F(w))^{-(n-d)} f(x), \quad w < x, \end{aligned}$$

which is reported in Asgharzadeh et al. (2015b).

Remark 12.2. By proceeding similar to the derivation of the PLF under Type-I hybrid censoring, predictive likelihood functions can be obtained under the other hybrid censoring plans. In particular, one has to consider the expression in (12.7) with w appropriately chosen from Table 4.7. As a result, this illustrates that the computational problem is completely solved by discussing the problem for the Type-II censoring case and the data $\boldsymbol{x}_{d-1}$, w, d, where $w \in \{x_d, T\}$.

Prediction for exponential distributions

Prediction of future observations based on a Type-I hybrid censored sample $X_{1:n}, \ldots, X_{D_{\text{I}}:n}$ with observed sample size D_{I} of exponentially distributed lifetimes has been first considered by Ebrahimi (1992). In this case, the predictive likelihood is given by

$$\begin{aligned} \mathscr{L}(x_{d+s}, \vartheta \mid \boldsymbol{x}_d, d) = &\frac{n!}{(s-1)!(n-d-s)!} \\ &\times \left(e^{-w/\vartheta} - e^{-x_{d+s}/\vartheta}\right)^{s-1} \vartheta^{-d-1} e^{-(\sigma+(n-d-s+1)x_{d+s})/\vartheta}, \end{aligned}$$

with $\sigma = \sum_{j=1}^{d} x_j$. This case is included in Asgharzadeh et al. (2015b) choosing the shape parameter of the Weibull distribution equal to one. As a result one gets for given ϑ the ML predictor

$$\widehat{\pi}_{\mathsf{ML}}(\boldsymbol{X}_{D_{\mathsf{I}}}, D_{\mathsf{I}}) = W - \vartheta \log\Big(1 - \frac{s-1}{n-D_{\mathsf{I}}}\Big)$$

where

$$W = T\mathbb{1}_{\{1,\dots,m-1\}}(D_{\mathsf{I}}) + X_{m:n}\mathbb{1}_{\{m\}}(D_{\mathsf{I}}) = X_{m:n} \wedge T.$$

In view of Remark 12.2, the problem can be solved along the lines of that for order statistics when ϑ is unknown. Therefore, the solution can be obtained directly from the results of Kaminsky and Rhodin (1985) and Ebrahimi (1992), that is,

$$\widehat{\vartheta} = \frac{1}{D_{\mathsf{I}}+1}\Big(\sum_{j=1}^{D_{\mathsf{I}}} X_{j:n} + (n-D_{\mathsf{I}})W\Big), \quad \widehat{\pi}_{\mathsf{ML}}(\boldsymbol{X}_{D_{\mathsf{I}}}, D_{\mathsf{I}}) = W - \widehat{\vartheta} \log\Big(1 - \frac{s-1}{n-D_{\mathsf{I}}}\Big), \quad (12.9)$$

that is, the parameter ϑ is replaced by its predictive MLE (see also Raqab and Nagaraja, 1995; Kaminsky and Nelson, 1998). As pointed out by Ebrahimi (1992), the BUP of $X_{D_{\mathsf{I}}+s:n}$ for known ϑ is given by

$$\widehat{\pi}_{\mathsf{BU}}(\boldsymbol{X}_{D_{\mathsf{I}}}, D_{\mathsf{I}}) = W + \vartheta \sum_{j=D_{\mathsf{I}}+1}^{D_{\mathsf{I}}+s} \frac{1}{n-j+1},$$

which has the same functional form as the BUP under Type-II censoring (cf. Kaminsky and Nelson, 1975). In case of an unknown parameter ϑ, replacing ϑ by an appropriate estimator (e.g., $\widehat{\vartheta}$ as in (12.9)) yields a reasonable predictor of $X_{D_{\mathsf{I}}+s:n}$ (see also Ebrahimi, 1992).

Asgharzadeh et al. (2015b) have considered conditional mean prediction as proposed in Raqab and Nagaraja (1995) for Weibull distributions with known parameters. Thus, exponential distributions are included as particular case leading to the conditional mean predictor

$$\widehat{\pi}_{\mathsf{CM}}(\boldsymbol{X}_{D_{\mathsf{I}}}, D_{\mathsf{I}}) = W - \vartheta \log\Big(1 - \mathrm{med}_{s,n-D_{\mathsf{I}}-s+1}\Big)$$

where $\mathrm{med}_{\alpha,\beta}$ denotes the median of a $Beta(\alpha,\beta)$-distribution. Replacing ϑ by an appropriate estimator, a predictor for unknown ϑ results.

Prediction for Weibull distributions

For $Weibull(\vartheta,\beta)$-distributions, Asgharzadeh et al. (2015b) discussed point prediction under Type-I hybrid censoring. From (12.7), the predictive likelihood function is given by

$$\mathscr{L}(x_{d+s}, \vartheta, \beta \mid \boldsymbol{x}_d, d) = \frac{n!}{(s-1)!(n-d-s)!}\beta^{d+1}\vartheta^{-d-1}\left(e^{-w^\beta/\vartheta} - e^{-x_{d+s}^\beta/\vartheta}\right)^{s-1}$$
$$\times x_{d+s}^{\beta-1}\left(\prod_{j=1}^{d} x_j\right)^{\beta-1} \exp\left\{-\sum_{j=1}^{d} x_j^\beta/\vartheta - (n-d-s+1)x_{d+s}^\beta/\vartheta\right\}$$

with w taken from Table 4.7. Proceeding as above, they found for known parameters that the maximum likelihood predictor is given by

$$\widehat{\pi}_{\mathsf{ML}}(\boldsymbol{X}_{D_\mathsf{I}}, D_\mathsf{I}) = \left(W^\beta - \vartheta \log\left(1 - \frac{(s-1)\beta W^\beta}{\beta(\vartheta D_\mathsf{I} - \sum_{j=1}^{D_\mathsf{I}} X_{j:n}^\beta) + \vartheta}\right)\right)^{1/\beta} \tag{12.10}$$

Replacing the parameters by the predictive MLE yields the maximum likelihood predictor for unknown parameters. If we choose $\beta = 1$, i.e., exponentially distributed lifetimes, and replace in (12.10) ϑ by the predictive maximum likelihood estimator given in (12.9), we get the corresponding predictor in case of the exponential distribution.

The BUP of $X_{D_\mathsf{I}+s:n}$ for known ϑ and β is given by

$$\widehat{\pi}_{\mathsf{BU}}(\boldsymbol{X}_{D_\mathsf{I}}, D_\mathsf{I}) = E(X_{D_\mathsf{I}+s:n} \mid \boldsymbol{X}, D_\mathsf{I}) \tag{12.11}$$
$$= s\binom{n-D_\mathsf{I}}{s}\vartheta^{1/\beta}\sum_{j=1}^{s-1}(-1)^{s-j-1}\binom{s-1}{j}\frac{e^{(n-D_\mathsf{I}-j)W/\vartheta}}{(n-D_\mathsf{I}-j)^{1+1/\beta}}\Gamma\left(1+\frac{1}{\beta}; \frac{(n-D_\mathsf{I}+-j)W^\beta}{\vartheta}\right),$$

where

$$\Gamma(\alpha, x) = \int_x^\infty t^{\alpha-1}e^{-t}\,dt, \quad x \geq 0,$$

denotes the incomplete gamma function (see Asgharzadeh et al., 2015b). If the parameters are supposed unknown, they have to be replaced by appropriate estimators, e.g., the MLEs or AMLEs. Finally, a CMP of $X_{D_\mathsf{I}+s:n}$ is presented as

$$\widehat{\pi}_{\mathsf{BU}}(\boldsymbol{X}_{D_\mathsf{I}}, D_\mathsf{I}) = \left(W^\beta - \vartheta \log\left(1 - \mathrm{med}_{s,n-D_\mathsf{I}-s+1}\right)\right)^{1/\beta}$$

where $\mathrm{med}_{\alpha,\beta}$ denotes the median of a *Beta*(α, β)-distribution. Replacing ϑ and β by appropriate estimators, a predictor for unknown parameters results.

Asgharzadeh et al. (2015b) proposed also Bayesian predictors for $X_{D_\mathsf{I}+s:n}$ given the informative sample $X_{1:n}, \ldots, X_{D_\mathsf{I}:n}, D_\mathsf{I}$. They assumed independent gamma priors for the parameters, that is, $\beta \sim \Gamma(b_1, a_1)$, $\lambda = 1/\vartheta \sim \Gamma(b_2, a_2)$. Then, the joint posterior density function $\pi^*(\cdot \mid \boldsymbol{x}_d, d)$ is given by

$$\pi^*(\beta, \lambda \mid \boldsymbol{x}_d, d) \propto g_1(\lambda \mid \beta, \boldsymbol{x}_d, d) \cdot g_2(\beta \mid \boldsymbol{x}_d, d),$$

where $g_1(\cdot \mid \beta, \boldsymbol{x}_d, d)$ is the density function of a $\Gamma(\sum_{j=1}^{d} x_j^{\beta} + (n-d)w^{\beta} + b_2, d + a_2)$-distribution and $g_2(\cdot \mid \boldsymbol{x}_d, d)$ is a proper density function with

$$g_2(\beta \mid \boldsymbol{x}_d, d) \propto \frac{e^{-b_1\beta}\beta^{d+a_1-1}}{\left(\sum_{j=1}^{d} x_j^{\beta} + (n-d)w^{\beta} + b_2\right)^{d+a_2}} \prod_{j=1}^{d} x_j^{\beta-1}.$$

From (12.2), the predictive density function is given by

$$f_{d+s}(t \mid \boldsymbol{x}_d, d) = \int f_{\beta,\lambda}^{X_{s+d:n}|X_{d:n}=x_d, D_1=d}(t)\pi^*(\beta, \lambda \mid \boldsymbol{x}_d, d)d\beta\, d\lambda, \quad \boldsymbol{x}_d \in \mathbb{R}^d, t \in \mathbb{R}.$$

According to (12.4), the Bayes predictor w.r.t. square loss is given by

$$\widehat{\pi}_{\mathsf{BA}}(\boldsymbol{X}_{D_1}, D_1) = \int_T^{\infty} y f_s(y \mid \boldsymbol{X}_{D_1}, D_1)dy.$$

Under LINEX loss function (see p. 101), one gets

$$\widehat{\pi}_{\mathsf{BA;L}}(\boldsymbol{X}_{D_1}, D_1) = -\frac{1}{a}\log\left(\int_T^{\infty} e^{-ay} f_s(y \mid \boldsymbol{X}_{D_1}, D_1)dy\right).$$

Asgharzadeh et al. (2015b) showed that $g_2(\cdot \mid \boldsymbol{x}_d, d)$ is log-concave so that a general simulation method due to Devroye (1984) can be used to simulate data from this distribution. Following Geman and Geman (1984), they proposed the following procedure to generate samples from the conditional posterior density function:

(1) Generate β_1 from $g_2(\cdot \mid \boldsymbol{x}_d, d)$ using the method developed by Devroye (1984).

(2) Generate λ_1 from $g_1(\cdot \mid \beta_1, \boldsymbol{x}_d, d)$

(3) Repeat steps (1) and (2) N times and obtain the sample $(\beta_1, \lambda_1), \ldots, (\beta_N, \lambda_N)$.

For further details, see Asgharzadeh et al. (2015b). Then, using the Gibbs sampling approach, the simulation consistent estimator of the predictive density function $f_{d+s}(\cdot \mid \boldsymbol{x}_d, d)$ is defined by

$$\widehat{f_{d+s}}(t \mid \boldsymbol{x}_d, d) = \sum_{j=1}^{N} f_{d+s}(t \mid \beta_j, \lambda_j, \boldsymbol{x}_d, d)$$

where $f_{d+s}(\cdot \mid \beta_j, \lambda_j, \boldsymbol{x}_d, d)$ denotes the conditional density function of $X_{d+s:n}$ given the informative sample and the parameters $\beta_j, \lambda_j, 1 \le j \le N$. Then, $\widehat{\pi}_{\mathsf{BA}}(\boldsymbol{x}_d, d)$ and $\widehat{\pi}_{\mathsf{BA;L}}(\boldsymbol{x}_d, d)$ are obtained by

$$\widehat{\pi}_{\mathsf{BA}}(\boldsymbol{x}_d, d) = \frac{1}{N}\sum_{j=1}^{N} I(\boldsymbol{x}_d, d; \beta_j, \lambda_j), \quad \widehat{\pi}_{\mathsf{BA;L}}(\boldsymbol{x}_d, d) = \frac{1}{N}\sum_{j=1}^{N} J(\boldsymbol{x}_d, d; \beta_j, \lambda_j).$$

$I(\boldsymbol{x}_d, d; \beta_j, \lambda_j)$ can be taken from (12.11) whereas $J(\boldsymbol{x}_d, d; \beta_j, \lambda_j)$ is obtained from

$$J(\boldsymbol{x}_d, d; \beta, \lambda) = E(e^{-cX_{d+s:n}} \mid \boldsymbol{x}_d, d) = \int_w^\infty e^{-cy} f_{d+s}(y \mid \beta, \lambda, \boldsymbol{x}_d, d) dy.$$

The integral can be evaluated by analogy with (12.11) (see Asgharzadeh et al., 2015b; c denotes the parameter of the LINEX loss function as given on p. 101).

Prediction for other distributions

As has been pointed out at the beginning of Section 12.2.2, point prediction can be carried out in the same way by considering the respective hybrid censored informative sample $X_{1:n}, \ldots, X_{D_{\mathsf{HCS}}:n}$ and the effectively observed sample size D_{HCS} induced by the hybrid censoring scheme. In particular, an analogous approach can be applied to progressive hybrid censoring but, so far, we are not aware of respective results.

Similar results for generalized exponential distributions under Type-I hybrid censoring have been established by Valiollahi et al. (2017). Kayal et al. (2017) and Kayal et al. (2018) obtained analogous findings for Burr XII distribution and inverted exponentiated Rayleigh distributions, respectively. A general account to distributions with a cumulative distribution function $F_{\boldsymbol{\theta}}$ of the form

$$F_{\boldsymbol{\theta}}(t) = 1 - \exp\{-\Lambda_{\boldsymbol{\theta}}(t)\}, \tag{12.12}$$

with cumulative hazard function $\Lambda_{\boldsymbol{\theta}}$ is investigated in Valiollahi et al. (2019). Particular choices of $\Lambda_{\boldsymbol{\theta}}$ yield exponential, Weibull, Pareto, and Burr Type XII distributions. Khan and Mitra (2021) discussed exponential–logarithmic distributions proposed in Tahmasbi and Rezaei (2008). Bayesian prediction for the lognormal distribution and generalized half normal distribution has been studied in Singh and Tripathi (2016) and Sultana and Tripathi (2018), respectively. Log-logistic distributions under Type-I hybrid censoring with possible left censoring are considered in Abou Ghaida and Baklizi (2022). Further distributions discussed are generalized Lindley distribution (Singh et al., 2016) and Poisson-Exponential distributions (Monfared et al., 2022).

12.2.3 Point prediction for other hybrid censoring schemes

A Type-II hybrid censored informative sample from generalized exponential distributions has been studied in Valiollahi et al. (2017) and Sen et al. (2018b). Shafay (2016a) studied Bayesian one- and two-sample prediction given a generalized Type-II hybrid censored informative sample for the cumulative distribution function in (12.12).

Conditional median prediction for a unified hybrid-censored Burr Type XII distribution has been considered in Panahi and Sayyareh (2015). Bayesian prediction under unified hybrid censoring with lifetimes from the cumulative distribution function given in (12.12) is discussed in Mohie El-Din et al. (2017).

Prediction of times to failure of progressively censored units based on Type-I and Type-II progressive hybrid censored data has been studied in Ameli et al. (2018) (for progressively Type-II censored data, see, e.g., Balakrishnan and Cramer, 2014, Chapter 16; Balakrishnan and Rao, 1997; Basak et al., 2006; Basak and Balakrishnan, 2009). Mohie El-Din et al. (2019) presented results on one- and two sample Bayesian prediction for Gompertz distribution under generalized progressive hybrid censoring. Sen et al. (2018c) and Sen et al. (2020) studied the problem for log-normal and generalized exponential distributions, respectively. For Burr III distributions and Kumaraswamy distributions under progressive type-I hybrid censoring, we refer to Singh et al. (2019) and Sultana et al. (2020), respectively.

12.3. Interval prediction

In interval prediction, statistical intervals are constructed that will contain the future observation with a specified probability (see Hahn et al., 2017, Sections 2.3 and 15.5). In the following, we will present results on such *prediction intervals* in presence of hybrid censoring.

12.3.1 Classical prediction intervals

A prediction interval based on Type-I hybrid censored exponentially distributed lifetimes has been presented in Ebrahimi (1992) using the conditional cumulative distribution function in (12.13) belonging to the conditional density function presented in (12.8). For an exponential distribution with parameter ϑ, the cumulative distribution function is given by

$$\begin{aligned} F_s(y \mid \boldsymbol{x}_d, d) &= F^{X_{d+s:n}|X_{1:n},\ldots,X_{d:n},D}(y \mid \boldsymbol{x}_d, d) \\ &= s\binom{n-d}{s}\vartheta^{-1}\int_w^y \left(e^{-w/\vartheta} - e^{-x/\vartheta}\right)^{s-1} e^{-(n-d-s+1)x/\vartheta+(n-d)w/\vartheta}\,dx \\ &= s\binom{n-d}{s}\vartheta^{-1}\int_0^{y-w} \left(1 - e^{-x/\vartheta}\right)^{s-1} e^{-(n-d-s+1)x/\vartheta}\,dx, \quad y \ge w. \end{aligned} \tag{12.13}$$

Then, for ϑ known, a (conditional) confidence interval (y_1, y_2) with level $1-\alpha$ is defined by the solution of the equations

$$F_s(y_1 \mid \boldsymbol{x}_d, d) = \frac{\alpha}{2}, \quad F_s(y_2 \mid \boldsymbol{x}_d, d) = 1 - \frac{\alpha}{2}.$$

If ϑ is supposed unknown, a suitable estimator can be used to construct an approximate confidence interval.

Interval prediction for Type-I and Type-II hybrid censored data from generalized exponential distributions has been studied in Valiollahi et al. (2017). The construction is based on the distribution of the random variable

$$Z = 1 - \frac{1-(1-e^{-\lambda Y})^{\beta}}{1-(1-e^{-\lambda W})^{\beta}}$$

which, given the informative sample, has a *Beta*$(s, n - D_{\mathsf{HCS}} - s + 1)$-distribution. Thus, they considered Z as a pivotal quantity and constructed a prediction interval $(\eta(\alpha/2), \eta(1-\alpha/2))$ with level $1-\alpha$ where

$$\eta(t) = -\frac{1}{\lambda}\log\left(1 - \left(1 - [1 - \xi_{s,n-D_{\mathsf{HCS}}-s+1;t}][1-(1-e^{-\lambda W})^{\beta}]\right)^{1/\beta}\right), \qquad t \in (0,1),$$

and $\xi_{s,n-D_{\mathsf{HCS}}-s+1;t}$ denotes the t-quantile of the *Beta*$(s, n - D_{\mathsf{HCS}} - s + 1)$-distribution. When the parameters are supposed unknown, they have to be estimated and plugged into the above formula. Then, of course, the respective interval is an approximate prediction interval. The same approach had earlier been applied in Asgharzadeh et al. (2015b) for Weibull lifetimes. In this case, the function η has to be replaced by

$$\eta_W(t) = -\frac{1}{\lambda}\log\left(\left[1 - \xi_{s,n-D_{\mathsf{HCS}}-s+1;t}\right]e^{-\lambda T^{\beta}}]\right)^{1/\beta}\right), \qquad t \in (0,1).$$

Since the *Beta*$(s, n - D_{\mathsf{HCS}} - s + 1)$-distribution is unimodal, Valiollahi et al. (2017) proposed a second construction leading to highest conditional density prediction limits where the bounds are given by $(\eta^*(\omega_1), \eta^*(\omega_2))$ with level $1-\alpha$ where

$$\eta^*(\omega) = -\frac{1}{\lambda}\log\left(1 - \left(1 - [1-\omega][1-(1-e^{-\lambda W})^{\beta}]\right)^{1/\beta}\right), \qquad \omega \in (0,1),$$

and ω_1 and ω_2 are the solutions of the equations

$$\int_{\omega_1}^{\omega_2} f_{Beta(s,n-D_{\mathsf{HCS}}-s+1)}(t)\,dt = 1-\alpha, \qquad f_{Beta(s,n-D_{\mathsf{HCS}}-s+1)}(\omega_1) = f_{Beta(s,n-D_{\mathsf{HCS}}-s+1)}(\omega_2).$$

12.3.2 Bayesian prediction intervals

AL-Hussaini (1999) provided a number of references on the applications of Bayesian prediction in different areas of applied statistics. Bayesian prediction intervals are obtained from the Bayes predictive density function f_{θ}^* given in (12.2) and (12.3) for the one- and two-sample situation, respectively. Bayesian prediction bounds are based on the predictive cumulative distribution function $F_{\theta}^*(\cdot \mid \boldsymbol{x})$ defined by

$$F_{\theta}^*(y \mid \boldsymbol{x}) = \int_{-\infty}^{y} f_{\theta}^*(t \mid \boldsymbol{x})\,dt, \qquad y > 0.$$

A $(1-\alpha)$-prediction interval is then defined by the bounds b_l and b_u satisfying the equations

$$F_{\theta}^*(b_l \mid \boldsymbol{x}) = \frac{\alpha}{2}, \quad F_{\theta}^*(b_u \mid \boldsymbol{x}) = 1 - \frac{\alpha}{2}.$$

Several authors developed Bayesian prediction intervals based on different forms of hybrid censored data by considering the cumulative distribution function of the lifetimes given by (12.12) with differentiable $\Lambda_{\boldsymbol{\theta}}$. In particular, the following hybrid censoring schemes are discussed in this framework

Type-I hybrid censoring: Shafay and Balakrishnan (2012)
Type-II hybrid censoring: Balakrishnan and Shafay (2012)
Generalized Type-I hybrid censoring: Shafay (2017)
Generalized Type-II hybrid censoring: Shafay (2016a)
Unified hybrid censoring: Mohie El-Din et al. (2017)

Following the suggestion of AL-Hussaini (1999), they considered the following conjugate prior of $\boldsymbol{\theta}$

$$\pi_{\delta}(\boldsymbol{\theta}) \propto \zeta(\boldsymbol{\theta}, \delta) e^{-\eta(\boldsymbol{\theta};\delta)}, \tag{12.14}$$

where $\boldsymbol{\theta} \in \Theta$ is the vector of parameters of the distribution under consideration and δ is the vector of hyperparameters.

Based on the prior in (12.14), the above mentioned authors developed a general procedure for determining the Bayesian prediction intervals for both one-sample and two-sample problems based on different forms of hybrid censoring schemes. They have illustrated their procedure for the one-parameter exponential and the two-parameter Pareto distributions for which the conjugate priors exist. In all cases considered, the Bayesian prediction intervals for the one-sample problem can not be obtained in explicit forms. For this reason, they used the Markov Chain Monte Carlo (MCMC) technique to compute predictive intervals in this case, and showed that the MCMC technique performs very well. In the two-sample case, however, the Bayesian prediction intervals can be derived in explicit forms.

Remark 12.3. Bayesian prediction intervals are studied for other lifetime distributions and various (progressive) hybrid censoring schemes in the literature. The following list provides a selection of references for some lifetime distributions: Weibull distributions (Asgharzadeh et al., 2015b), Rayleigh distribution (Asgharzadeh and Azizpour, 2016), inverted exponentiated Rayleigh distribution (Kayal et al., 2018), flexible Weibull distribution (Sharma, 2017), generalized exponential distributions (Valiollahi et al., 2017; Sen et al., 2018b, 2020), lognormal distribution (Singh and Tripathi, 2016; Sen et al., 2018c), generalized Lindley distribution (Singh et al., 2016), Burr III distribution (Singh et al., 2019), Burr XII distribution (Ateya and Mohammed, 2018), Gompertz distribution (Mohie El-Din et al., 2019), Lomax distribution (Asl et al., 2018), Poisson-exponential distribution (Monfared et al., 2022), Kumaraswamy distribution (Sultana et al., 2020), exponential-logarithmic distribution (Khan and Mitra, 2021).

CHAPTER 13

Adaptive progressive hybrid censoring

Contents

13.1. Introduction

A crucial assumption in the design of a progressively censored experiment is that the censoring plan $\mathscr{R} = (R_1, \dots, R_m)$ is prefixed. However, although this assumption is assumed in the standard model, it may not be satisfied in real life experiments since the experimenter may change the censoring numbers during the experiment (for whatever reasons). Therefore, it is desirable to have a model that takes into account such an adaptive process. Such a model has been proposed by Ng et al. (2009) who introduced a (prefixed) threshold parameter $T > 0$ as a control parameter in their life-testing experiment. Given some prefixed censoring plan $\mathscr{R} = (R_1, \dots, R_m)$, this plan is adapted after step (cf. (7.3))

$$D_{\mathsf{I}} = \begin{cases} 0, & X_{1:m:n}^{\mathscr{R}} > T \\ j-1, & X_{j-1:m:n}^{\mathscr{R}} \le T < X_{j:m:n}^{\mathscr{R}}, 2 \le j \le m \\ m, & X_{m:m:n}^{\mathscr{R}} \le T \end{cases} = \sum_{j=1}^{m} \mathbb{1}_{(-\infty, T]}(X_{j:m:n}^{\mathscr{R}}).$$

(see (7.3)) such that no further (progressive) censoring is carried out until the m-th failure time has been observed. At this time, the experiment is Type-II right censored and all surviving objects will be removed from the test.[1] Hence, the censoring plan is changed at the progressive censoring step $D_{\mathsf{I}} + 1$, i.e., at the first observed failure time exceeding the threshold T. Note that an adaption of the censoring plan is only possible when $D_{\mathsf{I}} < m - 1$, that is, there are at least two future censoring times available in the life test. Therefore, given an initially planned censoring plan $\mathscr{R} = (R_1, \dots, R_m)$, the observed sample is given by

$$\begin{cases} X_{1:m:n}^{\mathscr{R}}, \dots, X_{m:m:n}^{\mathscr{R}}, & \text{if } D_{\mathsf{I}} \ge m-1, \\ X_{1:m:n}^{\mathscr{R}}, \dots, X_{D_{\mathsf{I}}:m:n}^{\mathscr{R}}, X_{D_{\mathsf{I}}+1:m:n}^{\mathscr{R}^*}, \dots, X_{m:m:n}^{\mathscr{R}^*}, & \text{if } D_{\mathsf{I}} < m-1, \end{cases}$$

[1] So far, we notice that the adaption of the censoring plan is controlled by the random counter D_{I}.

Hybrid Censoring Know-How
https://doi.org/10.1016/B978-0-12-398387-9.00021-0

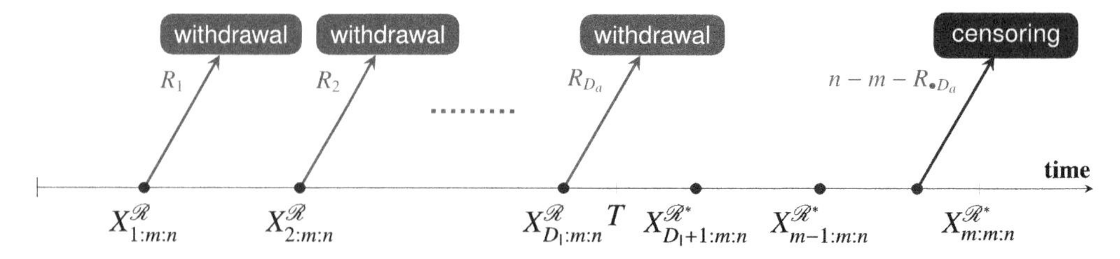

Figure 13.1 Adaptive progressive Type-II censoring as introduced in the Ng-Kundu-Chan model when $X^{\mathscr{R}}_{m:m:n} > T$.

where the effectively applied censoring plan $\mathscr{R}^*$ is given by the adapted censoring plan

$$\mathscr{R}^* = \mathscr{R}^*(D_I) = (R_1, \ldots, R_{D_I}, 0^{*m-D_I-1}, n - m - R_{\bullet D_I}), \quad R_{\bullet D_I} = \sum_{i=1}^{D_I} R_i. \tag{13.1}$$

Therefore, as long as the failures occur before time T, the initially planned progressive censoring plan is employed. After passing time T, no items are withdrawn at all except for the last failure time when all remaining surviving units are removed.

Remark 13.1. For $D_I = d \in \{0, \ldots, m\}$, let $\mathscr{R}_d$ denote $\mathscr{R}^*$ conditionally on $D_I = d$. Note that this yields the particular cases $\mathscr{R}_0 = (0^{*m-1}, n - m)$, that is, right censoring, and $\mathscr{R}_m = \mathscr{R}$, that is, the original censoring plan is realized.

Fig. 13.1 depicts the generation procedure of adaptive progressively Type-II censored order statistics in the Ng-Kundu-Chan model. After exceeding T, the censoring numbers except for the final (right) censoring number R_m^* are defined to be zero. This means that, after T, the next successive $m - D_I - 1$ failures are observed. The experiment is terminated when m failures have been observed. If $T > X^{\mathscr{R}}_{m-1:m:n}$, no adaption is carried out, that is $\mathscr{R} = \mathscr{R}^*$.

The Ng-Kundu-Chan model has also been called *adaptive Type-II progressive hybrid censoring scheme* since the threshold T is used to adapt the censoring plan to the data. However, this censoring scheme remains a pure Type-II progressive censoring with a change in the censoring plan as has been illustrated in Cramer and Iliopoulos (2010) and Balakrishnan and Cramer (2014), Chapter 6 (see also Cramer and Iliopoulos, 2015 for a more general account).

13.2. Adaptive Type-II progressive hybrid censoring

As has been pointed out in Cramer and Iliopoulos (2010), the Ng-Kundu-Chan model is a particular case of adaptive progressive Type-II censoring. Therefore, the general results established by Cramer and Iliopoulos (2010) can be directly applied to this setting, that is, to adaptive Type-II progressive hybrid censoring.

Defining by $\Gamma_i^* = \sum_{k=i}^m (R_k^* + 1)$, $1 \le i \le m$, the number of objects available in the life test before the i-th censoring step, we get

$$\Gamma_i^* = \begin{cases} \gamma_i, & 1 \le i \le D_{\mathsf{I}} \\ n - i + 1 - R_{\bullet D_{\mathsf{I}}}, & D_{\mathsf{I}} + 1 \le i \le m \end{cases} = \begin{cases} \gamma_i, & 1 \le i \le D_{\mathsf{I}} \\ \gamma_{D_{\mathsf{I}}+1} + D_{\mathsf{I}} - i + 1, & D_{\mathsf{I}} + 1 \le i \le m \end{cases}.$$

Notice that $\Gamma_1^*, \ldots, \Gamma_m^*$ are random variables which depend on the random variable D_{I}. In particular, their joint probability mass function is given by

$$\begin{aligned} \Pr(\Gamma_i^* = k_i, 1 \le i \le m) &= \sum_{j=0}^m \Pr(\Gamma_i^* = k_i, 1 \le i \le m \mid D_{\mathsf{I}} = j) \Pr(D_{\mathsf{I}} = j) \\ &= \Pr(D_{\mathsf{I}} = j) \prod_{i=1}^j \mathbb{1}_{\{\gamma_i\}}(k_i) \prod_{i=j+1}^m \mathbb{1}_{\{\gamma_{j+1}+j-i+1\}}(k_i). \end{aligned}$$

The probability mass function of D_{I} can be taken from Lemma 7.1.

Now, as has been shown in Cramer and Iliopoulos (2010) (see also Balakrishnan and Cramer, 2014, Theorem 6.1.2), the normalized spacings of adaptively Type-II progressive hybrid censored data from an exponential distribution are independent.

Theorem 13.2. *Let $X_{1:m:n}^{\mathscr{R}^*}, \ldots, X_{m:m:n}^{\mathscr{R}^*}$ be adaptively progressively Type-II censored order statistics generated from a two-parameter exponential distribution $Exp(\mu, \vartheta)$. Then, the normalized spacings*

$$S_{1,m} = \Gamma_1^*(X_{1:m:n}^{\mathscr{R}^*} - \mu), \quad S_{i,m} = \Gamma_i^*(X_{j:m:n}^{\mathscr{R}^*} - X_{j-1:m:n}^{\mathscr{R}^*}), \quad j = 2, \ldots, m,$$

are independent $Exp(\vartheta)$-distributed random variables. Moreover,

$$\mathsf{TTT} = \sum_{i=1}^m S_{i,m} = \sum_{i=1}^m (R_i^* + 1)(X_{j:m:n}^{\mathscr{R}^*} - \mu) \sim \Gamma(\vartheta, m).$$

Remark 13.3. Notice that the normalization is important because $\Gamma_1^*, \ldots, \Gamma_m^*$ are random variables depending on the failure times, that is, on D_{I}. Therefore, $\Gamma_1^*, \ldots, \Gamma_m^*$ and $X_{1:m:n}^{\mathscr{R}^*}, \ldots, X_{m:m:n}^{\mathscr{R}^*}$ are not independent in general. However, since $\Gamma_1^* = \gamma_1 = n$ is constant, the minimum $X_{1:m:n}^{\mathscr{R}^*}$ of the data and the (normalized) spacings $S_{2,m}, \ldots, S_{m,m}$ are independent.

For a general cumulative distribution function F and threshold T, we get

$$\begin{aligned} &\Pr(X_{i:m:n}^{\mathscr{R}^*} \le x_i, 1 \le i \le m, D_{\mathsf{I}} = 0) \\ &\quad = \Pr(T < X_{1:m:n}^{\mathscr{R}_0} \le x_1, X_{i:m:n}^{\mathscr{R}_0} \le x_i, 2 \le i \le m) \\ &\quad = \Pr(T < X_{1:n} \le x_1, X_{i:n} \le x_i, 2 \le i \le m) \end{aligned}$$

$$= \Pr(X_{i:n} \le x_i, 1 \le i \le m) - \Pr(X_{1:n} \le x_1 \wedge T, X_{i:n} \le x_i, 2 \le i \le m)$$
$$= F_{1,\dots,m:n}(x_1, x_2, \dots, x_m) - F_{1,\dots,m:n}(x_1 \wedge T, x_2, \dots, x_m).$$

Analogously,

$$\begin{aligned}\Pr(X_{i:m:n}^{\mathscr{R}^*} \le x_i, 1 \le i \le m, D_{\mathsf{I}} = m) &\\ &= \Pr(X_{i:m:n}^{\mathscr{R}_m} \le x_i, 1 \le i \le m-1, X_{m:m:n}^{\mathscr{R}_m} \le x_m \wedge T)\\ &= F_{1,\dots,m:m:n}^{\mathscr{R}}(x_1, \dots, x_{m-1}, x_m \wedge T).\end{aligned}$$

For $j = 2, \dots, m-1$, we find

$$\begin{aligned}\Pr(X_{i:m:n}^{\mathscr{R}^*} \le x_i, 1 \le i \le m, D_{\mathsf{I}} = j) &\\ &= \Pr(X_{i:m:n}^{\mathscr{R}_j} \le x_i, 1 \le i \le j, X_{i:m:n}^{\mathscr{R}^*} \le x_i, j+1 \le i \le m, X_{j:m:n}^{\mathscr{R}_j} \le T < X_{j+1:m:n}^{\mathscr{R}_j})\\ &= \Pr(X_{i:m:n}^{\mathscr{R}_j} \le x_i, i \in \{1, \dots, j-1, j+2, \dots, m\}, X_{j:m:n}^{\mathscr{R}_j} \le x_j \wedge T,\\ &\qquad\qquad T < X_{j+1:m:n}^{\mathscr{R}_j} \le x_{j+1})\\ &= F_{1,\dots,m:m:n}^{\mathscr{R}_j}(x_1, \dots, x_j \wedge T, x_{j+1}, \dots, x_m)\\ &\qquad - F_{1,\dots,m:m:n}^{\mathscr{R}_j}(x_1, \dots, x_j \wedge T, x_{j+1} \wedge T, x_{j+2}, \dots, x_m).\end{aligned}$$

Thus, for $x_1 \le \dots \le x_j \le T \le x_{j+1} \le \dots \le x_m$, we get the density function

$$f^{X_{i:m:n}^{\mathscr{R}^*}, 1 \le i \le m, D_{\mathsf{I}}}(\boldsymbol{x}_m, j) = f^{X_{i:m:n}^{\mathscr{R}_j}, 1 \le i \le m, D_{\mathsf{I}}}(\boldsymbol{x}_m, j) = \begin{cases} f_{1,\dots,m:n}(\boldsymbol{x}_m), & j = 0\\ f_{1,\dots,m:m:n}^{\mathscr{R}_j}(\boldsymbol{x}_m), & j \in \{1, \dots, m-1\}\\ f_{1,\dots,m:m:n}^{\mathscr{R}}(\boldsymbol{x}_m), & j = m \end{cases}.$$

Summing up, we find for $x_1 \le \dots \le x_m$ the expression

$$f^{X_{i:m:n}^{\mathscr{R}^*}, 1 \le i \le m, D_{\mathsf{I}}}(\boldsymbol{x}_m, j) = f_{1,\dots,m:m:n}^{\mathscr{R}_j}(\boldsymbol{x}_m) \mathbb{1}_{(-\infty, T]}(x_j) \mathbb{1}_{[T, \infty)}(x_{j+1}), \tag{13.2}$$

which illustrates that the joint density function of $X_{i:m:n}^{\mathscr{R}^*}$, $1 \le i \le m$, and D_{I} is the density function of a progressively Type-II censored sample with censoring plan $\mathscr{R}_j$.

Remark 13.4. The distribution of $X_{i:m:n}^{\mathscr{R}^*}$, $1 \le i \le m$, is generally a mixture which is seen from (13.2) to be (writing $x_0 = -\infty$, $x_{m+1} = \infty$)

$$f^{X_{i:m:n}^{\mathscr{R}^*}, 1 \le i \le m}(\boldsymbol{x}_m) = \sum_{j=0}^{m} f_{1,\dots,m:m:n}^{\mathscr{R}_j}(\boldsymbol{x}_m) \mathbb{1}_{(-\infty, T]}(x_j) \mathbb{1}_{[T, \infty)}(x_{j+1}).$$

13.3. Inference for adaptive Type-II progressive hybrid censored data

As mentioned above, adaptive Type-II progressive hybrid censoring can be seen as adaptive progressive Type-II censoring according to the Ng-Kundu-Chan model. Thus, we assume to observe the failure times $X_{1:m:n}^{\mathscr{R}^*}, \ldots, X_{m:m:n}^{\mathscr{R}^*}$ (with effectively applied censoring plan $\mathscr{R}^*$) and the random counter D_{I} determining the censoring plan $\mathscr{R}^*$.

Let $F_{\boldsymbol{\theta}}$, $\boldsymbol{\theta} \in \Theta \subseteq \mathbb{R}^q$, be an absolutely continuous cumulative distribution function with density function $f_{\boldsymbol{\theta}}$. The data are given by the sample $\boldsymbol{x}_m$, d. Then, from (13.2), the likelihood function is given by

$$\mathscr{L}(\boldsymbol{\theta} \mid \boldsymbol{x}_m, d) = f_{1,\ldots,m:m:n;\boldsymbol{\theta}}^{\mathscr{R}_j}(\boldsymbol{x}_m)\mathbb{1}_{(-\infty,T]}(x_j)\mathbb{1}_{[T,\infty)}(x_{j+1}) \propto f_{1,\ldots,m:m:n;\boldsymbol{\theta}}^{\mathscr{R}_j}(\boldsymbol{x}_m). \tag{13.3}$$

As an immediate result, we get from (13.3) the following theorem (see also Cramer and Iliopoulos, 2010).

Theorem 13.5. *Let $\widehat{\boldsymbol{\theta}} = \widehat{\boldsymbol{\theta}}(\boldsymbol{X}^{\mathscr{R}})$ be the maximum likelihood estimator of $\boldsymbol{\theta}$ when $\mathscr{R}$ is a prefixed progressive censoring plan. Then, using the effectively applied censoring plan $\mathscr{R}^*$, the maximum likelihood estimator of $\boldsymbol{\theta}$ under adaptive Type-II progressive hybrid censoring is given by $\widehat{\boldsymbol{\theta}}^* = \widehat{\boldsymbol{\theta}}(\boldsymbol{X}^{\mathscr{R}^*})$.*

Theorem 13.5 shows that, for the computation of the maximum likelihood estimator, it does not matter whether the censoring plan has been prefixed in advance or adaptively adjusted in the censoring process. Having an expression of the MLE under progressive Type-II censoring, we can replace the censoring plan by the observed one $\mathscr{R}^*$ and get the desired MLE under adaptive Type-II progressive hybrid censoring. Therefore, the forms of the estimators are identical using the effectively applied censoring plan $\mathscr{R}^*$. However, the distribution of the estimators may be different due to the adaptive process. An exception is given by these cases, where Theorem 6.1.2 in Balakrishnan and Cramer (2014) applies, that is, the distribution of the estimator $\widehat{\boldsymbol{\theta}}(\boldsymbol{X}^{\mathscr{R}})$ does not depend on the censoring plan.

Example 13.6. Suppose $X_j \sim Exp(\mu, \vartheta)$, $1 \le j \le n$. Then, for $\mu = 0$ and a censoring plan $\mathscr{R}^*$, the maximum likelihood estimator of ϑ is given by

$$\widehat{\vartheta} = \frac{1}{m}\sum_{j=1}^{m}(R_j^* + 1)X_{j:m:n}^{\mathscr{R}^*}.$$

Then, Theorem 13.2 yields $2m\widehat{\vartheta}/\vartheta \sim \chi_{2m}^2$ so that the assumptions of Theorem 6.1.2 in Balakrishnan and Cramer (2014) are satisfied. In particular, this shows $E_\vartheta \widehat{\vartheta} = \vartheta$ and $\mathrm{Var}_\vartheta\, \widehat{\vartheta} = \vartheta^2/m$. As for progressive Type-II censoring, $\widehat{\vartheta}$ is an unbiased estimator of ϑ. It can be shown that it is consistent and that $\sqrt{m}(\widehat{\vartheta} - \vartheta)/\vartheta$ and $\sqrt{m}(\widehat{\vartheta} - \vartheta)/\widehat{\vartheta}$ are asymptotically standard normal. A two-sided confidence interval for ϑ is given by

$$\Big[\frac{2m\widehat{\vartheta}}{\chi^2_{2m,1-\alpha/2}}, \frac{2m\widehat{\vartheta}}{\chi^2_{2m,\alpha/2}}\Big].$$

For an unknown location parameter, the maximum likelihood estimators are given by

$$\widehat{\mu} = X^{\mathscr{R}^*}_{1:m:n} \quad \text{and} \quad \widehat{\vartheta} = \frac{1}{m}\sum_{j=2}^{m} \Gamma^*_j (X^{\mathscr{R}^*}_{j:m:n} - X^{\mathscr{R}^*}_{j-1:m:n})$$

which, by Theorem 13.2, are independent with distributions $Exp(\mu, \vartheta/n)$ and $\Gamma(\vartheta/m, m-1)$ respectively. Hence, confidence intervals and statistical tests can be constructed as for progressive Type-II censoring (see, e.g., Balakrishnan and Cramer, 2014, Section 17.1).

Remark 13.7. Ng et al. (2009) calculated the maximum likelihood estimator of λ provided that the baseline distribution is an $Exp(1/\lambda)$-distribution, that is,

$$\begin{aligned}\widehat{\lambda} &= \frac{m}{\sum_{j=1}^{D_1}(R_j+1)X^{\mathscr{R}}_{1:m:n} + \sum_{j=D_1+1}^{m-1} X^{\mathscr{R}^*}_{j:m:n} + \big(n - m - \sum_{j=1}^{D_1} R_j\big)X^{\mathscr{R}^*}_{m:m:n}} \\ &= \frac{m}{\sum_{j=1}^{m}(R^*_j+1)X^{\mathscr{R}^*}_{j:m:n}},\end{aligned}$$

where the random variable $J = D_1$ denotes the change point in the censoring procedure and $\mathscr{R}^*$ is the censoring plan given in (13.1). Clearly, this result can be directly deduced from Theorem 13.5 (see also Example 13.6). Moreover, Ng et al. (2009) sketched several approaches to construct confidence intervals for λ including conditional inference, normal approximations, likelihood-ratio based inference, bootstrapping, and Bayesian inference.

From Theorem 13.5 and the likelihood function in (13.3), it follows that computation of maximum likelihood estimators under progressive Type-II censoring and under adaptive Type-II progressive hybrid censoring can be seen as the same problem (as long as the effectively applied censoring plan has been observed). Thus, from a computational point of view, the problem is solved by discussing the optimization under progressive Type-II censoring.

A similar argument applies to Bayesian inference. For instance, assuming some prior distribution $\pi_{\boldsymbol{a}}$ of the $\boldsymbol{\theta}$ with some hyperparameter $\boldsymbol{a}$ and considering the likelihood in (13.3), the posterior distribution is given by

$$\pi_{\boldsymbol{a}}(\boldsymbol{\theta} \mid \boldsymbol{x}_m) = \frac{f^{\mathscr{R}_j}_{1,\dots,m:m:n;\boldsymbol{\theta}}(\boldsymbol{x}_m) \cdot \pi_{\boldsymbol{a}}(\boldsymbol{\theta})}{\int f^{\mathscr{R}_j}_{1,\dots,m:m:n;\xi}(\boldsymbol{x}_m) \cdot \pi_{\boldsymbol{a}}(\xi)d\xi}.$$

As a result, the posterior mean, posterior median, or posterior mode equal those obtained under progressive Type-II censoring with censoring plan $\mathscr{R}_j$. In this regard, many results on Bayesian inference obtained under progressive Type-II censoring can be easily applied to the adaptive Type-II progressive hybrid censoring scheme. For information, we summarize some of these results.

As mentioned above, Ng et al. (2009) discussed exponential lifetimes under adaptive progressive hybrid censoring. Lin et al. (2009) considered this censoring model for Weibull lifetimes. Using a log-transformation of the data, they transformed the data to a location-scale family of extreme value distributions for which they derived the maximum likelihood estimates of the transformed parameters. However, writing the problem in terms of the effectively applied censoring plan $\mathscr{R}^*$ given in (13.1), it follows that the likelihood function equals that given in Balakrishnan et al. (2004a). Therefore, the maximum likelihood estimates can be directly taken from Balakrishnan et al. (2004a) which is also clear from Theorem 13.5. The same argument applies to the derivation of approximate maximum likelihood estimates.

In view of this connection to progressive Type-II censoring, we do not present further details here. For further reading, we refer to the following publications dealing with several topics in this direction: Liang (2014) (Bayesian sampling plans for exponential distribution), Ye et al. (2014) (extreme value distribution), Nassar and Abo-Kasem (2016a) (Burr XII distribution), Nassar et al. (2018) (Weibull distribution), Panahi and Asadi (2021) (Burr III distribution), Nassar et al. (2017) (inverse Weibull distribution), Panahi and Moradi (2020) (inverted exponentiated Rayleigh distribution), Elshahhat and Nassar (2021) (Hjorth distribution), Elshahhat and Elemary (2021) (Xgamma distribution), Hemmati and Khorram (2011), Liu and Gui (2020) (competing risks).

13.4. Adaptive Type-I progressive hybrid censoring

By imitating the construction of progressive Type-II hybrid censoring, Lin and Huang (2012) introduced a censoring model called *adaptive Type-I progressive hybrid censoring*. In this model, the experiment is terminated latest at a maximum test duration T but if the m-th failure time $X_{m:n^*:n}$ has been observed before T and if some objects are still remaining in the life test, the monitoring of failures is continued until T but the censoring plan is adapted in such a way that no further progressive censoring occurs until T. Moreover, the experiment is stopped at T so that the experiment is Type-I censored. As in the sampling situation in (7.6), we start the illustration of the model with the progressively Type-II censored sample $X_{1:n^*:n}^{\mathscr{R}}, \ldots, X_{n^*:n^*:n}^{\mathscr{R}}$ and associated censoring plan $\mathscr{R}$.

As a matter of fact, the adaptive Type-I progressive hybrid censoring scheme can be seen as a Type-I censored version of a progressively Type-II censored sample. Notice that in the original paper by Lin and Huang (2012) the censoring plan is adapted depending on whether $X^{\mathscr{R}}_{m:n^*:n} < T$ or not. In this model, the threshold T has two roles. First, it is involved in the decision whether the censoring plan is adapted or not. Secondly, it bounds the test duration. However, since the censoring plan is adapted only after the m-th failure, we can consider the modified (or extended) sample

$$X^{\mathscr{R}^*}_{1:m^*:n}, \dots, X^{\mathscr{R}^*}_{m:m^*:n}, X^{\mathscr{R}^*}_{m+1:m^*:n}, \dots, X^{\mathscr{R}^*}_{m^*:m^*:n} \tag{13.4}$$

by design with censoring plan

$$\mathscr{R}^* = (R_1, \dots, R_{m-1}, 0^{*\gamma_m}) \tag{13.5}$$

and $m^* = m + \gamma_m - 1$ maximum observed failure times. Then,

$$(X^{\mathscr{R}}_{1:m^*:n}, \dots, X^{\mathscr{R}}_{m:m^*:n}) = (X^{\mathscr{R}^*}_{1:m^*:n}, \dots, X^{\mathscr{R}^*}_{m:m^*:n}),$$

that is, the first m failures are the same in both models! Imposing Type-I censoring to the progressively Type-II censored sample given in (13.4) yields exactly the data given in Lin and Huang (2012). Thus, an adaption of the censoring plan is not necessary at all. In fact, the corresponding γ's associated with the censoring plan $\mathscr{R}^*$ are given by

$$\gamma_j^* = \begin{cases} \gamma_j, & j = 1, \dots, m-1 \\ \gamma_m + m - j, & j = m, \dots, m + \gamma_m - 1 \end{cases}. \tag{13.6}$$

Furthermore, comparing the likelihood function in (7.9) with the γ's taken from (13.6) illustrates that they coincide with the expression given in Lin and Huang (2012, Section 2.1). Furthermore, it should be noticed that Lin and Huang (2012) use the notation $X_{i:n}$ for the lifetimes observed after T. But, due to the progressive censoring before T, these are not really order statistics. Its just a notation.

As a consequence, results obtained under adaptive Type-I progressive hybrid censoring can be obtained directly from the results for Type-I progressive hybrid censoring using the censoring plan given in (13.5).

Remark 13.8. Due to the above comments, we do not present further details here. For further reading, we refer to the following publications dealing with several topics on adaptive Type-I progressive hybrid censoring: Lin et al. (2012) (Weibull distribution), Liang (2014) (Bayesian sampling plans for exponential distribution), Okasha et al. (2021) (Lomax distribution), Nassar and Dobbah (2020) (bathtub-shaped distribution), Almarashi et al. (2022) (Nadarajah–Haghighi distribution), Nassar and Ashour (2014), Ashour and Nassar (2014), Ashour and Jones (2017), Okasha and Mustafa (2020) (competing risks).

Appendix

A. Geometrical objects

Definition A.1 (Simplex). Let $a_1, \ldots, a_n > 0$ be positive values. Then, the set

$$\mathcal{S}_n(a_1, \ldots, a_n) = \left\{ \boldsymbol{w}_n = (w_1, \ldots, w_n) \,\middle|\, w_j \geq 0, 1 \leq j \leq n, \sum_{j=1}^{n} a_j w_j \leq 1 \right\} \subseteq \mathbb{R}^n$$

is called simplex with vertices $\mathbf{0}, a_1\boldsymbol{e}_1, \ldots, a_n\boldsymbol{e}_n$. $\boldsymbol{e}_1, \ldots, \boldsymbol{e}_n$ denote the canonical basis of the n-dimensional Euclidean space $\mathbb{R}^n$.

$\mathcal{S}_n = \mathcal{S}_n(1^{*n}) \subseteq \mathbb{R}^n$ is called standard simplex in $\mathbb{R}^n$.

For $Exp(\mu, \vartheta)$-distributions and Type-I progressive hybrid censoring with parameter $\gamma_1 > \cdots > \gamma_d > 0$, the support of normalized spacings (given $D_{\mathrm{I}} = d$) is given by (see also (4.24))

$$\mathcal{W}_d^{\leq}(T) = \left\{ \boldsymbol{x}_d \mid x_j \geq 0, 1 \leq j \leq d, \sum_{j=1}^{d} \frac{x_j}{\gamma_j} \leq T - \mu \right\}$$

which is a simplex $\mathcal{S}_d(1/[\gamma_1(T-\mu)], \ldots, 1/[\gamma_d(T-\mu)])$.

For $\mu = 0$, $\alpha_j = 1 - \gamma_{d+1}/\gamma_j$, $\beta_j = 1 - \gamma_d/\gamma_j$, $1 \leq j \leq d$, the set

$$\mathcal{M}_{d-1}^{(s)} = \left\{ \boldsymbol{w}_{d-1} \,\middle|\, w_j \geq 0, 1 \leq j \leq d-1, s - \gamma_d T \leq \sum_{j=1}^{d-1} \beta_j w_j, \sum_{j=1}^{d-1} \alpha_j w_j \leq s - \gamma_{d+1} T \right\},$$

where $s \in [\gamma_{d+1}T, \gamma_1 T]$, denotes a polyhedron in $\mathbb{R}^{d-1}$ which is used to describe the support of the joint distribution of spacings and the total time on test under Type-I (progressive) hybrid censoring. As pointed out in Cramer and Balakrishnan (2013), it holds with $s^* = s - \gamma_{d+1}T$

$$\mathcal{M}_{d-1}^{(s)} = \mathcal{S}_n(\alpha_1/s^*, \ldots, \alpha_{d-1}/s^*) \cap \mathcal{H}_{d-1}^{(s)}$$

where

$$\mathcal{H}_{d-1}^{(s)} = \left\{ \boldsymbol{w}_{d-1} \,\middle|\, s - \gamma_d T \leq \sum_{j=1}^{d-1} \beta_j w_j \right\},$$

denotes a hyperplane.

Therefore, the volume of $\mathcal{M}_{d-1}^{(s)}$ can be obtained from results obtained in Gerber (1981) and Cho and Cho (2001). It is related to B-spline functions (see Appendix C) by

$$\text{Volume}(\mathcal{M}_{d-1}^{(s)}) = \left(\frac{T^d(\gamma_d - \gamma_{d+1})}{d!} \prod_{j=1}^{d-1} \gamma_j \right) B_{d-1}(s \mid \gamma_{d+1}T, \ldots, \gamma_1 T), \ s \in [\gamma_{d+1}T, \gamma_1 T].$$

The connection of B-splines to the volume of intersection of a simplex and a hyperplane was observed first by Curry and Schoenberg (1966); see also de Boor (1976), Dahmen and Micchelli (1986), and Górny and Cramer (2019a).

B. Distributions

The following parametrization of the distributions is used throughout the monograph. The density functions are given only on the support of the corresponding distribution.

Definition B.1 (Uniform distribution). The uniform distribution *Uniform*(a, b), with parameters $a, b \in \mathbb{R}$, $a < b$, is defined by the density function

$$f(t) = \frac{1}{b-a}, \qquad a < t < b.$$

For $a = 0$ and $b = 1$, it defines the standard uniform distribution *Uniform*$(0, 1)$.

Definition B.2 (Beta distribution). The beta distribution *Beta*(α, β), with parameters $\alpha, \beta > 0$, is defined by the density function

$$f(t) = \frac{1}{\mathsf{B}(\alpha, \beta)} t^{\alpha-1}(1-t)^{\beta-1}, \quad 0 < t < 1,$$

where $\mathsf{B}(\alpha, \beta) = \frac{\Gamma(\alpha+\beta)}{\Gamma(\alpha)\Gamma(\beta)}$ denotes the complete beta function and $\Gamma(\cdot)$ denotes the gamma function.

Definition B.3 (Power distribution). The power distribution *Power*(α), with parameter $\alpha > 0$, is a particular beta distribution *Beta*$(\alpha, 1)$.

Definition B.4 (Reflected power distribution). The reflected power distribution *RPower*(β), with parameter $\beta > 0$, is a particular beta distribution *Beta*$(1, \beta)$.

Definition B.5 (Exponential distribution). The two-parameter exponential distribution *Exp*(μ, ϑ), with parameters $\mu \in \mathbb{R}$ and $\vartheta > 0$, is defined by the density function

$$f(t) = \frac{1}{\vartheta} e^{-\frac{t-\mu}{\vartheta}}, \qquad t > \mu.$$

For $\mu = 0$, it defines the exponential distribution *Exp*(ϑ).

Definition B.6 (Weibull distribution). The three-parameter Weibull distribution *Weibull*(μ, ϑ, β), with parameters $\mu \in \mathbb{R}$ and $\vartheta, \beta > 0$, is defined by the density function

$$f(t) = \frac{\beta}{\vartheta}(t-\mu)^{\beta-1} e^{-\frac{(t-\mu)^\beta}{\vartheta}}, \quad t > \mu.$$

(1) For $\mu = 0$, it defines the two-parameter Weibull distribution *Weibull*(ϑ, β). Sometimes the Weibull distribution is used with the parametrization *Weibull*(ϑ^β, β) which corresponds to the density function

$$f(t) = \frac{\beta}{\vartheta}\left(\frac{t}{\vartheta}\right)^{\beta-1} e^{-(t/\vartheta)^\beta}, \quad t > 0. \tag{B.1}$$

(2) For $\beta = 2$, the distribution *Weibull*$(\mu, \vartheta, 2)$ is known as two-parameter Rayleigh distribution.

Definition B.7 (Gamma distribution). The gamma distribution $\Gamma(\vartheta, \beta)$, with parameters $\vartheta, \beta > 0$, is defined by the density function

$$f(t) = \frac{1}{\Gamma(\beta)\vartheta^\beta} t^{\beta-1} e^{-\frac{t}{\vartheta}}, \quad t > 0,$$

where $\Gamma(\cdot)$ denotes the gamma function. The respective density function is denoted by $f_{\Gamma(\vartheta,\beta)}$.

Definition B.8 (Inverse Gamma distribution). The inverse gamma distribution $I\Gamma(\alpha, \beta)$, with parameters $\alpha, \beta > 0$, is defined by the density function

$$f(t) = \frac{\alpha^\beta}{\Gamma(\beta)} t^{-\beta-1} e^{-\frac{\alpha}{t}}, \quad t > 0,$$

where $\Gamma(\cdot)$ denotes the gamma function.

Definition B.9 (χ^2-distribution). The χ^2-distribution χ^2_n, with degrees of freedom $n \in \mathbb{N}$, is a particular gamma distribution $\Gamma(2, \frac{n}{2})$.

Definition B.10 (F-distribution). The F-distribution $F_{n,m}$, with parameters $n, m \in \mathbb{N}$, is defined by the density function

$$f(t) = \frac{\Gamma\left(\frac{n+m}{2}\right)}{\Gamma\left(\frac{n}{2}\right)\Gamma\left(\frac{m}{2}\right)} \left(\frac{n}{m}\right)^{\frac{n}{2}} \frac{t^{\frac{n}{2}-1}}{(1+\frac{n}{m}t)^{\frac{n+m}{2}}}, \quad t > 0,$$

where $\Gamma(\cdot)$ denotes the gamma function.

Definition B.11 (Pareto-distribution). The Pareto distribution *Pareto*(α), with parameter $\alpha > 0$, is defined by the density function

$$f(t) = \frac{\alpha}{t^{\alpha+1}}, \quad t \geq 1.$$

Definition B.12 (Lomax-distribution). The Lomax distribution *Lomax*(α), with parameter $\alpha > 0$, is defined by the density function

$$f(t) = \frac{\alpha}{(t+1)^{\alpha+1}}, \quad t \geq 0.$$

Definition B.13 (Extreme value distribution (Type-I)). The extreme value distribution (Type-I) $EV_{\text{I}}(0, 1)$ is defined by the density function

$$f(t) = e^{t-e^t}, \quad t \in \mathbb{R}.$$

Definition B.14 (Extreme value distribution (Type-II)/Gumbel distribution). The extreme value distribution (Type-II) or Gumbel distribution $EV_{\text{II}}(0, 1)$ is defined by the density function

$$f(t) = e^{-t-e^{-t}}, \quad t \in \mathbb{R}.$$

Definition B.15 (Laplace distribution). The Laplace distribution *Laplace*(μ, σ), with parameters $\sigma > 0$, $\mu \in \mathbb{R}$, is defined by the density function

$$f(t) = \frac{1}{\sigma} e^{-\frac{|t-\mu|}{\sigma}}, \quad t \in \mathbb{R}. \tag{B.2}$$

For $\mu = 0$ and $\sigma = 1$, it defines the standard Laplace distribution *Laplace*(0, 1).

Definition B.16 (Logistic distribution (Type-I)). The logistic distribution is defined by the density function

$$f(t) = \frac{e^t}{(1+e^t)^2}, \quad t \in \mathbb{R}.$$

Definition B.17 (Normal distribution). The normal distribution $N(\mu, \sigma^2)$, with parameters $\sigma^2 > 0$, $\mu \in \mathbb{R}$, is defined by the density function

$$f(t) = \frac{1}{\sqrt{2\pi\sigma^2}} e^{-\frac{(t-\mu)^2}{2\sigma^2}}, \quad t \in \mathbb{R}.$$

For $\mu = 0$ and $\sigma^2 = 1$, it defines the standard normal distribution $N(0, 1)$.

Definition B.18 (Log-normal distribution). The log-normal distribution *log-N*(μ, σ), with parameters $\sigma > 0$, $\mu \in \mathbb{R}$, is defined by the density function

$$f(t) = \frac{1}{\sqrt{2\pi\sigma^2}\, t} e^{-\frac{(\log t-\mu)^2}{2\sigma^2}}, \quad t > 0.$$

Definition B.19 (Binomial distribution). The binomial distribution *bin*(n, p), with parameters $n \in \mathbb{N}$, $p \in [0, 1]$, is defined by the probability mass function

$$f(k) = \binom{n}{k} p^k (1-p)^{n-k}, \quad k \in \{0, \ldots, n\}.$$

Definition B.20. A *Dirichlet*$(a_1, \ldots, a_{n+1})$-distribution with parameters $a_1, \ldots, a_{n+1}$ is defined by the density function

$$f(w_1, \ldots, w_n) = \frac{\Gamma(a_1 + \cdots + a_{n+1})}{\Gamma(a_1)\Gamma(a_2)\cdots\Gamma(a_{n+1})} w_1^{a_1-1} \cdots w_n^{a_n-1} \left(1 - \sum_{i=1}^{n} w_i\right)^{a_{n+1}-1},$$

$$0 < w_1, \ldots, w_n < 1, \ \sum_{i=1}^{n} w_i < 1.$$

For $(a_1, \ldots, a_{n+1}) = (1^{*n+1})$, the *Dirichlet*$(1^{*n+1})$-distribution is known as the uniform distribution on the standard simplex $\mathcal{S}_n = \mathcal{S}_n(1^{*n})$.

C. B-splines and divided differences

As first pointed out by Cramer and Balakrishnan (2013), B-splines play an interesting role in the representation of density functions and cumulative distribution functions under hybrid censoring schemes. Therefore, we present some background information and references for further reading on this topic. For a more detailed discussion, we refer to Chapter 3 in Górny (2017). Since its discussion by Schoenberg (1946) and Curry and Schoenberg (1947), B-splines have been extensively studied in many areas and numerous results and extensions have been provided (see, e.g., Curry and Schoenberg, 1966; de Boor, 1976, 2001, 2005; Schumaker, 2007). Connections to probability and statistics have been established in various directions. Its relation to urn models has been pointed out by Goldman (1988a,b) whereas Dahmen and Micchelli (1986) and Karlin et al. (1986) established connections between multivariate B-splines and Dirichlet density functions. Dahmen and Micchelli (1986) and Ignatov and Kaishev (1989) found that the density function of a linear combination of order statistics based on standard uniform random variables can be expressed as a B-spline function.

Originally, polynomial splines are used in approximation problems (see de Boor, 2001; Schumaker, 2007). For instance, they exhibit desirable smoothness properties, can be easily handled from a mathematical point of view, and have nice numerical features. Usually, they are defined via divided differences so that we start with a definition of these quantities. Notice that there are various approaches to introduce divided differences.

Definition C.1. Let $t_0, \ldots, t_d \in \mathbb{R}$ be a pairwise difference knot sequence and $g : \mathbb{R} \to \mathbb{R}$ be a function. Then, the divided difference of g w.r.t. the knot sequence $t_0, \ldots, t_d$, $d \in \mathbb{N}$, is denoted by $[t_0, \ldots, t_d; g(\cdot)]$. The divided differences are iteratively defined by

$$[t_0, \ldots, t_d; g(\cdot)] = \frac{[t_1, \ldots, t_d; g(\cdot)] - [t_0, \ldots, t_{d-1}; g(\cdot)]}{t_d - t_0}, \quad d \geq 1,$$

$$[t_j; g(\cdot)] = g(t_j), \quad j = 0, \ldots, d.$$

Divided differences can also be defined for knot sequences not necessarily pairwise different provided that g is sufficiently smooth (see de Boor, 2001; Schumaker, 2007). For instance, assuming $t_0 = \cdots = t_d$, one gets

$$[t_0, \ldots, t_d; g(\cdot)] = \frac{1}{d!} \frac{\partial^d}{\partial x^d} g(x)\Big|_{x=t_0}.$$

Since in our discussion the knots usually are strictly ordered, we restrict ourselves to this particular case. However, many results are also true for this more general setting.

For pairwise distinct knots $t_0, \ldots, t_d$, the divided difference $[t_0, \ldots, t_d; g(\cdot)]$ can be calculated via the sum (cf. Schumaker, 2007, p. 46)

$$[t_0, \ldots, t_d; g(\cdot)] = \sum_{j=0}^{d} \frac{g(t_j)}{\prod_{i=0, i \neq j}^{d} (t_j - t_i)}. \tag{C.1}$$

For further reading on divided differences, we refer to de Boor (2001, Chapter 1) and Schumaker (2007, Section 2.7). A compact formula for $[t_0, \ldots, t_d; g(\cdot)]$ with multiple knots $t_0, \ldots, t_d$ and sufficiently smooth g can be found in Soltani and Roozegar (2012, Lemma 2.1).

For a knot sequence $t_0, \ldots, t_d$ (not all knots are equal), the Curry-Schoenberg B-spline of degree $d-1$ is defined by the divided difference applied to the function $g_d(\cdot) = (\cdot - x)_+^{d-1}$ for some $x \in \mathbb{R}$. $(z)_+ = \max\{z, 0\}$ denotes the positive part of a real value $z \in \mathbb{R}$. Then, the B-spline function is defined by (see, e.g., de Boor, 2001, p. 87)

$$B_{d-1}(x \mid t_0, \ldots, t_d) = d\,[t_0, \ldots, t_d; (\cdot - x)_+^{d-1}], \quad x \in \mathbb{R}. \tag{C.2}$$

The B-spline is not affected by the order of the knot sequence so that without loss of generality one can assume an ordered sequence. In particular, for $t_0 \leq \cdots \leq t_d$, $t_0 \neq t_d$, the B-spline equals zero outside the interval $[t_0, t_d]$, that is,

$$B_{d-1}(x \mid t_0, \ldots, t_d) = 0, \quad x \notin [t_0, t_d],$$

(cf. Schumaker, 2007, p. 121). This property is quite useful in the analysis of hybrid censoring models since we do not have to take care about the support of the respective functions. Furthermore, assuming $t_0 < t_1 = \cdots = t_d$, the corresponding B-spline simplifies considerably, that is,

$$B_{d-1}(x \mid t_0, \ldots, t_d) = d\,\frac{(x - t_0)^{d-1}}{(t_1 - t_0)^d} \mathbb{1}_{[t_0, t_1)}(x), \quad x \in \mathbb{R}, \tag{C.3}$$

where $\mathbb{1}_{[t_0, t_1)}$ denotes the indicator function. Identity (C.3) can be easily verified by a well known recurrence relation for B-splines (cf., e.g., de Boor, 2001, p. 90).

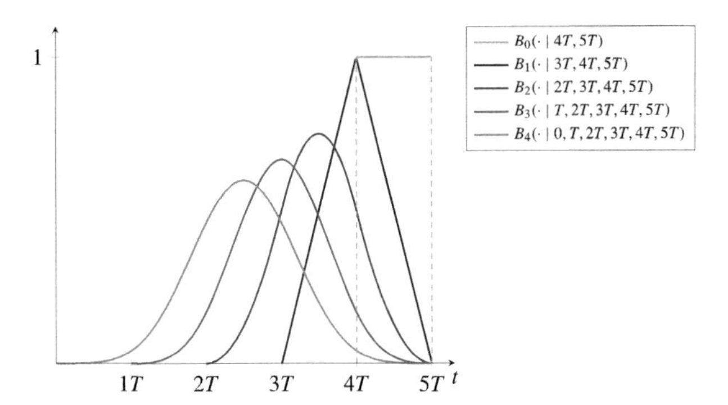

Figure C.1 Plots of B-splines for equally spaced knots and $n = 5$, $\gamma_i = n - i + 1$, $i \in \{1, \dots, 5\}$.

Under hybrid censoring, the knot sequence is often given by the increasingly ordered sequence $\gamma_{d+1}T < \cdots < \gamma_1 T$ where $\gamma_1, \gamma_2, \dots$ denote parameters related to (progressively type-II censored) order statistics and $T > 0$ denotes the time threshold. Thus, we get from (C.1) and (C.2) that the univariate B-Spline B_{d-1} of degree $d-1$ with knots $\gamma_{d+1}T < \cdots < \gamma_1 T$ exhibits the representation

$$B_{d-1}(s \mid \gamma_{d+1}T, \dots, \gamma_1 T) = \frac{(-1)^d}{T^d} d \sum_{i=1}^{d+1} a_{i,d+1}[\gamma_i T - s]_+^{d-1}, \quad s \in [\gamma_{d+1}T, \gamma_1 T], \tag{C.4}$$

where $a_{i,d+1}$ is defined in Theorem 2.53. Plots of B-splines for equally spaced knots and $n = 5$, $\gamma_i = n - i + 1$, $i \in \{1, \dots, 5\}$ are depicted in Fig. C.1

Representations for particular knot sequences are useful in the discussion of hybrid censoring schemes. Notice that the γ_i's correspond to the model parameters in progressive censoring. By choosing $\gamma_i = n - i + 1$, $1 \le i \le n$, this reduces to the model of order statistics. In this case, we get the following expressions:

$$B_0(s \mid a, b) = \frac{1}{b-a} \mathbb{1}_{[a,b)}(s), \quad a < b, s \in \mathbb{R} \tag{C.5}$$

$$\begin{aligned} &B_{d-1}(s \mid (n-d)T, \dots, nT) \\ &\quad = \frac{1}{T^d (d-1)!} \sum_{i=0}^{d} (-1)^i \binom{d}{i} [(n-i)T - s]_+^{d-1}, \\ &B_{d-1}(s \mid 0, (n-d+1)T, \dots, nT) \\ &\quad = \frac{1}{T^d (d-1)!} \sum_{i=1}^{d} (-1)^{i-1} \binom{d}{i-1} \frac{d-i+1}{n-i+1} [(n-i+1)T - s]_+^{d-1} \\ &\qquad + (-1)^d \frac{d(n-d)!}{n! T^d} [-s]_+^{d-1}. \end{aligned}$$

Bibliography

Abo-Eleneen, Z.A., 2007. Fisher information and optimal schemes in progressive Type-II censored samples. Model Assisted Statistics and Applications 2, 153–163.

Abo-Eleneen, Z.A., 2008. Fisher information in type II progressive censored samples. Communications in Statistics. Theory and Methods 37, 682–691.

Abo-Eleneen, Z.A., 2011. The entropy of progressively censored samples. Entropy 13 (2), 437–449.

Abo-Kasem, O.E., Elshahhat, A., 2021. Analysis of two Weibull populations under joint progressively hybrid censoring. Communications in Statistics. Simulation and Computation, 1–22. In press.

Abo-Kasem, O.E., Elshahhat, A., 2022. A new two sample generalized Type-II hybrid censoring scheme. American Journal of Mathematical and Management Sciences 41 (2), 170–184.

Abo-Kasem, O.E., Nassar, M., Dey, S., Rasouli, A., 2019. Classical and Bayesian estimation for two exponential populations based on joint Type-I progressive hybrid censoring scheme. American Journal of Mathematical and Management Sciences 38, 2325–8454.

Abou Ghaida, W.R., Baklizi, A., 2022. Prediction of future failures in the log-logistic distribution based on hybrid censored data. International Journal of System Assurance Engineering and Management, 1–9.

Abushal, T.A., Soliman, A.A., Abd-Elmougod, G.A., 2022. Inference of partially observed causes for failure of Lomax competing risks model under type-II generalized hybrid censoring scheme. Alexandria Engineering Journal 61 (7), 5427–5439.

Adell, J., Sangüesa, C., 2005. Approximation by B-spline convolution operators. A probabilistic approach. Journal of Computational and Applied Mathematics 174, 79–99.

Agarwal, G.G., Dalpatadu, R.J., Singh, A.K., 2002. Linear functions of uniform order statistics and B-splines. Communications in Statistics. Theory and Methods 31, 181–192.

Ahmadi, M.V., Doostparast, M., Ahmadi, J., 2015. Statistical inference for the lifetime performance index based on generalised order statistics from exponential distribution. International Journal of Systems Science 46, 1094–1107.

AL-Hussaini, E.K., 1999. Predicting observables from a general class of distributions. Journal of Statistical Planning and Inference 79 (1), 79–91.

Al-Zahrani, B., Gindwan, M., 2014. Parameter estimation of a two-parameter Lindley distribution under hybrid censoring. International Journal of System Assurance Engineering and Management 5 (4), 628–636.

Alizadeh Noughabi, H., 2017. Testing exponentiality based on Kullback—Leibler information for progressively Type II censored data. Communications in Statistics. Simulation and Computation 46 (10), 7624–7638.

Alizadeh Noughabi, H., Chahkandi, M., 2018. Testing the validity of the exponential model for hybrid Type-I censored data. Communications in Statistics. Theory and Methods 47 (23), 5770–5778.

Alizadeh Noughabi, H., Mohtashami Borzadaran, G.R., 2020. An updated review of goodness of fit tests based on entropy. Journal of the Iranian Statistical Society 19 (2).

Alma, O.G., Arabi Belaghi, R., 2016. On the estimation of the extreme value and normal distribution parameters based on progressive type-II hybrid-censored data. Journal of Statistical Computation and Simulation 86 (3), 569–596.

Almarashi, A.M., Algarni, A., Okasha, H., Nassar, M., 2022. On reliability estimation of Nadarajah–Haghighi distribution under adaptive type-I progressive hybrid censoring scheme. Quality and Reliability Engineering International 38 (2), 817–833.

Almohaimeed, B., 2017. On the entropy of progressive hybrid censoring schemes. Applied Mathematics & Information Sciences 11 (6), 1811–1814.

Ameli, S., Rezaie, M., Ahmadi, J., 2018. Prediction of times to failure of censored units in progressive hybrid censored samples for the proportional hazards family. Journal of Statistical Research of Iran 14 (2), 131–155.

Anderson, T.W., Darling, D.A., 1954. A test of goodness of fit. Journal of the American Statistical Association 49 (268), 765–769.

Arabi Belaghi, R., Noori Asl, M., 2019. Estimation based on progressively type-I hybrid censored data from the Burr XII distribution. Statistical Papers 60, 761–803.

Arnold, B.C., Balakrishnan, N., Nagaraja, H.N., 1992. A First Course in Order Statistics. Wiley, New York.

Arnold, B.C., Balakrishnan, N., Nagaraja, H.N., 2008. A First Course in Order Statistics. Society for Industrial and Applied Mathematics (SIAM), Philadelphia.

Arnold, B.C., Becker, A., Gather, U., Zahedi, H., 1984. On the Markov property of order statistics. Journal of Statistical Planning and Inference 9, 147–154.

Asgharzadeh, A., 2009. Approximate MLE for the scaled generalized exponential distribution under progressive type-II censoring. Journal of the Korean Statistical Society 38 (3), 223–229.

Asgharzadeh, A., Azizpour, M., 2016. Bayesian inference for Rayleigh distribution under hybrid censoring. International Journal of System Assurance Engineering and Management 7, 239–249.

Asgharzadeh, A., Kazemi, M., 2014. Stress-strength reliability of exponential distribution based on hybrid censored samples. In: The 12-th Iranian Statistical Conference (ISC12), pp. 26–32.

Asgharzadeh, A., Kazemi, M., Kundu, D., 2015a. Estimation of $P(X > Y)$ for Weibull distribution based on hybrid censored samples. International Journal of System Assurance Engineering and Management 8 (S1), 489–498.

Asgharzadeh, A., Kazemi, M., Kuş, C., 2013. Analysis of the hybrid censored data from the logistic distribution. Journal of Probability and Statistical Science 11 (2), 183–198.

Asgharzadeh, A., Valiollahi, R., Kundu, D., 2015b. Prediction for future failures in Weibull distribution under hybrid censoring. Journal of Statistical Computation and Simulation 85 (4), 824–838.

Ashour, S.K., Jones, P.W., 2017. Inference for Weibull distribution under adaptive Type-I progressive hybrid censored competing risks data. Communications in Statistics. Theory and Methods 46 (10), 4756–4773.

Ashour, S.K., Nassar, M.M., 2014. Analysis of generalized exponential distribution under adaptive Type-II progressive hybrid censored competing risks data. International Journal of Advanced Statistics and Probability 2 (2), 108–113.

Asl, M.N., Belaghi, R.A., Bevrani, H., 2018. Classical and Bayesian inferential approaches using Lomax model under progressively type-I hybrid censoring. Journal of Computational and Applied Mathematics 343, 397–412.

Ateya, S., Mohammed, H., 2018. Prediction under Burr-XII distribution based on generalized Type-II progressive hybrid censoring scheme. Journal of the Egyptian Mathematical Society 26 (3), 491–508.

Azizpour, M., Asgharzadeh, A.a., 2018. Inference for the type-II generalized logistic distribution with progressive hybrid censoring. Journal of Statistical Research of Iran 14 (2), 189–217.

Bagdonavičius, V., 1978. Testing the hypothesis of additive accumulation of damages. Probability Theory and Its Applications 23, 403–408.

Bagdonavičius, V., Nikulin, M., 2002. Accelerated Life Models: Modeling and Statistical Analysis. Chapman & Hall/CRC Press, Boca Raton/Florida.

Bai, X., Shi, Y., Liu, Y., Liu, B., 2018. Reliability estimation of multicomponent stress-strength model based on copula function under progressively hybrid censoring. Journal of Computational and Applied Mathematics 344, 100–114.

Balakrishnan, N., 1983. Empirical power study of a multi-sample test of exponentiality based on spacings. Journal of Statistical Computation and Simulation 18, 265–271.

Balakrishnan, N., 1985. Order statistics from the half logistic distribution. Journal of Statistical Computation and Simulation 20 (4), 287–309.

Balakrishnan, N., 1989a. Approximate maximum likelihood estimation of the mean and standard deviation of the normal distribution based on type II censored samples. Journal of Statistical Computation and Simulation 32 (3), 137–148.

Balakrishnan, N., 1989b. Recurrence relations among moments of order statistics from two related sets of independent and non-identically distributed random variables. Annals of the Institute of Statistical Mathematics 41, 323–329.

Balakrishnan, N., 1990. Approximate maximum likelihood estimation for a generalized logistic distribution. Journal of Statistical Planning and Inference 26 (2), 221–236.

Balakrishnan, N. (Ed.), 1992. Handbook of the Logistic Distribution. Dekker, New York.

Balakrishnan, N., 2007. Progressive censoring methodology: an appraisal (with discussions). Test 16, 211–296.
Balakrishnan, N., 2009. A synthesis of exact inferential results for exponential step-stress models and associated optimal accelerated life-tests. Metrika 69, 351–396.
Balakrishnan, N., Aggarwala, R., 2000. Progressive Censoring: Theory, Methods, and Applications. Birkhäuser, Boston.
Balakrishnan, N., Alam, F.M.A., 2019. Maximum likelihood estimation of the parameters of a multiple step-stress model from the Birnbaum-Saunders distribution under time-constraint: a comparative study. Communications in Statistics. Simulation and Computation 48 (5), 1535–1559.
Balakrishnan, N., Basu, A.P. (Eds.), 1995. The Exponential Distribution: Theory, Methods, and Applications. Taylor & Francis, Newark.
Balakrishnan, N., Beutner, E., Kamps, U., 2008a. Order restricted inference for sequential k-out-of-n systems. Journal of Multivariate Analysis 99 (7), 1489–1502.
Balakrishnan, N., Burkschat, M., Cramer, E., 2008b. Best linear equivariant estimation and prediction in location-scale families. Sankhyā B 70, 229–247.
Balakrishnan, N., Burkschat, M., Cramer, E., Hofmann, G., 2008c. Fisher information based progressive censoring plans. Computational Statistics & Data Analysis 53, 366–380.
Balakrishnan, N., Chan, P.S., 1992. Estimation for the scaled half logistic distribution under Type II censoring. Computational Statistics & Data Analysis 13 (2), 123–141.
Balakrishnan, N., Childs, A., Chandrasekar, B., 2002a. An efficient computational method for moments of order statistics under progressive censoring. Statistics & Probability Letters 60 (4), 359–365.
Balakrishnan, N., Cohen, A.C., 1991. Order Statistics and Inference: Estimation Methods. Academic Press, Boston.
Balakrishnan, N., Cramer, E., 2008. Progressive censoring from heterogeneous distributions with applications to robustness. Annals of the Institute of Statistical Mathematics 60, 151–171.
Balakrishnan, N., Cramer, E., 2014. The Art of Progressive Censoring. Applications to Reliability and Quality. Birkhäuser, New York.
Balakrishnan, N., Cramer, E., 2023. Progressive censoring methodology: a review. In: Pham, H. (Ed.), Springer Handbook of Engineering Statistics, 2 edition. Springer, New York. In press.
Balakrishnan, N., Cramer, E., Iliopoulos, G., 2014. On the method of pivoting the cdf for exact confidence intervals with illustration for exponential mean under life-test with time constraints. Statistics & Probability Letters 89, 124–130.
Balakrishnan, N., Cramer, E., Kamps, U., 2005. Relation for joint densities of progressively censored order statistics. Statistics 39 (6), 529–536.
Balakrishnan, N., Cramer, E., Kamps, U., Schenk, N., 2001. Progressive type II censored order statistics from exponential distributions. Statistics 35, 537–556.
Balakrishnan, N., Cutler, C.D., 1996. Maximum likelihood estimation of Laplace parameters based on Type-II censored samples. In: Morrison, D.F., Nagaraja, H.N., Sen, P.K. (Eds.), Statistical Theory and Applications. Papers in Honor of Herbert A. David. Springer, Berlin, pp. 145–151.
Balakrishnan, N., Davies, K.F., 2013. Pitman closeness results for Type-I censored data from exponential distribution. Statistics & Probability Letters 83 (12), 2693–2698.
Balakrishnan, N., Davies, K.F., Keating, J.P., 2009a. Pitman closeness of order statistics to population quantiles. Communications in Statistics. Simulation and Computation 38 (4), 802–820.
Balakrishnan, N., Davies, K.F., Keating, J.P., Mason, R.L., 2010. Simultaneous closeness among order statistics to population quantiles. Journal of Statistical Planning and Inference 140 (9), 2408–2415.
Balakrishnan, N., Dembińska, A., 2008. Progressively Type-II right censored order statistics from discrete distributions. Journal of Statistical Planning and Inference 138, 845–856.
Balakrishnan, N., Dembińska, A., 2009. Erratum to 'Progressively Type-II right censored order statistics from discrete distributions' [J. Statist. Plann. Inference 138 (2008) 845–856]. Journal of Statistical Planning and Inference 139, 1572–1574.
Balakrishnan, N., Govindarajulu, Z., Balasubramanian, K., 1993. Relationships between moments of two related sets of order statistics and some extensions. Annals of the Institute of Statistical Mathematics 45 (2), 243–247.

Balakrishnan, N., Habibi Rad, A., Arghami, N.R., 2007a. Testing exponentiality based on Kullback-Leibler information with progressively Type-II censored data. IEEE Transactions on Reliability 56, 301–307.

Balakrishnan, N., Han, D., 2008. Exact inference for a simple step-stress model with competing risks for failure from exponential distribution under Type-II censoring. Journal of Statistical Planning and Inference 138 (12), 4172–4186.

Balakrishnan, N., Han, D., Iliopoulos, G., 2011. Exact inference for progressively Type-I censored exponential failure data. Metrika 73, 335–358.

Balakrishnan, N., Iliopoulos, G., 2009. Stochastic monotonicity of the mle of exponential mean under different censoring schemes. Annals of the Institute of Statistical Mathematics 61, 753–772.

Balakrishnan, N., Iliopoulos, G., 2010. Stochastic monotonicity of the MLEs of parameters in exponential simple step-stress models under Type-I and Type-II censoring. Metrika 72, 89–109.

Balakrishnan, N., Iliopoulos, G., Keating, J.P., Mason, R.L., 2009b. Pitman closeness of sample median to population median. Statistics & Probability Letters 79 (16), 1759–1766.

Balakrishnan, N., Kamps, U., Kateri, M., 2012. A sequential order statistics approach to step-stress testing. Annals of the Institute of Statistical Mathematics 64, 303–318.

Balakrishnan, N., Kannan, N., Lin, C.T., Wu, S.J.S., 2004a. Inference for the extreme value distribution under progressive Type-II censoring. Journal of Statistical Computation and Simulation 74 (1), 25–45.

Balakrishnan, N., Kateri, M., 2008. On the maximum likelihood estimation of parameters of Weibull distribution based on complete and censored data. Statistics & Probability Letters 78, 2971–2975.

Balakrishnan, N., Kocherlakota, S., 1985. On the double Weibull distribution: order statistics and estimation. Sankhyā: The Indian Journal of Statistics, Series B (1960-2002) 47 (2), 161–178.

Balakrishnan, N., Kundu, D., 2013. Hybrid censoring: models, inferential results and applications. Computational Statistics & Data Analysis 57, 166–209.

Balakrishnan, N., Kundu, D., 2019. Birnbaum-Saunders distribution: a review of models, analysis, and applications. Applied Stochastic Models in Business and Industry 35, 4–49.

Balakrishnan, N., Kundu, D., Ng, H.K.T., Kannan, N., 2007b. Point and interval estimation for a simple step-stress model with Type-II censoring. Journal of Quality Technology 39 (1), 35–47.

Balakrishnan, N., Lin, C.-T., 2003. On the distribution of a test for exponentiality based on progressively type-II right censored spacings. Journal of Statistical Computation and Simulation 73 (4), 277–283.

Balakrishnan, N., Ling, M.H., 2013. Expectation maximization algorithm for one shot device accelerated life testing with Weibull lifetimes, and variable parameters over stress. IEEE Transactions on Reliability 62 (2), 537–551.

Balakrishnan, N., Mi, J., 2003. Existence and uniqueness of the MLEs for normal distribution based on general progressively type-II censored samples. Statistics & Probability Letters 64 (4), 407–414.

Balakrishnan, N., Ng, H.K.T., Kannan, N., 2002b. A test of exponentiality based on spacings for progressively type-II censored data. In: Huber-Carol, C., Balakrishnan, N., Nikulin, M., Mesbah, M. (Eds.), Goodness-of-Fit Tests and Model Validity. Birkhäuser, Boston, MA, pp. 89–111.

Balakrishnan, N., Ng, H.K.T., Kannan, N., 2004b. Goodness-of-fit tests based on spacings for progressively type-ii censored data from a general location-scale distribution. IEEE Transactions on Reliability 53, 349–356.

Balakrishnan, N., Papadatos, N., 2002. The use of spacings in the estimation of a scale parameter. Statistics & Probability Letters 57 (2), 193–204.

Balakrishnan, N., Rao, C.R., 1997. Large-sample approximations to the best linear unbiased estimation and best linear unbiased prediction based on progressively censored samples and some applications. In: Panchapakesan, S., Balakrishnan, N. (Eds.), Advances in Statistical Decision Theory and Applications. Birkhäuser, Boston, MA, pp. 431–444.

Balakrishnan, N., Rao, C.R., 1998a. Order statistics: a historical perspective. In: Balakrishnan, N., Rao, C.R. (Eds.), Order Statistics: Theory & Applications. In: Handbook of Statistics, vol. 16. Elsevier, Amsterdam, pp. 3–24. Chapter 1.

Balakrishnan, N., Rao, C.R. (Eds.), 1998b. Order Statistics: Applications. Handbook of Statistics, vol. 17. Elsevier, Amsterdam.

Balakrishnan, N., Rao, C.R. (Eds.), 1998c. Order Statistics: Theory & Methods. Handbook of Statistics, vol. 16. Elsevier, Amsterdam.

Balakrishnan, N., Rao, C.R., 2003. Some efficiency properties of best linear unbiased estimators. Journal of Statistical Planning and Inference 113 (2), 551–555.

Balakrishnan, N., Rasouli, A., 2008. Exact likelihood inference for two exponential populations under joint Type-II censoring. Computational Statistics & Data Analysis 52, 2725–2738.

Balakrishnan, N., Rasouli, A., Farsipour, N., 2008d. Exact likelihood inference based on an unified hybrid censored sample from the exponential distribution. Journal of Statistical Computation and Simulation 78 (5), 475–488.

Balakrishnan, N., Sandhu, R.A., 1995. A simple simulational algorithm for generating progressive Type-II censored samples. American Statistician 49, 229–230.

Balakrishnan, N., Sandhu, R.A., 1996. Best linear unbiased and maximum likelihood estimation for exponential distributions under general progressive type-II censored samples. Sankhyā B 58, 1–9.

Balakrishnan, N., Shafay, A.R., 2012. One- and two-sample Bayesian prediction intervals based on Type-II hybrid censored data. Communications in Statistics. Theory and Methods 41 (9), 1511–1531.

Balakrishnan, N., Su, F., 2015. Exact likelihood inference for k exponential populations under joint Type-II censoring. Communications in Statistics. Simulation and Computation 44 (3), 591–613.

Balakrishnan, N., Su, F., Liu, K.-Y., 2015. Exact likelihood inference for k exponential populations under joint progressive type-II censoring. Communications in Statistics. Simulation and Computation 44 (4), 902–923.

Balakrishnan, N., Varadan, J., 1991. Approximate mles for the location & scale parameters of the extreme value distribution with censoring. IEEE Transactions on Reliability 40 (2), 146–151.

Balakrishnan, N., Wong, K., 1991. Approximate MLEs for the location and scale parameters of the half-logistic distribution with type-II right-censoring. IEEE Transactions on Reliability 40 (2), 140–145.

Balakrishnan, N., Wong, K.H.T., 1994. Best linear unbiased estimation of location and scale parameters of the half-logistic distribution based on Type II censored samples. American Journal of Mathematical and Management Sciences 14 (1–2), 53–101.

Balakrishnan, N., Xie, Q., 2007a. Exact inference for a simple step-stress model with Type-I hybrid censored data from the exponential distribution. Journal of Statistical Planning and Inference 137 (11), 3268–3290.

Balakrishnan, N., Xie, Q., 2007b. Exact inference for a simple step-stress model with Type-II hybrid censored data from the exponential distribution. Journal of Statistical Planning and Inference 137 (8), 2543–2563.

Balakrishnan, N., Xie, Q., Kundu, D., 2009c. Exact inference for a simple step-stress model from the exponential distribution under time constraint. Annals of the Institute of Statistical Mathematics 61 (1), 251–274.

Balakrishnan, N., Zhang, L., Xie, Q., 2009d. Inference for a simple step-stress model with Type-I censoring and lognormally distributed lifetimes. Communications in Statistics. Theory and Methods 38 (10), 1690–1709.

Balakrishnan, N., Zhu, X., 2014. On the existence and uniqueness of the maximum likelihood estimates of the parameters of Birnbaum–Saunders distribution based on Type-I, Type-II and hybrid censored samples. Statistics 48 (5), 1013–1032.

Balakrishnan, N., Zhu, X., 2016. Exact likelihood-based point and interval estimation for Laplace distribution based on Type-II right censored samples. Journal of Statistical Computation and Simulation 86 (1), 29–54.

Balasooriya, U., 1995. Failure-censored reliability sampling plans for the exponential distribution. Journal of Statistical Computation and Simulation 52 (4), 337–349.

Balasooriya, U., Balakrishnan, N., 2000. Reliability sampling plans for the lognormal distribution, based on progressively censored samples. IEEE Transactions on Reliability 49, 199–203.

Balasooriya, U., Saw, S.L.C., Gadag, V., 2000. Progressively censored reliability sampling plans for the Weibull distribution. Technometrics 42, 160–167.

Banerjee, A., Kundu, D., 2008. Inference based on Type-II hybrid censored data from a Weibull distribution. IEEE Transactions on Reliability 57 (2), 369–378.

Banerjee, B., Pradhan, B., 2018. Kolmogorov–Smirnov test for life test data with hybrid censoring. Communications in Statistics. Theory and Methods 47 (11), 2590–2604.

Baratpour, S., Habibi Rad, A., 2012. Testing goodness-of-fit for exponential distribution based on cumulative residual entropy. Communications in Statistics. Theory and Methods 41 (8), 1387–1396.

Baratpour, S., Habibi Rad, A., 2016. Exponentiality test based on the progressive type II censoring via cumulative entropy. Communications in Statistics. Simulation and Computation 45 (7), 2625–2637.

Barlevy, G., Nagaraja, H.N., 2015. Properties of the vacancy statistic in the discrete circle covering problem. In: Choudhary, P.K., Nagaraja, C.H., Ng, H.K.T. (Eds.), Ordered Data Analysis, Modeling and Health Research Methods – In Honor of H.N. Nagaraja's 60th Birthday. Springer, New York, pp. 73–86.

Barlow, R.E., Madansky, A., Proschan, F., Scheuer, E.M., 1968. Statistical estimation procedures for the 'burn-in' process. Technometrics 10 (1), 51–62.

Bartholomew, D.J., 1957. A problem in life testing. Journal of the American Statistical Association 52 (279), 350–355.

Bartholomew, D.J., 1963. The sampling distribution of an estimate arising in life testing. Technometrics 5 (3), 361–374.

Basak, I., Balakrishnan, N., 2009. Predictors of failure times of censored units in progressively censored samples from normal distribution. Sankhyā 71-B, 222–247.

Basak, I., Basak, P., Balakrishnan, N., 2006. On some predictors of times to failure of censored items in progressively censored samples. Computational Statistics & Data Analysis 50 (5), 1313–1337.

Basu, A.P., 1995. Accelerated life testing with applications. In: Balakrishnan, N., Basu, A.P. (Eds.), The Exponential Distribution: Theory, Methods and Applications. Gordon and Breach, Newark, NJ, pp. 377–383.

Basu, S., Singh, S.K., Singh, U., 2018. Bayesian inference using product of spacings function for progressive hybrid Type-I censoring scheme. Statistics 52 (2), 345–363.

Basu, S., Singh, S.K., Singh, U., 2019. Estimation of inverse Lindley distribution using product of spacings function for hybrid censored data. Methodology and Computing in Applied Probability 21 (4), 1377–1394.

Bayoud, H.A., 2014. Bayesian inference for exponential lifetime models based on type-II hybrid censoring. Journal of Statistical Theory and Applications 13 (2), 175.

Beg, M.A., 1980. Estimation of $Pr(Y < X)$ for exponential-family. IEEE Transactions on Reliability R29, 158–159.

Berkson, J., Gage, R.P., 1952. Survival curve for cancer patients following treatment. Journal of the American Statistical Association 47 (259), 501–515.

Bhattacharya, R., 2020. Implementation of compound optimal design strategy in censored life-testing experiment. Test 29 (4), 1029–1050.

Bhattacharya, R., Aslam, M., 2019. Design of variables sampling plans based on lifetime-performance index in presence of hybrid censoring scheme. Journal of Applied Statistics 46 (16), 2975–2986.

Bhattacharya, R., Aslam, M., 2020. Generalized multiple dependent state sampling plans in presence of measurement data. IEEE Access 8, 162775–162784.

Bhattacharya, R., Pradhan, B., 2017. Computation of optimum Type-II progressively hybrid censoring schemes using variable neighborhood search algorithm. Test 26 (4), 802–821.

Bhattacharya, R., Pradhan, B., Dewanji, A., 2014a. Optimum life testing plans in presence of hybrid censoring: a cost function approach. Applied Stochastic Models in Business and Industry 30 (5), 519–528.

Bhattacharya, R., Pradhan, B., Dewanji, A., 2015. Computation of optimum reliability acceptance sampling plans in presence of hybrid censoring. Computational Statistics & Data Analysis 83, 91–100.

Bhattacharya, R., Pradhan, B., Dewanji, A., 2016. On optimum life-testing plans under Type-II progressive censoring scheme using variable neighborhood search algorithm. Test 25 (2), 309–330.

Bhattacharya, R., Saha, B.N., Farías, G.G., Balakrishnan, N., 2020. Multi-criteria-based optimal life-testing plans under hybrid censoring scheme. Test 29, 430–453.

Bhattacharya, S., Pradhan, B., Kundu, D., 2014b. Analysis of hybrid censored competing risks data. Statistics 48 (5), 1138–1154.

Bhattacharyya, G.K., 1995. Inferences under two-sample and multi-sample situations. In: Balakrishnan, N., Basu, A.P. (Eds.), The Exponential Distribution. Gordon and Breach, Amsterdam, pp. 93–118.

Bhattacharyya, G.K., Mehrotra, K.G., 1981. On testing equality of two exponential distributions under combined type II censoring. Journal of the American Statistical Association 76 (376), 886–894.

Birnbaum, Z.W., 1956. On a use of Mann-Whitney statistics. In: Proceedings of the Third Berkeley Symposium on Mathematical Statistics and Probability (Berkeley, CA, 1954/55), volume I: Contributions to the Theory of Statistics. University of California Press, Berkeley, CA, pp. 13–17.
Birnbaum, Z.W., Saunders, S.C., 1958. A statistical model for life-length of materials. Journal of the American Statistical Association 53 (281), 151–160.
Blyth, C.R., 1972. Some probability paradoxes in choice from among random alternatives: rejoinder. Journal of the American Statistical Association 67 (338), 379–381.
Boag, J.W., 1949. Maximum likelihood estimates of the proportion of patients cured by cancer therapy. Journal of the Royal Statistical Society, Series B, Methodological 11 (1), 15–53.
Brascamp, H.J., Lieb, E.H., 1975. Some inequalities for gaussian measures and the long-range order of the one-dimensional plasma. In: Arthurs, A.M. (Ed.), Functional Integration and Its Applications, Proceedings of the Conference on Functional Integration, Cumberland Lodge, England. Clarendon Press, Oxford, pp. 1–14.
Buonocore, A., Pirozzi, E., Caputo, L., 2009. A note on the sum of uniform random variables. Statistics & Probability Letters 79 (19), 2092–2097.
Burkschat, M., 2008. On optimality of extremal schemes in progressive Type-II censoring. Journal of Statistical Planning and Inference 138, 1647–1659.
Burkschat, M., 2009. Systems with failure-dependent lifetimes of components. Journal of Applied Probability 46 (4), 1052–1072.
Burkschat, M., Cramer, E., 2012. Fisher information in generalized order statistics. Statistics 46, 719–743.
Burkschat, M., Cramer, E., Górny, J., 2016. Type-I censored sequential k-out-of-n systems. Applied Mathematical Modelling 40 (19–20), 8156–8174.
Burkschat, M., Navarro, J., 2013. Dynamic signatures of coherent systems based on sequential order statistics. Journal of Applied Probability 50 (1), 272–287.
Casella, G., Berger, R.L., 2002. Statistical Inference, 2 edition. Duxbury Press, Boston.
Celeux, G., Chauveau, D., Diebolt, J., 1996. Stochastic versions of the EM algorithm: an experimental study in the mixture case. Journal of Statistical Computation and Simulation 55 (4), 287–314.
Celeux, G., Dieboldt, J., 1985. The sem algorithm: a probabilistic teacher algorithm derived from the EM algorithm for the mixture problem. Computational Statistics Quarterly 2, 73–82.
Çetinkaya, Ç., 2020. Inference based on Type-II hybrid censored data from a Pareto distribution. Journal of Reliability and Statistical Studies 13, 253–264.
Çetinkaya, Ç., 2021. Reliability estimation of a stress-strength model with non-identical component strengths under generalized progressive hybrid censoring scheme. Statistics 55 (2), 250–275.
Chakrabarty, J.B., Chowdhury, S., Roy, S., 2021. Optimum reliability acceptance sampling plan using Type-I generalized hybrid censoring scheme for products under warranty. International Journal of Quality and Reliability Management 38 (3), 780–799.
Chan, P., Ng, H.K.T., Su, F., 2015. Exact likelihood inference for the two-parameter exponential distribution under Type-II progressively hybrid censoring. Metrika 78, 747–770.
Chandrasekar, B., Balakrishnan, N., 2002. On a multiparameter version of Tukey's linear sensitivity measure and its properties. Annals of the Institute of Statistical Mathematics 54 (4), 796–805.
Chandrasekar, B., Childs, A., Balakrishnan, N., 2004. Exact likelihood inference for the exponential distribution under generalized Type-I and Type-II hybrid censoring. Naval Research Logistics 51 (7), 994–1004.
Chen, G., Balakrishnan, N., 1995. A general purpose approximate goodness-of-fit test. Journal of Quality Technology 27, 154–161.
Chen, J., Chou, W., Wu, H., Zhou, H., 2004. Designing acceptance sampling schemes for life testing with mixed censoring. Naval Research Logistics 51 (4), 597–612.
Chen, L.-S., Liang, T., Yang, M.-C., 2021. Optimal curtailed Bayesian sampling plans for exponential distributions with Type-I hybrid censored samples. Communications in Statistics. Simulation and Computation 50 (3), 764–777.
Chen, M.-H., Shao, Q.-M., 1999. Monte Carlo estimation of Bayesian credible and HPD intervals. Journal of Computational and Graphical Statistics 8 (1), 69–92.
Chen, S., Bhattacharyya, G.K., 1988. Exact confidence bounds for an exponential parameter under hybrid censoring. Communications in Statistics. Theory and Methods 17, 1857–1870.

Chen, S., Gui, W., 2021. Order restricted classical and Bayesian inference of a multiple step-stress model from two-parameter Rayleigh distribution under Type I censoring. Communications in Statistics. Theory and Methods 0 (0), 1–31.

Cheng, C., Zhao, H., 2016. Exact inferences of the two-parameter exponential distribution and Pareto distribution with hybrid censored data. Pacific Journal of Applied Mathematics 8 (1), 65–88.

Childs, A., Balakrishnan, N., 1997. Maximum likelihood estimation of Laplace parameters based on general Type-II censored examples. Statistical Papers 38 (3), 343–349.

Childs, A., Balakrishnan, N., Chandrasekar, B., 2012. Exact distribution of the MLEs of the parameters and of the quantiles of two-parameter exponential distribution under hybrid censoring. Statistics 46, 441–458.

Childs, A., Chandrasekar, B., Balakrishnan, N., 2008. Exact likelihood inference for an exponential parameter under progressive hybrid censoring schemes. In: Vonta, F., Nikulin, M., Limnios, N., Huber-Carol, C. (Eds.), Statistical Models and Methods for Biomedical and Technical Systems. Birkhäuser, Boston, pp. 323–334.

Childs, A., Chandrasekar, B., Balakrishnan, N., Kundu, D., 2003. Exact likelihood inference based on Type-I and Type-II hybrid censored samples from the exponential distribution. Annals of the Institute of Statistical Mathematics 55 (2), 319–330.

Cho, Y., Cho, E., 2001. The volume of simplices clipped by a half space. Applied Mathematics Letters 14 (6), 731–735.

Cho, Y., Sun, H., Lee, K., 2014. An estimation of the entropy for a Rayleigh distribution based on doubly-generalized type-II hybrid censored samples. Entropy 16, 3655–3669.

Cho, Y., Sun, H., Lee, K., 2015a. Estimating the entropy of a Weibull distribution under generalized progressive hybrid censoring. Entropy 17, 102–122.

Cho, Y., Sun, H., Lee, K., 2015b. Exact likelihood inference for an exponential parameter under generalized progressive hybrid censoring scheme. Statistical Methodology 23, 18–34.

Christensen, R., 2020. Plane Answers to Complex Questions: The Theory of Linear Models, 5 edition. Springer, New York.

Cohen, A.C., 1949. On estimating the mean and standard deviation of truncated normal distributions. Journal of the American Statistical Association 44 (248), 518–525.

Cohen, A.C., 1950. Estimating the mean and variance of normal populations from singly truncated and doubly truncated samples. The Annals of Mathematical Statistics 21 (4), 557–569.

Cohen, A.C., 1955. Maximum likelihood estimation of the dispersion parameter of a chi-distributed radial error from truncated and censored samples with applications to target analysis. Journal of the American Statistical Association 50 (272), 1122–1135.

Cohen, A.C., 1959. Simplified estimators for the normal distribution when samples are singly censored or truncated. Technometrics 1 (3), 217–237.

Cohen, A.C., 1961. Tables for maximum likelihood estimates: singly truncated and singly censored samples. Technometrics 3 (4), 535–541.

Cohen, A.C., 1963. Progressively censored samples in life testing. Technometrics 5, 327–329.

Cohen, A.C., 1991. Truncated and Censored Samples. Theory and Applications. Marcel Dekker, New York.

Cohen, A.C., 1995. MLEs under censoring and truncation and inference. In: Balakrishnan, N., Basu, A.P. (Eds.), The Exponential Distribution: Theory, Methods, and Applications. Gordon and Breach, Newark, NJ, pp. 33–51.

Constantine, K., Tse, S.K., Karson, M., 1986. Estimation of $P(X < Y)$ in the gamma case. Communications in Statistics. Simulation and Computation 15, 365–388.

Cover, T.M., Thomas, J.A., 2006. Elements of Information Theory, 2 edition. John Wiley & Sons, Hoboken, NJ.

Cox, D., 1959. The analysis of exponentially distributed life-times with two types of failure. Journal of the Royal Statistical Society, Series B 21, 411–421.

Cramer, E., 2001. Inference for stress-strength systems based on Weinman multivariate exponential samples. Communications in Statistics. Theory and Methods 30, 331–346.

Cramer, E., 2003. Contributions to Generalized Order Statistics. Habilitationsschrift. University of Oldenburg, Oldenburg, Germany.

Cramer, E., 2004. Logconcavity and unimodality of progressively censored order statistics. Statistics & Probability Letters 68, 83–90.
Cramer, E., 2006. Dependence structure of generalized order statistics. Statistics 40, 409–413.
Cramer, E., 2016. Sequential order statistics. In: Balakrishnan, N., Brandimarte, P., Everitt, B., Molenberghs, G., Piegorsch, W., Ruggeri, F. (Eds.), Wiley StatsRef: Statistics Reference Online. John Wiley & Sons, Ltd.
Cramer, E., 2021. Ordered and censored lifetime data in reliability: an illustrative review. WIREs: Computational Statistics, e1571.
Cramer, E., Bagh, C., 2011. Minimum and maximum information censoring plans in progressive censoring. Communications in Statistics. Theory and Methods 40, 2511–2527.
Cramer, E., Balakrishnan, N., 2013. On some exact distributional results based on Type-I progressively hybrid censored data from exponential distributions. Statistical Methodology 10 (1), 128–150.
Cramer, E., Burkschat, M., Górny, J., 2016. On the exact distribution of the MLEs based on Type-II progressively hybrid censored data from exponential distributions. Journal of Statistical Computation and Simulation 86 (10), 2036–2052.
Cramer, E., Davies, K., 2018. Restricted optimal progressive censoring. Communications in Statistics. Simulation and Computation 47, 1216–1239.
Cramer, E., Ensenbach, M., 2011. Asymptotically optimal progressive censoring plans based on Fisher information. Journal of Statistical Planning and Inference 141, 1968–1980.
Cramer, E., Górny, J., Laumen, B., 2021. Multi-sample progressive Type-I censoring of exponentially distributed lifetimes. Communications in Statistics. Theory and Methods 22, 5285–5313.
Cramer, E., Herle, K., Balakrishnan, N., 2009. Permanent expansions and distributions of order statistics in the INID case. Communications in Statistics. Theory and Methods 38, 2078–2088.
Cramer, E., Iliopoulos, G., 2010. Adaptive progressive Type-II censoring. Test 19, 342–358.
Cramer, E., Iliopoulos, G., 2015. Adaptive progressive censoring. In: Choudhary, P.K., Nagaraja, C.H., Ng, H.K.T. (Eds.), Ordered Data Analysis, Modeling and Health Research Methods – In Honor of H.N. Nagaraja's 60th Birthday. Springer, New York, pp. 73–86.
Cramer, E., Kamps, U., 1996. Sequential order statistics and k-out-of-n systems with sequentially adjusted failure rates. Annals of the Institute of Statistical Mathematics 48, 535–549.
Cramer, E., Kamps, U., 1997. The UMVUE of $P(X < Y)$ based on Type-II censored samples from Weinman multivariate exponential distributions. Metrika 46, 93–121.
Cramer, E., Kamps, U., 1998. Sequential k-out-of-n systems with Weibull components. Economic Quality Control 13, 227–239.
Cramer, E., Kamps, U., 2001a. Estimation with sequential order statistics from exponential distributions. Annals of the Institute of Statistical Mathematics 53, 307–324.
Cramer, E., Kamps, U., 2001b. Sequential k-out-of-n systems. In: Balakrishnan, N., Rao, C.R. (Eds.), Advances in Reliability. In: Handbook of Statistics, vol. 20. Elsevier, Amsterdam, pp. 301–372. Chapter 12.
Cramer, E., Kamps, U., 2003. Marginal distributions of sequential and generalized order statistics. Metrika 58, 293–310.
Cramer, E., Lenz, U., 2010. Association of progressively Type-II censored order statistics. Journal of Statistical Planning and Inference 140 (2), 576–583.
Cramer, E., Navarro, J., 2015. Progressive Type-II censoring and coherent systems. Naval Research Logistics 62, 512–530.
Cramer, E., Navarro, J., 2016. The progressive censoring signature of coherent systems. Applied Stochastic Models in Business and Industry 32 (5), 697–710.
Cramer, E., Tamm, M., 2014. On a correction of the scale MLE for a two-parameter exponential distribution under progressive Type-I censoring. Communications in Statistics. Theory and Methods 43, 4401–4414.
Cramer, E., Tran, T.-T.-H., 2009. Generalized order statistics from arbitrary distributions and the Markov chain property. Journal of Statistical Planning and Inference 139, 4064–4071.
Crowder, M.J., 2001. Classical Competing Risks. Chapman & Hall/CRC, Boca Raton, FL.
Curry, H., Schoenberg, I., 1966. On Pólya frequency functions IV: the fundamental spline functions and their limits. Journal D'analyse Mathématique 17, 71–107.

Curry, H.B., Schoenberg, I.J., 1947. On spline distributions and their limits – the Pólya distribution functions. Bulletin of the American Mathematical Society 53, 1114.
D'Agostino, R.B., Stephens, M.A. (Eds.), 1986. Goodness-of-Fit Techniques. Marcel Dekker, New York.
Dahmen, K., Burkschat, M., Cramer, E., 2012. A- and D-optimal progressive Type-II censoring designs based on Fisher information. Journal of Statistical Computation and Simulation 82, 879–905.
Dahmen, W., Micchelli, C.A., 1986. Statistical encounters with *B*-splines. In: Function Estimates. Arcata, Calif., 1985. Amer. Math. Soc., Providence, RI, pp. 17–48.
Das, B., Nag, A.S., 2002. A test of exponentiality in life-testing against Weibull alternatives under hybrid censoring. Calcutta Statistical Association Bulletin 52 (1–4), 371–380.
David, H.A., 1973. Concomitants of order statistics. Bulletin of the International Statistical Institute 45, 295–300.
David, H.A., 1995. First (?) occurrence of common terms in mathematical statistics. American Statistician 49 (2), 121–133.
David, H.A., Nagaraja, H.N., 1998. Concomitants of order statistics. In: Balakrishnan, N., Rao, C.R. (Eds.), Order Statistics: Theory & Methods. In: Handbook of Statistics, vol. 16. Elsevier, Amsterdam, pp. 487–513.
David, H.A., Nagaraja, H.N., 2003. Order Statistics, 3rd edition. John Wiley & Sons, Hoboken, New Jersey.
Davies, K.F., 2021. Pitman closeness results for Type-I hybrid censored data from exponential distribution. Journal of Statistical Computation and Simulation 91 (1), 58–80.
de Boor, C., 1976. Splines as linear combinations of B-splines. A survey. In: Lorentz, G.G., Chui, C.K., Schumaker, L.L. (Eds.), Approximation Theory II. Academic Press, New York, pp. 1–47.
de Boor, C., 2001. A Practical Guide to Splines, revised edition. Springer, Berlin, New York.
de Boor, C., 2005. Divided differences. Surveys in Approximation Theory 1, 46–69.
Deemer, W.L., Votaw, D.F., 1955. Estimation of parameters of truncated or censored exponential distributions. The Annals of Mathematical Statistics 26 (3), 498–504.
Dembo, A., Cover, T.M., 1991. Information theoretic inequalities. IEEE Transactions on Information Theory 37, 1501–1518.
Dempster, A.P., Laird, N.M., Rubin, D.B., 1977. Maximum likelihood from incomplete data via the EM algorithm. Journal of the Royal Statistical Society, Series B 39, 1–38.
Devroye, L., 1984. A simple algorithm for generating random variates with a log-concave density. Computing 33 (3), 247–257.
Dey, S., Pradhan, B., 2014. Generalized inverted exponential distribution under hybrid censoring. Statistical Methodology 18, 101–114.
Dodson, B., 2006. The Weibull Analysis Handbook. ASQ Quality Press.
Doostparast, M., Ahmadi, M.V., Ahmadi, J., 2013. Bayes estimation based on joint progressive type II censored data under LINEX loss function. Communications in Statistics. Simulation and Computation 42 (8), 1865–1886.
Döring, M., Cramer, E., 2019. On the power of goodness-of-fit tests for the exponential distribution under progressive Type-II censoring. Journal of Statistical Computation and Simulation 89, 2997–3034.
Draper, N., Guttman, I., 1987. Bayesian analysis of hybrid life tests with exponential failure times. Annals of the Institute of Statistical Mathematics 39 (1), 219–225.
Du, K., Wang, M., Lu, T., Sun, X., 2021. Estimation based on hybrid censored data from the power Lindley distribution. Communications in Statistics. Simulation and Computation, 1–19. In press.
Dube, S., Pradhan, B., Kundu, D., 2011. Parameter estimation of the hybrid censored lognormal distribution. Journal of Statistical Computation and Simulation 81 (3), 275–287.
Dufour, R., Maag, U.R., 1978. Distribution results for modified Kolmogorov-Smirnov statistics for truncated or censored. Technometrics 20 (1), 29–32.
Ebrahimi, N., 1986. Estimating the parameters of an exponential distribution from a hybrid life test. Journal of Statistical Planning and Inference 14 (2), 255–261.
Ebrahimi, N., 1988. Determining the sample size for a hybrid life test based on the cost function. Naval Research Logistics 35 (1), 63–72.
Ebrahimi, N., 1992. Prediction intervals for future failures in the exponential distribution under hybrid censoring. IEEE Transactions on Reliability 41 (1), 127–132.

Ebrahimi, N., Soofi, E., Zahedi, H., 2004. Information properties of order statistics and spacings. IEEE Transactions on Information Theory 50 (1), 177–183.
Efron, B., Johnstone, I.M., 1990. Fisher's information in terms of the hazard rate. The Annals of Statistics 18 (1), 38–62.
Elshahhat, A., Elemary, B.R., 2021. Analysis for xgamma parameters of life under Type-II adaptive progressively hybrid censoring with applications in engineering and chemistry. Symmetry 13 (11), 2112.
Elshahhat, A., Nassar, M., 2021. Bayesian survival analysis for adaptive Type-II progressive hybrid censored Hjorth data. Computational Statistics 36 (3), 1965–1990.
Emam, W., Sultan, K.S., 2021. Bayesian and maximum likelihood estimations of the Dagum parameters under combined-unified hybrid censoring. Mathematical Biosciences and Engineering 18 (3), 2930–2951.
Epstein, B., 1954. Truncated life tests in the exponential case. The Annals of Mathematical Statistics 25, 555–564.
Epstein, B., 1960a. Estimation from life test data. Technometrics 2 (4), 447–454.
Epstein, B., 1960b. Statistical life test acceptance procedures. Technometrics 2 (4), 435–446.
Epstein, B., 1960c. Statistical techniques in life testing. Technical report. National Technical Information Service, US Department of Commerce, Washington, DC.
Epstein, B., Sobel, M., 1953. Life testing. Journal of the American Statistical Association 48, 486–502.
Epstein, B., Sobel, M., 1954. Some theorems relevant to life testing from an exponential distribution. The Annals of Mathematical Statistics 25, 373–381.
Escobar, L.A., Meeker, W.Q., 1986. Algorithm as 218: elements of the Fisher information for the smallest extreme value distribution and censored data. Journal of the Royal Statistical Society. Series C. Applied Statistics 35 (1), 80–86.
Escobar, L.A., Meeker, W.Q., 2001. The asymptotic equivalence of the Fisher information matrices for type I and type II censored data from location-scale families. Communications in Statistics. Theory and Methods 30 (10), 2211–2225.
Fairbanks, K., Madsen, R., Dykstra, R., 1982. A confidence interval for an exponential parameter from a hybrid life test. Journal of the American Statistical Association 77 (377), 137–140.
Feizjavadian, S.H., Hashemi, R., 2015. Analysis of dependent competing risks in the presence of progressive hybrid censoring using Marshall–Olkin bivariate Weibull distribution. Computational Statistics & Data Analysis 82, 19–34.
Feller, W., 1968. An Introduction to Probability Theory and Its Applications, volume I, 3 edition. Wiley, New York.
Fernández, A.J., 2005. Progressively censored variables sampling plans for two-parameter exponential distributions. Journal of Applied Statistics 32 (8), 823–829.
Fertig, K.W., Mann, N.R., 1980. Life-test sampling plans for two-parameter Weibull populations. Technometrics 22 (2), 165–177.
Fischer, T., Balakrishnan, N., Cramer, E., 2008. Mixture representation for order statistics from inid progressive censoring and its applications. Journal of Multivariate Analysis 99, 1999–2015.
Ganguly, A., Kundu, D., 2016. Analysis of simple step-stress model in presence of competing risks. Journal of Statistical Computation and Simulation 86 (10), 1989–2006.
Garg, R., Kumar, K., 2021. On estimation of $P(Y > X)$ for generalized inverted exponential distribution based on hybrid censored data. Statistica 81 (3), 335–361.
Geman, S., Geman, D., 1984. Stochastic relaxation, Gibbs distributions, and the Bayesian restauration of images. IEEE Transactions on Pattern Analysis and Machine Intelligence 6, 721–741.
Gerber, L., 1981. The volume cut off a simplex by a half-space. Pacific Journal of Mathematics 94, 311–314.
Gertsbakh, I., 1995. On the Fisher information in type-I censored and quantal response data. Statistics & Probability Letters 23 (4), 297–306.
Ghitany, M.E., Al-Jarallah, R.A., Balakrishnan, N., 2013. On the existence and uniqueness of the MLEs of the parameters of a general class of exponentiated distributions. Statistics 47, 605–612.
Ghitany, M.E., Alqallaf, F., Balakrishnan, N., 2014. On the likelihood estimation of the parameters of Gompertz distribution based on complete and progressively Type-II censored samples. Journal of Statistical Computation and Simulation 84, 1803–1812.
Goel, R., Krishna, H., 2022. Statistical inference for two Lindley populations under balanced joint progressive type-II censoring scheme. Computational Statistics 37 (1), 263–286.

Goldberger, A.S., 1962. Best linear unbiased prediction in the generalized linear regression model. Journal of the American Statistical Association 57 (298), 369–375.

Goldman, R.N., 1988a. Urn models and B-splines. Constructive Approximation 4 (1), 265–288.

Goldman, R.N., 1988b. Urn models, approximations, and splines. Journal of Approximation Theory 54 (1), 1–66.

Górny, J., 2017. A New Approach to Hybrid Censoring. PhD thesis. RWTH Aachen University, Aachen, Germany.

Górny, J., Cramer, E., 2016. Exact likelihood inference for exponential distributions under generalized progressive hybrid censoring schemes. Statistical Methodology 29, 70–94.

Górny, J., Cramer, E., 2018a. Exact inference for a new flexible hybrid censoring scheme. Journal of the Indian Society for Probability and Statistics 19 (1), 169–199.

Górny, J., Cramer, E., 2018b. Modularization of hybrid censoring schemes and its application to unified progressive hybrid censoring. Metrika 81, 173–210.

Górny, J., Cramer, E., 2019a. A volume based approach to establish B-spline based expressions for density functions and its application to progressive hybrid censoring. Journal of the Korean Statistical Society 38, 340–355.

Górny, J., Cramer, E., 2019b. From B-spline representations to gamma representations in hybrid censoring. Statistical Papers 60, 1119–1135.

Górny, J., Cramer, E., 2019c. Type-I hybrid censoring of uniformly distributed lifetimes. Communications in Statistics. Theory and Methods 48, 412–433.

Górny, J., Cramer, E., 2020a. On exact inferential results for a simple step-stress model under a time constraint. Sankhya. Series B 82 (2), 201–239.

Górny, J., Cramer, E., 2020b. Type-I hybrid censoring of multiple samples. Journal of Computational and Applied Mathematics 366, 112404.

Gouno, E., Balakrishnan, N., 2001. Step-stress accelerated life test. In: Balakrishnan, N., Rao, C.R. (Eds.), Advances in Reliability. In: Handbook of Statistics, vol. 20. North Holland, Amsterdam, pp. 623–639.

Govindarajulu, Z., 1963. Relationships among moments of order statistics in samples from two related populations. Technometrics 5 (4), 514–518.

Goyal, T., Rai, P.K., Maurya, S.K., 2020. Bayesian estimation for GDUS exponential distribution under type-I progressive hybrid censoring. Annals of Data Science 7 (2), 307–345.

Gradshteyn, I.S., Ryzhik, I.M., 2007. Table of Integrals, Series, and Products, 7 edition. Academic Press, Boston.

Graybill, F.A., 1983. Matrices with Applications in Statistics, 2 edition. Wadsworth, Belmont.

Green, P.J., Łatuszyński, K., Pereyra, M., Robert, C.P., 2015. Bayesian computation: a summary of the current state, and samples backwards and forwards. Statistics and Computing 25 (4), 835–862.

Guilbaud, O., 2001. Exact non-parametric confidence intervals for quantiles with progressive type-II censoring. Scandinavian Journal of Statistics 28, 699–713.

Guilbaud, O., 2004. Exact non-parametric confidence, prediction and tolerance intervals with progressive type-II censoring. Scandinavian Journal of Statistics 31 (2), 265–281.

Gupta, A.K., 1952. Estimation of the mean and standard deviation of a normal population from a censored sample. Biometrika 39 (3/4), 260–273.

Gupta, P.K., Singh, B., 2013. Parameter estimation of Lindley distribution with hybrid censored data. International Journal of System Assurance Engineering and Management 4 (4), 378–385.

Gupta, R.D., Gupta, R.C., Sankaran, P.G., 2004. Some characterization results based on factorization of the (reversed) hazard rate function. Communications in Statistics. Theory and Methods 33 (12), 3009–3031.

Gupta, R.D., Kundu, D., 1998. Hybrid censoring schemes with exponential failure distribution. Communications in Statistics. Theory and Methods 27 (12), 3065–3083.

Gupta, R.D., Kundu, D., 1999. Generalized exponential distributions. Australian & New Zealand Journal of Statistics 41 (2), 173–188.

Gupta, R.D., Kundu, D., 2001. Exponentiated exponential family: an alternative to gamma and Weibull distributions. Biometrical Journal 43 (1), 117–130.

Gupta, R.D., Kundu, D., 2002. Generalized exponential distribution; statistical inferences. Journal of Statistical Theory and Applications 1, 101–118.

Gupta, R.D., Kundu, D., 2006. On the comparison of Fisher information of the Weibull and GE distributions. Journal of Statistical Planning and Inference 136 (9), 3130–3144.
Gupta, R.D., Kundu, D., 2007. Generalized exponential distribution: existing results and some recent developments. Journal of Statistical Planning and Inference 137 (11), 3537–3547.
Habibi Rad, A., Izanlo, M., 2011. An EM algorithm for estimating the parameters of the generalized exponential distribution under unified hybrid censored data. Journal of Statistical Research of Iran 8 (2), 149–162.
Habibi Rad, A., Yousefzadeh, F., 2010. Analysis of hybrid censored data from the lognormal distribution. Journal of Statistical Research of Iran 7 (1).
Habibi Rad, A., Yousefzadeh, F., 2014. Inference based on unified hybrid censored data from a Weibull distribution. In: Athens: ATINER'S Conference Paper Series, STA2014-1258.
Habibi Rad, A., Yousefzadeh, F., Balakrishnan, N., 2011. Goodness-of-fit test based on Kullback-Leibler information for progressively Type-II censored data. IEEE Transactions on Reliability 60, 570–579.
Hahn, G.J., Meeker, W.Q., Escobar, L.A., 2017. Statistical Intervals: A Guide for Practitioners. John Wiley & Sons, New York.
Hald, A., 1949. Maximum likelihood estimation of the parameters of a normal distribution which is truncated at a known point. Skandinavisk Aktuarietidskrift 1949 (1), 119–134.
Han, D., Balakrishnan, N., 2010. Inference for a simple step-stress model with competing risks for failure from the exponential distribution under time constraint. Computational Statistics & Data Analysis 54 (9), 2066–2081.
Hansen, P., Mladenović, N., 1999. An introduction to variable neighborhood search. In: Voß, S., Martello, S., Osman, I.H., Roucairol, C. (Eds.), Meta-Heuristics: Advances and Trends in Local Search Paradigms for Optimization. Kluwer Academic Publishers, New York, pp. 433–458.
Harris, T.E., Meier, P., Tukey, J.W., 1950. Timing of the distribution of events between observations; a contribution to the theory of follow-up studies. Human Biology 22 (4), 249–270.
Harter, H.L., 1978. MTBF confidence bounds based on MIL-STD-781C fixed-length test results. Journal of Quality Technology 10 (4), 164–169.
Hashemi, R., Azar, J., 2011. Analysis of competing risks in the Burr XII Model in presence of progressive hybrid censoring. International Mathematical Forum 6 (62), 3069–3078.
Hemmati, F., Khorram, E., 2011. Bayesian analysis of the adaptive Type-II progressively hybrid censoring scheme in presence of competing risks. In: Proc. ICCS-11, pp. 181–194.
Hemmati, F., Khorram, E., 2013. Statistical analysis of the log-normal distribution under type-II progressive hybrid censoring schemes. Communications in Statistics. Simulation and Computation 42 (1), 52–75.
Hosono, Y.O.H., Kase, S., 1981. Design of single sampling plans for doubly exponential characteristics. In: Lenz, H.J., Wetherill, G.B., Wilrich, P.T. (Eds.), Frontiers in Quality Control. Physica Verlag, Vienna, pp. 94–112.
Huang, W., Yang, K., 2010. A new hybrid censoring scheme and some of its properties. Tamsui Oxford Journal of Mathematical Sciences 26, 355–367.
Huber-Carol, C., Balakrishnan, N., Nikulin, M., Mesbah, M. (Eds.), 2002. Goodness-of-Fit Tests and Model Validity. Birkhäuser, Boston, MA.
Hyun, S., Lee, J., Yearout, R., 2016. Parameter estimation of Type-I and Type-II hybrid censored data from the log-logistic distribution. Industrial and Systems Engineering Review 4 (1), 37–44.
Ignatov, A.G., Kaishev, V.K., 1985. B-splines and linear combinations of uniform order statistics. Technical report.
Ignatov, Z.G., Kaishev, V.K., 1989. A probabilistic interpretation of multivariate b-splines and some applications. Serdica 15, 91–99.
Iliopoulos, G., 2015. On exact confidence intervals in a competing risks model with generalized hybrid type-i censored exponential data. Journal of Statistical Computation and Simulation 85 (14), 2953–2961.
Iliopoulos, G., Balakrishnan, N., 2009. Conditional independence of blocked ordered data. Statistics & Probability Letters 79, 1008–1015.
Iliopoulos, G., Balakrishnan, N., 2011. Exact likelihood inference for Laplace distribution based on Type-II censored samples. Journal of Statistical Planning and Inference 141 (3), 1224–1239.
Ishii, G., 1978. Tokeiteki yosoku (Statistical prediction). In: Basic Sugaku, vol. 7. Gendai-Sugakusha, Tokyo, Japan (in Japanese).

Ismail, A.A., 2014. Likelihood inference for a step-stress partially accelerated life test model with Type-I progressively hybrid censored data from Weibull distribution. Journal of Statistical Computation and Simulation 84 (11), 2486–2494.

Jansen, M., Górny, J., Cramer, E., 2022. Exact likelihood inference for an exponential parameter under a multi-sample Type-II progressive hybrid censoring model. American Journal of Mathematics and Management Science 41, 101–127.

Jeon, Y.E., Kang, S.-B., 2021. Estimation of the Rayleigh distribution under unified hybrid censoring. Austrian Journal of Statistics 50 (1), 59–73.

Jeong, H.-S., Park, J.-I., Yum, B.-J., 1996. Development of r, t hybrid sampling plans for exponential lifetime distributions. Journal of Applied Statistics 23 (6), 601–608.

Jeong, H.-S., Yum, B.-J., 1995. Type-I censored life test plans in the exponential case. Communications in Statistics. Simulation and Computation 24 (1), 187–205.

Jia, X., Guo, B., 2017. Exact inference for exponential distribution with multiply Type-I censored data. Communications in Statistics. Simulation and Computation 46 (9), 7210–7220.

Jia, X., Guo, B., 2018. Inference on the reliability of Weibull distribution by fusing expert judgements and multiply Type-I censored data. In: 2018 IEEE International Systems Engineering Symposium (ISSE), pp. 1–7.

Jia, X., Nadarajah, S., Guo, B., 2018. Exact inference on Weibull parameters with multiply Type-I censored data. IEEE Transactions on Reliability 67 (2), 432–445.

Joarder, A., Krishna, H., Kundu, D., 2011. Inferences on Weibull parameters with conventional type-I censoring. Computational Statistics & Data Analysis 55 (1), 1–11.

Johnson, N.L., Kotz, S., Balakrishnan, N., 1994. Continuous Univariate Distributions, vol. 1, 2 edition. John Wiley & Sons, New York.

Johnson, N.L., Kotz, S., Kemp, A.W., 2005. Univariate Discrete Distributions, 3 edition. John Wiley & Sons, Hoboken, NJ.

Johnson, R.A., Mehrotra, K.G., 1972. Locally most powerful rank tests for the two-sample problem with censored data. The Annals of Mathematical Statistics 43 (3), 823–831.

Jones, M., Balakrishnan, N., 2021. Simple functions of independent beta random variables that follow beta distributions. Statistics & Probability Letters 170, 109011.

Kaminsky, K.S., Nelson, P.I., 1975. Best linear unbiased prediction of order statistics in location and scale families. Journal of the American Statistical Association 70, 145–150.

Kaminsky, K.S., Nelson, P.I., 1998. Prediction of order statistics. In: Balakrishnan, N., Rao, C.R. (Eds.), Applications. In: Handbook of Statistics, vol. 17. Elsevier, Amsterdam, pp. 431–450.

Kaminsky, K.S., Rhodin, L., 1985. Maximum likelihood prediction. Annals of the Institute of Statistical Mathematics 37, 507–517.

Kamps, U., 1995a. A concept of generalized order statistics. Journal of Statistical Planning and Inference 48, 1–23.

Kamps, U., 1995b. A Concept of Generalized Order Statistics. Teubner, Stuttgart.

Kamps, U., 2016. Generalized order statistics. In: Balakrishnan, N., Brandimarte, P., Everitt, B., Molenberghs, G., Piegorsch, W., Ruggeri, F. (Eds.), Wiley StatsRef: Statistics Reference Online. John Wiley & Sons, Ltd.

Kamps, U., Cramer, E., 2001. On distributions of generalized order statistics. Statistics 35, 269–280.

Kang, S.B., Cho, Y.S., Choi, S.H., 2001. Approximate maximum likelihood estimation for the three-parameter Weibull distribution. The Korean Communications in Statistics 8, 209–217.

Karlin, S., Micchelli, C.A., Rinott, Y., 1986. Multivariate splines: a probabilistic perspective. Journal of Multivariate Analysis 20 (1), 69–90.

Kateri, M., Balakrishnan, N., 2008. Inference for a simple step-stress model with Type-II censoring, and Weibull distributed lifetimes. IEEE Transactions on Reliability 57 (4), 616–626.

Kateri, M., Kamps, U., Balakrishnan, N., 2010. Multi-sample simple step-stress experiment under time constraints. Statistica Neerlandica 64 (1), 77–96.

Kayal, T., Tripathi, Y.M., Kundu, D., Rastogi, M.K., 2022. Statistical inference of Chen distribution based on Type-I progressive hybrid censored samples. Statistics, Optimization and Information Computing 10 (2), 627–642.

Kayal, T., Tripathi, Y.M., Rastogi, M.K., 2018. Estimation and prediction for an inverted exponentiated Rayleigh distribution under hybrid censoring. Communications in Statistics. Theory and Methods 47 (7), 1615–1640.

Kayal, T., Tripathi, Y.M., Rastogi, M.K., Asgharzadeh, A., 2017. Inference for Burr XII distribution under Type I progressive hybrid censoring. Communications in Statistics. Simulation and Computation 46 (9), 7447–7465.

Keating, J.P., Mason, R.L., Sen, P.K., 1993. Pitman's Measure of Closeness: A Comparison of Statistical Estimators. SIAM, Society for Industrial and Applied Mathematics, Philadelphia, PA.

Kelly, D., Smith, C., 2011. Bayesian Inference for Probabilistic Risk Assessment. Springer.

Khan, R.A., Mitra, M., 2021. Estimation issues in the exponential-logarithmic model under hybrid censoring. Statistical Papers 62 (1), 419–450.

Kim, M., Yum, B.-J., 2011. Life test sampling plans for Weibull distributed lifetimes under accelerated hybrid censoring. Statistical Papers 52 (2), 327–342.

Klenke, A., 2014. Probability Theory: A Comprehensive Course, 2 edition. Springer, Berlin.

Kocherlakota, S., Balakrishnan, N., 1986. One- and two-sided sampling plans based on the exponential distribution. Naval Research Logistics Quarterly 33 (3), 513–522.

Kohansal, A., 2020. Bayesian and classical estimation of $R = P(X < Y)$ based on Burr type XII distribution under hybrid progressive censored samples. Communications in Statistics. Theory and Methods 49 (5), 1043–1081.

Kohansal, A., Nadarajah, S., 2019. Stress–strength parameter estimation based on type-II hybrid progressive censored samples for a Kumaraswamy distribution. IEEE Transactions on Reliability 68, 1296–1310.

Kohansal, A., Rezakhah, S., 2013. Testing exponentiality based on Rényi entropy with progressively type-II censored data.

Kohansal, A., Rezakhah, S., Khorram, E., 2015. Parameter estimation of Type-II hybrid censored weighted exponential distribution. Communications in Statistics. Simulation and Computation 44, 1273–1299.

Koley, A., 2018. Some Contributions to Exponential Failure Data Modeling. PhD thesis. Department of Mathematics and Statistics, Indian Institute of Technology Kanpur, Kanpur, India.

Koley, A., Kundu, D., 2017. On generalized progressive hybrid censoring in presence of competing risks. Metrika 80 (4), 401–426.

Koley, A., Kundu, D., Ganguly, A., 2017. Analysis of Type-II hybrid censored competing risks data. Statistics: A Journal of Theoretical and Applied Statistics, 1304–1325.

Kong, F., 1998. Parameter estimation under multiply type-II censoring. In: Balakrishnan, N., Rao, C.R. (Eds.), Order Statistics: Applications, vol. 17. North-Holland, Amsterdam, pp. 315–335.

Kotz, S., Kozubowski, T.J., Podgórski, K., 2001. The Laplace Distribution and Generalizations. A Revisit with Applications to Communications, Economics, Engineering, and Finance. Birkhäuser, Boston.

Kotz, S., Lumelskii, Y., Pensky, M., 2003. The Stress-Strength Model and Its Generalizations – Theory and Applications. World Scientific Publications, Singapore.

Koziol, J.A., Byar, D.P., 1975. Percentage points of the asymptotic distributions of one and two sample K-S statistics for truncated or censored data. Technometrics 17 (4), 507–510.

Kullback, S., 1959. Information Theory and Statistics. Wiley, New York.

Kundu, D., 2007. On hybrid censored Weibull distribution. Journal of Statistical Planning and Inference 137 (7), 2127–2142.

Kundu, D., Basu, S., 2000. Analysis of incomplete data in presence of competing risks. Journal of Statistical Planning and Inference 87 (2), 221–239.

Kundu, D., Ganguly, A., 2017. Analysis of Step-Stress Models. Academic Press Inc., London, UK.

Kundu, D., Gupta, R.D., 2006. Estimation of $P[Y < X]$ for Weibull distributions. IEEE Transactions on Reliability 55 (2), 270–280.

Kundu, D., Gupta, R.D., 2007. Analysis of hybrid life-tests in presence of competing risks. Metrika 65 (2), 159–170.

Kundu, D., Joarder, A., 2006a. Analysis of Type-II progressively hybrid censored competing risks data. Journal of Modern Applied Statistical Methods 5 (1), 152–170.

Kundu, D., Joarder, A., 2006b. Analysis of Type-II progressively hybrid censored data. Computational Statistics & Data Analysis 50 (10), 2509–2528.

Kundu, D., Kannan, N., Balakrishnan, N., 2004. Analysis of progressively censored competing risks data. In: Balakrishnan, N., Rao, C.R. (Eds.), Advances in Survival Analysis. In: Handbook of Statistics, vol. 23. Elsevier, Amsterdam, pp. 331–348.

Kundu, D., Koley, A., 2017. Interval estimation of the unknown exponential parameter based on time truncated data. American Journal of Mathematical and Management Sciences 36, 188–195.

Kundu, D., Pradhan, B., 2009. Estimating the parameters of the generalized exponential distribution in presence of hybrid censoring. Communications in Statistics. Theory and Methods 38, 2030–2041.

Kundu, D., Samanta, D., Ganguly, A., Mitra, S., 2013. Bayesian analysis of different hybrid & progressive life tests. Communications in Statistics. Simulation and Computation 42 (9), 2160–2173.

Laumen, B., 2017. Progressive Censoring and Stage Life Testing. PhD thesis. RWTH Aachen, Aachen, Germany.

Laumen, B., Cramer, E., 2015. Inference for the lifetime performance index with progressively Type-II censored samples from gamma distributions. Economic Quality Control 30, 59–73.

Laumen, B., Cramer, E., 2019. Progressive censoring with fixed censoring times. Statistics 53, 569–600.

Lawless, J.F., 2003. Statistical Models and Methods for Lifetime Data, 2 edition. Wiley, New York.

Lee, K., 2017. Estimation of entropy of the inverse Weibull distribution under generalized progressive hybrid censored data. Journal of the Korean Data and Information Science Society 28 (3), 659–668.

Lee, K., 2020. Estimation of the entropy with generalized type I hybrid censored Weibull data. Journal of the Korean Data and Information Science Society 31 (3), 687–697.

Lee, K., 2021. Estimating the parameters of the Weibull distribution under generalized type II hybrid censoring. Journal of the Korean Data and Information Science Society 32 (4), 905–915.

Lee, K., Sun, H., Cho, Y., 2016a. Exact likelihood inference of the exponential parameter under generalized type ii progressive hybrid censoring. Journal of the Korean Statistical Society 45, 123–136.

Lee, K.-J., Lee, J.-I., Park, C.-K., 2016b. Analysis of generalized progressive hybrid censored competing risks data. Journal of the Korean Society of Marine Engineering 40 (2), 131–137.

Lee, W., Lee, K., 2019. Goodness-of-fit tests based on generalized Lorenz curve for progressively Type II censored data from a location-scale distributions. Communications for Statistical Applications and Methods AB 26 (2), 191–203.

Lehmann, E.L., Casella, G., 1998. Theory of Point Estimation, 2 edition. Springer, New York.

Lehmann, E.L., Romano, J.P., 2005. Testing Statistical Hypotheses, 3 edition. Springer, Berlin.

Liang, T., 2014. Designing Bayesian sampling plans with adaptive progressive hybrid censored samples. Advances in Statistics 2014, 198696.

Liang, T., Yang, M.-C., 2013. Optimal Bayesian sampling plans for exponential distributions based on hybrid censored samples. Journal of Statistical Computation and Simulation 83 (5), 922–940.

Lieberman, G.J., Resnikoff, G.J., 1955. Sampling plans for inspection by variables. Journal of the American Statistical Association 50 (270), 457–516.

Lieblein, J., 1955. On moments of order statistics from the Weibull distribution. The Annals of Mathematical Statistics 26, 330–333.

Lim, J., Park, S., 2007. Censored Kullback-Leibler information and goodness-of-fit test with Type II censored data. Journal of Applied Statistics 34 (9), 1051–1064.

Lin, C., Huang, Y., 2012. On progressive hybrid censored exponential distribution. Journal of Statistical Computation and Simulation 82 (5), 689–709.

Lin, C.-T., Chou, C.-C., Huang, Y.-L., 2012. Inference for the Weibull distribution with progressive hybrid censoring. Computational Statistics & Data Analysis 56 (3), 451–467.

Lin, C.-T., Hsu, Y.-Y., Lee, S.-Y., Balakrishnan, N., 2019. Inference on constant stress accelerated life tests for log-location-scale lifetime distributions with type-I hybrid censoring. Journal of Statistical Computation and Simulation 89 (4), 720–749.

Lin, C.-T., Huang, Y.-L., Balakrishnan, N., 2008. Exact Bayesian variable sampling plans for the exponential distribution based on Type-I and Type-II hybrid censored samples. Communications in Statistics. Simulation and Computation 37 (6), 1101–1116.

Lin, C.-T., Huang, Y.-L., Balakrishnan, N., 2010a. Corrections on "Exact Bayesian variable sampling plans for the exponential distribution based on Type-I and Type-II hybrid censored samples". Communications in Statistics. Simulation and Computation 39 (7), 1499–1505.

Lin, C.-T., Huang, Y.-L., Balakrishnan, N., 2010b. Exact Bayesian variable sampling plans for exponential distribution under Type-I censoring. In: Huber, C., Limnios, N., Mesbah, M., Nikulin, M. (Eds.), Mathematical Methods in Survival Analysis, Reliability and Quality of Life. ISTE, pp. 151–162.

Lin, C.-T., Huang, Y.-L., Balakrishnan, N., 2011. Exact Bayesian variable sampling plans for the exponential distribution with progressive hybrid censoring. Journal of Statistical Computation and Simulation 81 (7), 873–882.

Lin, C.-T., Huang, Y.-L., Balakrishnan, N., 2013. Correction on 'Exact Bayesian variable sampling plans for the exponential distribution with progressive hybrid censoring, J. Stat. Comput. Simul. 81, 873–882, 2011'. Journal of Statistical Computation and Simulation 83, 402–404.

Lin, C.-T., Ng, H.K.T., Chan, P.S., 2009. Statistical inference of type-II progressively hybrid censored data with Weibull lifetimes. Communications in Statistics. Theory and Methods 38, 1710–1729.

Lin, Y.-P., Liang, T., Huang, W.-T., 2002. Bayesian sampling plans for exponential distribution based on Type I censoring data. Annals of the Institute of Statistical Mathematics 54 (1), 100–113.

Lindley, D., 1980. Approximate Bayesian methods. Trabajos de Estadistica Y de Investigacion Operativa 31, 223–245.

Ling, L., Xu, W., Li, M., 2009. Parametric inference for progressive Type-I hybrid censored data on a simple step-stress accelerated life test model. Mathematics and Computers in Simulation 79 (10), 3110–3121.

Littel, A.S., 1952. Estimation of the T-year survival rate from follow-up studies over a limited period of time. Human Biology 24 (2), 87–116.

Liu, S., Gui, W., 2020. Estimating the parameters of the two-parameter Rayleigh distribution based on adaptive Type II progressive hybrid censored data with competing risks. Mathematics 8 (10).

Liu, S., Gui, W., 2021. Statistical inference for bathtub-shaped distribution based on generalized progressive hybrid censored data. Communications in Statistics. Theory and Methods, 1–24. In press.

Lloyd, E.H., 1952. Least-squares estimation of location and scale parameters using order statistics. Biometrika 39 (1/2), 88–95.

Lodhi, C., Tripathi, Y.M., Wang, L., 2021. Inference for a general family of inverted exponentiated distributions with partially observed competing risks under generalized progressive hybrid censoring. Journal of Statistical Computation and Simulation 91 (12), 2503–2526.

Louis, T.A., 1982. Finding the observed information when using the em algorithm. Journal of the Royal Statistical Society, Series B 44, 226–233.

Ly, A., Marsman, M., Verhagen, J., Grasman, R.P.P.P., Wagenmakers, E.-J., 2017. A tutorial on Fisher information. Journal of Mathematical Psychology 80, 40–55.

Mahto, A.K., Lodhi, C., Tripathi, Y.M., Wang, L., 2022a. Inference for partially observed competing risks model for Kumaraswamy distribution under generalized progressive hybrid censoring. Journal of Applied Statistics 40 (8), 2064–2092.

Mahto, A.K., Tripathi, Y.M., Sultana, F., Rastogi, M.K., 2022b. Statistical inference for a Gumbel type-II distribution under hybrid censoring. Journal of Statistical Computation and Simulation 92, 2290–2316.

Malmquist, S., 1950. On a property of order statistics from a rectangular distribution. Skandinavisk Aktuarietidskrift 33, 214–222.

Mann, N.R., 1967. Results on location and scale parameter estimation with application to the extreme-value distribution. Technical Report ARL 67-0023. Aerospace Research Labs., Wright-Patterson Air Force Base, Ohio, AD 653575.

Mann, N.R., Fertig, K.W., 1973. Tables for obtaining Weibull confidence bounds and tolerance bounds based on best linear invariant estimates of parameters of the extreme-value distribution. Technometrics 15 (1), 87–101.

Mao, S., Liu, B., Shi, Y., 2021. Statistical inference for a simple step stress model with competing risks based on generalized Type-I hybrid censoring. Journal of Systems Science and Information 9 (5), 533–548.

Mao, S., Shi, Y., 2020. Likelihood inference under generalized hybrid censoring scheme with competing risks. Chinese Quarterly Journal of Mathematics 31 (2), 178–188.

Mao, S., Shi, Y.-M., Sun, Y.-D., 2014. Exact inference for competing risks model with generalized type-I hybrid censored exponential data. Journal of Statistical Computation and Simulation 84 (11), 2506–2521.

Mao, S., Shi, Y.-m., Wang, X.-l., 2017. Exact inference for joint Type-I hybrid censoring model with exponential competing risks data. Acta Mathematicae Applicatae Sinica, English Series 33 (3), 645–658.

Marohn, F., 2002. A characterization of generalized Pareto distributions by progressive censoring schemes and goodness-of-fit tests. Communications in Statistics. Theory and Methods 31 (7), 1055–1065.
Maswadah, M., 2022. Improved maximum likelihood estimation of the shape-scale family based on the generalized progressive hybrid censoring scheme. Journal of Applied Statistics 49, 2825–2844.
Mathai, A.M., 1993. A Handbook of Generalized Special Functions for Statistical and Physical Sciences. Clarendon Press, Oxford.
McLachlan, G.J., Krishnan, T., 2008. The EM Algorithm and Extensions, 2 edition. Wiley-Interscience.
Meeker, W.Q., Escobar, L.A., 1998. Statistical Methods for Reliability Data. John Wiley & Sons, New York.
Meeker, W.Q., Hahn, G.J., 1985. How to plan accelerated life tests. In: ASQC Basic References in Quality Control: Statistical Techniques, vol. 10. The American Society for Quality Control, Milwaukee.
Meeter, C.A., Meeker, W.Q., 1994. Optimum accelerated life tests with a nonconstant scale parameter. Technometrics 36 (1), 71–83.
Mehrotra, K.G., Bhattacharyya, G.K., 1982. Confidence intervals with jointly type-II censored samples from two exponential distributions. Journal of the American Statistical Association 77 (378), 441–446.
Mehrotra, K.G., Johnson, R., Bhattacharyya, G., 1979. Exact Fisher information for censored samples and the extended hazard rate functions. Communications in Statistics. Theory and Methods 8 (15), 1493–1510.
MIL-STD-781-C, 1977. Reliability Design Qualification and Production Acceptance Tests: Exponential Distribution. U.S. Government Printing Office, Washington, DC.
Mirjalili, M., Torabi, H., Nadeb, H., Baferki, S.F., 2016. Stress-strength reliability of exponential distribution based on Type-I progressively hybrid censored samples. Journal of Statistical Research of Iran 13 (1), 89–105.
Mohie El-Din, M.M., Nagy, M., Abu-Moussa, M.H., 2019. Estimation and prediction for Gompertz distribution under the generalized progressive hybrid censored data. Annals of Data Science 6, 673–705.
Mohie El-Din, M.M., Nagy, M., Shafay, A.R., 2017. Statistical inference under unified hybrid censoring scheme. Journal of Statistics Applications & Probability 6 (1), 149–167.
Mokhtari, E.B., Rad, A.H., Yousefzadeh, F., 2011. Inference for Weibull distribution based on progressively type-ii hybrid censored data. Journal of Statistical Planning and Inference 141 (8), 2824–2838.
Mondal, S., Kundu, D., 2019a. Bayesian inference for Weibull distribution under the balanced joint type-II progressive censoring scheme. American Journal of Mathematical and Management Sciences 39, 56–74.
Mondal, S., Kundu, D., 2019b. Exact inference on multiple exponential populations under a joint type-II progressive censoring scheme. Statistics 53 (6), 1329–1356.
Mondal, S., Kundu, D., 2019c. Point and interval estimation of Weibull parameters based on joint progressively censored data. Sankhya. Series B 81, 1–25.
Mondal, S., Kundu, D., 2020. On the joint Type-II progressive censoring scheme. Communications in Statistics. Theory and Methods 49, 958–976.
Monfared, M.M., Belaghi, R.A., Behzadi, M.H., Singh, S., 2022. Estimation and prediction based on type-I hybrid censored data from the Poisson-Exponential distribution. Communications in Statistics. Simulation and Computation 51 (5), 2560–2585.
Montgomery, D.C., 2013. Statistical Quality Control. A Modern Introduction, 7 edition. John Wiley & Sons, Hoboken, NJ.
Morabbi, H., Razmkhah, M., 2009. Entropy of hybrid censoring schemes. Journal of Statistical Research of Iran 6, 161–176.
Nadarajah, S., 2011. The exponentiated exponential distribution: a survey. AStA Advances in Statistical Analysis 95 (3), 219–251.
Nagaraja, H.N., 1994. Tukey's linear sensitivity and order statistics. Annals of the Institute of Statistical Mathematics 46 (4), 757–768.
Nagaraja, H.N., Abo-Eleneen, Z.A., 2003. Fisher information in order statistics. Pakistan Journal of Statistics 19 (1), 161–173.
Nagy, M., Sultan, K.S., Abu-Moussa, M.H., 2021. Analysis of the generalized progressive hybrid censoring from Burr Type-XII lifetime model. AIMS Mathematics 6 (9), 9675–9704.
Nassar, M., Abo-Kasem, O., 2016a. Estimation of Burr Type XII parameters under adaptive Type-II progressive hybrid censoring scheme. International Journal of Engineering and Applied Sciences 9 (1), 1–11.

Nassar, M., Abo-Kasem, O., 2016b. Estimation of the inverse Weibull parameters under adaptive type-II progressive hybrid censoring scheme. Journal of Computational and Applied Mathematics.

Nassar, M., Abo-Kasem, O., Zhang, C., Dey, S., 2018. Analysis of Weibull distribution under adaptive Type-II progressive hybrid censoring scheme. Journal of the Indian Society for Probability and Statistics 19, 25–65.

Nassar, M., Ashour, S.K., 2014. Analysis of exponential distribution under adaptive Type-I progressive hybrid censored competing risks data. Pakistan Journal of Statistics and Operation Research 10 (2), 229–245.

Nassar, M., Dobbah, S.A., 2020. Analysis of reliability characteristics of bathtub-shaped distribution under adaptive Type-I progressive hybrid censoring. IEEE Access 8, 181796–181806.

Nassar, M., Nassr, S., Dey, S.A., 2017. Analysis of Burr Type-XII distribution under step stress partially accelerated life tests with Type-I and adaptive Type-II progressively hybrid censoring schemes. Annals of Data Science 4 (2), 227–248.

Navarro, J., 2021. Introduction to System Reliability Theory. Springer.

Nelson, W., 1980. Accelerated life testing – step-stress models and data analyses. IEEE Transactions on Reliability R-29 (2), 103–108.

Nelson, W., 1982. Applied Life Data Analysis. Wiley, New York.

Nelson, W., 1990. Accelerated Testing: Statistical Models, Test Plans, and Data Analyses. John Wiley & Sons, Hoboken, NJ.

Nelson, W., Meeker, W.Q., 1978. Theory for optimum accelerated censored life tests for Weibull and extreme value distributions. Technometrics 20 (2), 171–177.

Ng, H.K.T., Chan, P.S., Balakrishnan, N., 2002. Estimation of parameters from progressively censored data using EM algorithm. Computational Statistics & Data Analysis 39 (4), 371–386.

Ng, H.K.T., Chan, P.S., Balakrishnan, N., 2004. Optimal progressive censoring plans for the Weibull distribution. Technometrics 46 (4), 470–481.

Ng, H.K.T., Kundu, D., Balakrishnan, N., 2006. Point and interval estimation for the two-parameter Birnbaum-Saunders distribution based on Type-II censored samples. Computational Statistics & Data Analysis 50 (11), 3222–3242.

Ng, H.K.T., Kundu, D., Chan, P.S., 2009. Statistical analysis of exponential lifetimes under an adaptive Type-II progressive censoring scheme. Naval Research Logistics 56, 687–698.

Nikulin, M.S., Chimitova, E.V., 2017. Chi-Squared Goodness-of-Fit Tests for Censored Data. John Wiley & Sons.

Noori Asl, M., Arabi Belaghi, R., Bevrani, H., 2017. On Burr XII distribution analysis under progressive Type-II hybrid censored data. Methodology and Computing in Applied Probability 19 (2), 665–683.

Okasha, H., Lio, Y., Albassam, M., 2021. On reliability estimation of Lomax distribution under adaptive Type-I progressive hybrid censoring scheme. Mathematics 9 (22).

Okasha, H., Mustafa, A., 2020. E-Bayesian estimation for the Weibull distribution under adaptive Type-I progressive hybrid censored competing risks data. Entropy 22 (8).

Pakyari, R., 2021. Goodness-of-fit testing based on Gini index of spacings for progressively Type-II censored data. Communications in Statistics. Simulation and Computation, 1–10. In press.

Pakyari, R., Balakrishnan, N., 2012. A general purpose approximate goodness-of-fit test for progressively Type-II censored data. IEEE Transactions on Reliability 61, 238–244.

Pakyari, R., Balakrishnan, N., 2013. Goodness-of-fit tests for progressively Type-II censored data from location-scale distributions. Journal of Statistical Computation and Simulation 83, 167–178.

Panahi, H., 2017. Estimation methods for the generalized inverted exponential distribution under Type II progressively hybrid censoring with application to spreading of micro-drops data. Communications in Mathematics and Statistics 5 (2), 159–174.

Panahi, H., Asadi, S., 2021. On adaptive progressive hybrid censored Burr type III distribution: application to the nano droplet dispersion data. Quality Technology & Quantitative Management 18 (2), 179–201.

Panahi, H., Moradi, N., 2020. Estimation of the inverted exponentiated Rayleigh distribution based on adaptive Type II progressive hybrid censored sample. Journal of Computational and Applied Mathematics 364, 112345.

Panahi, H., Sayyareh, A., 2015. Estimation and prediction for a unified hybrid-censored Burr Type XII distribution. Journal of Statistical Computation and Simulation 86 (1), 55–73.

Park, S., 1994. Fisher Information in Order Statistics. PhD thesis. University of Chicago, Chicago, IL.
Park, S., 1995. The entropy of consecutive order statistics. IEEE Transactions on Information Theory.
Park, S., 1996. Fisher information in order statistics. Journal of the American Statistical Association 91 (433), 385–390.
Park, S., 2003. On the asymptotic Fisher information in order statistics. Metrika 57 (1), 71–80.
Park, S., 2005. Testing exponentiality based on Kullback-Leibler information with the Type-II censored data. IEEE Transactions on Reliability 54, 22–26.
Park, S., 2014. On Kullback-Leibler information of order statistics in terms of the relative risk. Metrika 77, 609–616.
Park, S., 2016. On the Kullback–Leibler information of hybrid censored data. Communications in Statistics. Theory and Methods 45 (15), 4486–4493.
Park, S., Balakrishnan, N., 2009. On simple calculation of the Fisher information in hybrid censoring schemes. Statistics & Probability Letters 79, 1311–1319.
Park, S., Balakrishnan, N., 2012. A very flexible hybrid censoring scheme and its Fisher information. Journal of Statistical Computation and Simulation 82 (1), 41–50.
Park, S., Balakrishnan, N., Kim, S.W., 2011. Fisher information in progressive hybrid censoring schemes. Statistics: A Journal of Theoretical and Applied Statistics 45 (6), 623–631.
Park, S., Balakrishnan, N., Zheng, G., 2008. Fisher information in hybrid censored data. Statistics & Probability Letters 78 (16), 2781–2786.
Park, S., Pakyari, R., 2015. Cumulative residual Kullback-Leibler information with the progressively Type-II censored data. Statistics & Probability Letters 106, 287–294.
Park, S., Shin, M., 2014. Kullback–Leibler information of a censored variable and its applications. Statistics 48 (4), 756–765.
Park, S., Zheng, G., 2004. Equal Fisher information in order statistics. Sankhyā 66 (1), 20–34.
Parsi, S., Bairamov, I., 2009. Expected values of the number of failures for two populations under joint Type-II progressive censoring. Computational Statistics & Data Analysis 53 (10), 3560–3570.
Parsi, S., Ganjali, M., Farsipour, N.S., 2011. Conditional maximum likelihood and interval estimation for two Weibull populations under joint Type-II progressive censoring. Communications in Statistics. Theory and Methods 40 (12), 2117–2135.
Pintilie, M., 2006. Competing Risks: A Practical Perspective. Wiley, Hoboken, NJ.
Pitman, E.J.G., 1937. The "closest" estimates of statistical parameters. Mathematical Proceedings of the Cambridge Philosophical Society 33, 212–222.
Prajapat, K., Koley, A., Mitra, S., Kundu, D., 2021. An optimal Bayesian sampling plan for two-parameter exponential distribution under Type-I hybrid censoring. Sankhya. Series A.
Prajapati, D., Mitra, S., Kundu, D., 2019a. Decision theoretic sampling plan for one-parameter exponential distribution under Type-I and Type-I hybrid censoring schemes. In: Lio, Y., Ng, H.K.T., Tsai, T.-R., Chen, D.-G. (Eds.), Statistical Quality Technologies: Theory and Practice. Springer International Publishing, Cham, pp. 183–210.
Prajapati, D., Mitra, S., Kundu, D., 2019b. A new decision theoretic sampling plan for Type-I and Type-I hybrid censored samples from the exponential distribution. Sankhya. Series B 81 (2), 251–288.
Prajapati, D., Mitra, S., Kundu, D., 2020. Bayesian sampling plan for the exponential distribution with generalized Type-II hybrid censoring scheme. Communications in Statistics. Simulation and Computation, 1–32. In press.
Prakash, G., 2020. Pareto distribution under hybrid censoring: some estimation. Journal of Modern Applied Statistical Methods 19 (Article 16), eP3004.
Prékopa, A., 1973. On logarithmic concave measures and functions. Acta Scientiarum Mathematicarum 34, 335–343.
Proschan, F., 1963. Theoretical explanation of observed decreasing failure rate. Technometrics 5 (3), 375–383.
Pundir, P.S., Gupta, P.K., 2018. Stress-strength reliability of two-parameter bathtub-shaped lifetime model based on hybrid censored samples. Journal of Statistics & Management Systems 21 (7), 1229–1250.
Rao, C.R., 1973. Linear Statistical Inference and Its Applications, 2 edition. John Wiley & Sons, Hoboken, NJ.
Raqab, M., Nagaraja, H., 1995. On some predictors of future order statistics. Metron 53, 185–204.

Raqab, M.Z., Bdair, O.M., Rastogi, M.K., Al-aboud, F.M., 2021. Inference for an exponentiated half logistic distribution with application to cancer hybrid censored data. Communications in Statistics. Simulation and Computation 50 (4), 1178–1201.
Rasouli, A., Balakrishnan, N., 2010. Exact likelihood inference for two exponential populations under joint progressive Type-II censoring. Communications in Statistics. Theory and Methods 39 (12), 2172–2191.
Rastogi, M.K., Tripathi, Y.M., 2011. Estimating a parameter of Burr type XII distribution using hybrid censored observations. International Journal of Quality and Reliability Management 28 (8), 885–893.
Rastogi, M.K., Tripathi, Y.M., 2013a. Estimation using hybrid censored data from a two-parameter distribution with bathtub shape. Computational Statistics & Data Analysis 67, 268–281.
Rastogi, M.K., Tripathi, Y.M., 2013b. Inference on unknown parameters of a Burr distribution under hybrid censoring. Statistical Papers 54 (3), 619–643.
Rényi, A., 1953. On the theory of order statistics. Acta Mathematica Academiae Scientiarum Hungaricae 4, 191–231.
Rezapour, M., Alamatsaz, M.H., Balakrishnan, N., 2013a. On properties of dependent progressively Type-II censored order statistics. Metrika 76, 909–917.
Rezapour, M., Alamatsaz, M.H., Balakrishnan, N., Cramer, E., 2013b. On properties of progressively Type-II censored order statistics arising from dependent and non-identical random variables. Statistical Methodology 10 (1), 58–71.
Rüschendorf, L., 1985. Two remarks on order statistics. Journal of Statistical Planning and Inference 11, 71–74.
Salah, M., Ahmed, E., Alhussain, Z., Ahmed, H., El-Morshedy, M., Eliwa, M., 2021. Statistical inferences for type-II hybrid censoring data from the alpha power exponential distribution. PLoS ONE 16 (1), e0244316.
Salem, A.M., Abo-Kasem, O.E., 2011. Estimation for the parameters of the exponentiated Weibull distribution based on progressive hybrid censored samples. International Journal of Contemporary Mathematical Sciences 6 (35), 1713–1724.
Samanta, D., Kundu, D., 2021. Bayesian inference of a dependent competing risk data. Journal of Statistical Computation and Simulation 91 (15), 3069–3086.
Santner, T., Tenga, R., 1984. Testing goodness of fit to the increasing failure rate family with censored data. Naval Research Logistics 31 (4), 631–646.
Sarhan, A.E., Greenberg, B.G., 1956. Estimation of location and scale parameters by order statistics from singly and doubly censored samples. The Annals of Mathematical Statistics 27 (2), 427–451.
Schneider, H., 1986. Truncated and Censored Samples from Normal Populations. Marcel Dekker, New York.
Schneider, H., 1989. Failure-censored variables-sampling plans for lognormal and Weibull distributions. Technometrics 31 (2), 199–206.
Schoenberg, I.J., 1946. Contributions to the problem of approximation of equidistant data by analytic functions, Part A: on the problem of smoothing or graduation, a first class of analytic approximation formulas. Quarterly of Applied Mathematics 4, 45–99.
Schumaker, L., 2007. Spline Functions: Basic Theory, 3 edition. Cambridge University Press, Cambridge, NY.
Sedyakin, N.M., 1966. On one physical principle in reliability theory. Technical Cybernetics 3, 80–87 (in Russian).
Sen, T., Bhattacharya, R., Tripathi, Y.M., Pradhan, B., 2018a. Generalized hybrid censored reliability acceptance sampling plans for the Weibull distribution. American Journal of Mathematical and Management Sciences 37 (4), 324–343.
Sen, T., Bhattacharya, R., Tripathi, Y.M., Pradhan, B., 2020. Inference and optimum life testing plans based on Type-II progressive hybrid censored generalized exponential data. Communications in Statistics. Simulation and Computation 49 (12), 3254–3282.
Sen, T., Pradhan, B., Tripathi, Y.M., Bhattacharya, R., 2018b. Fisher information in generalized progressive hybrid-censored data. Statistics: A Journal of Theoretical and Applied Statistics 52 (5), 1025–1039.
Sen, T., Singh, S., Tripathi, Y.M., 2018c. Statistical inference for lognormal distribution with type-I progressive hybrid censored data. American Journal of Mathematical and Management Sciences 38 (1), 70–95.

Sen, T., Tripathi, Y.M., Bhattacharya, R., 2018d. Statistical inference and optimum life testing plans under Type-II hybrid censoring scheme. Annals of Data Science 5 (4), 679–708.

Seo, J.I., Kim, Y., 2017. Robust Bayesian estimation of a two-parameter exponential distribution under generalized Type-I progressive hybrid censoring. Communications in Statistics. Simulation and Computation 46 (7), 5795–5807.

Seo, J.I., Kim, Y., 2018. Robust Bayesian analysis for exponential parameters under generalized Type-II progressive hybrid censoring. Communications in Statistics. Theory and Methods 47 (9), 2259–2277.

Shafay, A., Balakrishnan, N., Abdel-Aty, Y., 2014. Bayesian inference based on a jointly type-II censored sample from two exponential populations. Journal of Statistical Computation and Simulation 84 (11), 2427–2440.

Shafay, A.R., 2016a. Bayesian estimation and prediction based on generalized Type-II hybrid censored sample. Journal of Statistical Computation and Simulation 86 (10), 1970–1988.

Shafay, A.R., 2016b. Exact inference for a simple step-stress model with generalized Type-I hybrid censored data from the exponential distribution. Communications in Statistics. Simulation and Computation 45, 181–206.

Shafay, A.R., 2017. Bayesian estimation and prediction based on generalized type-i hybrid censored sample. Communications in Statistics. Theory and Methods 46 (10), 4870–4887.

Shafay, A.R., 2022. Exact likelihood inference for two exponential populations under joint Type-II hybrid censoring scheme. Applied Mathematics & Information Sciences 16, 389–401.

Shafay, A.R., Balakrishnan, N., 2012. One- and two-sample Bayesian prediction intervals based on Type-I hybrid censored data. Communications in Statistics. Simulation and Computation 41 (1), 65–88.

Shapiro, S.S., Wilk, M.B., 1972. An analysis of variance test for the exponential distribution (complete samples). Technometrics 14 (2), 355–370.

Sharma, V.K., 2017. Estimation and prediction for Type-II hybrid censored data follow flexible Weibull distribution. Statistica 77 (4), 386–414.

Shi, Y., Wu, M., 2016. Statistical analysis of dependent competing risks model from Gompertz distribution under progressively hybrid censoring. SpringerPlus 5 (1), 1745.

Shoaee, S., Khorram, E., 2016. Statistical inference of $R = P(X < Y)$ for Weibull distribution under Type-II progressively hybrid censored data. Journal of Statistical Computation and Simulation 86 (18), 3815–3834.

Singh, B., Goel, R., 2018. Reliability estimation of modified Weibull distribution with Type-II hybrid censored data. Iranian Journal of Science and Technology, Transactions A: Science 42 (3), 1395–1407.

Singh, B., Gupta, P.K., Sharma, V.K., 2014. On type-II hybrid censored Lindley distribution. Statistics Research Letters 3, 58–62.

Singh, D.P., Lodhi, C., Tripathi, Y.M., Wang, L., 2021. Inference for two-parameter Rayleigh competing risks data under generalized progressive hybrid censoring. Quality and Reliability Engineering International 37 (3), 1210–1231.

Singh, S., Belaghi, R.A., Asl, M.N., 2019. Estimation and prediction using classical and Bayesian approaches for Burr III model under progressive type-I hybrid censoring. International Journal of System Assurance Engineering and Management 10 (4), 746–764.

Singh, S., Tripathi, Y.M., 2016. Bayesian estimation and prediction for a hybrid censored lognormal distribution. IEEE Transactions on Reliability 65 (2), 782–795.

Singh, S.K., Singh, U., Sharma, V.K., 2016. Estimation and prediction for Type-I hybrid censored data from generalized Lindley distribution. Journal of Statistics & Management Systems 19 (3), 367–396.

Soltani, A.R., Roozegar, R., 2012. On distribution of randomly ordered uniform incremental weighted averages: divided difference approach. Statistics & Probability Letters 82 (5), 1012–1020.

Soofi, E.S., 2000. Principal information theoretic approaches. Journal of the American Statistical Association 95 (452), 1349–1353.

Spinelli, J.J., Stephens, M.A., 1987. Tests for exponentiality when origin and scale parameters are unknown. Technometrics 29 (4), 471–476.

Spurrier, J.D., Wei, L.J., 1980. A test of the parameter of the exponential distribution in the type I censoring case. Journal of the American Statistical Association 75 (370), 405–409.

Stephens, M.A., 1986. Tests based on EDF statistics. In: D'Agostino, R.B., Stephens, M.A. (Eds.), Goodness-of-Fit Techniques. Marcel Dekker, New York, pp. 97–193.

Su, F., 2013. Exact likelihood inference for multiple exponential populations under joint censoring. PhD thesis. McMaster University, Hamilton, ON.
Su, F., Balakrishnan, N., Zhu, X., 2018. Exact likelihood-based point and interval estimation for lifetime characteristics of Laplace distribution based on hybrid type-I and type-II censored data. In: Tez, M., von Rosen, D. (Eds.), Trends and Perspectives in Linear Statistical Inference. Springer International Publishing, Cham, pp. 203–239.
Su, F., Zhu, X., 2016. Exact likelihood inference for two exponential populations based on a joint generalized Type-I hybrid censored sample. Journal of Statistical Computation and Simulation 86 (7), 1342–1362.
Sukhatme, P.V., 1937. Test of significance for samples of the χ^2-population with two degrees of freedom. Annual of Eugenics 8, 52–56.
Sultan, K.S., Emam, W., 2021. The combined-unified hybrid censored samples from Pareto distribution: estimation and properties. Applied Sciences 11 (13).
Sultana, F., Tripathi, Y.M., 2018. Estimation and prediction for the generalized half normal distribution under hybrid censoring. Journal of Testing and Evaluation 48 (2), 20170721.
Sultana, F., Tripathi, Y.M., Rastogi, M.K., Wu, S.-J., 2018. Parameter estimation for the Kumaraswamy distribution based on hybrid censoring. American Journal of Mathematical and Management Sciences 37 (3), 243–261.
Sultana, F., Tripathi, Y.M., Wu, S.-J., Sen, T., 2020. Inference for Kumaraswamy distribution based on Type I progressive hybrid censoring. Annals of Data Science.
Sundberg, R., 2001. Comparison of confidence procedures for type I censored exponential lifetimes. Lifetime Data Analysis 7 (4), 393–413.
Tahmasbi, R., Rezaei, S., 2008. A two-parameter lifetime distribution with decreasing failure rate. Computational Statistics & Data Analysis 52 (8), 3889–3901.
Takada, Y., 1981. Relation of the best invariant predictor and the best unbiased predictor in location and scale families. The Annals of Statistics 9 (4), 917–921.
Takada, Y., 1991. Median unbiasedness in an invariant prediction problem. Statistics & Probability Letters 12 (4), 281–283.
Tang, L.C., 2003. Multiple steps step-stress accelerated. In: Pham, H. (Ed.), Handbook of Reliability Engineering. Springer, New York, pp. 441–455.
Tanner, M.A., 1993. Tools for Statistical Inference. Methods for the Exploration of Posterior Distributions and Likelihood Functions, 2 edition. Springer, New York.
Teitler, S., Rajagopal, A.K., Ngai, K.L., 1986. Maximum entropy and reliability distributions. IEEE Transactions on Reliability 35 (4), 391–395.
Thomas, D.R., Wilson, W.M., 1972. Linear order statistic estimation for the two-parameter Weibull and extreme value distributions from Type-II progressively censored samples. Technometrics 14, 679–691.
Tian, Y., Yang, A., Li, E., Tian, M., 2016. Parameters estimation for mixed generalized inverted exponential distributions with type-II progressive hybrid censoring. Hacettepe Journal of Mathematics and Statistics 47, 1023–1039.
Tian, Y., Zhu, Q., Tian, M., 2015. Estimation for mixed exponential distributions under type-II progressively hybrid censored samples. Computational Statistics & Data Analysis 89, 85–96.
Tiku, M., 1980. Goodness of fit statistics based on the spacings of complete or censored samples. Australian Journal of Statistics 22, 260–275.
Tomer, S.K., Panwar, M.S., 2015. Estimation procedures for Maxwell distribution under type-I progressive hybrid censoring scheme. Journal of Statistical Computation and Simulation 85 (2), 339–356.
Tong, H., 1974. A note on the estimation of $Pr(Y < X)$ in the exponential case. Technometrics 16, 625. Correction Technometrics 17, 395 (1975).
Tong, H., 1975a. Errata: a note on the estimation of $Pr\{Y < X\}$ in the exponential case. Technometrics 17 (3), 395.
Tong, H., 1975b. Letter to the editor. Technometrics 17, 393.
Torabi, H., Mirjalili, S.M., Nadeb, H., 2017. A new simple and powerful normality test for progressively Type-II censored data. arXiv:1704.06787.
Tsiatis, A., 1975. A nonidentifiability aspect of the problem of competing risks. Proceedings of the National Academy of Sciences of the United States of America 72, 20–22.

Tu, J., Gui, W., 2020. Bayesian inference for the Kumaraswamy distribution under generalized progressive hybrid censoring. Entropy 22 (9), 1032.

Tukey, J.W., 1965. Which part of the sample contains the information? Proceedings of the National Academy of Sciences of the United States of America 53, 127–134.

Valiollahi, R., Asgharzadeh, A., Kundu, D., 2017. Prediction of future failures for generalized exponential distribution under Type-I or Type-II hybrid censoring. Brazilian Journal of Probability and Statistics 31 (1), 41–61.

Valiollahi, R., Asgharzadeh, A., Raqab, M.Z., 2019. Prediction of future failures times based on Type-I hybrid censored samples of random sample sizes. Communications in Statistics. Simulation and Computation 48 (1), 109–125.

van Bentum, T., Cramer, E., 2019. Stochastic monotonicity of MLEs of the mean for exponentially distributed lifetimes under sequential hybrid censoring. Statistics & Probability Letters 148, 1–8.

Viveros, R., Balakrishnan, N., 1994. Interval estimation of parameters of life from progressively censored data. Technometrics 36, 84–91.

Voinov, V., Nikulin, M., Balakrishnan, N., 2013. Chi-Squared Goodness of Fit Tests with Applications. Academic Press, San Diego, CA.

Volovskiy, G., 2018. Likelihood-Based Prediction in Models of Ordered Data. PhD thesis. RWTH Aachen, Aachen, Germany.

Volterman, W., Davies, K.F., Balakrishnan, N., 2013a. Simultaneous Pitman closeness of progressively Type-II right-censored order statistics to population quantiles. Statistics 47, 439–452.

Volterman, W., Davies, K.F., Balakrishnan, N., 2013b. Two-sample Pitman closeness comparison under progressive Type-II censoring. Statistics 47, 1305–1320.

Wang, B.X., 2008. Goodness-of-fit test for the exponential distribution based on progressively type-ii censored sample. Journal of Statistical Computation and Simulation 78 (2), 125–132.

Wang, F.-K., 2014. Using BBPSO algorithm to estimate the Weibull parameters with censored data. Communications in Statistics. Simulation and Computation 43 (10), 2614–2627.

Wang, K., Gui, W., 2021. Statistical inference of generalized progressive hybrid censored step-stress accelerated dependent competing risks model for Marshall-Olkin bivariate Weibull distribution. Statistics 55 (5), 1058–1093.

Wang, L., 2018. Inference for Weibull competing risks data under generalized progressive hybrid censoring. IEEE Transactions on Reliability 67 (3), 998–1007.

Wang, L., Li, H., 2022. Inference for exponential competing risks data under generalized progressive hybrid censoring. Communications in Statistics. Simulation and Computation 51 (3), 1255–1271.

Wang, L., Tripathi, Y.M., Lodhi, C., 2020. Inference for Weibull competing risks model with partially observed failure causes under generalized progressive hybrid censoring. Journal of Computational and Applied Mathematics 368, 112537.

Wang, Y., He, S., 2005. Fisher information in censored data. Statistics & Probability Letters 73, 199–206.

Wilks, S.S., 1962. Mathematical Statistics. Wiley, New York.

Wong, K.M., Chen, S., 1990. The entropy of ordered sequences and order statistics. IEEE Transactions on Information Theory 36 (2), 276–284.

Wu, M., Shi, Y., Sun, Y., 2017. Statistical analysis for competing risks model from a Weibull distribution under progressively hybrid censoring. Communications in Statistics. Theory and Methods 46 (1), 75–86.

Wu, M., Shi, Y., Wang, Y., 2016. E-Bayesian estimation for competing risk model under progressively hybrid censoring. Journal of Systems Engineering and Electronics 27 (4), 936–944.

Wu, S.-F., 2010. Interval estimation for the two-parameter exponential distribution under progressive censoring. Quality and Quantity 44, 181–189.

Xie, Q., Balakrishnan, N., Han, D., 2008. Exact inference and optimal censoring scheme for a simple step-stress model under progressive Type-II censoring. In: Balakrishnan, N. (Ed.), Advances in Mathematical and Statistical Modeling. Birkhäuser, Boston, pp. 107–137.

Xiong, C., 1998. Inferences on a simple step-stress model with type-II censored exponential data. IEEE Transactions on Reliability 47 (2), 142–146.

Yadav, A.S., Saha, M., Singh, S.K., Singh, U., 2019a. Bayesian estimation of the parameter and the reliability characteristics of xgamma distribution using Type-II hybrid censored data. Life Cycle Reliability and Safety Engineering 8 (1), 1–10.

Yadav, A.S., Singh, S.K., Singh, U., 2019b. Bayesian estimation of stress-strength reliability for Lomax distribution under type-ii hybrid censored data using asymmetric loss function. Life Cycle Reliability and Safety Engineering 8 (3), 257–267.

Yang, M.-C., Chen, L.-S., Liang, T., 2017. Optimal Bayesian variable sampling plans for exponential distributions based on modified type-II hybrid censored samples. Communications in Statistics. Simulation and Computation 46 (6), 4722–4744.

Ye, Z., Ng, H.K.T., 2014. On analysis of incomplete field failure data. Annals of Applied Statistics 8 (3), 1713–1727.

Ye, Z.-S., Chan, P.-S., Xie, M., Ng, H.K.T., 2014. Statistical inference for the extreme value distribution under adaptive Type-II progressive censoring schemes. Journal of Statistical Computation and Simulation 84 (5), 1099–1114.

Yeh, L., 1988. Bayesian approach to single variable sampling plans. Biometrika 75 (2), 387–391.

Yeh, L., 1994. Bayesian variable sampling plans for the exponential distribution with Type I censoring. The Annals of Statistics 22 (2), 696–711.

Zhang, Y., Meeker, W.Q., 2005. Bayesian life test planning for the Weibull distribution with given shape parameter. Metrika 61 (3), 237–249.

Zheng, G., 2001. A characterization of the factorization of hazard function by the Fisher information under Type-II censoring with application to the Weibull family. Statistics & Probability Letters 52 (3), 249–253.

Zheng, G., Balakrishnan, N., Park, S., 2009. Fisher information in ordered data: a review. Statistics and Its Interface 2, 101–113.

Zheng, G., Gastwirth, J.L., 2000. Where is the Fisher information in an ordered sample? Statistica Sinica 10 (4), 1267–1280.

Zheng, G., Park, S., 2004. On the Fisher information in multiply censored and progressively censored data. Communications in Statistics. Theory and Methods 33 (8), 1821–1835.

Zheng, G., Park, S., 2005. Another look at life testing. Journal of Statistical Planning and Inference 127 (1–2), 103–117.

Zhu, T., 2020. Statistical inference of Weibull distribution based on generalized progressively hybrid censored data. Journal of Computational and Applied Mathematics 371, 112705.

Zhu, T., 2021a. Goodness-of-fit tests for progressively Type-II censored data: application to the engineering reliability data from continuous distribution. Quality Engineering 33 (1), 128–142.

Zhu, T., 2021b. A new approach for estimating parameters of two-parameter bathtub-shaped lifetime distribution under modified progressive hybrid censoring. Quality and Reliability Engineering International 37 (5), 2288–2304.

Zhu, X., 2015. Inference for Birnbaum-Saunders, Laplace and some related distributions under censored data. PhD thesis. McMaster University, Hamilton, ON, Canada.

Zhu, X., Balakrishnan, N., 2016. Exact inference for Laplace quantile, reliability, and cumulative hazard functions based on type-II censored data. IEEE Transactions on Reliability 65 (1), 164–178.

Zhu, X., Balakrishnan, N., 2017. Exact likelihood-based point and interval estimation for lifetime characteristics of Laplace distribution based on a time-constrained life-testing experiment. In: Adhikari, A., Adhikari, M.R., Chaubey, Y.P. (Eds.), Mathematical and Statistical Applications in Life Sciences and Engineering. Springer Singapore, Singapore, pp. 327–372.

Zhu, X., Balakrishnan, N., Feng, C., Ni, J., Yu, N., Zhou, W., 2020. Exact likelihood-ratio tests for joint type-II censored exponential data. Statistics 54 (3), 636–648.

Zhu, X., Balakrishnan, N., Saulo, H., 2019. On the existence and uniqueness of the maximum likelihood estimates of parameters of Laplace Birnbaum-Saunders distribution based on Type-I, Type-II and hybrid censored samples. Metrika 82 (7), 759–778.

Index

Q

R

S

9780123983879